# Combustion Science and Engineering

# Combustion Science and Engineering

Edited by **Alden Whitley**

**NY**RESEARCH
P R E S S

New York

Published by NY Research Press,
23 West, 55th Street, Suite 816,
New York, NY 10019, USA
www.nyresearchpress.com

**Combustion Science and Engineering**
Edited by Alden Whitley

International Standard Book Number: 978-1-63238-510-9 (Hardback)

Printed in the United States of America.

# Contents

# Preface

This book has been a concerted effort by a group of academicians, researchers and scientists, who have contributed their research works for the realization of the book. This book has materialized in the wake of emerging advancements and innovations in this field. Therefore, the need of the hour was to compile all the required researches and disseminate the knowledge to a broad spectrum of people comprising of students, researchers and specialists of the field.

This book elucidates the concepts and innovative models around prospective developments with respect to combustion science and engineering. Combustion refers to a series of exothermic redox chemical reactions between fuel and oxidant which results in production of heat, light and energy. Combustion engineering deals with applying the rules of science and engineering in order to form solutions for problems related to combustion and fire, in various fields. Combustion engineering is used in cars, gas ovens, fossil fuel power plants, steam engines, etc. This book includes topics that are of utmost significance and bound provide incredible insights to readers. Different approaches, evaluations, methodologies and advanced studies on this subject have been included in this text. Scientists, engineers, researchers, students and all related to this field will benefit alike from this text.

At the end of the preface, I would like to thank the authors for their brilliant chapters and the publisher for guiding us all-through the making of the book till its final stage. Also, I would like to thank my family for providing the support and encouragement throughout my academic career and research projects.

**Editor**

# Combustion Characteristics of Butane Porous Burner for Thermoelectric Power Generation

**K. F. Mustafa,[1] S. Abdullah,[1] M. Z. Abdullah,[2] and K. Sopian[1]**

[1]Department of Mechanical and Materials Engineering, Faculty of Engineering and Built Environment,
Universiti Kebangsaan Malaysia (UKM), 43600 Bangi, Selangor, Malaysia
[2]School of Mechanical Engineering, Universiti Sains Malaysia Engineering Campus, Seri Ampangan,
14300 Nibong Tebal, Penang, Malaysia

Correspondence should be addressed to K. F. Mustafa; mekhairil@usm.my

Academic Editor: Kalyan Annamalai

The present study explores the utilization of a porous burner for thermoelectric power generation. The porous burner was tested with butane gas using two sets of configurations: single layer porcelain and a stacked-up double layer alumina and porcelain. Six PbSnTe thermoelectric (TE) modules with a total area of $54\,cm^2$ were attached to the wall of the burner. Fins were also added to the cold side of the TE modules. Fuel-air equivalence ratio was varied between the blowoff and flashback limit and the corresponding temperature, current-voltage, and emissions were recorded. The stacked-up double layer negatively affected the combustion efficiency at an equivalence ratio of 0.20 to 0.42, but single layer porcelain shows diminishing trend in the equivalence ratio of 0.60 to 0.90. The surface temperature of a stacked-up porous media is considerably higher than the single layer. Carbon monoxide emission is independent for both porous media configurations, but moderate reduction was recorded for single layer porcelain at lean fuel-air equivalence ratio. Nitrogen oxides is insensitive in the lean fuel-air equivalence ratio for both configurations, even though slight reduction was observed in the rich region for single layer porcelain. Power output was found to be highly dependent on the temperature gradient.

## 1. Introduction

The merits of utilizing thermoelectric (TE) devices in various energy conversion systems have led to their application in many engineering fields. Coupled with strong drives for research and developments, numerous scientists have shown appreciable interest in these devices owing to their unique and easy method of transforming thermal energy into electricity. The working principle of TE devices is practically simple, relying primarily on the temperature difference between the hot and cold sides of the TE modules for electricity power generation. The generated temperature difference agitated the electron charge carrier and thereby resulted in the flow of electric current in the circuit. Of practical significance is that the modules are generally quiet and highly reliable, and their compactness allows easy integration into a burner or furnace.

Thermal to electric energy transformation that pertains to thermoelectric system takes place when the heat produced from the combustion products in the burner flows through the TE modules. The temperature gradient between the hot combustion gases and relatively cold ambient triggers the electricity flow through the module and into the circuit. In general, the burners for TE applications can be fired using many types of hydrocarbon fuels, either liquid or gaseous. Qiu and Hayden [1], for instance, studied the thermoelectric power generation using natural gas-fired burner. The electricity generated was used to power the electrical components for residential heating system. Their work was further extended to investigate the beneficial effect of including an air recuperator and additional TE modules in the system, Qiu and Hayden [2]. The apparent advantage of an air recuperator was evident with an improved electrical efficiency reported in their later work. A number of useful applications of TE modules have also been disseminated by other notable investigators, for example, in a cook stove by Champier et al. [3], and exhaust gas extraction for vehicular applications, as shown by Yang and Stabler [4], Kim et al. [5], and Deng et al. [6]. With regards to butane gas, despite its popularity as household portable burner, considerable researches on it remained largely shallow. Posthill et al. [7] endeavored on the application of butane gas for TE generator by demonstrating a mini combustor as a heat source for silicon germanium (SiGe) TE modules. Much earlier, Rahman and Shuttleworth [8] devised an experiment of TE applications using butane gas for powering laptop computer. Other than these, Yoshida et al. [9] attempted a catalytic microcombustor fired by butane and hydrogen gas in TE power generation. Although the abovementioned studies did much to elucidate the TE power generation covering wide spectrum of applications, combustion and emission aspects of the burner itself are generally neglected. Lack of understanding and inadequate information from the works cited previously highlighted the unintended voids left by these authors in omitting the burner performance and gas emission from the burner. This however opens up new avenues for researchers to realize that much effort is still needed to cover these aspects to regard TE modules as a good candidate for TE power generation. The present work is therefore initiated to expound more succinctly the overall performance of the TE power generation system, by including the thermal characteristics of the burner.

Porous media combustion is one of the newer techniques invented to achieve combustion stability with concomitant reduction in emissions. The essential feature of porous media combustion fundamentally pertains to heterogeneous combustion between the solid matrix with its void filled with fluids. Howell et al. [10] expounded that the combustion of fuel and air mixture in the porous matrix is rigorously heated via enhanced convective mode as the reactants flow through the interstitial voids in the matrix. An enhanced heat transfer mechanism between the combustion products and porous structure is beneficial in allowing greater control of flame stability with improved radiant energy. With better flame stability, fuel-air equivalence ratio can be widened and thereby the burner can be operated at leaner equivalence ratio. This will then be potentially translated into an improved overall efficiency and hazardous exhaust gases associated with emission products can be ameliorated.

Considerable research effort has also been expended in devising stacked-up porous media to obtain flame stability for both liquid and gaseous fuels. Kerosene fuel combustion using silicone-carbide-coated carbon-carbon (C-C) foam was comprehensively tested by Periasamy et al. [11], Vijaykant and Agrawal [12], and Periasamy and Gollahalli [13]. Jugjai and Polmart [14] made use of alumina spheres for kerosene fuel evaporation enhancement in two-section porous burner. Combustion of natural gas in silicon carbide coated C-C porous material was elucidated by Marbach and Agrawal [15]. Smucker and Ellzey [16] elicited the merits of yttria stabilized zirconia in stretching the operating range of fuel-air equivalence ratio of propane and methane in a two-section porous burner. The findings exposited by these researchers generally agree with the apparent advantages of incorporating the stacked-up porous media in a burner. However, coupling the inherent advantages of a porous burner for TE power

generation requires comprehensive temperature and burner efficiency and these remain largely unexplored. The published work of Hanamura et al. [17] was primarily restricted towards numerical aspects of superadiabatic combustion in investigating the use of porous element for TE power generation. In-depth understanding and sufficient knowledge have not been generated since, and this forms the thrust and motivation for designing the present work.

The primary focus of the study undertaken was to evaluate the electricity generation using thermoelectric modules (TEM) from the combustion of butane gas in a porous burner. The study covers two types of porous media configurations, using single layer porcelain and double layer alumina and porcelain. The aim was further narrowed to assess the thermoelectric power generation at several ranges of operating fuel-air equivalence ratio. Combustion characteristics are discussed in terms of the temperature profiles, combustion efficiency, and the emissions generated as the products of combustion. Results yielded in the study are analyzed for both sets of experiments and presented to demonstrate the feasibility of utilizing stacked-up porous media for TE power generation system.

## 2. Materials and Methods

*2.1. Experimental Setup and Procedure.* The design concept in this work is to integrate a porous burner operating on butane gas with TE modules for the generation of electricity. TE modules allow temperature difference between two dissimilar conductors to produce voltage via Seebeck effect. The electricity power generation system consists essentially of a hexagonal burner and six PbSnTe TEM cells ($3 \, cm \times 3 \, cm$) attached to the side wall of the burner.

The stainless steel burner was fabricated in the common machine workshop in the School of Mechanical Engineering, Universiti Sains Malaysia (USM). Stainless steel with a thickness of 3 mm was chosen owing to the fact that it has lower thermal conductivity ($16.2 \, W/mK$). Low thermal conductivity ensures high temperature encapsulation for the burner. The burner can be divided into two sections: (1) the main combustion chamber to house the porous media, and (2) a base premixed chamber. The main combustion chamber was designed to be hexagonal in shape with 4.5 cm width for all sides of the burner. The choice of the hexagonal shape was solely made for the flexible placement and easy positioning of the TEM cells even though no geometric optimization was carried out for the dimensions of the burner. The hexagonal shape burner was the primary combustion zone and filled with two types of porous media; the top layer is alumina ($Al_2O_3$) and the bottom layer is porcelain. Alumina has a thickness of 12.7 mm with 8 pores per cm (8 ppcm) and 85% porosity. It was chosen owing to its high working temperature, thermal shock resistance, and low pressure drop. The bottom layer is porcelain with a thickness of 15.0 mm having 16 ppcm and 86% porosity. Both porous media were carefully placed inside the chamber and cemented with special glue at all contacting edges of the burner. A tight fit between the porous media and burner is critical because unwanted gaps between the two could cause the flame to develop and propagate around the porous media. When the burner was operated with double layer porous media, alumina was carefully placed on top of porcelain so that there was no air space between the two.

The base premixed chamber is a small hexagonal chamber which was designed to increase the residence time of the fuel and air mixture prior to the entrance of the main combustion zone. This chamber was also fabricated using stainless steel and welded together to the bottom part of the primary combustion zone of the burner to create a single piece. The height of the chamber is 8 cm. The bottom end of the chamber forms a fuel feeding side which was connected to the fuel supply pipe from the butane gas container. All TEM cells were positioned equispaced around the wall of the primary combustion zone and fixed in their position to the wall of the main combustion zone by a thermal pad. The thickness of the thermal pad is negligible compared to the overall dimensions of the combustion zone. This implies that the temperature fluctuation across the pad can be safely neglected without significant influence on the overall heat transfer across the TE modules. There are six TEM cells with a total area of 54 $cm^2$. The thermoelectric elements in the module are made from PbSnTe doped in either p- or n-type semiconductor properties. Since electricity generation using TEM cells is strongly influenced by the temperature gradient between the hot and cold surfaces, six steel fins were added to the ambient side TE modules to enhance cooling. The terminals of all TEM cells were electrically connected in series.

The working fuel used throughout the entire experiment is butane (chemical formula $C_4H_{10}$). The butane gas is kept in a container with a capacity of 230 g similar to the one used in a commercially available portable cooking burner. It was tightly secured in its place with a manual locking device and put in the horizontal plane housing. The fuel releasing knob can be set from the minimum (zero fuel flow rate) to the maximum (maximum fuel flow rate) and it was connected to the opening of the butane fuel supply. During the experiment, the fuel releasing knob was adjusted and only used to approximate the butane gas supplied, since the fuel flow rate was precisely metered using Vogtlin flow regulator GCR-C9KA BA20 (Switzerland). The flow meter was calibrated in the range of 0–2.000 liter per minute (lpm) with flow accuracy of up to 0.005 lpm. It was supplied through reinforced plastic fuel tubing with an internal diameter of 4 mm. The pipe was connected to the entrance of the premixed chamber of the burner. It should be noted that the air admission into the premixed chamber was done in the artificial conditions of inducement via fuel entrainment. Since the fuel feeding pipe sits only few millimeters from the mouth belly of the premixed chamber, sufficiently high gas velocity of butane created the air inducement and entrained the air into the premixed chamber.

Throughout this study, experiments were conducted by varying the fuel flow rate to demonstrate the changing values of the fuel-air equivalence ratio ($\phi$) on the electricity power generation. Two sets of experiments were conducted, with single layer porcelain porous medium and double layer alumina and porcelain porous media. In a double layer configuration, alumina was stacked on top of porcelain and

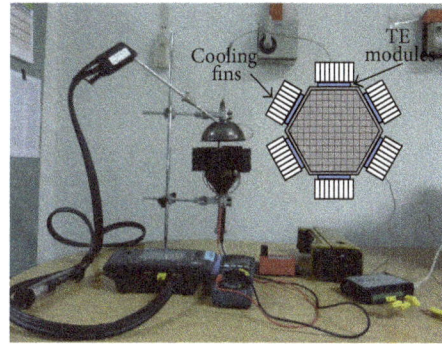

FIGURE 1: Experimental setup without the thermal imager.

cemented at the edges to eliminate any air gaps between the two porous media. The entire experiment was conducted by varying the fuel-air equivalence ratio from the leanest to the richest. The leanest equivalence ratio was not necessarily the theoretical value for butane, but it was rather observed based on the flame stability during combustion with porous media. The stable flame is defined as one that is entirely contained within or on the porous medium for a given fuel and air flow rate and remained steady (Periasamy and Gollahalli [13]). Even though combustion was successfully initiated for all intended values of fuel-air equivalence ratio, it could not be stabilized for periods longer than 15 minutes without the flame either flashing back at lower airflows and igniting the spray or blowing out the downstream end of the porous media at higher flow rates.

For each set, the emission level of carbon monoxide (CO) and nitrogen oxides ($NO_x$) was recorded using a portable combustion analyzer CA-CALC 6203 suitable for quasicontinuous measurement of combustion products. The data were later saved on a PC as a Microsoft Excel worksheet. The probe tip of the combustion analyzer was positioned 10 cm from the top surface of the porous media and fixed at the central position of the porous chamber. The surface temperature of the porous media was imaged using Fluke Ti27 9 Hz thermal imager which provides accurate surface temperature distribution. The surface temperature was only captured after the emission gases were recorded by the combustion analyzer. Since both apparatuses were in identically positioned on top of the porous burner, the readings were done alternately. The temperatures of the fins attached to the thermoelectric modules were also captured using the thermal imager. The uncertainty in temperature measurement is ±2°C. The current-voltage readings were determined using Sanwa Digital Multimeter CD771 with series connection for all terminals. Figure 1 depicts the actual experimental setup without the thermal imager and Figure 2 represents the schematic diagram. The detail dimensions of the burner are shown in Figure 3.

*2.2. Quality of the Experimental Data.* Experiments for both sets of burner configurations were repeated thrice to ensure the repeatability and the reliability of the measured data. For all experimental data, the mean value ($\overline{X}$) and the standard deviation ($S_X$) are expressed as follows [18, 19]:

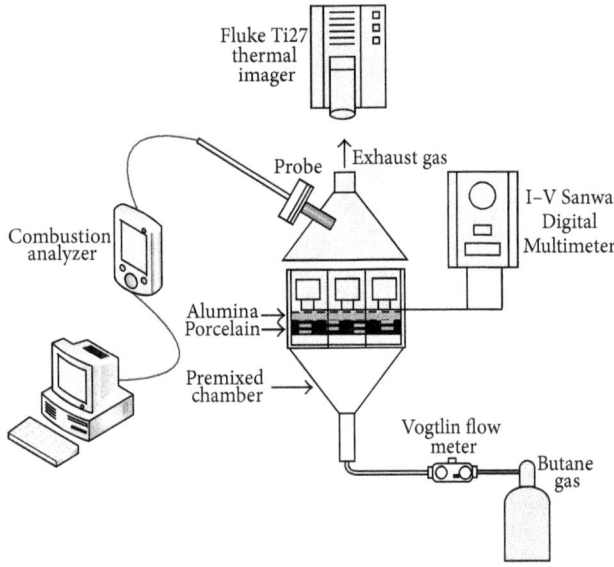

FIGURE 2: Schematic diagram including the thermal imager. The thermal imager was positioned on top and in the center of the porous burner.

$$\overline{X} = \frac{1}{n}\sum_{i=1}^{n} X_i$$

$$S_X = \left[ \frac{1}{n-1}\sum_{i=1}^{n}\left(X_i - \overline{X}\right)^2 \right]^{1/2}. \tag{1}$$

Uncertainty is defined as [18, 19]

$$\text{Uncertainty} = \frac{\text{Standard deviation}}{\text{Mean value}} \times 100\%. \tag{2}$$

The values of $\overline{X}$, $S_X$, and uncertainty for the measured parameters are tabulated in Tables 1 and 2. The experimental data are tabulated for the surface temperature of the porous media, mass flow rate of butane ($m_f$), voltage ($V$), current ($I$), carbon monoxide (CO), and nitrogen oxides ($NO_x$).

The measured experimental data are reliable since the maximum uncertainty was only 2.8%, which can be regarded as very low. The uncertainty of the measured data is shown throughout the range of fuel-air equivalence ratio for single and layer and double layer are shown in Tables 3 and 4, respectively.

The maximum uncertainty for single layer porcelain and double layer porcelain and alumina is 2.8% and 2.2%, respectively.

### 2.3. Numerical Investigation.

The numerical procedure adopted in this study is used to determine the influence of porous material properties on the thermal characteristics of the burner. The governing equations for mass, solid energy, gas energy, and gas species are used [16]:

$$\frac{\partial \left(\rho_g \varepsilon\right)}{\partial t} + \frac{\partial \left(\rho_g \varepsilon u\right)}{\partial x} = 0, \tag{3}$$

$$\rho_g C_g \varepsilon \frac{\partial T_g}{\partial t} + \rho_g C_g \varepsilon u + \Sigma \rho \varepsilon Y_i V_i C_{gi} \frac{\partial T_g}{\partial x} + \varepsilon \Sigma \dot{\omega}_i h_i W_i$$
$$= \varepsilon \frac{\partial}{\partial x}\left(\left(k_g + \rho C_p D_{II}^d\right)\frac{\partial T_g}{\partial x}\right) - h_v\left(T_g - T_s\right), \tag{4}$$

$$\rho_s C_s \frac{\partial T_s}{\partial t} = k_s \frac{\partial^2 T_s}{\partial x^2} + h_v\left(T_g - T_s\right) - \frac{\partial q_r}{\partial x}, \tag{5}$$

$$\rho_g \varepsilon \frac{\partial Y_i}{\partial t} + \rho_g \varepsilon u \frac{\partial Y_i}{\partial x} + \frac{\partial}{\partial x}\left(\rho \varepsilon Y_i V_i\right) - \varepsilon \dot{\omega}_i W_i = 0, \tag{6}$$

where $\varepsilon$ is porosity, $\rho$ is gas density, $u$ is gas velocity, $t$ is time, $C_g$ is the specific heat of gas, $T_g$ is temperature of the gas, $x$ is the distance, $Y_i$, $V_i$, $C_{gi}$, $\omega_i$, $h_i$, and $W_i$ are the mass fraction, diffusion velocity, specific heat, molar rate of production, molar enthalpy, and molecular weight of the $i$th species, respectively, $k_g$ is gas thermal conductivity, $h_v$ is the volumetric heat transfer coefficient between the porous media and the gas, $T_s$, $C_s$, and $k_s$ are the temperature, specific heat, and effective thermal conductivity of the porous medium, respectively, and $q_r$ is radiant heat flux in the $x$ direction [16].

## 3. Results and Discussion

### 3.1. Surface and Submerged Temperature Distributions.

The computed surface and submerged temperature distributions against porous material thickness at various fuel-air equivalence ratios are shown in Figure 4. In all cases, the submerged temperature distributions increase as the thickness of the porous material increases. In the transition zone of the burner ($x = 15.0$ mm), noticeable increase in the submerged temperature was observed. This has the effect on stabilizing the combustion zone in the porous burner. A change in the pore size has significantly affected the downstream temperature distributions, as the computed temperature gradually increases until the top layer of the double layer porcelain and alumina. The predicted and the experimental values are tabulated in the inset of Figure 4. It can be seen that the predicted and experimental values of the surface temperatures agree well within 5% range. The sensitivity of the combustion zone in the submerged layer of the porous alumina towards the flow velocity is thought to be the contributing factor in the experimental values. It can also be seen in Figure 4 that the influence of fuel-air equivalence ratio on the submerged temperature profiles is only marginal, since no obvious pattern can be interpreted from the figure. Unlike the free flame combustion, combustion in porous media is relatively complex, and enriching the fuel-air equivalence ratio is not normally accompanied by an increase in the submerged temperature distributions.

### 3.2. Combustion Efficiency.

Fuel-air equivalence ratio is defined as the ratio between the actual fuel-air mixture and the stoichiometric fuel-air mixture. For butane ($C_4H_{10}$), using the stoichiometric combustion equation, it can be

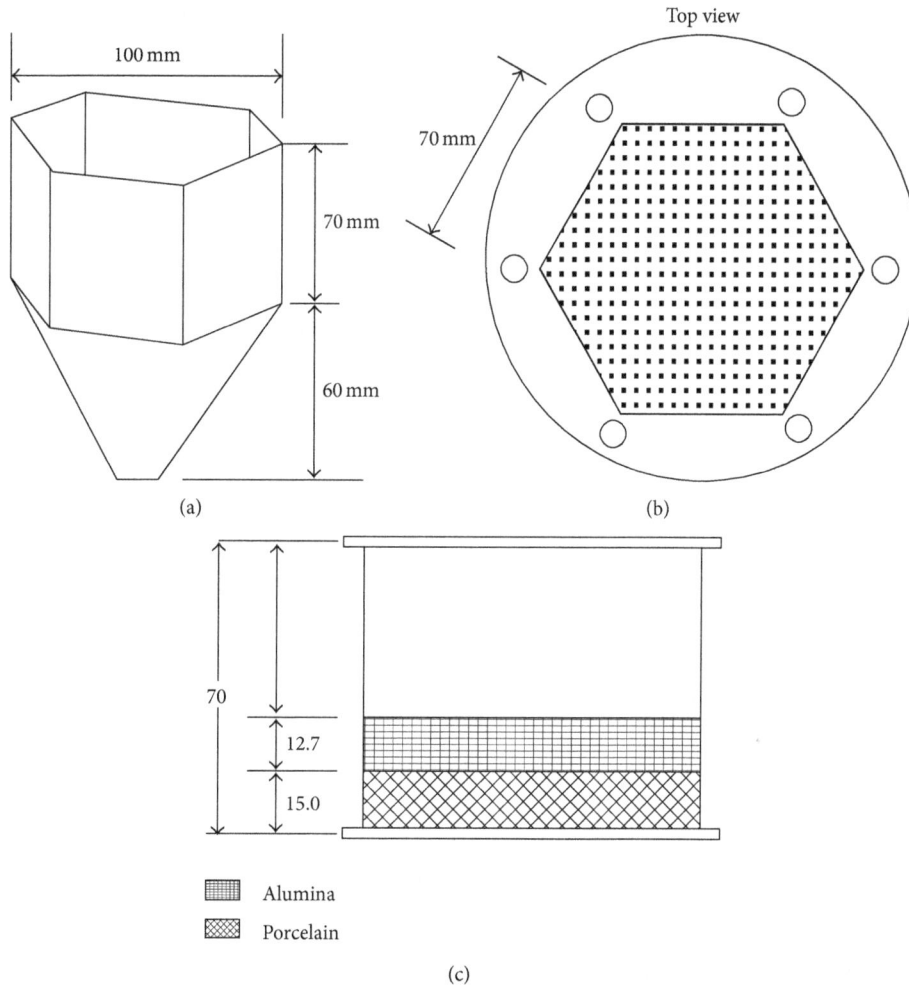

FIGURE 3: Burner drawing and dimensions: (a) isometric view, (b) top view, and (c) side view without the premixing chamber. All dimensions are in mm and figures are not drawn to scale.

TABLE 1: $\overline{X}$, $S_X$, and uncertainty for single layer porcelain.

| Measured parameters | Surface temperature (°C) | $\dot{m}_f$ L/min | V Volt | I Amp | CO ppm | $NO_x$ ppm |
|---|---|---|---|---|---|---|
| Mean value, $\overline{X}$ | 384.3 | 1.105 | 9.96 | 0.077 | 828 | 24 |
| Standard deviation, $S_X$ | 2.9 | 0.020 | 0.12 | 0.00216 | 12 | 0 |
| Uncertainty | 0.7% | 1.8% | 1.2% | 2.8% | 1.4% | 0.0% |

derived from chemical composition that the stoichiometric fuel-air mixture is 0.065. In our study, the fuel-air equivalence ratio, $\phi$ was varied by adjusting the amount of butane gas supplied into the burner. Experiments were conducted for both single layer (using porcelain only) and double layer alumina and porcelain. For a meaningful comparison, the fuel-air equivalence ratio was intended to be identical for both configurations of porous media, but it was evidently observed during the experiment that combustion instability has demarcation effect on the range of fuel-air equivalence ratio covered in the experiment. For single layer porcelain, the "rich" fuel-air equivalence ratio was extended until the combustion reached the critical flashback region, where

the associated flashback triggered the onset of flame shifting towards the premixed chamber and extinguished. This flashback point for single layer porcelain was found to be approximately at fuel-air equivalence ratio of 0.90. For double layer alumina and porcelain, the maximum fuel-air equivalence ratio was marginally lower compared to single layer porcelain and the value of 0.62 was recorded in this experiment. The sensitivity of the flame stabilization towards fuel-air equivalence ratio was also apparent in the lean region with the minimum values of 0.34 and 0.20 for single and double layer porous media, respectively. In the rich region of equivalence ratio, flashback is the primary issue, but the lean region is moderately dominated to some extent by

TABLE 2: $\overline{X}$, $S_X$, and uncertainty for double layer porcelain and alumina.

| Measured parameters | Surface temperature (°C) | $\dot{m}_f$ L/min | V Volt | I Amp | CO ppm | $NO_x$ ppm |
|---|---|---|---|---|---|---|
| Mean value, $\overline{X}$ | 576.7 | 0.880 | 11.03 | 0.083 | 643 | 32 |
| Standard deviation, $S_X$ | 5.2 | 0.015 | 0.20 | 0.001 | 10 | 0 |
| Uncertainty | 0.9% | 1.7% | 1.8% | 1.1% | 1.6% | 0.0% |

TABLE 3: Uncertainty of the measured data for single layer porcelain in the range of fuel-air equivalence ratio.

| Fuel-air equivalence ratio | Uncertainty | | | | | |
|---|---|---|---|---|---|---|
| | Surface temperature (°C) | $\dot{m}_f$ L/min | V Volt | I Amp | CO ppm | $NO_x$ ppm |
| 0.38 | 0.8% | 2.1% | 1.1% | 2.2% | 1.1% | 1.0% |
| 0.39 | 1.0% | 2.2% | 1.4% | 2.8% | 1.1% | 1.0% |
| 0.41 | 0.9% | 1.8% | 1.2% | 2.3% | 1.6% | 1.0% |
| 0.50 | 0.6% | 1.5% | 1.1% | 2.1% | 1.2% | 0.0% |
| 0.51 | 0.6% | 1.2% | 1.2% | 0.8% | 1.1% | 0.0% |
| 0.58 | 0.7% | 0.9% | 1.5% | 1.6% | 1.6% | 0.0% |
| 0.60 | 0.8% | 2.4% | 0.9% | 1.7% | 1.2% | 0.0% |
| 0.62 | 1.1% | 2.3% | 1.0% | 1.8% | 1.1% | 1.0% |
| 0.65 | 1.2% | 2.6% | 1.0% | 2.5% | 1.4% | 2.0% |
| 0.67 | 1.5% | 2.1% | 0.9% | 2.8% | 1.3% | 0.0% |
| 0.95 | 1.% | 1.2% | 0.8% | 2.6% | 1.3% | 0.0% |

blowoff phenomenon. Blowoff occurs at low fuel flow rate and the diminishing fuel-air equivalence ratio exacerbated the flame stability. Even though the reduced stability of the flame was quite subtle, it was visually evident when the flame began to extinguish at the top surface of the porous media during the experiment. The beneficial effect of stacking up the porous media with different pore size can be noticed at leaner equivalence ratio. The lean limit with double layer porous media suggests that a slight improvement in the flame stability with concomitant reduction in the equivalence ratio occurs in the burner. This was confirmed by an earlier work of Hsu et al. [20] in their experiment using stacked porous ceramic burner with premixed methane gas. Without making quantitative comparison due to the difference in a fuel, the observed flame stability at lean limit of equivalence ratio in this study was remarkably similar to the reported work of Hsu et al. [20]. The transition from large (alumina) to small (porcelain) pore size porous media also seems to enhance the flame stability by reducing the flame speed and combustion intensity. To illustrate the effectiveness of the thermal energy conversion in our study, the combustion efficiency is plotted against fuel-air equivalence ratio and shown in Figure 5. The combustion efficiency is calculated using the following expression [21]:

$$\eta_{comb} = 1 - \frac{\dot{m}_{net} X_k H_{coal} + \dot{n} y_{CO} H_{CO} + \dot{n}\bar{c}_p (T - T_o) + UA (T - T_o)}{\dot{m}_f Q_{net}},$$

(7)

where $\dot{m}_{net}$ is the mass flow rate discharge from the combustor and $X_k$ is the carbon content in the discharged solid particles. The first term on the right-hand side the numerator in (7) is the loss due to the carbon content in the discharge mass, the second term is the loss due to CO content, the third term is the loss in the flue gas, and the fourth term in the heat loss in the wall of the combustor.

It is observed that the combustion efficiency for both single and double layer porous media fluctuates between 58% and 73% in the entire range of fuel-air equivalence ratio. The general trend in the figure suggests that single layer porcelain gives marginally higher combustion efficiency compared to double layer alumina and porcelain. A maximum combustion efficiency of 73% is recorded at fuel-air equivalence ratio of 0.52 for single layer porcelain. On the other hand, a maximum combustion efficiency of 68% is attained for double layer porous media at fuel-air equivalence ratio of 0.57. It has been reported by Charoensuk and Lapirattanakun [22] that it is possible to achieve combustion efficiency in excess of 80% in a stacked porous combustor. However, the said value is attained when the CO emission is low and the burner is incorporated with a staged air supply. The surface temperature profile at maximum combustion efficiency for single layer porcelain is shown in Figure 6.

In a burner application, the combustion efficiency is generally governed by the ratio between the theoretical and actual amount of fuel-air mixture [23]. It can also be interpreted as the ratio of the useful heat to the amount of heat input in the burner. For single layer porcelain, when the fuel-air equivalence ratio is gradually enriched, the combustion efficiency deteriorates and diminishes to an

TABLE 4: Uncertainty of the measured data for the double layer porcelain and alumina in the range of fuel-air equivalence ratio.

| Fuel-air equivalence ratio | Uncertainty | | | | | |
| | Surface temperature (°C) | $\dot{m}_f$ L/min | $V$ Volt | $I$ Amp | CO ppm | $NO_x$ ppm |
| --- | --- | --- | --- | --- | --- | --- |
| 0.20 | 0.8% | 2.1% | 1.9% | 0.9% | 1.7% | 0.0% |
| 0.30 | 1.1% | 2.2% | 1.8% | 1.2% | 1.8% | 0.0% |
| 0.35 | 0.9% | 2.0% | 1.6% | 1.1% | 1.6% | 0.0% |
| 0.42 | 1.2% | 1.7% | 1.9% | 1.1% | 1.5% | 1.0% |
| 0.57 | 0.9% | 1.8% | 2.0% | 1.1% | 1.5% | 1.0% |
| 0.59 | 1.0% | 1.9% | 2.1% | 1.3% | 1.4% | 0.0% |
| 0.60 | 1.2% | 1.7% | 2.2% | 1.4% | 1.7% | 0.0% |
| 0.61 | 1.4% | 1.9% | 2.1% | 1.2% | 1.9% | 1.0% |

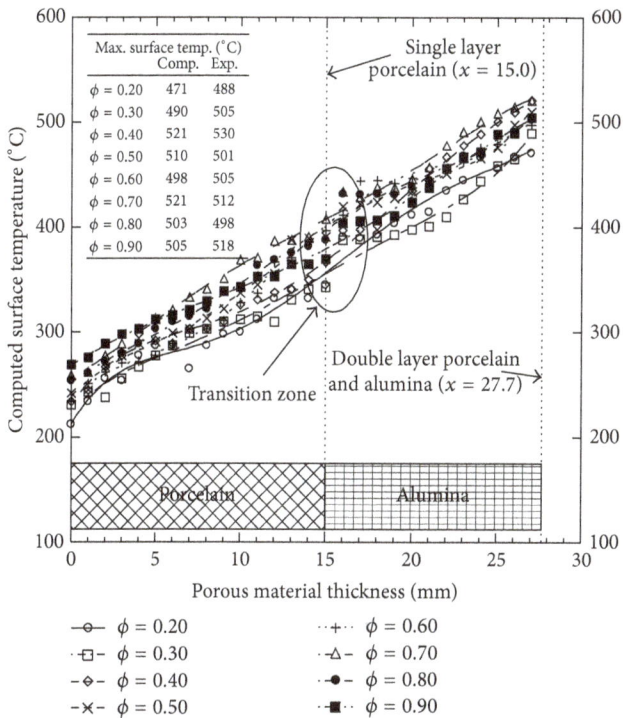

FIGURE 4: Computed surface temperature distributions against porous material thickness at various fuel-air equivalence ratios.

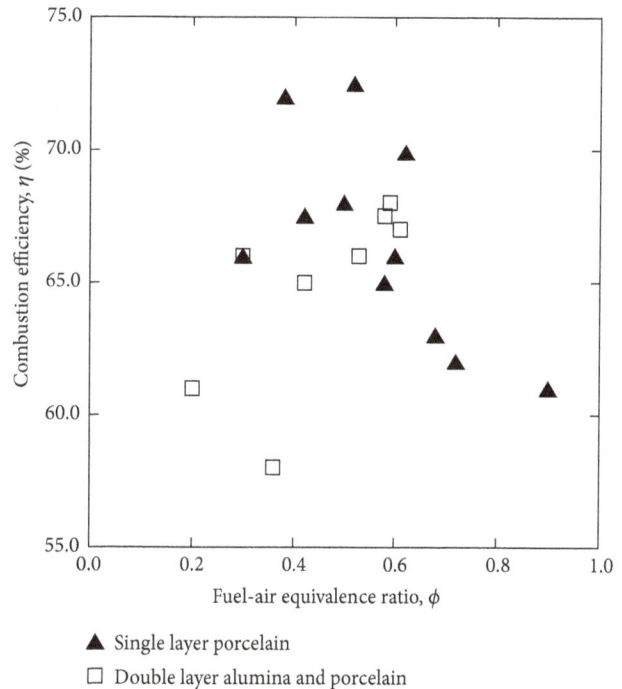

FIGURE 5: Combustion efficiency against fuel-air equivalence ratio.

approximately 60%. The reduced combustion efficiency is supported by the temperature profile images captured in the experiment using the thermal imager. The augmentation in the fuel-air equivalence ratio towards the rich region (Figures 7(a), 7(b), and 7(c)) is accompanied by a reduction in the maximum surface temperature recorded by the imager. Since the wall temperature of the burner only shows minor fluctuations (maximum 5°C), there is little, if any, effect of this temperature on the overall heat transfer. This implies that as the amount of fuel supplied is increased, lack of oxygen prevents complete combustion in the burner and contributed towards the reduced combustion efficiency. Furthermore, since the conduction and convection heat transfer mechanisms are dominant, an improved heat transfer at higher surface temperature for the porous media is evident.

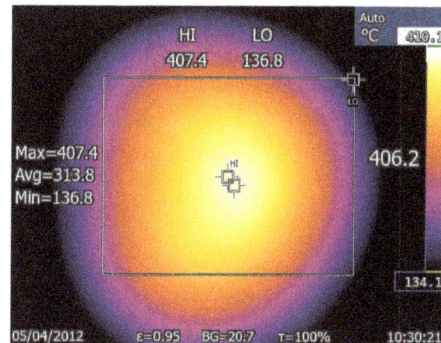

FIGURE 6: Surface temperature profile for single layer porcelain at fuel-air equivalence ratio of 0.52, maximum combustion efficiency 73%.

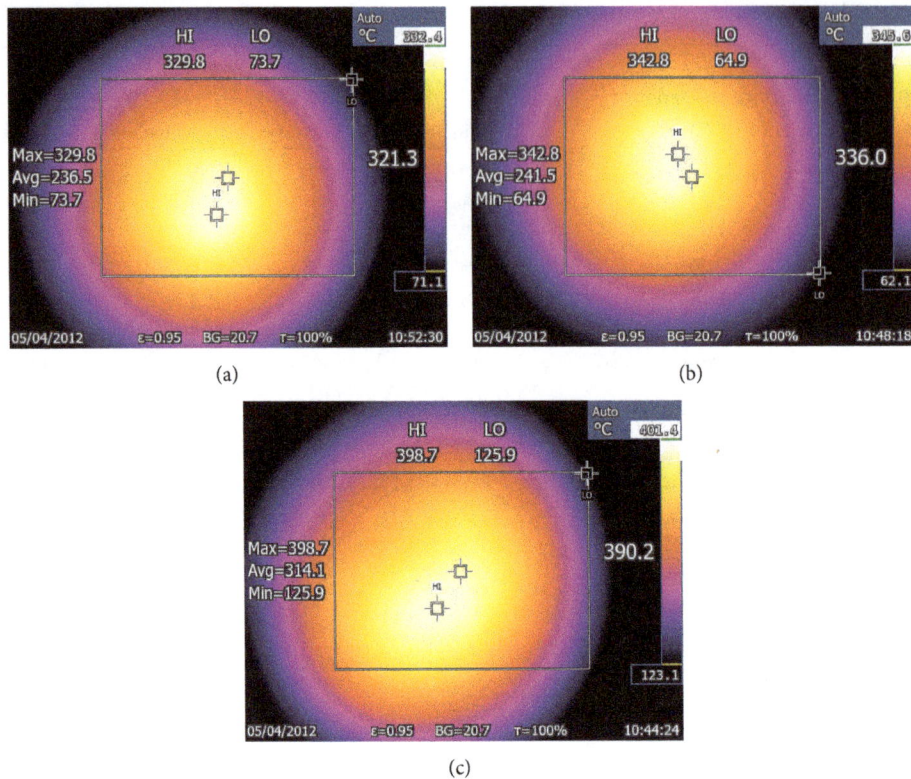

FIGURE 7: Surface temperature profile for single layer porcelain at various fuel-air equivalence ratios (a) $\phi = 0.90$, (b) $\phi = 0.75$, and (c) $\phi = 0.68$.

For double layer alumina and porcelain porous media, the deleterious effect of a stacked porous media is evident, as the combustion efficiency is markedly lower at lean fuel-air equivalence ratio.

The temperature profile images for double layer porous media in the lean region of fuel-air equivalence ratio are shown in Figures 8(a) and 8(b). The figures are temperature profiles for two reference points of fuel-air equivalence ratio ($\phi = 0.30$ and $\phi = 0.35$). These points represent the lean region of fuel-air equivalence ratio for double layer alumina and porcelain porous media. The surface temperature at these points is considerably higher than the surface temperature of the single layer porcelain only. Furthermore, the wall temperature for double layer is slightly higher than the single layer porcelain, but the maximum temperature fluctuations were less than 10°C. The concomitant reduction in combustion efficiency is remarkably peculiar, because it suggests that the temperature difference is not the only governing factor for heat transfer mechanism at these regions of fuel-air equivalence ratio.

For single layer porcelain, it has been delineated that in the rich region of fuel-air equivalence ratio the conduction and convective heat transfer are pronounced. However, stacking up the porous media with bigger pore size alumina (8 ppcm) appears to substantially increase the surface temperature but adversely affect the combustion efficiency. This can be partly explained by considering the role of pores in the matrix of the porous media. As the flame propagates downstream towards bigger size pore, the magnitudes of the

turbulence intensity are higher. As the intensity increases, reaction rates and turbulent flame speed is higher in alumina. This causes the flame temperature to be significantly higher in this section. However, since the combustion efficiency is negatively affected, it is postulated here that the mechanics of flow in the pores of porous media matrix could have contributed towards the diminishing combustion efficiency.

3.3. CO Emission. The emission level of nitrogen oxides ($NO_x$) and carbon monoxide (CO) were measured using combustion analyzer CA-CALC 6203. The probe tip of the combustion analyzer was aligned in the horizontal position to be in the center of the burner with 10 cm vertical distance from the top surface of the porous media. It seemed reasonable to suppose that the vertical distance of 10 cm from the top surface of the porous media to the probe tip was sufficient to ensure uniformity across the entire section of the measured plane of combustion surface. All emission readings have the uncertainty of ±5 ppm. Figure 9 shows the carbon monoxide (CO) emission level (ppm) against fuel-air equivalence ratio, $\phi$.

This will subsequently reduce the reaction rates and turbulent speed in the double layer alumina and porcelain greater than the single layer porcelain only. Figure 9 reveals that CO emission level for stacked alumina and porcelain is flatter until the fuel-air equivalence ratio of about 0.55 before it steeply increases as the equivalence ratio approaches the rich region. Emissions of CO are in the range of 400–800 ppm from the lean limit of 0.20 to 0.55 and markedly increase

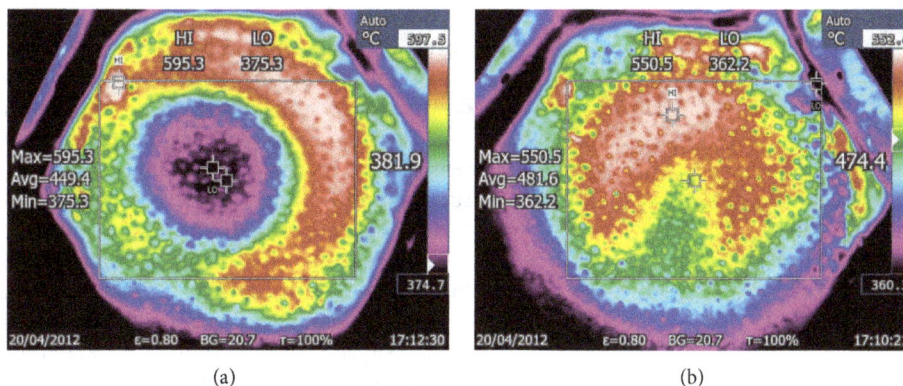

(a)                                                                                   (b)

FIGURE 8: Surface temperature profile for double layer alumina and porcelain at various fuel-air equivalence ratio: (a) $\phi = 0.30$ and (b) $\phi = 0.35$.

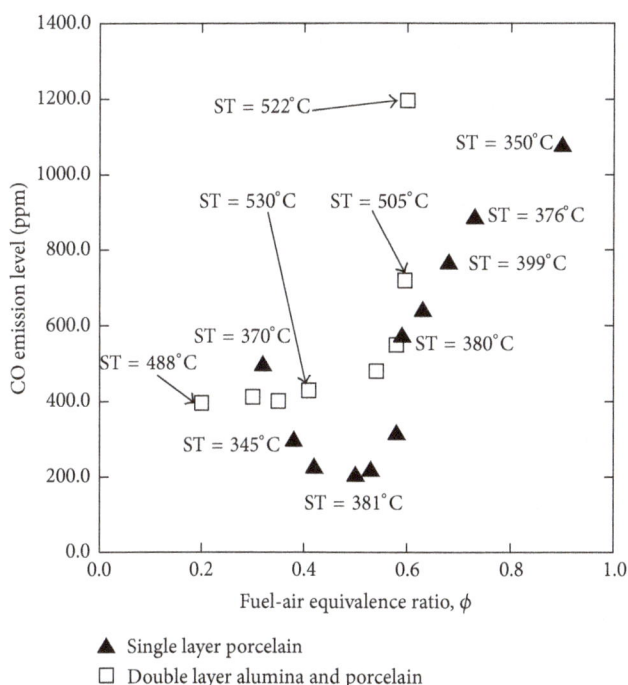

▲ Single layer porcelain
□ Double layer alumina and porcelain

FIGURE 9: CO emission level (ppm) against fuel-air equivalence ratio, $\phi$ for both single (porcelain) and double layer (alumina and porcelain) porous media.

to more than 1200 ppm when the fuel-air equivalence ratio was extended towards rich mixture. It has been elucidated earlier that the incorporation of stacked porous media with different pore size reduces the reaction rates and turbulent speed, but flame stability was moderately enhanced in the lean region of combustion. Furthermore, CO emission is also a useful indicator of the completeness of combustion. Low level of CO is normally created when the combustion is most complete and negligible amount is generated when the fuel is completely burned. Substantial amount of CO emission recorded in this study suggested a deleterious impact of porous media combustion using butane gas as primary fuel. The fuel premixing chamber designed in our investigation was primarily intended to increase the mixture residence

time prior to combustion, by obviating the need of an air compressor which would have been externally driven by an external source. Since the fuel nozzle sits only few millimeters from the opening of the premixed chamber, the air induction was mainly achieved via entrainment to the premixed chamber. However, since the amount of CO level is appreciably high, it can be postulated that the concept of premixing the butane gas with entrained air brought certain degree of shortcomings which has impaired the combustion efficiency. Figure 9 also illustrates that the amount of CO generated for single layer porcelain is equally high, with the maximum value comparable to the double layer alumina and porcelain. The measured value is fairly moderate at around 500 ppm for lean fuel-air equivalence ratio but gradually reduces as the mixture is enriched. There is a fair degree of scatter lying above the minimum value of about 200 ppm and shows increasing trend when the mixture was continually enriched. The surface temperature is also shown in Figure 8. Those highlighted in the figure represent vital surface temperatures at extreme ends of fuel-air equivalence ratio and other points which have temperature difference of about ±20°C with contiguous measuring points. Evidently, when comparison is made for both single and double layer porous media, the surface temperatures do not greatly affect the amount of CO emission recorded across the range of equivalence ratio in the experiment. However, the difference in the surface temperature of double layer alumina and porcelain is apparent, with maximum temperature difference of about 200°C. By arranging the smaller pore size porcelain upstream of the flow, the finer porous medium structure (16 ppcm) creates greater flow resistance compared to alumina (8 ppcm). The turbulence intensity is smaller in small pore size, as described by Hall and Hiatt [24]. As soon as the flow enters alumina with bigger pore size, the turbulence intensity increases and the flow propagates downstream towards the top surface of alumina. This in particular dictates that concomitant increase in the turbulence intensity is thought to be wholly beneficial for the significant increment in the measured temperature of the double layer porous media.

3.4. $NO_x$ Emission. Figure 10 represents the $NO_x$ emission level against fuel-air equivalence ratio for both single layer

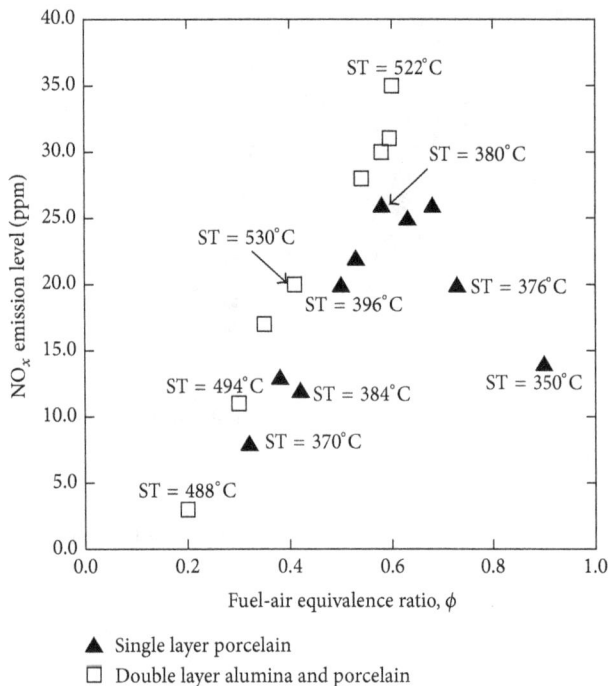

FIGURE 10: $NO_x$ emission level (ppm) against fuel-air equivalence ratio, $\phi$ for both single (porcelain) and double layer (alumina and porcelain) porous media.

porcelain and double layer alumina and porcelain. The surface temperatures at selected points of fuel-air equivalence ratio are also highlighted in the figure for both single and double layer porous media. The data plotted in Figure 10 clearly shows a linear correlation for double layer alumina and porcelain porous media in the ranges of fuel-air equivalence ratio investigated. For single layer porcelain, an increasing trend of the $NO_x$ level is apparent and exhibits maxima at fuel-air equivalence ratio of about 0.60 before the level diminishes towards the rich limit of fuel-air equivalence ratio. Furthermore, essential features in the figure indicate that the level of $NO_x$ emission is clearly higher for double layer porous media compared to single layer porcelain at all operating ranges of fuel-air equivalence ratio. Maximum level of $NO_x$ for single layer and double layer porous media is 27 ppm and 35 ppm, respectively. The sensitivity of $NO_x$ formation is susceptible by two factors: the surface temperature and fuel-air equivalence ratio. When the surface temperature of the combustion zone is high, the amount of $NO_x$ emission recorded is generally high, as can be shown by comparing the surface temperature of the single and double layer porous media. This reflects the dependency of the $NO_x$ formation on the surface temperature.

For double layer porous media, as we sweep from the lean to the rich limit of the equivalence ratio, the level of $NO_x$ increases accordingly. This general trend implies that by increasing the fuel flow rate as the fuel-air equivalence ratio is enriched, the amount of input energy is also increased (since the input energy is the product of fuel flow rate and the calorific value of the fuel). It seems acceptable to

suppose that by premixing greater amount of fuel with air, once the combustion stabilizes in the upstream section of the porous media, the flame propagates towards larger pore downstream porous media. In the larger pore porous media (alumina), combustion intensity increases and the surface temperature increases accordingly. However, the maximum surface temperature recorded in our study is not at the richest fuel-air equivalence ratio but shifted slightly towards leaner fuel-air equivalence ratio (approximately 0.57). This shows the caution required to draw a direct conclusion based on this observation alone, because the difficulty arises owing to the complexity and lack of proper understanding of the underlying physics of $NO_x$ formation. This needs to be corroborated further by analyzing the exact temperature distribution to understand the detail mechanism of $NO_x$ formation. The single layer porcelain yields lower surface temperature as all data scatters lying below the measured surface temperature of the double layer porous media. The maximum temperature for the single layer porcelain (399°C) is 89°C lower than the lowest surface temperature of the double layer porous media (488°C). It is also evident from Figure 9 that the single layer porcelain allows fuel-air equivalence ratio to be enriched to about 0.90 before flashback occurs in the combustion zone. Interestingly, the maximum $NO_x$ emission occurs almost at the same fuel-air equivalence ratio of double layer porous media. It then gradually decreases as the fuel-air equivalence ratio was enriched. Since single layer porous media consists of smaller pore size compared to double layer porous media, there is no transition of the combustion intensity throughout the entire section of the porous media. This could in particular dictate that the path taken by the combustion flow gases does not suffer from adverse combustion intensity changes, as would have taken place in the double layer porous media. Temperature distribution is much more uniform and lower, which inferred the significantly lower amount of $NO_x$ obtained in this section of porous media.

### 3.5. Temperature Difference versus Fuel-Air Equivalence Ratio.
The plot of temperature difference against fuel-air equivalence ratio is shown in Figure 11. The temperature difference shown in the figure is based on the temperature difference of surface temperature and the wall temperature for each calculated points of fuel-air equivalence ratio. The wall temperature refers to the average fins temperature attached to the thermoelectric modules. Temperature difference is a more meaningful parameter than the exact surface temperature of the porous media since the electricity generated in thermoelectric modules works on Seebeck effect, which is strongly dependent on temperature difference generated from a burner. Figure 11 shows that the temperature difference for double layer porous media is higher than the single layer porous media in the range of fuel-air equivalence ratio investigated. The maximum temperature difference for double layer porous media is 459°C (at fuel-air equivalence ratio of 0.57) and for single layer porous media is 400°C (at fuel-air equivalence ratio of 0.67). For double layer porous media, a change in the fuel-air equivalence ratio does not create perceptible change in the measured temperature difference. However, closer inspection in the figure reveals

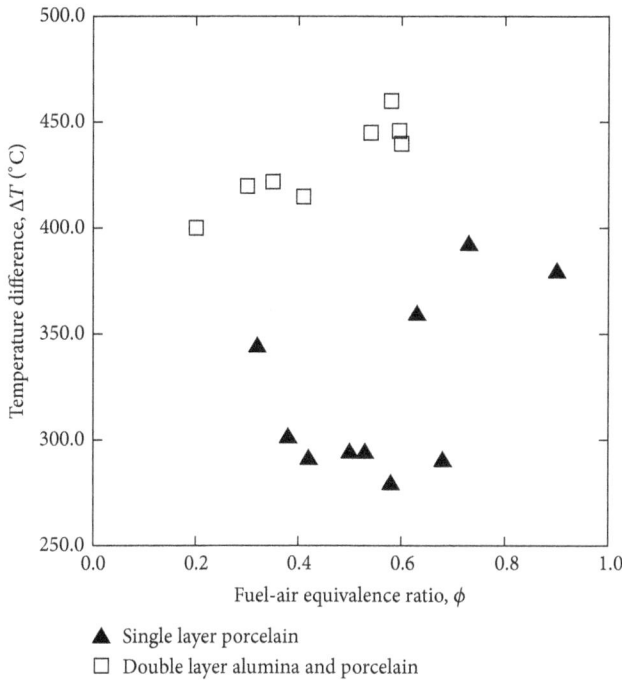

FIGURE 11: Temperature difference (°C) against fuel-air equivalence ratio, $\phi$ for both single (porcelain) and double layer (alumina and porcelain) porous media.

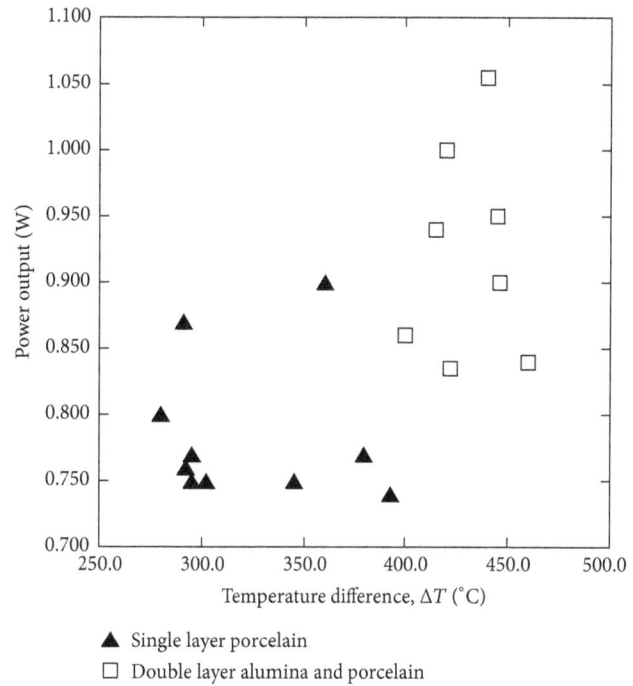

FIGURE 12: Temperature difference (°C) against power output (W) for both single porcelain and double layer alumina and porcelain porous media.

that there are few scatters lying in the temperature difference region of approximately 450°C when the fuel-air equivalence ratio is nearing the flashback (rich) region. This is the highest temperature difference for the ranges of fuel-air equivalence ratio varied in this study. It has been expounded earlier that flashback is characterized by the sudden reverse flow towards the upstream section of the stacked porous media at rich spectrum of fuel-air equivalence ratio. However, since the thermal conductivity of alumina (40 W/mK) is greater than the thermal conductivity of porcelain (only 1.5 W/mK), a new thermal equilibrium for double layer alumina and porcelain is significantly longer to attain compared to single layer porous media. Heat feedback from the porcelain to the alumina occurs and resulted in higher temperature difference generated in the burner. For single layer porous media using porcelain, low thermal conductivity and smaller pore size have adversely affected the temperature difference obtained, as illustrated by the lower temperature difference when comparison is made with double layer porous media. Combustion intensity is much more rigorous in bigger pore size alumina than in the smaller pore porcelain. On the other hand, the calmer combustion intensity in the smaller pore porcelain has enhanced combustion stability, but at the expense of the lower temperature difference obtained in the section. In addition, when the fuel-air equivalence ratio generally recedes towards the lean mixture, temperature difference is lowest, owing to the compounding effect of reduced combustion intensity and low energy input from the supplied butane gas.

*3.6. Electric Power Output from Thermoelectric Modules.* The temperatures difference between the porous media surface, voltage ($V$), and current ($A$) are generated from the raw data

of this study and they are tabulated in Table 5 below. The data tabulated in Table 5, in particular voltage and current can be extracted to give power output (Watt) and this is plotted against the temperature difference as shown in Figure 12.

The electric power (Watt) generated from this study is found to be marginally higher for double layer alumina and porcelain compared to single layer porcelain. In general, it is observed that an increase in the temperature difference resulted in higher power output produced by the system. The lowest power obtained using the system was found to be about 0.750 W for single layer porcelain and the highest is about 1.05 W, obtained in the double layer porous media. Six PbSnTe thermoelectric modules used in the system were thermally connected in parallel and electrically connected in series. Each leg of the thermoelectric modules was connected in series and the final pair of legs forms the terminal for voltage and current measurement. The current and voltage output were measured using the digital multimeter. The wall temperature of the burner was measured using Fluke Ti27 9 Hz thermal imager, without encountering significant fluctuations in the measured temperature throughout the entire experiment. It was also observed that the wall temperature consistently pulsated between 3°C and 5°C and they were thought to have very little influence on the outcome of the overall temperature difference attained in this study.

*3.7. Feasibility of the Proposed Burner for TE Power Generation.* A porous burner for TE power generation could be viewed as an alternative power device. The primary feature of our designed is a strong dependence of the thermal and electrical variables on the fuel-air equivalence ratio. A change

TABLE 5

| Double layer alumina and porcelain | | | Single layer porcelain | | |
|---|---|---|---|---|---|
| Temperature difference (°C) | Voltage (V) | Current (A) | Temperature difference (°C) | Voltage (V) | Current (A) |
| 401 | 10.64 | 0.081 | 345 | 10.02 | 0.080 |
| 459 | 10.35 | 0.082 | 310 | 9.98 | 0.087 |
| 441 | 10.45 | 0.085 | 295 | 10.11 | 0.076 |
| 441 | 11.30 | 0.085 | 341 | 10.55 | 0.070 |
| 414 | 11.46 | 0.086 | 307 | 10.19 | 0.075 |
| 414 | 10.85 | 0.077 | 306 | 9.45 | 0.08 |
| 410 | 11.10 | 0.084 | 289 | 9.75 | 0.083 |
| 434 | 12.10 | 0.087 | 361 | 10 | 0.090 |
| | | | 399 | 10.05 | 0.074 |
| | | | 376 | 9.78 | 0.076 |
| | | | | 9.65 | 0.071 |

in fuel-air equivalence ratio is associated with the change in chemical energy of the fuel. This leads to various thermal and electrical parameters of the burner, which could be optimized for specific TE power generation. The surface temperature of a double layer porous burner is considerably higher than the single layer throughout the entire range of fuel-air equivalence ratio. However, the combustion efficiency for a double layer is marginally lower than the single layer. This indicates that operating the single layer burner in the lean region is beneficial compared to the double layer. Furthermore, the system is temperature dependent, and the maximum permissible operating temperature of the TE cells must be strictly adhered. It is also shown that the CO and $NO_x$ is generally low in the lean region of combustion. For CO, this reduction is due to completeness of combustion, and, for $NO_x$, the observed trend is fundamentally related to the temperature profiles. The emission findings accord well with the surface temperatures of the burner in the lean operating region. It is also interesting to note that the double layer porous burner yields greater power output throughout the range of fuel-air equivalence ratio. However, the calculated power is very low, leading to a reduced overall efficiency of the system. Therefore, if the system improvement were needed in the current setup, the use of high performance semiconductor materials must be chosen for future TE power generation. It is also important to note that butane is used in our current study. If portability of the fuel used is of the utmost priority, our present setup with butane gas has been demonstrated to be feasible. However, other commercially available fuels, such as propane or methane can be employed without major hardware modifications.

## 4. Conclusion

An experimental study has been conducted to evaluate the characteristics of a porous burner for thermoelectric (TE) power generation. Two types of configurations were assessed: double layer porous burner composed of alumina and porcelain and single layer porcelain only. The characteristics of the burner are presented in terms of the combustion efficiency, surface temperature, and the emission level. The electricity was generated using six PbSnTe thermoelectric

(TE) modules, which were attached to the wall of the burner. The surface temperature for the double layer porous media is significantly higher than the single layer. The flow transition from the smaller pore size porcelain (16 ppcm) to the higher pore size alumina (8 ppcm) has contributed to the higher surface temperature recorded for the double layer porous media. In the range of fuel-air equivalence ratio investigated, for double layer porous media, $NO_x$ emission increases linearly and peaked at fuel-air equivalence ratio of 0.60. Similar trend is also observed for single layer porcelain, but level decreased when the fuel-air equivalence ratio was extended further towards the rich region. The amount of CO emission is generally high owing to the complex flow mechanism as the flame propagates downstream from the smaller pore size porcelain to bigger pore size alumina. The electric power generated is calculated based on the current and voltage produced from the TE modules. The values are generally dependent on the temperature difference between the burner and the wall, with higher power generated at greater temperature difference.

## Conflict of Interests

The authors declare that there is no conflict of interests regarding the publication of this paper.

## Acknowledgment

This work was supported by Ministry of Higher Education, Malaysia, under the Fundamental Research Grant Scheme (FRGS) (Grant no. 6071236).

## References

[1] K. Qiu and A. C. S. Hayden, "Integrated thermoelectric generator and application to self-powered heating systems," in *Proceedings of the 25th International Conference on Thermoelectrics (ICT '06)*, pp. 198–203, Vienna, Austria, August 2006.

[2] K. Qiu and A. C. S. Hayden, "A natural-gas-fired thermoelectric power generation system," *Journal of Electronic Materials*, vol. 38, no. 7, pp. 1315–1319, 2009.

[3] D. Champier, J. P. Bedecarrats, M. Rivaletto, and F. Strub, "Thermoelectric power generation from biomass cook stoves," *Energy*, vol. 35, no. 2, pp. 935–942, 2010.

[4] J. Yang and F. R. Stabler, "Automotive applications of thermoelectric materials," *Journal of Electronic Materials*, vol. 38, no. 7, pp. 1245–1251, 2009.

[5] S.-K. Kim, B.-C. Won, S.-H. Rhi, S.-H. Kim, J.-H. Yoo, and J.-C. Jang, "Thermoelectric power generation system for future hybrid vehicles using hot exhaust gas," *Journal of Electronic Materials*, vol. 40, no. 5, pp. 778–783, 2011.

[6] Y. D. Deng, W. Fan, K. Ling, and C. Q. Su, "A 42-V electrical and hybrid driving system based on a vehicular waste-heat thermoelectric generator," *Journal of Electronic Materials*, vol. 41, no. 6, pp. 1698–1705, 2012.

[7] J. Posthill, A. Reddy, E. Siivola et al., "Portable power sources using combustion of butane and thermoelectrics," in *Proceedings of the 24th International Conference on Thermoelectrics (ICT '05)*, pp. 520–523, Clemson, SC, USA, June 2005.

[8] M. M. Rahman and R. Shuttleworth, "Thermoelectric power generation for battery charging," in *Proceedings of the International Conference on Energy Management and Power Delivery (EMPD '95)*, pp. 186–191, November 1995.

[9] K. Yoshida, S. Tanaka, S. Tomonari, D. Satoh, and M. Esashi, "High-energy density miniature thermoelectric generator using catalytic combustion," *Journal of Microelectromechanical Systems*, vol. 15, no. 1, pp. 195–203, 2006.

[10] J. R. Howell, M. J. Hall, and J. L. Ellzey, "Combustion of hydrocarbon fuels within porous inert media," *Progress in Energy and Combustion Science*, vol. 22, no. 2, pp. 121–145, 1996.

[11] C. Periasamy, S. K. Sankara-Chinthamony, and S. R. Gollahalli, "Experimental evaluation of evaporation enhancement with porous media in liquid-fueled burners," *Journal of Porous Media*, vol. 10, no. 2, pp. 137–150, 2007.

[12] S. Vijaykant and A. K. Agrawal, "Liquid fuel combustion within silicon-carbide coated carbon foam," *Experimental Thermal and Fluid Science*, vol. 32, no. 1, pp. 117–125, 2007.

[13] C. Periasamy and S. R. Gollahalli, "Experimental investigation of kerosene spray flames in inert porous media near lean extinction," *Energy and Fuels*, vol. 25, no. 8, pp. 3428–3436, 2011.

[14] S. Jugjai and N. Polmart, "Enhancement of evaporation and combustion of liquid fuels through porous media," *Experimental Thermal and Fluid Science*, vol. 27, no. 8, pp. 901–909, 2003.

[15] T. L. Marbach and A. K. Agrawal, "Experimental study of surface and interior combustion using composite porous inert media," *Journal of Engineering for Gas Turbines and Power*, vol. 127, no. 2, pp. 307–313, 2005.

[16] M. T. Smucker and J. L. Ellzey, "Computational and experimental study of a two-section porous burner," *Combustion Science and Technology*, vol. 176, no. 8, pp. 1171–1189, 2004.

[17] K. Hanamura, T. Kumano, and Y. Iida, "Electric power generation by super-adiabatic combustion in thermoelectric porous element," *Energy*, vol. 30, no. 2-4, pp. 347–357, 2005.

[18] W.-H. Chen, C.-Y. Liao, C.-I. Hung, and W.-L. Huang, "Experimental study on thermoelectric modules for power generation at various operating conditions," *Energy*, vol. 45, no. 1, pp. 874–881, 2012.

[19] K. Mustafa, S. Abdullah, M. Abdullah, K. Sopian, and A. Ismail, "Experimental investigation of the performance of a liquid fuel-fired porous burner operating on kerosene-vegetable cooking oil (VCO) blends for micro-cogeneration of thermoelectric power," *Renewable Energy*, vol. 74, pp. 505–516, 2015.

[20] P.-F. Hsu, W. D. Evans, and J. R. Howell, "Experimental and numerical study of premixed combustion within nonhomogeneous porous ceramics," *Combustion Science and Technology*, vol. 90, no. 1–4, pp. 149–172, 1993.

[21] A. Gungor, "Analysis of combustion efficiency in CFB coal combustors," *Fuel*, vol. 87, no. 7, pp. 1083–1095, 2008.

[22] J. Charoensuk and A. Lapirattanakun, "On flame stability, temperature distribution and burnout of air-staged porous media combustor firing LPG with different porosity and excess air," *Applied Thermal Engineering*, vol. 31, no. 16, pp. 3125–3141, 2011.

[23] T. D. Eastop and A. McConkey, *Applied Thermodynamics for Engineering Technologies*, Prentice Hall, London, UK, 1993.

[24] M. J. Hall and J. P. Hiatt, "Exit flows from highly porous media," *Physics of Fluids*, vol. 6, no. 2, pp. 469–479, 1994.

# Influence of Sorbent Characteristics on Fouling and Deposition in Circulating Fluid Bed Boilers Firing High Sulfur Indian Lignite

**Selvakumaran Palaniswamy,[1] M. Rajavel,[1] A. Leela Vinodhan,[1] B. Ravi Kumar,[1] A. Lawrence,[1] and A. K. Bakthavatsalam[2]**

[1] *Bharat Heavy Electricals Limited, Tiruchirappalli, Tamil Nadu 620 014, India*
[2] *National Institute of Technology, Tiruchirappalli, Tamil Nadu 620015, India*

Correspondence should be addressed to Selvakumaran Palaniswamy; pskumaran9454@gmail.com

Academic Editor: Michael Fairweather

125 MWe circulating fluidized bed combustion (CFBC) boiler experienced severe fouling in backpass of the boiler leading to obstruction of gas flow passage, while using high sulfur lignite with sorbent, calcium carbonate, to capture sulfur dioxide. Optical microscopy of the hard deposits showed mainly anhydrite ($CaSO_4$) and absence of intermediate phases such as calcium oxide or presence of sulfate rims on decarbonated limestone. It is hypothesized that loose unreacted calcium oxides that settle on tubes are subjected to recarbonation and further extended sulfation resulting in hard deposits. Foul probe tests were conducted in selected locations of backpass for five different compositions of lignite, with varied high sulfur and ash contents supplied from the mines along with necessary rates of sorbent limestone to control $SO_2$, and the deposits build-up rate was determined. The deposit build-up was found increasing, with increase in ash content of lignite, sorbent addition, and percentage of fines in limestone. Remedial measures and field modifications to dislodge deposits on heat transfer surfaces, to handle the deposits in ash conveying system, and to control sorbent fines from the milling circuit are explained.

## 1. Backdrop

India with growing energy consumption is looking at utilizing all its potential energy resources in the most economic and environmentally sustainable manner. Coal will continue to be the major energy source in India due to its availability. Per capita consumption of electricity and GDP growth has direct relation, and energy intensity in developing countries like India is comparatively more than the developed world, and the gap between supply and demand is ever increasing. The demand for all forms of energy is expected to increase substantially in the foreseeable future and is expected to get doubled by 2030. Although coal would continue to be a major energy source in India due to its availability, lignite is fast emerging as an alternate source of fuel for electricity generation. In India, the total lignite potential is 4177 million tonnes. Indian lignites have a typical analytical range of ash content of 15 to 35%, sulfur content of 1.0 to 7.0%, and moisture content of 10 to 45%. The varieties found in Gujarat and Rajasthan region have moderate to high sulfur (1 to 7%) content. It has become an economic necessity to use these lignites for power generation in view of spurt in energy demand, with $SO_2$ emission controlled. Circulating fluid bed combustion (CFBC) technology is employed considering the impurities, moisture, ash, and sulfur content and wide variations in lignite. Hence, the share of lignite-based pit head thermal projects in Gujarat and Rajasthan is increasing. The size of CFB boilers in India using lignite has reached already over 250 MWe and set to increase above 500 MWe, and that underlines the importance.

Slagging, fouling, and ash deposition are major problems experienced in PF boilers. In contrast, agglomeration of bed particles in fluidized bed combustion system is considered as a primary operational issue. Interaction and coalescence of bed particles and ash *(sintering)* are considered to be the principal sources of agglomeration in CFB boilers employing bed material and fuel ash as binary system. Choking/blocking in fuel path is another peculiar operational problem experienced worldwide in CFB boilers, firing pet-coke, low rank coals, and biofuels [1].

Lignite mineralogy greatly influences combustion behavior. Agglomeration and clogging/blocking are experienced due to sintering of lignite ash with limestone (sorbent) at lower temperature regime in which CFB boilers operate (640–960°C). At this low temperature range, the extensive knowledge built with respect to slagging, fouling, and corrosion phenomenon occurring at higher temperatures in pulverized fuel combustion may not be applicable. In CFB boilers, ash sintering contributes to deposit formation in cyclone, return leg, and postcyclone flue gas channel (backpass) [1]. In operating units, rapid sintering lead to heavy agglomerate formation, which finally inhibited circulation in dense phase areas (such as seal pot) and in the backpass. Understanding the sintering behavior of fuel is required for resolving such problems.

Over the past decades, designers and operators of fluidized beds have been concentrating on developing the CFBC technology by establishing the optimum operating conditions and troubleshooting associated with refractory and so forth. Due attention has not been paid to understand the limestone characteristics that are important for efficient capture of $SO_2$. *Present work describes influence of limestone and its grain size in blocking/clogging of cyclone and hard deposits in second pass of CFB boiler during combustion of high sulfur lignite with high ash content (20 to 30%) in CFB units in Giral, Rajasthan state of India.*

## 2. Operational Issues

High sulfur content lignite, available at Giral, Barmer District, and Rajasthan state, is used as fuel. These lignites had posed several operational issues during initial stage of commissioning and stabilization. High sulfur in the lignite needs high limestone feed rates to control emissions. High limestone feed rates caused huge quantities of backpass deposits, which led to obstruction of gas flow passage. Despite providing steam soot blowers for clearing the deposits obstruction of gas flow increased with increase in limestone feed rate.

*2.1. Cyclone Standpipe Blockage.* During commissioning, ash holdup occurred in cyclone standpipe at low loads of about 20 to 40 MW. Ash analysis of the hold-up material is carried out.

*2.2. Backpass Fouling.* Sulfur dioxide emitted during combustion is absorbed in situ by adding limestone of size less than 1.0 mm. The CFB boiler experienced fouling in superheater/reheater (SH/RH) coils while adding required quantity of limestone (Figure 1). Heavy and rapid deposit buildup has been experienced on the flue gas side of the heat transfer tubes. Deposit buildup was most severe at low temperature superheater (LTSH)-SH 1B tube bank. Also, growth of ash deposit in final stage reheater tube bank was observed during the initial period of operation. These deposits increased gas-side pressure drop and in turn increased loading of induced draught (ID) fans, with high current, causing boiler trips.

Consequently, CFB boiler was required to be operated with less quantity of limestone which resulted in more sulfur dioxide emissions. The fouling took place mostly in LTSH coils of backpass which is placed between reheater and economizer. Due to fouling in the backpass, fly ash particles collected in hoppers of economizer and in other zones got sintered during intermittent storage. Nonoperation of soot blowers (SB) and water ingress while starting soot blowing caused cakes formation. Dislodgement of such cakes leads to difficulty in ash evacuation. Deashing system pump was chocked often, due to sintered particles (lumps) formed due to water ingress.

## 3. Experiments: Laboratory and Field

*3.1. Lignite.* Six samples (sample 1 to sample 6) of high sulfur lignite collected from Giral/Rajasthan/India (covering a range of high sulfur content) are considered for the present study of backpass fouling propensity of the high sulfur fuels in CFB boiler. All the fuel samples are prepared in accordance with ASTM-D 2013. The as-received solid fuels are crushed to pass a number 4 sieve (4.75 mm) and then air dried until the loss in weight is not more than 0.1% per hour. Air dried samples are again crushed to pass a number 72 mesh (212 microns). Samples of sizes less than 72 mesh are used for analyses of proximate, ultimate, and calorific values. Adequate quantity of ash of each fuel is generated using proximate analyses at 750°C for further analyses of chemical composition, ash fusion temperature. The proximate, ultimate, and gross calorific values of the samples were carried out using TGA 701 proximate Analyzer (LECO), Elemental analyzer Vario EL III, and PARR Isoperibol Bomb Calorimeter, respectively. The chemical composition of ashes was carried out by ICP- AES, Perkin Elmer.

*3.2. Limestone.* The sorbents are characterized based on the $CaCO_3$ content, particle size distribution of the parent sorbent, and a relative sulfation reactivity parameter [3]. Calcium utilization, in general, increases as the sorbents particle size decreases. *As the particle size distribution of the feed sorbent changes in a CFB due to attrition, it is taken for granted that the feed size distribution of limestone (input) is not as important as the resultant sorbent size distribution in the boiler.* On the contrary, mathematical model results show that sulfur capture efficiency is related to particle attrition/fragmentation of sorbent inventory in addition to input particle size distribution to the performance of circulating fluidized bed CFB combustors [4]. The physical and chemical properties of a sorbent are important when evaluating for use in CFB application. Sorbents although chemically similar, may have different sulfation performance. Extensive literature studies on process of desulfurization in CFBC show that sorbent conversion degree is dependent not only on residence time in combustor but also on its porosity, pore structure and pore size distribution [5]. The detailed analyses of Indian

(a)

(b)

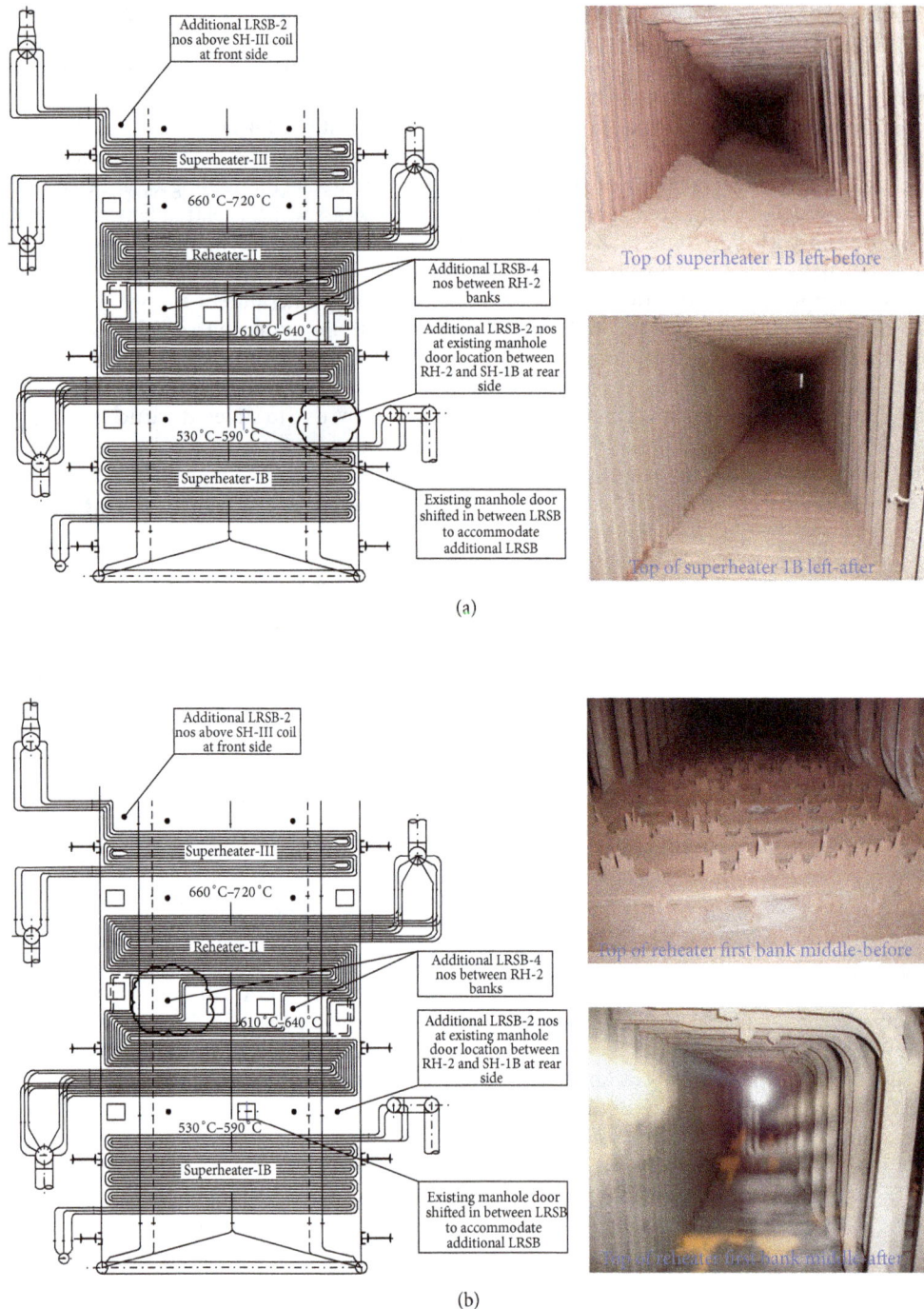

FIGURE 1: Deposits in superheater/reheater coils before and after introduction of high pressure soot blowers and location of additional soot blowers in backpass.

limestones-chemical composition, calcium and magnesium carbonate contents, that are used in CFB were performed using Inductively Coupled Plasma-Atomic Emission Spectroscopy (ICP-AES) Perkin Elmer Optima 2000 DU and using Inductively Coupled Plasma-Mass Spectroscopy (ICP-MS) Perkin Elmer. *Sulfation of limestones of different size fractions showed that sorbent requirement (g of sorb/g of sulfur) is less for finer size fractions* [6].

### 3.3. Deposit Sampling Using Probes and Field Experiment

*3.3.1. Deposit Probes.* Field experiment using deposit probes is taken up, as the wide range of characterization of the selected limestones with respect to their potential difference as desulfurisation agents in CFBC boilers yielded no definitive evidence of the fouling and deposition faced in the operating units.

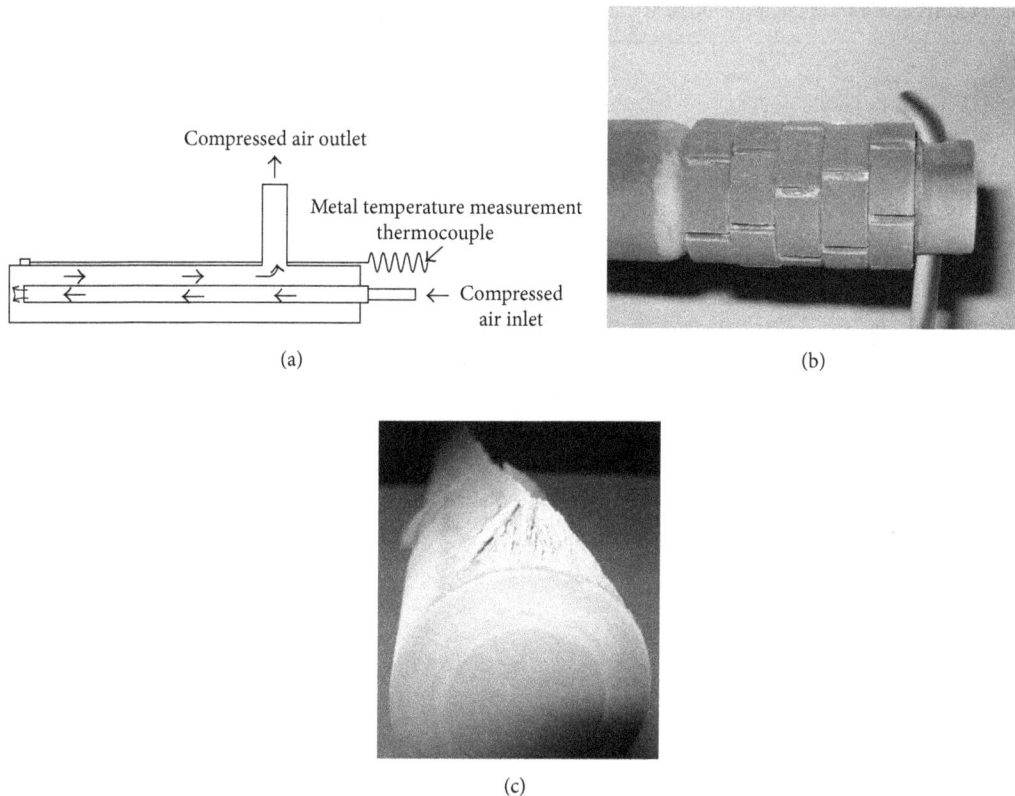

FIGURE 2: (a) Schematic sketch of probe to collect fouling samples, (b) steel probe with rings [2], and (c) foul probe with deposits.

A deposit probe is a good tool for finding out the mechanisms of deposit formation. Air cooled deposit probes of type Figure 2 was used for sampling of deposits, which are equipped with detachable rings [2]. The temperature of the probe can be controlled by varying flow rate of pressurized air. For each test, a new probe/ring is used and the weight of the probe/ring is checked before and after exposure. Taking into account exposure time, a rate of deposit buildup (g/(m² h)) can be calculated. Deposited probes/rings are stored for analysis.

Deposits were collected from three different locations in the backpass after SH-1B, in between RH-2 bundles, and after RH-2 (Figure 3). Chemical composition analysis of the probe deposits is carried out. The sieve analysis of deposits shows significant share of particles smaller than 50 μm size. It was clear that addition of limestone significantly increased the formation of hard deposits compared to firing only lignite, that is, without any limestone.

*3.3.2. Particle Size Distribution of Injected Lime.* The sieve analysis of collected deposits showed that these deposits were built up mainly by fine lime particles injected into furnace. Figure 4 shows distribution of the particle size for two samples done by wet sieving. The share of particles smaller than 50 μm size indicated that fine fractions were higher than envisaged during design (0 to 5% less than 50 μm). *Earlier researchers have shown that the particle size distribution of sorbent could significantly affect deposit formation rate* [2].

## 4. Results and Discussions

Analyses of proximate, ultimate, and gross calorific value and chemical composition of ashes for the seven lignite samples are listed in Table 1. Analysis of chemical composition of the hold-up material in the cyclone standpipe is furnished in Table 2. Detailed limestone analyses-chemical composition, calcium and magnesium carbonate contents for the Indian limestones that are used in CFB, are furnished in Table 3. Fouling probe test condition/measurement details are furnished and the chemical composition analysis of the probe deposits is furnished in Table 4. Mineralogy of the probe deposits as determined by XRD is furnished in Table 5.

*4.1. Correlation with Conventional Ash Deposition Indices.* Various conventional indices, based upon ash chemistry, have been calculated as indicators of slagging and fouling propensity [7]. Values for the following indices, for the high sulfur lignite samples 1 to 7 are given in Table 1:

Silica ratio = $SiO_2/(SiO_2 + Fe_2O_3 + CaO + MgO) * 100$

Base/acid ratio = $(Fe_2O_3 + CaO + MgO + Na_2O + K_2O)/(SiO_2 + Al_2O_3 + TiO_2)$

Iron index = $Fe_2O_3 * B/A$

Iron/calcium ratio = $Fe_2O_3/CaO$

Iron + calcium in ash = $Fe_2O_3 + CaO$

TABLE 1: Proximate, ultimate, chemical composition of ash, ash fusion temperatures, and ash deposition indices of high sulfur lignite.

| Sample ID | Sample 1 Lignite Giral | Sample 2 Lignite Giral | Sample 3 Lignite Giral | Sample 4 Lignite Giral | Sample 5 Lignite Giral | Sample 6 Standpipe blockage Giral |
|---|---|---|---|---|---|---|
| Proximate analysis (wt % on air dried basis) | | | | | | |
| Moisture | 11.8 | 10.0 | 29.6 | 29.1 | 15.0 | 9.6 |
| Volatile matter | 37.5 | 29.5 | 27.8 | 28.4 | 33.7 | 37.8 |
| Ash | 18.6 | 34.5 | 15.6 | 13.9 | 18.7 | 26.8 |
| Fixed carbon | 32.1 | 26.0 | 27.0 | 28.6 | 32.6 | 25.8 |
| Gross calorific value Cal/g | 4865 | 3445 | 3645 | 4059 | 4720 | 4030 |
| Ultimate (wt % on air dried basis) | | | | | | |
| Carbon | 51.6 | 38.5 | 35.3 | 39.5 | 49.1 | 41.0 |
| Hydrogen | 3.8 | 2.5 | 2.6 | 2.6 | 3.3 | 4.0 |
| Nitrogen | 0.6 | 0.6 | 0.9 | 0.8 | 0.7 | 0.6 |
| Sulfur | 6.94 | 5.5 | 4.1 | 4.7 | 6.70 | 4.0 |
| Chemical composition of ash (wt %) | | | | | | |
| $SiO_2$ | 25.9 | 39.2 | 41.0 | 36.2 | 25.1 | 34.1 |
| $Al_2O_3$ | 12.6 | 27.5 | 22.0 | 17.7 | 14.2 | 14.8 |
| $Fe_2O_3$ | 28.8 | 16.5 | 21.4 | 25.7 | 26.4 | 11.9 |
| $TiO_2$ | 1.3 | 2.1 | 2.3 | 2.4 | 1.5 | 1.6 |
| CaO | 8.3 | 4.2 | 3.7 | 5.1 | 6.6 | 3.3 |
| MgO | 3.3 | 2.1 | 2.2 | 2.2 | 3.1 | 1.3 |
| $Na_2O$ | 7.2 | 1.4 | 1.7 | 2.8 | 8.3 | 4.1 |
| $K_2O$ | 0.3 | 0.6 | 0.4 | 0.4 | 0.3 | 0.2 |
| $SO_3$ | 11.0 | 6.2 | 5.7 | 7.2 | 13.7 | 28.7 |
| Ash fusion temperatures °C (oxidizing atmosphere) | | | | | | |
| Temperatures | 1 | 2 | 3 | 4 | 5 | 6 |
| Deformation T1 | >1152 | 1267 | 1275 | 1311 | >1152 | 1244 |
| Softening T2 | >1214 | 1290 | 1300 | 1321 | >1214 | 1260 |
| Hemisphere T3 | >1230 | 1307 | 1333 | 1364 | >1230 | >1300 |
| Fusion T4 | >1250 | 1377 | 1360 | 1385 | >1250 | >1300 |
| Ash deposition indices | | | | | | |
| Si ratio | 39.06 | *63.22* | *60.02* | *52.31* | 41.01 | *67.4* |
| Base/acid | 1.20 | *0.36* | *0.45* | *0.64* | 1.78 | *0.41* |
| Iron index | 34.56 | *5.93* | *9.63* | *16.45* | 47.0 | *4.9* |
| Fe/Ca | 3.47 | 3.93 | 5.78 | 5.04 | 4.0 | 3.6 |
| Fe + Ca | *37.1* | *20.7* | *25.1* | *30.8* | *33.0* | *15.2* |

TABLE 2: Cyclone outlet standpipe blockage—chemical composition of fuel* ash and clinkers.

| Material | $Na_2O$ | MgO | $Al_2O_3$ | $SiO_2$ | $SO_3$ | $P_2O_5$ | $K_2O$ | CaO | $Fe_2O_3$ | $TiO_2$ |
|---|---|---|---|---|---|---|---|---|---|---|
| Fuel ash—Table 1 sample 6 | 4.1 | 1.3 | 14.8 | 34.1 | 28.7 | — | 0.2 | 3.3 | 11.9 | 1.6 |
| Black clinker | 2.6 | 1.8 | 3.8 | 6.4 | 37.1 | 0.1 | 0.1 | 30.8 | 16.9 | 0.4 |
| Brown clinker | 2.4 | 1.6 | 4.5 | 7.2 | 29.7 | 0.3 | 0.4 | 31.5 | 21.8 | 0.6 |
| Grey clinker | 2.1 | 1.1 | 3.6 | 6.4 | 35.0 | 0.3 | 0.1 | 31.9 | 18.9 | 0.6 |

*Table 1 sample 6.

The interpretation of such ash deposition indices requires caution, as these have been developed for a particular range or type of coal, and influence of boiler design/operating conditions is not accounted. Ash chemistry indices do not count the mineralogical mode of occurrence of the elements of concern and mineral associations, both of which are equally important as the ash chemistry in determination of slagging and fouling. With the above limitations, it can be seen from Table 1 that the values for most of the common ash deposition indices suggest that the lignite samples would

TABLE 3: Elemental analysis—calcium and magnesium carbonate contents of limestones.

| Limestone sample ID | (1) SLPP | (2) Ariyalur | (3) NLC Barsingsar | (4) Kutch | (5) Giral Rajasthan |
|---|---|---|---|---|---|
| $Al_2O_3$ % | 4.26 | 1.72 | 0.74 | 2.78 | 1.98 |
| BaO % | 0.02 | 0.01 | 0.00 | 0.00 | 0.01 |
| CaO % | 38.6 | 48.4 | 52.1 | 45.0 | 47.3 |
| $Fe_2O_3T$ % | 12.32 | 2.27 | 0.28 | 1.63 | 0.79 |
| $K_2O$ % | 0.03 | 0.20 | 0.04 | 0.31 | 0.20 |
| MgO % | 0.89 | 0.35 | 0.37 | 1.24 | 0.71 |
| MnO % | 0.34 | 0.04 | 0.01 | 0.03 | 0.02 |
| $Na_2O$ % | 0.02 | 0.09 | 0.02 | 0.21 | 0.10 |
| $P_2O_5$ % | 0.13 | 0.16 | 0.07 | 0.08 | 0.08 |
| $SiO_2$ % | 6.38 | 4.75 | 2.05 | 6.96 | 6.81 |
| SrO % | 0.02 | 0.01 | 0.03 | 0.07 | 0.03 |
| $TiO_2$ % | 0.43 | 0.08 | 0.03 | 0.23 | 0.10 |
| LOI (900°C) | 34.4 | 39.4 | 41.4 | 39.4 | 38.6 |
| $CaCO_3$ g/100 g of stone | 70.52 | 88.67 | 95.7 | 82.04 | 87.30 |
| $MgCO_3$ g/100 g of stone | 1.9 | 0.74 | 0.8 | 2.67 | 1.54 |

TABLE 4: Deposit sampling using probes.

(a) Foul probe test conditions—position windward

| Test serial number | Gas temp. °C | Probe temp. °C | Exposure hours | Limestone tonnes/hr | $SO_2$ ppm | Rate of buildup g/m$^2$ hr | Lignite fired during test Giral sample numbers (Table 1) |
|---|---|---|---|---|---|---|---|
| 1 | 685 | 500 | 0.5 | 0 | >5000 | 62 | Sample number 2 |
| 2 | 635 | 500 | 0.5 | 0 | >5000 | 34 | Sample number 3 |
| 3 | 720 | 600 | 0.5 | 0 | >5000 | 73 | Sample number 2 |
| 4 | 680 | 500 | 2 | 5 | 1800 | 39 | Sample number 3 |
| 5 | 690 | 500 | 0.5 | 8 | 1800 | 27 | Sample number 4 |
| 6 | 700 | 500 | 2 | 12 | 1200 | 61 | Sample number 2 |

(b) Chemical composition of foul probe deposit samples

| Serial number | $Na_2O$ | MgO | $Al_2O_3$ | $SiO_2$ | $SO_3$ | $K_2O$ | CaO | $TiO_2$ | MnO | $Fe_2O_3$ |
|---|---|---|---|---|---|---|---|---|---|---|
| 1 | 3.3 | 3.5 | 12.2 | 20.5 | 18 | 0.3 | 11.2 | 2.3 | 0.1 | 28.6 |
| 2 | 3.2 | 2.3 | 16.2 | 29.9 | 8.0 | 0.4 | 4.6 | 1.5 | 0.2 | 33.7 |
| 3 | 4.5 | 3.8 | 15.9 | 25.2 | 15.0 | 0.4 | 9.3 | 2.1 | 0.2 | 23.7 |
| 4 | 0.7 | 1.1 | 5.1 | 7.7 | 36.8 | 0.0 | 38.4 | 0.7 | 0.0 | 9.5 |
| 5 | 0.8 | 0.9 | 4.1 | 6.5 | 39.6 | 0.0 | 39.0 | 0.5 | 0.0 | 8.6 |
| 6 | 0.7 | 0.9 | 4.8 | 7.3 | 37.8 | 0.1 | 39.9 | 0.6 | 0.0 | 7.9 |

have a high propensity to form ash deposits [8, 9]. The values in bold and italics indicate high propensity for ash deposition. Agglomeration can start well below the ash fusion temperatures in fluidized beds for lignite, and influence of $Na_2O$ (AFT decreases) and $Al_2O_3$ (AFT increases) on Turkish lignite was studied by earlier researchers [10].

4.2. Sulfation of Free Lime in Backpass of Boiler. The investigations of the deposit hardening phenomenon in the CFB boilers have been widely discussed as the occurrence of three types of deposit consolidation mechanisms [11, 12]. Two out of the three consolidation mechanisms result in increase in volume of free CaO rich zones in deposits. Fine sorbent

TABLE 5: Ash mineralogy—XRD.

| Lignite Giral sample 2 Table 1 | |
|---|---|
| Mineral matter | % present |
| Quartz ($SiO_2$) | 1.2 |
| Anorthite | 3.0 |
| Diopside | 2.5 |
| Maghemite | 3.9 |
| Hematite | 10.5 |
| Anhydrite | 78.4 |
| Hexahydrite | 0.5 |
| Total | 100.0 |

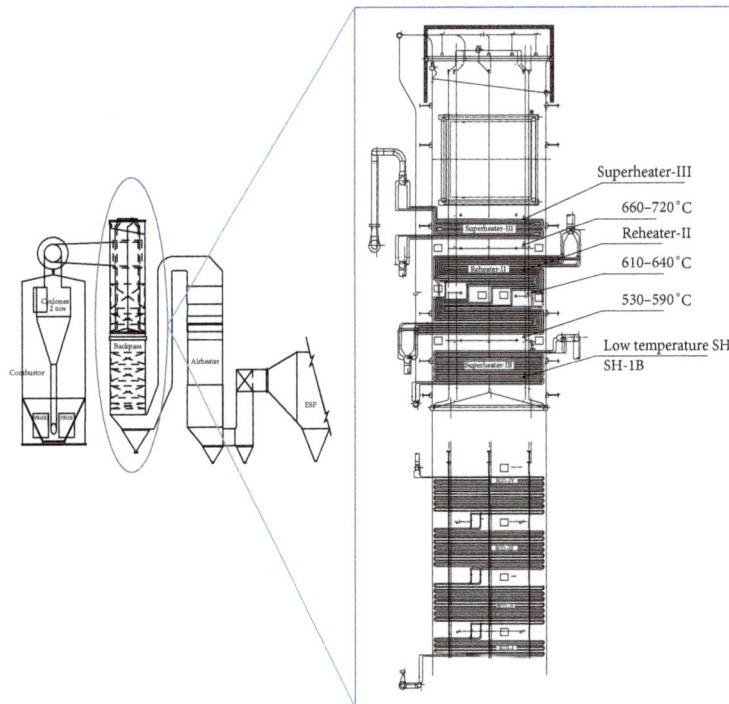

FIGURE 3: General arrangement of CFBC boiler and backpass.

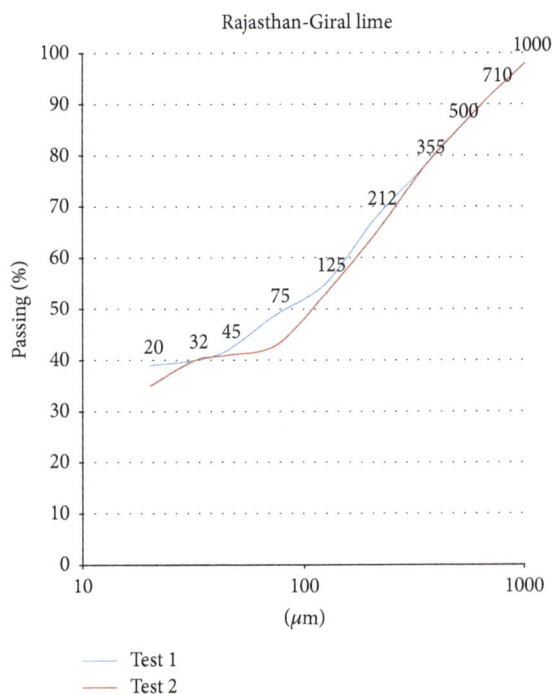

FIGURE 4: Shares of particles smaller than 50 $\mu$m in limestone samples.

particles, settled either on the tube surface or in the caverns on the "rough" surface of the old deposits, (Figure 5) are exposed to SO$_2$-containing flue gases. These sorbent particles are fine (i.e., not captured in the cyclone), and the majority

of particles are already calcined before entering the second pass of the boiler. During their residence on tube surfaces in the convective section, these particles undergo a continuous sulfation through an exothermic reaction (1). The sulfation process is described by the following overall reaction [2]:

$$CaO + SO_2 + \frac{1}{2}O_2 \longrightarrow CaSO_4 + 481\,kg/mol \qquad (1)$$

Further if the temperature of flue gas in vicinity of the sorbent particle is sufficiently high, then the local temperature of the deposits is likely to exceed the sintering temperature due to exothermic reaction and hence, as a result, the agglomeration could occur.

It had been shown by earlier researchers that the agglomeration can occur between 750 and 950°C via the second mechanism, the extended sulfation process [12]. The temperature for optimum sulfur capture is about 850°C [13]. *The issue to be understood is whether there exists an optimum temperature range for extended sulfation (long term)* [14]. Sulfation appears to be the dominant agglomeration mechanism in systems that use high sulfur fuel with calcium-based sorbents for low ash fuels like pet-coke [15]. The deposits are shown to be composed predominantly of CaSO$_4$ and in some cases almost pure CaSO$_4$ [16, 17]. Low temperature (down to 750°C), agglomeration mechanism may be via carbonation and then sulfation [18].

*Herein the fuel used is lignite having ash content ranging from 15 to 35% and the gas temperature range where the deposits occurred is from 600°C to 720°C.*

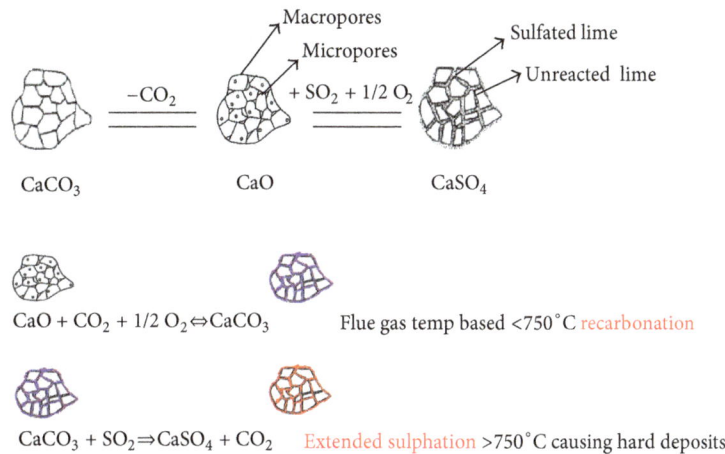

FIGURE 5: Consolidation mechanisms—sulfation of free lime.

In CFBC, sulfation is followed by carbonation of CaO and these reactions can be represented as follows [11]:

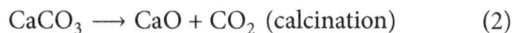

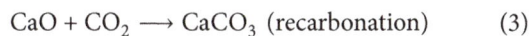

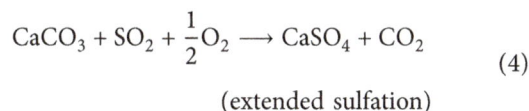

$$CaCO_3 \longrightarrow CaO + CO_2 \text{ (calcination)} \quad (2)$$

$$CaO + CO_2 \longrightarrow CaCO_3 \text{ (recarbonation)} \quad (3)$$

$$CaCO_3 + SO_2 + \frac{1}{2}O_2 \longrightarrow CaSO_4 + CO_2 \quad (4)$$

$$\text{(extended sulfation)}$$

Carbonation mechanism dominates between temperature range of 650 and 790°C at typical $CO_2$ partial pressures (15 kPa) in a CFB boiler, which is much faster than sulfation and is then followed by sulfation of the deposit.

A third possible mechanism thought to cause agglomeration is hydration followed by carbonation [12]. This type of fouling is not common in FBCs because they are normally operated at temperatures well above at which $Ca(OH)_2$ is stable under atmospheric conditions ($\leq 450°C$). The hydration reaction may be represented by the following equation:

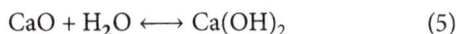

$$CaO + H_2O \longleftrightarrow Ca(OH)_2 \quad (5)$$

This must be followed by carbonation at temperatures below 450°C via the following reaction:

$$Ca(OH)_2 + CO_2 \longleftrightarrow CaCO_3 + H_2O \quad (6)$$

Traditional fouling mechanism due to presence of elements that are associated with ash softening or melting, in particular K, Na, and V, is not applicable for the fuels studied due to low levels of Na, K, and V present [19].

*4.3. Detailed Analysis of Ash Forming Matter in the Giral Lignite.* Giral lignite has high ash content, 15 to 35% (Table 1), which makes it unique, with respect to quantum of ash and the rate at which it was deposited at the backpass. The principal ash forming elements that play significant role in the fireside problems of the boiler, as indicated by mineralogy of the lignite (determined by XRD), are aluminum silicate (kaolinite minerals) and iron compounds (pyrite, $FeS_2$).

With no limestone addition, the flue gas was estimated to contain around 6,900 ppm $SO_2$ (with 6.1% sulfur in fuel and 3% $O_2$ in flue gases). With 12 t/h limestone addition, the corresponding emissions measured were 1400 ppm $SO_2$. The tests were conducted at site to study reactions of lime particles in flue gas to understand the formation of deposits containing various calcium compounds. The boiler load was varied by increasing the lignite feed and corresponding increase in the limestone to control the $SO_x$ level. The very fine limestone particles were calcined, and less than 50-micron level escaped out of the cyclone to backpass and settled over the superheater and reheater coils. As seen in Table 4 *chemical composition analysis indicates that adding limestone changes the whole chemistry of the deposits mainly from silicon-aluminum-iron-based deposits (samples 1 to 3) to calcium-based deposit (samples 4 to 6).* The calcium compounds present are mainly CaO, $CaCO_3$, and $CaSO_4$ as seen in XRD (Table 5).

The root cause of the fouling problem is carbonation and then sulfation reactions of the limestone particles. Loose limestone particles deposit sinter on surfaces and form hard deposits, particularly in flue gas temperature range around 500–700°C. As explained earlier, *it can be safely concluded, at Rajasthan-Giral, that recarbonation reaction is dominant in range of 650–750°C and the extended sulfation reaction (dominant in range of 750–850°C) leads to hardened deposits.*

Ash formed, due to combustion of high sulfur lignite, does not form (sticky or sintering) deposits without limestone addition. These hard deposits were formed due to fine calcined limestone particles ($<50\,\mu m$) that leave the cyclone. These particles settle on the superheater surfaces and react with $CO_2$ between 650 and 750°C leading to recarbonation and then with $SO_2$ between 750 and 850°C furthering extended sulfation, forming sintered and hard deposits (Figure 6). The hypothesis is that in CFBC, carbonation takes place as a dominant reaction forming calcium carbonate (at temperature range of 650 to 790°C) and then extended sulfation takes place between 750°C and 850°C. The environment of flue gas and exothermic reactions contributes to the conversion of the deposits already formed as calcium carbonate into calcium sulfate. The particles settle as deposits

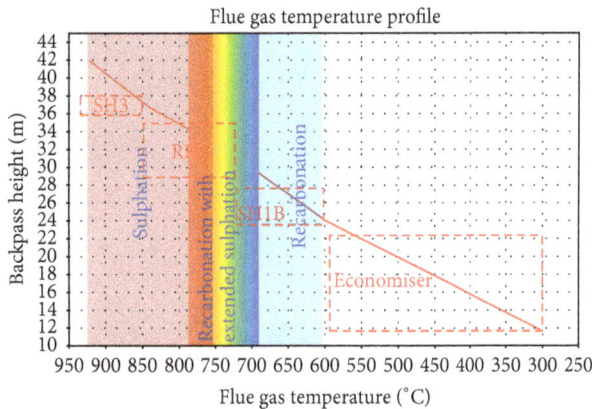

FIGURE 6: Recarbonation and extended sulfation range and location.

on the tube surface, continue their reaction journey and form as calcium sulfate.

*4.4. Optical Microscopy.* Optical microscopy of the deposit samples shows a layered structure (Figure 7) defined mainly by mineralogical variation, principally in anhydrite ($CaSO_4$) and iron oxides. Giral ashes are unusual in the occurrence of complete sulfation of the decarbonated limestone with no evidence of either the occurrence of intermediate phases such as calcium oxide or the presence of sulfate reaction rims (Figure 5) on decarbonated limestone [16, 17]. Reason for this unusual behavior is the high sulfur content of the Giral lignite which might have resulted in complete sulfation of the limestone. Additional factor is the greater proportion of fine particles in the milled Giral limestone which would react completely [6]. This observation is supported by the occurrence of fine anhydrite particles in the Giral backpass sample and a subsequent increase in grain size in the back end of the boiler, suggesting that winnowing of the fine particles has occurred in the hotter sections of the backpass.

# 5. Field Trials/Modifications and Improvement Carried out

*5.1. Standpipe Blockage.* The chemical compositions of the lignite (Table 1), cyclone ash (Table 2), and limestone (Table 3) were analyzed. During commissioning, cyclone standpipe choking due to clinkers (Figure 8) with low combustor temperature of less than 750°C was noticed. The analysis reveals that the composition does not vary much and contains mostly calcium oxide (CaO). The phenomenon of recarbonation of calcined limestone (CaO + $CO_2$ → $CaCO_3$) unreacted with sulphur dioxide was suspected, as a root cause for loose bonding of material at cyclone standpipe leading to blockage of cyclone [20]. This is reflected in the cyclone ash analysis by the presence of free lime (Table 2). The following steps were taken: (a) limestone feed size was checked with more sampling; (b) excessive limestone feed rate was reduced; (c) the operation procedure was revised to maintain higher combustor temperature before starting limestone addition; and (d) automatic pincing air

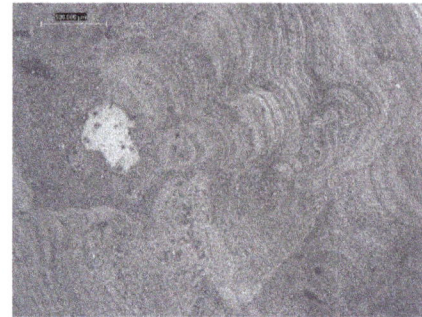

(a)

(b)

FIGURE 7: (a) Photomicrograph of superheater deposit. Reflected light images showing curvilinear layering. (b) Photomicrograph of anhydrite $CaSO_4$/iron oxide $Fe_2O_3$ layer—in transmitted polarised light-white anhydrite and dark brown iron oxide grains.

(a)

(b)

FIGURE 8: Cyclone outlet standpipe clinkers.

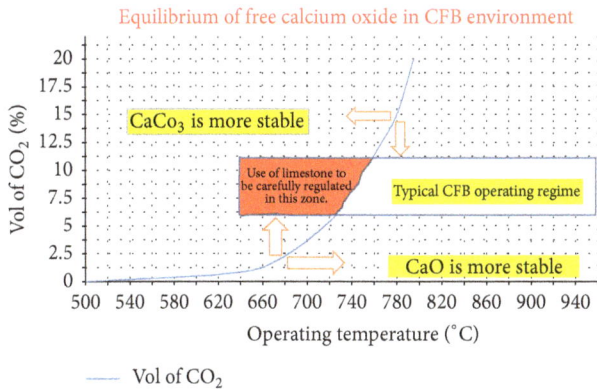

FIGURE 9: Recarbonation-prone regime for limestone addition.

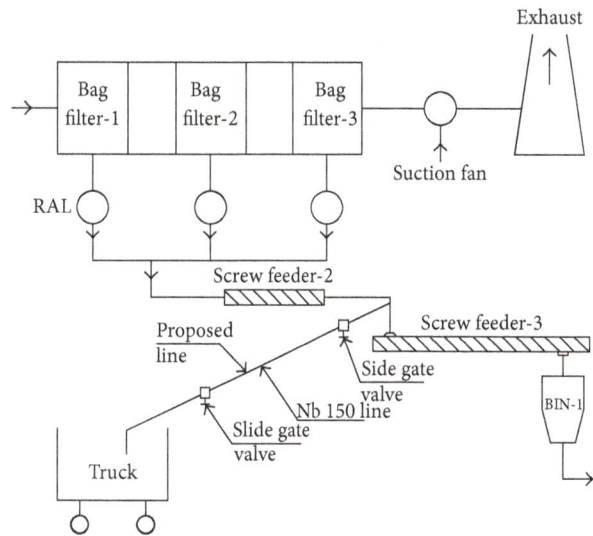

FIGURE 10: Lime mill arrangement for segregation of lime powder particles less than 50 microns.

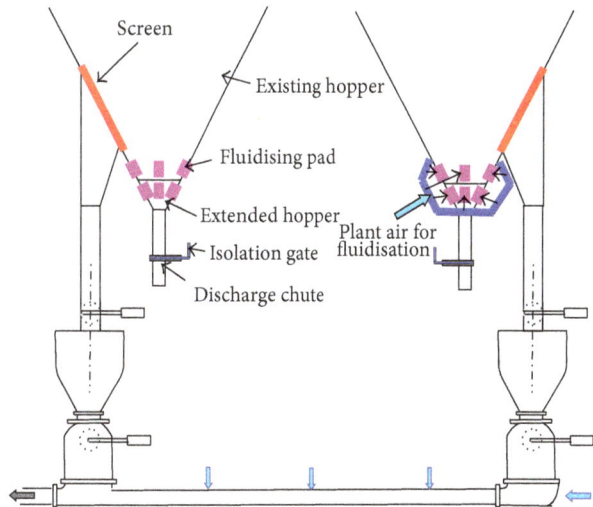

FIGURE 11: Modified arrangement of economizer hopper for removal of bigger particles.

arrangements at junction of the cyclone and standpipe to disturb the agglomeration were incorporated.

After incorporation of changes in operation procedure and with pincing air arrangements, the issue was resolved. The timing of pincing was reduced by maintaining temperature above regime of recarbonation at the cyclone standpipe. Figure 9 shows specific recommendations for avoiding, recarbonation-prone regime for limestone addition [20]. The curve denotes the limit of equilibrium of calcium compounds. As shown in the equilibrium diagram (Figure 8), $CaCO_3$ is stable on the left side of the line, whereas CaO is stable on the right side. In the field, CaO was found abundant because of excess limestone added to the furnace. When the temperature was reduced to recarbonation range, sticky carbonate causing agglomeration blocked (Figure 8) the cyclone standpipe.

*5.2. High Pressure Soot Blowing.* High pressure soot blowing was introduced in the final superheater (FSH), and reheater (RH) and in low temperature superheater (LTSH). After increase in soot blowing pressure from 10 to 20 kg/cm²g, deposits were completely eliminated. Deposits could be removed easily, nearer to the soot blower location, and deposits located away from lance, accumulated proportional to distance from soot blower. Because continuous soot blowing was needed to keep the boiler surfaces clean, additional soot blowers were introduced at selected locations as shown in Figure 1, and deposits were eliminated completely (Figure 1).

*5.3. Limestone Size Distribution.* Lignite without limestone addition caused little or *no hard deposit* buildup in the backpass of CFB boiler. The severity of the fouling (hard deposits) was clearly dependent on the amount of limestone addition. Deposits contained very small fines of less than $50\,\mu m$ size fractions. It was found that 30–40% of the feed limestone was smaller than $50\,\mu m$ (Figure 4). Both dry and wet sieving tests indicated fine fractions were higher than envisaged during design. (0 to 5% less than $50\,\mu m$). Excess quantity of fines $<50\,\mu m$ generated in the milling process was removed by providing a separate elimination line (Figure 10). In addition, the deashing arrangement was improved by introduction of

fluidizing pad at the discharge end and increase in diameter of discharge chute. A screen is provided inside hopper, close to the outlet chute, to separate ash particles below 6 mm into the ash evacuation system (Figure 11).

# 6. Conclusions

Sorbent limestone is used widely in CFB boilers effectively to control sulfur dioxide emissions. Hard deposits were formed in backpass of CFB boiler while using high sulfur Indian lignite and limestone sorbent to control $SO_2$. In addition large quantum of loose deposits caused severe blocking of the second pass. Unreacted calcium oxides that settled on heat transfer tubes at temperature between 650°C and 750°C were subjected to recarbonation and further extended sulfation which resulted in the hard deposits. Elimination of fines, less than 50 $\mu$m, in feed limestone could effectively reduce the hard deposits formation in backpass of CFB boiler. This confirms the finding of the previous studies carried out at other institutions firing high sulfur but low ash fuels. Rate of buildup of deposit and chemistry of deposits in backpass of CFB boiler were studied using special foul probes. The rate of buildup of deposit was proportional to the increase in ash content of lignite and sorbent feed rate. Solution to control the fouling in 125 MWe CFB boiler is to minimize the amount of free lime particles (CaO) in the system formed due to excess addition of fines in feed limestone (less than 50 $\mu$m). The fine fractions of limestone feed <50 $\mu$m coming out of milling circuit were removed by providing an elimination line.

Other CFB boiler operational issues faced, namely, cyclone standpipe blockage, cleaning the heat transfer surfaces deposited with huge quantum of loose ash, and ash evacuation to separate the large size deposits/particles, were effectively resolved through introduction of pincing air at the junction of cyclone and standpipe, high pressure (20 kg/cm$^2$g) soot blowing in selected locations, and incorporation of fluidizing pads and screens in ash hoppers, respectively.

Frequent soot blowing and provision of soot blowers at additional locations were *effective in clearing* the huge quantum of loose deposits.

## Abbreviations

| | |
|---|---|
| AFT: | Ash fusion temperature |
| ASTM: | American Society for Testing Materials |
| Al$_2$O$_3$: | Aluminum oxide |
| CaCO$_3$: | Calcium carbonate |
| CaO: | Calcium oxide |
| CaSO$_4$: | Calcium sulfate |
| CFBC: | Circulating fluidized bed combustion |
| GDP: | Gross domestic product |
| LTSH: | Low temperature superheater |
| LRSB: | Long retract soot blower |
| MWe: | Mega Watt electrical |
| RH: | Reheater |
| SH: | Superheater |
| SiO$_2$: | Silicon dioxide |
| SO$_2$: | Sulfur dioxide |
| TGA: | Thermogravimetric analysis |
| XRD: | X-ray diffraction. |

## Acknowledgment

The authors thank the Management of BHEL for the opportunity to present their views through this paper on this important topic. The views expressed in this paper are those of the authors and not necessarily those of BHEL.

## References

[1] A. Lawrence, V. Ilayaperumal, K. P. Dhandapani, S. V. Srinivasan, M. Muthukrishnan, and S. Sundarrajan, "A novel technique for characterizing sintering propensity of low rank fuels for CFBC boilers," *Fuel*, vol. 109, pp. 211–216, 2013.

[2] R. Kobyłecki, S. Gołąb, L. Krzemień, J. Tchórz, and Z. BisCzęstochowa, "Fouling in the back pass of a large scale CFBC," in *Proceedings of the 9th International Conference on Circulating Fluidized Beds*, 2008.

[3] S. V. Pisupati and A. W. Scaroni, "Sorbent characterizataion for FBC application," in *Proceedings of the 10th Annual Fluidized Bed Conference*, 1994.

[4] M. Fabio, S. Piero, S. Fabrizio, and U. Massimo, Sulfur uptake by Limestone based sorbent particles in CFBC: the influence of attrition / fragmentation on sorbent inventory and particle size distribution-CFB 10, 2011.

[5] M. Olas and R. Kobyłecki, BisZ—Simultaneous calcination and sulfation of limestone based sorbents in CFBC-effect of mechanical activation-CFB 9, 2009.

[6] S. J. Hari and V. P. Sarma, *A Study on Indian Limestones For Sulfur Capture-The EMS Energy Institute and John and Willie Leone Department of Energy Mineral Engineering*, The Pennsylvania State University, 2012.

[7] Common slagging and fouling indices, http://www.coaltech .com.au/LinkedDocuments/Slagging&Fouling.pdf.

[8] Rod Hatt, Coal Combustion, Inc.Correlating the slagging of a utility boiler with coal characteristics-http://65.163.62.71/ PDF%20Files/Corre Slag efc3.pdf.

[9] R. C. Attig and A. F. Duzy, "Coal ash deposition studies and application to boiler design," *Proceedings of American Power Conference*, vol. 31, pp. 290–300, 1969.

[10] H. Atakül, B. Hilmioğlu, and E. Ekinci, "The relationship between the tendency of lignites to agglomerate and their fusion characteristics in a fluidized bed combustor," *Fuel Processing Technology*, vol. 86, no. 12-13, pp. 1369–1383, 2005.

[11] E. J. Anthony, A. P. Iribarne, J. V. Iribarne, R. Talbot, L. Jia, and D. L. Granatstein, "Fouling in a 160 MWe FBC boiler firing coal and petroleum coke," *Fuel*, vol. 80, no. 7, pp. 1009–1014, 2001.

[12] E. J. Anthony, R. E. Talbot, L. Jia, and D. L. Granatstein, "Agglomeration and fouling in three industrial petroleum coke-fired CFBC boilers due to carbonation and sulfation," *Energy and Fuels*, vol. 14, no. 5, pp. 1021–1027, 2000.

[13] P. F. B. Hansen, K. Dam-Johansen, L. H. Bank, and K. Ostergaard, "Sulphur retention on limestone under fluidized bed combustion conditions. An experimental study," in *Proceedings of the 11th International Conference on Fluidized Bed Combustion*, pp. 73–82, April 1991.

[14] E. J. Anthony and D. L. Granatstein, "Sulfation phenomena in fluidized bed combustion systems," *Progress in Energy and Combustion Science*, vol. 27, no. 2, pp. 215–236, 2001.

[15] E. J. Anthony, A. P. Iribarne, and J. V. Iribarne, "A new mechanism for FBC agglomeration and fouling in 100 percent firing of petroleum coke," *Journal of Energy Resources Technology, Transactions of the ASME*, vol. 119, no. 1, pp. 55–61, 1997.

[16] E. J. Anthony, A. P. Iribarne, and J. V. Iribarne, "Fouling in a utility-scale CFBC boiler firing 100% petroleum coke," *Fuel Processing Technology*, vol. 88, no. 6, pp. 535–547, 2007.

[17] E. J. Anthony, L. Jia, and K. Laursen, "Strength development due to long term sulfation and carbonation/sulfation phenomena," *Canadian Journal of Chemical Engineering*, vol. 79, no. 3, pp. 356–366, 2001.

[18] E. J. Anthony and L. Jia, "Agglomeration and strength development of deposits in CFBC boilers firing high-sulfur fuels," *Fuel*, vol. 79, no. 15, pp. 1933–1942, 2000.

[19] E. J. Anthony, F. Preto, L. Jia, and J. V. Iribarne, "Agglomeration and fouling in petroleum coke-fired FBC boilers," *Journal of*

*Energy Resources Technology, Transactions of the ASME*, vol. 120, no. 4, pp. 285–292, 1998.

[20] M. Lakshminarasimhan, B. Ravikumar, A. Lawrence, and M. Muthukrishnan, High Sulfur Lignite Fired Large CFB Boilers: Design & Operating experience. International Conf.on Circulating Fluidized Beds and Fluidization Technology-CFB 10, 2011.

# 3

# Calculation of Spotting Particles Maximum Distance in Idealised Forest Fire Scenarios

**José C. F. Pereira, José M. C. Pereira, André L. A. Leite, and Duarte M. S. Albuquerque**

*IDMEC, Instituto Superior Tecnico, Universidade de Lisboa, Avenida Rovisco Pais, 1049-001 Lisbon, Portugal*

Correspondence should be addressed to José C. F. Pereira; jcfpereira@tecnico.ulisboa.pt

Academic Editor: Michael A. Delichatsios

Large eddy simulation of the wind surface layer above and within vegetation was conducted in the presence of an idealised forest fire by using an equivalent volumetric heat source. Firebrand's particles are represented as spherical particles with a wide range of sizes, which were located into the combustion volume in a random fashion and are convected in the ascending plume as Lagrangian points. The thermally thin particles undergo drag relative to the flow and moisture loss as they are dried and pyrolysis, char-combustion, and mass loss as they burn. The particle momentum, heat and mass transfer, and combustion governing equations were computed along particle trajectories in the unsteady 3D wind field until their deposition on the ground. The spotting distances are compared with the maximum spotting distance obtained with Albini model for several idealised line grass or torching trees fires scenarios. The prediction of the particle maximum spotting distance for a 2000 kW/m short grass fire compared satisfactorily with results from Albini model and underpredicted by 40% the results for a high intensity 50000 kW/m fire. For the cases of single and four torching trees the model predicts the maximum distances consistently but for slightly different particle diameter.

## 1. Introduction

Spot fires occur during wild forest fires when burning debris transported by the wind and convection column land far from the active fire source. Under such occurrence there is a probability to ignite another fire with dangerous consequences for fire brigades and firefighting which should be considered by the decision support systems for wildfire management and planning [1–3]. Many firebrand transport models have been developed; see the pioneers' works of Tarifa et al. [4], Lee and Hellman [5], and Albini [6]. Albini model predicts the maximum spotting distance [6–10] and has been included in several forest fire propagation models [11–17]. The computing time required to obtain predictions with the Albini model is much faster than real time, due to the inherent model based correlations. This is a great advantage over multidimensional Computational Fluid Dynamics (CFD) predictions; however it is believed that forest fire phenomena will benefit from CFD like for enclosure fire predictions; see, for example, [18–20].

The problem is of great complexity because a large number of random parameters are presented. The transport of firebrands involves several modelling difficulties: firstly the knowledge of the particle shape and size that lift off in the flame region; secondly the particle transport by unsteady wind and convection column fields; and last but not the least the probability to ignite a fire after particle landing. This work is only related to the transport of firebrand aiming to predict the particle maximum spotting distance. However it involves the coupled prediction of (i) the wind flow through and above vegetation; (ii) the fire source near region and convection column; and (iii) the particle heat and mass transfer and combustion along its trajectory.

The wind flow interaction with canopy trees has been extensively studied [21–25]. The vertical mean wind velocity displays an exponential profile type inside the forest and the turbulence levels are mainly due to turbulent kinetic energy production by shear at the canopy top, rather than by wake production by the individual elements. The flow within and above a forest is linked by turbulent motions, at larger scales relative to the forest depth, that are strongly intermittent in character [22, 26, 27]. Several attempts have been proposed to simulate the exchanges between a forest and the lower atmosphere by assembling averaged statistical turbulent models such as $k - \varepsilon$; see [28–31]. These models are inadequate in representing the intermittent character of the flow. Large eddy simulation (LES) explicitly simulates the dominant energetic turbulent scales resolved by the three-dimensional mesh [32–34] and consequently it was used in this work.

Due to the disparity of scales in the fire heat release region from those of the remaining computational domain, it is almost impossible to resolve and to predict the heat release rate, but it is to be given as an input parameter. The so-called Lagrangian thermal elements [35, 36] are used to model the fire release heat as they are convected about by the thermal induced motion. Under this assumption, the fire is a large collection of blobs carried along by the large-scale motion and the heat release rate associated with each element is represented by a simple function with a time scale determined from the plume correlations summarised by [37]. A far more simplified model is to prescribe a heat source either on the surface or in volume, usually approximated by Gaussian profiles [38] and by correlations of the flame height as a function of the firepower intensity. This was the procedure adopted to model the fire itself and the source was assumed stationary since the propagation fire velocity is small compared with the wind velocity.

The pioneers' works have employed the classical plume model approach of integral models to predict buoyant plumes in a cross flow responsible for firebrand lofting. Examples of developed simplified models are for initially axisymmetric jets [39, 40] and for buoyant plumes from fires [5, 7, 9] or integral plume models for line fires in a cross flow [41]. These models reduce the problem to a set of ordinary differential equations to be solved with an approximate expression or with an empirical fit to calculate the plume trajectory, width, velocity, and temperature. Three-dimensional field calculations of fire plumes have been extensively investigated; see, for example, [36, 38, 42]. In this work the latter approach is followed and a LES model was selected that takes into account explicitly the subgrid stresses and turbulent heat fluxes [43].

In the framework of time averaged turbulent flow modelling, the problem of the instantaneous velocity acting on the particle is usually treated under Lagrangian stochastic models; see [44] for a review. For the present purpose of the LES calculations during the firebrand trajectory it was assumed that the calculated instantaneous velocity field acts in the particle during the considered time step. There are three main ways on how to account with the with the wind interaction in the firebrands particles. The first way is to consider spherical particles with a wide range of sizes undergoing drag relative to the flow and moisture loss as they are dried and pyrolysis, char-combustion, and mass loss as they burn. Models to calculate flight paths of dispersed particles in a turbulent flow are well established and a spherical particle shape is assumed due to inherent difficulties to know the drag and momentum coefficients from other shapes; see, for example, [45]. The second way is to consider nonspherical particles shapes, cylinders or discs, under one-way coupling, meaning that the wind influences

the nonspherical particle orientation relative to the local wind obtained by momentum balances, but with prescribed drag and momentum coefficients as a function of relative particle orientation. The third one would correspond to the intrusive real body geometry of the nonspherical particle and the calculation (with Chimera or moving meshes) of their wake that may interfere with the particle itself during tumbling, fluttering, or chaotic free fall motion.

Recently, experimental apparatus has been constructed in order to generate a controlled size and mass distribution of glowing firebrands; see, for example, [46, 47], allowing studying the combustibility of firebrand material such as pine cones and scales and pieces of bark eucalypt; see [48]. Theoretical models for the drag coefficient of nonspherical particles are being established (see [49–51]) but the wide range of random shapes, sizes, and terminal velocities requires validation tests before practical use in reactive multidimensional calculations.

The combustion model of the woody, cellulose, or coal fuel particles commonly includes drying, pyrolysis, and char-oxidation processes. Their burning characteristics and diameter at landing are related to the potential for the firebrand to ignite the adjacent vegetation [52–55] and reviews of the modelling chemical and physical processes of wood and biomass pyrolysis have been presented; see, for example, [56–58]. Firebrand propagation prediction is based on either plume model, coupled fire-atmosphere, or semiempirical models to predict the fire spread; see, for example, [59–61]. Particles trajectories and spotting distances have been obtained for a wide range of idealised cases using these main assumptions about the fire source responsible for the convection column; see, for example, [55, 61–63].

Physics based on coupled fire-atmosphere models consider approximations of the governing equations from the fluid dynamics, the combustion, and the thermal degradation of solid fuel (see, e.g., [64–66]) aiming to preclude the use of existing simplified empirical wildfire models because they do not predict general fire behaviour; however the high-resolution and the high-fidelity combustion are not currently appropriate because of their computational cost. Several physics based on coupled fire-atmosphere studies have been conducted (see [67–72]) and some of these studies have been applied to the fire spotting problem. Among them [71] has considered particle combustion of cylindrical and disk-shaped firebrands for several geometrical parameters. Discs travel further than cylinders; also firebrands from canopy fires travel further than firebrands from surface fires. Depending on where the burning occurs, for example, the faces or around their circumference, this influences the firebrand lifetimes. In addition the simulations reveal that the coupled fire-atmosphere behaviour dominates the trajectories and landing patterns.

The main difficulty in the validation of the Albini model or of a CFD model, under real conditions, is that large field forest fires only show the spotting fires signature on the ground after the fire has been extinguished and it is unknown if they correspond to the particle maximum spotting distance. The particle responsible for the maximum spotting distance may not ignite a fire in opposition to the other particles that spot too much shorter distances. Consequently the intercomparison of different models may contribute to estimating the error bar of the spotting distances pattern.

The main objective of this work is to compare the maximum spotting distance obtained with the Albini model with the spotting particles maximum distance obtained with LES and firebrand combustion models. Therefore in this work a coupled solution of the three-dimensional velocity and temperature unsteady fields is obtained and for each time step the particles are allowed to burn during their convection. For each particle size the calculated spotting distances as well as their char, ash, and temperature allow one to obtain the maximum spotting distance for a prescribed fire. Two classes of fires are presented: grass fires and burning of trees. For both cases the predicted maximum spotting distances are compared with Albini model's results.

In the next section the models are briefly presented. The section of Results follows this, but prior to the firebrand transport results a LES benchmark test case was performed. It corresponds to the LES simulation of lower atmosphere with a homogeneous forest [22] to investigate the LES solution dependence on coarse grid resolution. Next, firebrand spotting is examined using a coupled fire/atmosphere LES (large eddy simulation) in which the processes of firebrand lofting, propagation, and deposition are connected. The idealised scenarios correspond to the Albini "spotting distance examples" [6] for short grass 2000 kW/m fire and wind-driven fire in chaparral 50000 kW/m fire. In addition torching trees were considered, based on the scenario given by [73], the first corresponding to a single Grand Fir tree and the second corresponding to four trees burning together. The paper closes with summary conclusions about the comparison of the maximum spotting distances.

## 2. Mathematical and Numerical Model

*2.1. Governing Equation.* The governing equations are the continuity Navier-Stokes and energy equations. The Boussinesq approximation is used and the equations include additional terms to account for the drag from the canopy trees and for the heat received by the air in contact with the vegetation. The filtered Navier-Stokes model equations can be expressed by

$$\frac{\partial \overline{u}_j}{\partial x_j} = 0, \tag{1}$$

$$\frac{\partial \overline{u}_i}{\partial t} + \frac{\partial \left(\overline{u}_j \overline{u}_i\right)}{\partial x_j} = \frac{1}{\rho}\frac{\partial \overline{p}}{\partial x_i} + \frac{\partial \overline{p}}{\partial x_j}\left(\nu\frac{\partial \overline{u}_i}{\partial x_j} - \overline{u_i'' u_j''}\right) + \beta g T \delta_{i3} \tag{2}$$

$$+ F_i, \quad i = 1, 2, 3,$$

$$\frac{D\overline{T}}{Dt} \equiv \frac{\partial \overline{T}}{\partial t} + \frac{\partial \left(\overline{u}_j \overline{T}\right)}{\partial x_j}$$

$$= \frac{\partial}{\partial x_j}\left(\mu\frac{\partial \overline{T}}{\partial x_j} - \overline{u_j'' T''}\right) + S_h + S_R. \tag{3}$$

Here $p$ is the pressure, $g$ is the gravitational acceleration, $\beta = -(\partial \rho/\partial T)_p/\rho$ is the volumetric expansion coefficient, and $n$ and $m$ are the constant molecular diffusivities of momentum and heat. The bar denotes the average over a computational grid cell and the double primes the deviations thereof. The Coriolis force has been excluded as it has little direct bearing on scales of motion for the domain considered of the order of 1 km.

*2.2. The Subgrid-Scale (SGS) Model.* There is a wide range of subgrid-scale (SGS) models as well as great knowledge of the modelling issues like gradient-diffusion hypothesis used in some of the large eddy simulations; see, for example, [74, 75]. For atmospheric flows the pioneer classical models (see, e.g., [33, 34]) have been being improved with models based on transport equations for the SGS stresses and fluxes. The second-order closure subgrid-scale equations model reported by [43] was selected for the present purpose to predict a plume at cross flow in the atmospheric surface layer. The model uses a transport equation for the subgrid-scale kinetic energy $\overline{E''} \equiv \overline{u_i''^2}/2$

$$\frac{D\overline{E''}}{Dt} = -\overline{u_i'' u_j''}\frac{\partial \overline{u}_i}{\partial x_j} + \beta g \overline{w'' T''} + \frac{\partial}{\partial x_i}\left[\frac{5}{3}lc_{3m}\overline{E''}^{1/2}\frac{\partial \overline{E''}}{\partial x_i}\right]$$

$$- c_{\varepsilon m}\frac{\overline{E''}^{3/2}}{l} - 2\frac{\overline{E''}}{\tau}. \tag{4}$$

The turbulent heat and momentum fluxes and their respective anisotropic components,

$$\overline{A_{ij}''} = \overline{u_i'' u_j''} - \frac{2}{3}\delta_{ij}\overline{E''}, \quad i = 1, 2, 3, \quad j = 1, 2, 3, \tag{5}$$

are determined from the following set of algebraically approximated second-order closure equations:

$$0 = -\left(1 - c_{Gm}\right)\frac{2}{3}\overline{E''}\left(\frac{\partial \overline{u}_i}{\partial x_i} + \frac{\partial \overline{u}_j}{\partial x_i}\right) + \left(1 - c_{Bm}\right)$$

$$\cdot \left[\beta g\left(\delta_{i3}\overline{u_j'' T''} + \delta_{j3}\overline{u_i'' T''} - \frac{2}{3}\delta_{ij}\overline{u_3'' T''}\right)\right] - c_{Rm} \tag{6a}$$

$$\cdot \frac{E''^{1/2}}{l}\overline{A_{ij}''},$$

$$0 = -\left(1 - c_{GT}\right)\frac{2}{3}\overline{E''}\frac{\partial \overline{T}}{\partial x_i} + \left(1 - c_{BT}\right)\beta g\overline{T''^2}\delta_{i3} - c_{RT} \tag{6b}$$

$$\cdot \frac{E''^{1/2}}{l}\overline{u_i'' T''},$$

$$0 = -2\overline{u_j'' T''}\frac{\partial \overline{T}}{\partial x_j} - c_{\varepsilon T}\frac{\overline{E''^{1/2} T''^2}}{l}. \tag{6c}$$

Equations (6a) to (6c) were solved explicitly as proposed by [43] yielding

$$\overline{A''_{ij}} = \frac{l}{c_{Rm}\overline{E''^{1/2}}}\left[\left(1 - c_{Gm}\right)\frac{2}{3}\overline{E''}\left(\frac{\partial \overline{u}_i}{\partial x_j} + \frac{\partial \overline{u}_j}{\partial x_i}\right) + \overline{B''_{ij}}\right], \quad (7a)$$

$$\overline{B''_{ij}} = \left(1 - c_{Bm}\right)\beta g\left(\overline{u''_j T''}\delta_{i3} + \overline{u''_i T''}\delta_{j3}\right.$$
$$\left. - \frac{2}{3}\delta_{ij}\overline{u''_3 T''}\right), \quad (7b)$$

$$\overline{u''_i T''} = -\frac{l}{c_{RT}\overline{E''^{1/2}}}\left[\left(1 - c_{GT}\right)\frac{2}{3}\overline{E''}\frac{\partial \overline{T}}{\partial x_i}\right.$$
$$\left. - \left(1 - c_{BT}\right)\beta g\overline{T''^2}\delta_{i3}\right], \quad (7c)$$

$$\overline{T''^2} = -\frac{2l}{c_{\varepsilon T}\overline{E''^{1/2}}}\left[\overline{u''_i T''}\frac{\partial \overline{T}}{\partial x_i}\right]. \quad (7d)$$

The coefficients of the SGS model are listed in Table 1 according to [43].

The length scale $l$ is prescribed as a function of height above the ground surface and of the mesh size $\Delta$:

$$l = \min\left(\Delta, c_l z\right), \quad \Delta = \frac{1}{3}\left(\Delta x + \Delta y + \Delta z\right). \quad (8)$$

The numerical stability enhanced procedures described by [76] were used to solve (4).

A second SGS model was also considered, the classic Smagorinsky [77] model, because it is used extensively in atmospheric boundary layer predictions. The simple Smagorinsky subgrid-scale model should be sufficient provided that the temporal and spatial resolution is fine enough to resolve a major fraction of the energetic scales; see, for example, [78]. These resolution requirements become computationally not feasible in higher resolutions and especially in proximity to the surface.

### 2.3. Additional Terms to Model the Forest.
The additional term ($F_i$) in (2) represents the drag force due to the canopy modelled as a porous body of uniform area density. The drag force was taken according to [18] as time dependent and equal to the product of the local foliage density (a function only of the vertical position), a constant drag coefficient $C_D = 0.15$, and the square of the local velocity such that the force $F_i$ the $x_i$-direction is given by

$$F_i = -C_D a V \overline{u}_i = -\frac{\overline{u}_i}{\tau}, \quad (9)$$

where $V$ is scalar speed. The drag coefficient is isotropic and the drag force directly opposes the local, instantaneous wind vector.

The $S_h$ term in (3) takes into account the heat source provided by the foliage that is heated by assuming that solar radiation penetrates the canopy and warms the foliage. The strength of the heat source, $S_h$, included in (3) is then the

TABLE 1: Coefficients of the SGS model.

| $c_{3m}$ | $c_{\varepsilon m}$ | $c_{Gm}$ | $c_{Bm}$ | $c_{Rm}$ | $c_{GT}$ | $c_{RT}$ | $c_{\varepsilon T}$ | $c_l$ |
|---|---|---|---|---|---|---|---|---|
| 0.2 | 0.845 | 0.55 | 0.55 | 3.5 | 0.5 | 1.63 | 2.02 | 0.845 |

vertical derivative of the upward kinematic vertical heat flux given by

$$Q(z) = Q(h)\exp\left(-\alpha F\right),$$
$$\text{and } F \text{ given by } F = \int_z^h L\,dz, \quad (10)$$

where $F$ is the downward cumulative leaf area index (nondimensional), which is an extinction coefficient and is taken to be 0.6. Only the test case corresponding to weakly convective conditions, $Q(h) = 0.005$, was considered corresponding to $W_k/U_k = 1.1$, where $W_k$ is the convective velocity scale [$W_k = (\beta g Q(h) 2h)^{1/3}$], and $h/l = -0.26$, where $l$ denotes the Monin-Obukhov length and the leaf area densities, $L$ (integration of the plant area density) considered as described by [43].

The last term in (3) represents the radiative heat source assuming a simplified heat loss:

$$S_R = \varepsilon\sigma\left(T^4 - T_0^4\right), \quad (11)$$

where $\sigma$ denotes the Stefan-Boltzmann constant and the emissivity was assumed equal to $\varepsilon = 0.25$.

The last term in (4) represents an additional dissipation process due to canopy drags in which $t$ is a time scale for the drag defined by (9). The term takes into account the removal of SGS kinetic energy by the action of drag on the assumption that wake motions are of even smaller scale than those making up the bulk of SGS kinetic energy.

An extra scalar transport equation to simulate the "smoke concentration" was added only for visualization proposes.

### 2.4. Firebrand Transport Model.
The motion of firebrands is studied by assuming that they behave as a point mass and the aerodynamic drag acts in the opposite direction to the motion of the centre of gravity from the firebrand. Considering only drag and gravity, the three-dimensional motion of a firebrand of mass $m$ moving at the velocity $u_p$ within the fire velocity $u_i$ is governed by the following system of differential equations:

$$m\frac{du_{pi}}{dt} = \frac{1}{2}C_D\rho A_p\left(u_i - u_{pi}\right)\left|u_i - u_{pi}\right| - m\cdot g, \quad (12a)$$

where the gravity force acts only in the vertical coordinate direction.

The drag coefficient $C_D$ depends on the particle shapes. Here a spherical shape is chosen for the firebrand because of the well-known dependence of the drag coefficient with Reynolds number:

$$C_D = \frac{24}{\text{Re}_p}\left(1 + 0.15\text{Re}_{pi}^{0.687}\right) \quad \text{for } \text{Re}_{pi} \le 1000, \quad (12b)$$

$$C_D = 0.44 \quad \text{for } \text{Re}_{pi} \ge 1000, \quad (12c)$$

where the particle Reynolds number is

$$\text{Re}_{pi} = \frac{\rho \left| u_i - u_{pi} \right| d_p}{\mu}. \tag{12d}$$

The particles for the test cases investigated have a range of a few millimetres and they are made of coal and cellulose. The rate of particle mass reduction due to water vaporisation, pyrolysis, and char combustion can be represented by the sum of the mass rates:

$$\frac{dm_p}{dt} = \frac{d\rho_p V_p}{dt} = \frac{dm_p^{H_2O}}{dt} + \frac{dm_p^{pyr}}{dt} + \frac{dm_p^{char}}{dt}. \tag{13}$$

During the drying and pyrolysis processes, the volume of the particle is assumed to remain constant and the mass equation reduces to

$$V_p \frac{d\rho_p}{dt} = \frac{dm_p^{H_2O}}{dt} + \frac{dm_p^{pyr}}{dt}. \tag{14}$$

Whilst during char combustion, the particle density is assumed to remain constant and the mass equation reduces to

$$\rho_p \frac{dV_p}{dt} = \frac{dm_p^{char}}{dt}. \tag{15}$$

The rates of mass loss due to dry in [79] and pyrolysis [80] can be deduced from the following Arrhenius-type laws:

$$\frac{dm_p^{pyr}}{dt} = -2.67 \times 10^9 \exp\left(\frac{-17921}{T_p}\right), \tag{16a}$$

$$\frac{dm_p^{H_2O}}{dt} = -6.0 \times 10^5 T_p^{-0.5} m_p^{H_2O} \exp\left(\frac{-6000}{T_p}\right). \tag{16b}$$

Following drying and pyrolysis, the particle mass change due to char combustion is given by [56]:

$$\frac{dm_p^{char}}{dt} = -28.0 \exp\left(\frac{-9646}{T_p}\right) \pi d_p^2 P_{O_2}^{0.5}, \tag{16c}$$

where $P_{O_2}$ is the pressure of oxygen adjacent to the particle.

By considering the assumption that particles are thermally thin, the temperature throughout any particle is uniform while it is being heated or in combustion. The particle heat conduct (convection) equation reduces to

$$\frac{d\left(c_p m_p T_p\right)}{\partial t} = \frac{dm_p^{H_2O}}{dt} Q^{vap} + \frac{dm_p^{pyr}}{dt} Q^{pyr}$$
$$+ \alpha_c \frac{dm_p^{char}}{dt} Q^{char} + Q_{con} + Q_{rad}, \tag{17a}$$

where $\alpha_c$ represents the fraction of the particle heat of char combustion, which is transferred to the particle. Here $\alpha_c = 0.3$. Water vaporisation and pyrolysis are endothermic processes ($Q^{vap} = 2.4\,\text{KJ/kg}$ and $Q^{pyr} = 0.418\,\text{KJ/kg}$) whilst

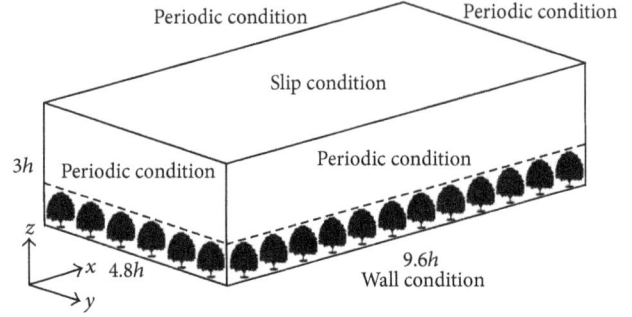

FIGURE 1: Computational domain of a deciduous forest.

char combustion is highly exothermic ($Q^{char} = -32.74\,\text{KJ/kg}$). The radiative heat transfer between the particle and the gas $Q_{rad}$ has been neglected here. Convective heat transfer is obtained for a sphere moving at the slip velocity through the gas:

$$Q_{con} = \frac{\text{Nu}\lambda}{d_p} A_p \left(T - T_p\right) - \text{Nu}\lambda \left(T - T_p\right) d_p, \tag{17b}$$

where the Nusselt number Nu is given by

$$\text{Nu} = 2 \left(1 + 0.345 \text{Re}_p^{1/2} \text{Pr}^{1/3}\right). \tag{17c}$$

The thermal conductivity $\lambda$ is the function of the mean temperature of gas and particle. The Lagrangian approach is used to calculate the particle motion. In this procedure the particle trajectories are integrated considering instantaneous gas mean velocity at each time step.

*2.5. Numerical Method and Boundary Conditions.* The discretization of the unsteady form of the momentum, energy, and scalar concentration equations was made with a finite volume method together with the QUICKEST discretization scheme [81] that is of 3rd-order accuracy in space and time. The explicit scheme uses quadratic Leith type of temporal discretization and upstream quadratic interpolation involving 21 grid points for the evaluation of the convective fluxes in three-dimensional problems. A Poisson equation was solved to ensure that the velocity field is divergence-free. The maximum Courant number used was equal to 0.5 and details about QUICKEST scheme are given in [82, 83]. Calculations were also obtained with another own developed finite volume code, SOL (see [84, 85]), that employs unstructured meshes and despite having second-order formal accuracy its order of accuracy is likely to be reduced by the use of flux limiters to avoid solution wiggles.

Figure 1 also shows the boundary conditions used in each plane of the block domain; in the streamwise direction a nonperiodic pressure gradient is adjusted in each time step in order to maintain a prescribed mean velocity, with periodic conditions in the planes normal to the streamwise direction.

The boundary conditions employed for the fire scenarios are antisymmetric conditions in the spanwise direction while at inlet a velocity profile was prescribed and at outlet a Sommerfeld wave extrapolation was considered to minimise reflection. At the bottom, a nonslip boundary condition is used in which the vertical fluxes of the horizontal momentum are evaluated from the Monin-Obukhov relationship according to [43]. At the top, free-slip boundary conditions are used for the horizontal velocity components and the vertical derivative of temperature and the vertical flux of SGS kinetic energy is set to zero. This artificial boundary condition may be interpreted as a strong inversion. For the homogeneous forest test case periodic boundary conditions are used in the spanwise direction.

## 3. Results

### 3.1. Wind Flow above and within a Model Forest

*3.1.1. Velocity Fields.* The computational domain extends over $9.6h \times 4.8h$ on the ground and over $3h$ on the vertical of a homogeneous deciduous forest as proposed by [22] (see Figure 1), with $h$ being the tree's height. A leaf area density vertical profile is considered with the distribution reported in the reference that corresponds to leaf area index of 5. A uniform mesh comprising $96 \times 48 \times 30$ mesh nodes was used by [22] that includes 10 grid nodes below the canopy height.

The predictions with the QUICKEST code employ a 5-node resolution of the tree's height corresponding to $50 \times 26 \times 17$ grid nodes. The coarser mesh was used because spotting distances in the forest fires cases require distances spanning a much wider domain, approximately two orders of magnitude of the tree's height, and for practical situations the resolution used by [22] would be unpractical.

Figure 2 shows the averaged streamwise velocity for a leaf area index equal to 5. The present predictions show the strong velocity deficit due to trees and the velocity variation with height is in satisfactory agreement with [22]. The predicted vertical profile of turbulence kinetic energy obtained with Smagorinsky is shown in Figure 3(a) and below the canopy the grid resolution is still not enough. But globally it follows in satisfactory agreement the turbulence kinetic energy reported by [22] as well as by the LES transport SGS equation for the subgrid fluxes used in QUICKEST. The subgrid-scale contribution is small and the trend is in satisfactory agreement with the reported values. The results obtained with QUICKEST show some discrepancies with those reported by [22] due to the lack of grid resolution.

Figure 3(b) shows the predicted turbulent shear stress for a leaf area index equal to 5. The nondimensionalised values, by its value at treetop height, show a satisfactory agreement in the rapid and almost linear decay inside the forest as demanded by the strong drag forces and above the canopy in near-linear fashion forced by the existence a deep mixed layer. Also as shown by [22] the SGS contribution to the Reynolds stress is small, indicating that it is generally a very small component of the total momentum flux. The contribution of the subgrid stress to the total shear stress is very small denoting that the large-scale motion is mainly responsible for the large values observed on the canopy and on the shear flow.

Figures 4(a) and 4(b) show a horizontal slice at the middle of the domain with velocity vectors ($U$ and $V$ components),

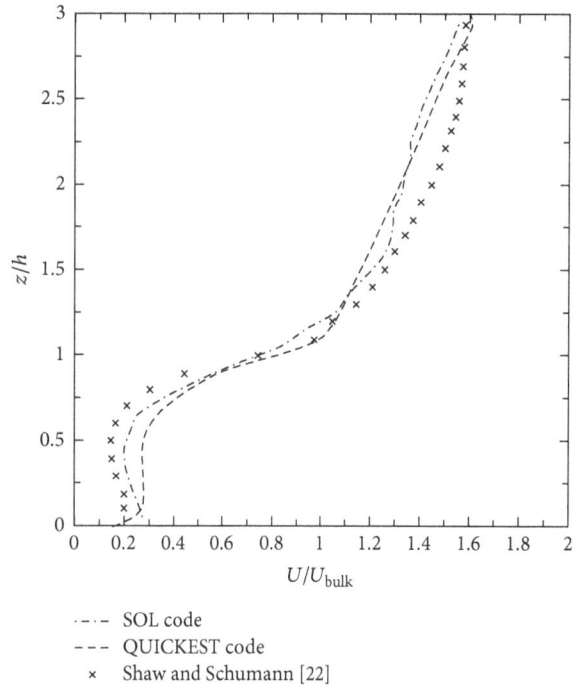

FIGURE 2: Spatial mean longitudinal velocity profiles for a leaf area index equal to 5.

together with contours of the spanwise and vertical velocity components, respectively. The figures show regions of updrafts and downdrafts due to the intermittent character of large-scale phenomena. Figure 4(c) shows a vertical slice contour plot of $U$ velocity with velocity vectors ($U$ and $W$ components), where the two distinct regions with high and low velocity due to the presence of the forest trees in the bottom of the domain can be observed.

This test case shows that the developed software code with the incorporation of the LES model reproduces satisfactorily the flow within and above the model forest. Although a coarse mesh was used the QUICKEST predictions are in satisfactory agreement with [22]. For the study related to spotting distance from a forest fire we have selected the QUICKEST model with SGS turbulent kinetic energy equation. This still state-of-the-art model should perform much better for the plume in cross flow than the SGS eddy diffusivity concept embodied in the Smagorinsky model for the treatment of the SGS heat fluxes.

*3.1.2. Particle Dispersion.* The homogeneous forest constitutes a good scenario to analyse the dispersion of particles that are released at a certain height above the canopy. Under wind periodic boundary conditions the particle will be transported by the unsteady wind and may experience several domain turnovers up to ground deposition. For this purpose the vertical dimension of the computational domain was doubled and particles were released at 50-metre height. At $t = 0$, the initial streamwise and vertical velocity components were assigned to be equal to local wind velocity and the terminal particle velocity, respectively. This scenario without the buoyant plume atmosphere interaction is only relevant

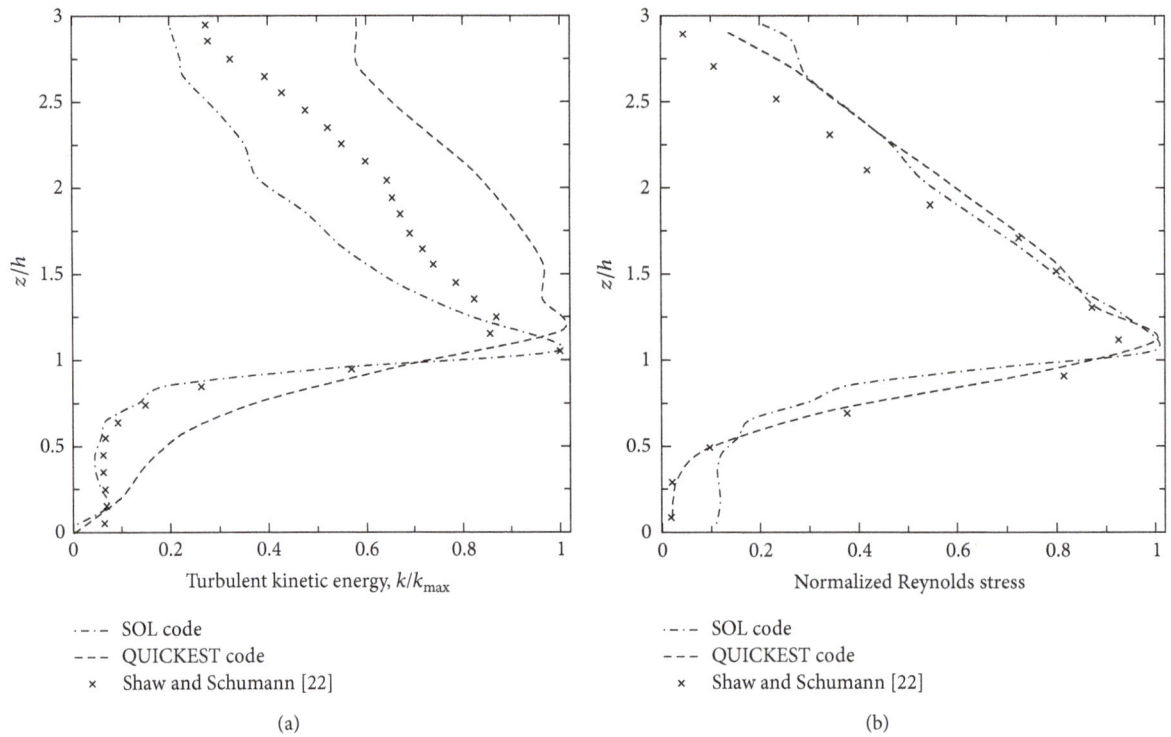

FIGURE 3: Vertical profile of the turbulent kinetic energy (a) and of the Reynolds stress (b) for a leaf area index equal to 5.

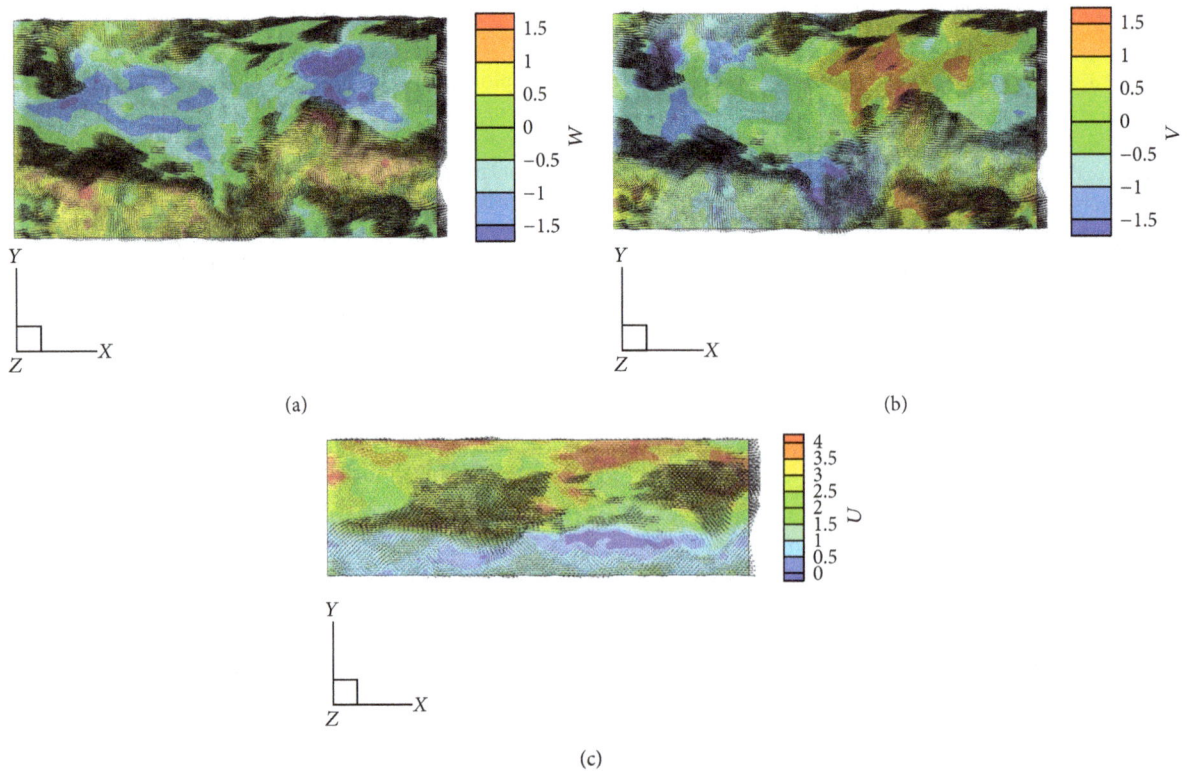

(a)

(b)

(c)

FIGURE 4: Instantaneous velocity contour plots with velocity vectors: (a) $W$ velocity with slice plane at $Z = 25$ m; (b) $V$ velocity, slice plane at $Z = 25$ m; and (c) $U$ velocity, slice plane at $Y = 48$ m.

(a)

(b)

(c)

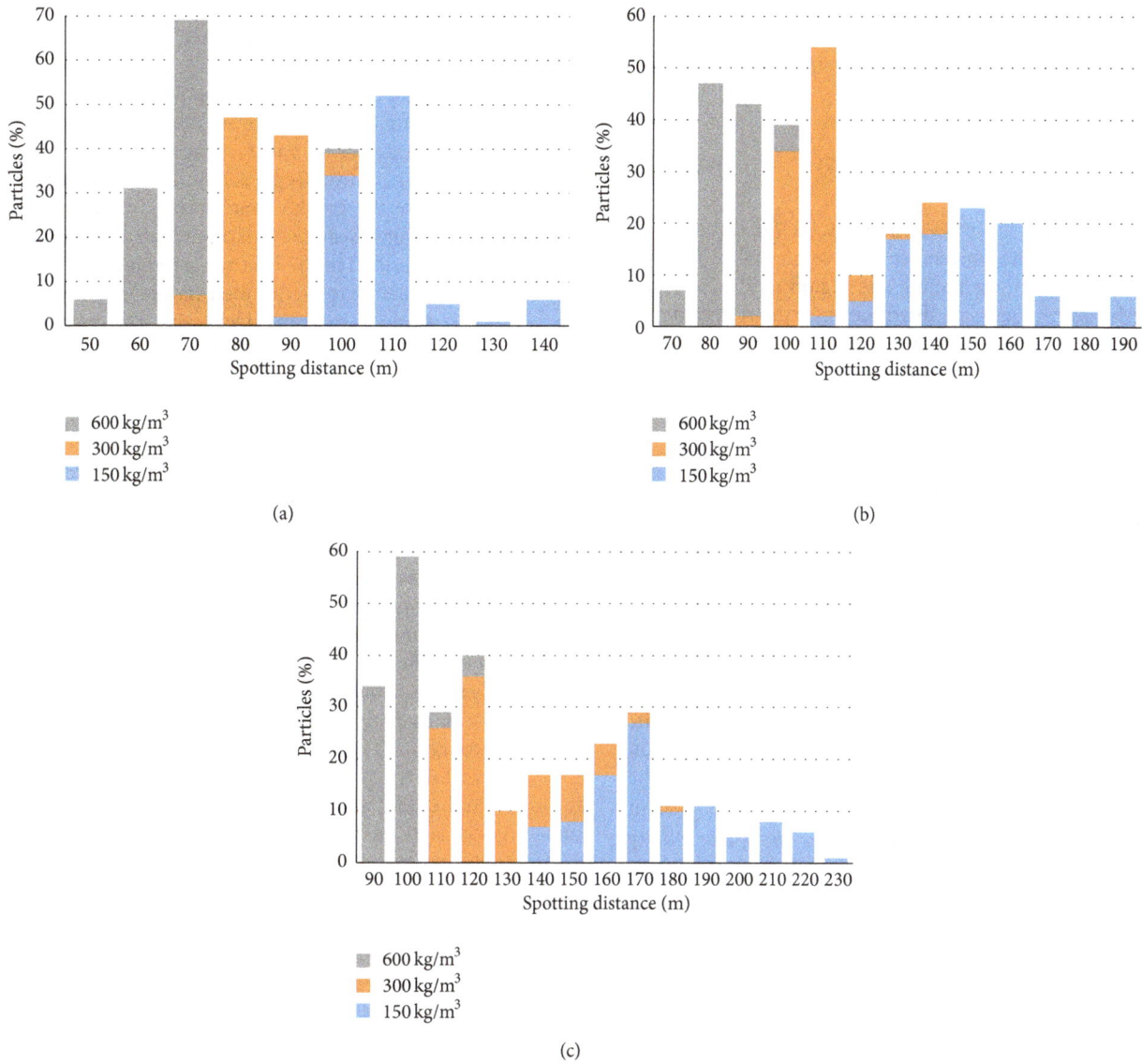

FIGURE 5: Spotting distances of the spherical particles for three different densities, 150, 300, and 600 kg/m$^3$, and three different drag coefficients, (a) Cd = 0.4; (b) Cd = 0.8; and (c) Cd = 1.2.

to simulating the particles final flight path, after leaving the convective column, and their interaction with the canopy dominated flow. In addition the wind velocity was 15 m/s to simulate strong winds occurring very often in large fires. The particles were assumed to be spherical with different drag coefficients, 0.4, 0.5, and 1.2, and densities, 150 kg/m$^3$, 300 kg/m$^3$, and 600 kg/m$^3$, in order to simulate different particle relaxation times.

Figures 5(a), 5(b), and 5(c) show the probabilities of spotting distances for the three drag coefficients, respectively. The light particles travel more and are strongly dependent on the drag coefficient. The high drag force communicates a higher particle momentum in the streamwise direction, and the particles with low density, 150 kg/m$^3$, have a maximum spotting distance corresponding to 140 m for Cd = 0.4 and reach up to 230 m for Cd = 1.2; the probability density curve

denotes the increase of the variability with the increase of drag. For the heavy particles with density 600 kg/m$^3$ the spotting distances change much less with the considered drag coefficients. It should be mentioned that despite the simplicity of the analysis one identifies a large variability, from 130 to 230 m for light particles with a realistic Cd = 1.2. The lifted-off particles that reach an altitude of 50 m, say by transient events of intense flammability of isolated torching trees, and begin their transport by the wind field outside the convection column may deposit with a large dispersion, creating a large uncertainty of secondary fires.

3.2. Spotting Distances from a Model Fire. In previous subsection quantitative or qualitative comparisons of the predictions were performed concluding that although coarse grids are used the numerical results are still consistent with the parameters that influence the transport of large diameter firebrands.

All the predictions that follow use the volumetric heat source strength calculated from

$$\int I\, ds = Q = \iiint_v q\, dv, \tag{18}$$

where $I$ stands for the power intensity, $ds$ for the line fire length, and $q$ for the power per unit volume. The integral was obtained in the prismatic flame region with base equal to flame height. A Gaussian distribution of the source depth $q$ function was assumed and adjustments were made in order to satisfy (17a), (17b), and (17c) on the computational domain.

*3.2.1. Line Grass Fire.* The first scenario corresponds to the examples given by [6, 86]. The problem consists in estimating the maximum spot fire distance from a heading fire in short grass, with intensity of 2000 kW/m when the wind speed at 10 m height is 5 m/s. The vegetation height is equal to 2.5 m and the vegetation area density corresponds to a triangular vertical distribution with maximum value of 0.5. The inlet velocity is a power law (1/7) and linear stratified 0.003 K/m above 30 m. The flame height was estimated from Rothermel correlations [41] to be 2.6 m height and the line fire length was 60 m. The volumetric heat source was allocated in the prismatic volume with equilateral triangular cross section of 3 m height (predicted by Rothermel correlations). The strength mean value was 49 kW/m$^3$ and on the ground 10 kW/m$^2$ to ensure the 2000 kW/m fire. The inlet velocity and temperature field were perturbed at each time step according to

$$T = T_0 + 0.1\xi\left(1 - \frac{y}{y^*}\right)T_*, \tag{19a}$$

$$U_i = U_i + 0.1\xi\left(1 - \frac{y}{y^*}\right)W_*, \tag{19b}$$

where $\xi$ is a Gaussian random number with zero mean and unit variance and the convective scale $W_* = (\beta g Q Z_{io})^{1/3}$ and $T_* = Q_s/W_*$. This procedure is required because no homogeneous directions can be considered and a velocity inlet profile is required for the phenomenon under consideration.

A grid of $64 \times 64 \times 64$ nodes discretizes the computational domain of $1000 \times 300$ metres in the horizontal plane and 250 metres in the vertical direction. The mesh is nonuniform distributed on the ground vicinity with 1 m resolution and on the line fire region; typically 8000 time iterations (16-minute real time) were performed. Particles were released with zero velocity inside the region where heat release was prescribed. Ten particle sizes classes were considered from 0.5 to 9.5 mm of diameter and for each class 50 particles were randomly distributed around the spanwise control volumes. The particles may experience ground deposit or burnout or exiting the domain and when this occurs, a new particle was inserted in the heat release region. A total of 500 particles were always tracked in each time step during 6000 time steps and the total number of particle ground depositions was

around 20000. The particle initial density was 200 kg/m$^3$ and the percentage in mass of char, ash, water, and pyrolysis was 70%, 20%, 1%, and 4%, respectively.

Figure 6(a) shows a three-dimensional view of the scalar smoke concentration field. The buoyant flow under cross flow originates two characteristic vortex structures and far away from the source the corotating vortex pair originates two plume tubes where the smoke concentration is maximum. For visualization purposes the smoke concentration equal to the unity was prescribed in the heat release volume.

The Albini model was programmed and their correlations for vegetation fires or torching trees give the maximum spotting distance and for the torching trees the model output includes also the particle diameter. However for vegetation fires Albini model does not predict the particle size diameter that corresponds to the maximum spotting distance. Another remark is concerned with the prescription of the fire and also its behaviour compared with a real situation where strong flame oscillations are present and taken into account by Albini. No effort was made to consider unsteady volumetric heat release sources. The predicted puffing of the plume could only affect very small particles.

Figure 6(b) shows the coordinates of the point where the particle attains the maximum altitude along their particle trajectory and the spotting distance of the corresponding particle. For each particle size class the maximum location depends strongly on its initial position. If the buoyant column does not capture the particle its trajectory is very short. As the total number of particles was around 20000 it is believed that a good statistical description of the source release positions was covered. According to Albini model the particle reaches the maximum altitude of 68 m for this scenario. From the present predictions this should correspond to particles around 2 mm diameter. Particles smaller than 1 mm burn during the flight according to the combustion model considered. Form Figure 6(b) it is possible to conclude that the particles trapped higher up in the plume are taken further downstream.

The results corresponding to the probability for ground deposition are shown in Figure 6(c). One should mention that the probability was evaluated by counting, for each class, the particles that deposit in each longitudinal slice of 2 m length and the number divided by the total of particles that deposit for each class. Consequently if a larger slice size was selected the probability would increase but with a similar distribution. Nevertheless Figure 6(c) shows that particles deposit close to the inlet due to large size or those with a small size released at a too low initial position or they fall out of the convection column. Figure 6(c) shows that relatively only few particles spot far than 200 metres. But there are differences in propagation distance among particles released randomly at the same height. The maximum predicted spotting distance was 400 metres for 1.5 and 2.5 mm spherical diameter. The Albini model predictions, corrected by [86], indicate that the maximum spotting distance is 450 m.

The second test case corresponds to a wind-driven fire in chaparral when the wind speed at 10 m height is 20 m/s and the fire intensity is 50000 kW/m. This corresponds to a severe surface fire with very high intensity and very strong

(a)

(b)

(c)

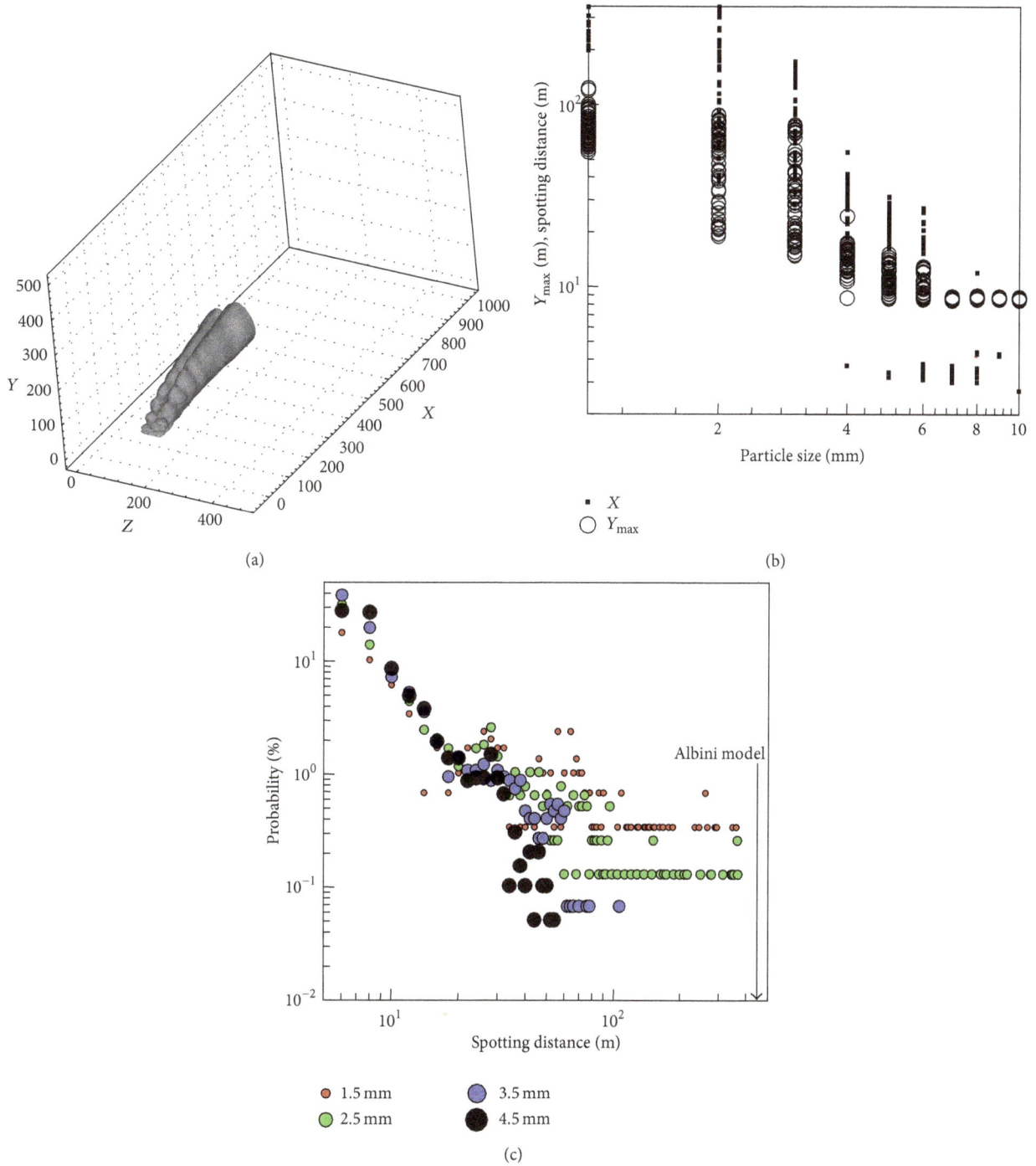

FIGURE 6: (a) Smoke scalar concentration for 20000 kW/m short grass fire; (b) maximum height and longitudinal distances during particle trajectory as a function of particle size; and (c) probability of particles spotting distances as a function of particle size.

wind. Large spotting distances might be expected in light of the results from the first example. Conditions similar to the previous example were used but the computational domain extends to 3 km in the longitudinal direction and by 500 m in both the spanwise and vertical directions. A $64 \times 64 \times 64$ coarse grid was used and the vegetation was assumed to have 4 m height and the flames were assumed to have 10 m height.

A volumetric heat release rate of 508 kW/m$^3$ was prescribed in the flame region of the 100-metre line fire.

Figure 7(a) show the three-dimensional view of smoke concentration up to 1 km from the source. Several distinct types of coherent vortical structures have been observed in our simulations. It is possible to identify the rolling up shear-layer ("hanging" the Kelvin-Helmholtz instability) like

(a)

(b)

(c)

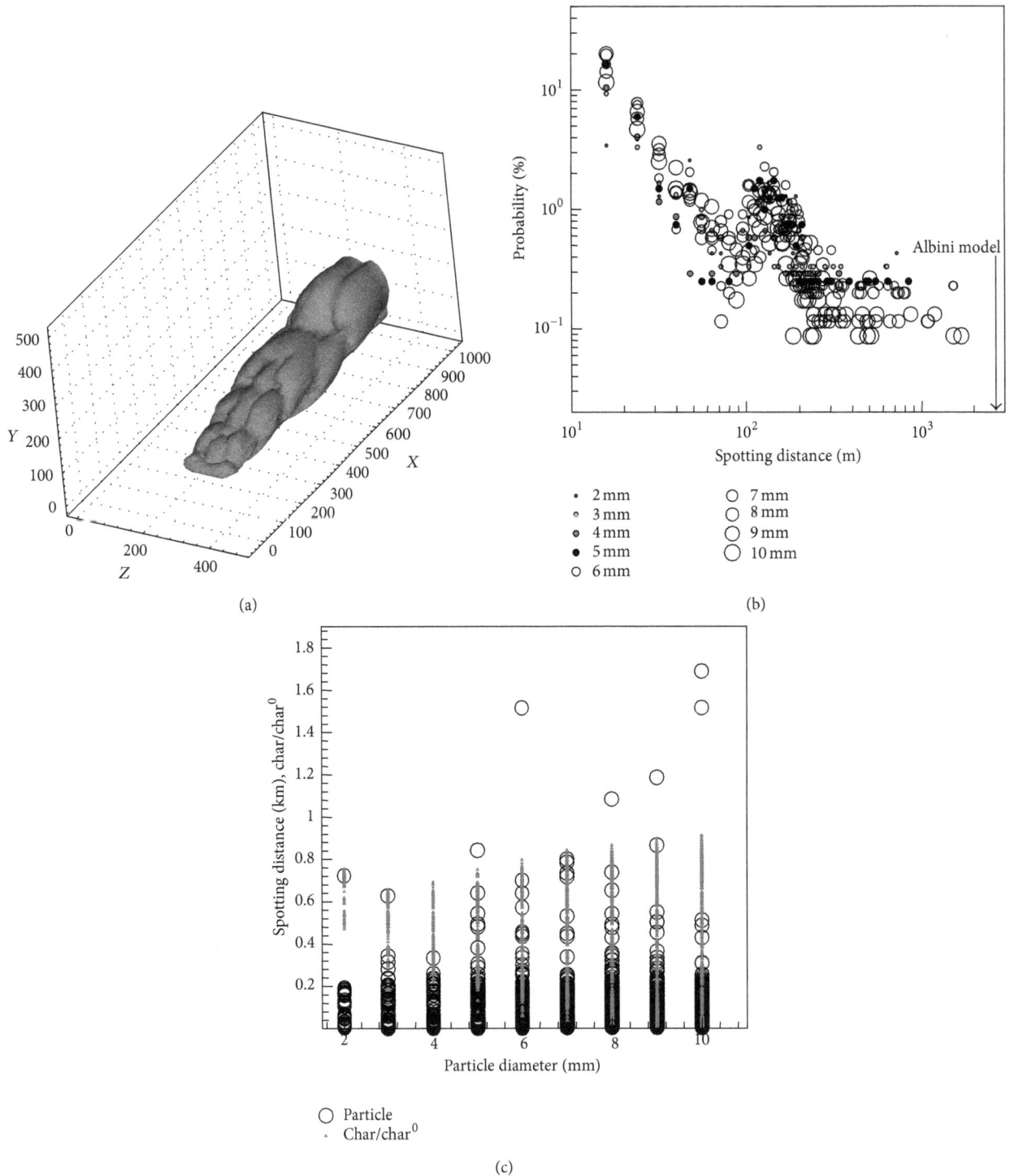

FIGURE 7: (a) Smoke scalar concentration for 50000 kW/m short grass fire; (b) probability of particles spotting distances as a function of particle size; and (c) particles spotting distances and char content as a function of particle size.

the one described by [87]. Due to the long line fire length the vortical structures corresponding to the counterrotating vortex air, plume did not yet bifurcate. It is also possible to distinguish the incipient edge vortices, at the windward corners of the line fuel surface and located near the ground surface. Finally in the wake, not visible in the figure, there

are wake vortices that interact with the ground for this strong wind.

Figure 7(b) shows the spotting distances probability as a function of particle size diameter valuated by the number of particles that deposit in 10 m longitudinal slice. Particles smaller than 2 mm diameter burn out during the flight. Due to the strong wind and convection column larger particles may be transported to larger distances without burning. Albini model predicts the maximum spotting distance at 2820 m. The present 3D predictions show spotting distances up to 1800 m that are smaller but under the present plume flow coherent structure the trapped particles will describe trajectories that could not be captured by simplified classic models. Firebrand motion is very sensitive to the variable flow field surrounding it that is governed by the fire-atmosphere dynamical system.

Figure 6(b) also shows that the distribution of spot fires decreases rapidly with distance downwind of the fires and can be described using a negative exponential function (see Figure 6(b)), while Figure 7(b) denotes around 100 m a slight increase followed by a rapid decrease. It may be due to the strong vertical flows.

Figure 7(c) shows the spotting distance, as well as the char content divided by initial char mass. One should observe that the ordinates axis displays two scales: (i) the spotting distance units in km and (ii) the char particle content at spotting divided by the initial particle content; this ratio is represented in the scale 0 to 1. Small particles deposit with high char content and large particles deposit with char percentages that range almost from zero (burnout) to 80%, due to different residence lifetimes along their trajectories. Particles with low char mean that their density is lower and consequently for the same drag are likely to be transported further downstream.

*3.2.2. Torching Trees.* The calculations correspond to the Rothermel example [73] that uses the Albini model version to calculate the maximum spotting distance from torching trees. The first example corresponds to a single Grand Fir torching tree of 50 m height in a forest with the other tree's 43 m height and the wind speed of 9 m/s at 6 m height. The heat of 38 kW/m$^3$ was released in a cone of 12 m diameter and 40-metre height above 10-metre ground level to which a power of 180 MW estimated from Rothermel correlations of burning speed and fuel mass available corresponds. However the author is not certain about the volumetric heat release of a Grand Fir torching tree under the wind velocity considered.

The instantaneous cross section planes of scalar smoke concentration display a horseshoe-like structure with counterrotating vortices. Figure 8(a) shows the maximum spotting distance for the single tree example. Albini model predicts the maximum spotting distance of 560 m and corresponding to a particle diameter of 4.9 mm. However Albini assumed the drag coefficient of an infinite cylindrical particle shape. A simple balance between drag and weight forces shows that a cylinder may have a diameter 1.9 times bigger than the spherical diameter. The predictions for spherical particles show that 3 mm particles may reach a maximum distance of 600 m and 2 mm particle 900 m and consequently the present predictions are in satisfactory agreement with Albini model.

Figure 8(b) shows three different quantities. The first is the spotting distance in km units. The second is the particle ratio of char mass at deposition to its initial value in a scale between 0 and 1 and the third is the particle average temperature relative to initial 973 K, also in the scale between

0 and 1. The predictions show that larger particles deposit with a higher temperature than the small ones due to the shorter flight time and less heat release to the cold ambient. The spreading of the particle char content increases with particle size. Figure 8(c) shows the dependence of char on particle spotting distance denoting that the particles that travel more have consumed most of the char as it would be expected, but the 10 mm particles have the highest char variability at deposition. This information will be valuable for probabilistic ignition models that account for vegetation, moister content, and so forth. The dependence of char content on temperature is shown in Figure 8(d) denoting that larger particles have a lower char content and a wide range of temperatures than small particles that keep the char at a high value and cool down very fast. This is consequence of the combustion model constants used and other values would produce different results, but no data about real spotting material combustion (leaves; fruits; etc.) was found.

The second example corresponds to four torching Grand Fir trees aligned in the spanwise direction with 12-metre spacing. Figure 9(a) shows the spotting distances. The Albini model predicts 780 m for a 7.4 mm cylindrical particle diameter. The present predictions suggest that the maximum spotting distance in between 700 and 800 m is reachable by 1.5 and 2.5 mm particles. The spotting distances and the corresponding particles char and temperature for four Grand Fir torching trees are represented in Figure 9(b). The average char content decreases with particle diameter, but its variability increases similarly to the previous example of a single torching tree.

## 4. Conclusion Remarks

In this work particle spotting distances were predicted with a coupled fire-atmosphere LES simulation. A volumetric heat source was used to simulate the fire and LES of the wind over and through homogeneous vegetation was conducted with the SGS kinetic model. This model was selected because it captures satisfactorily the mean flow and temperature effects that may be relevant to calculate particle spotting distances in different atmospheric boundary layer conditions: weakly unstable, neutral, and stratified.

The maximum spotting distance obtained with the Albini based correlation model is compared with the corresponding distance obtained from thousands of simulated particles in the 3D wind field. Classical examples were considered as benchmark test cases corresponding to vegetation line fires and torching trees reported by the pioneering researchers Rothermel and Albini.

The particle's initial position was prescribed randomly in all flame volumes and for the low intensity vegetation fire line the agreement between the two very different models is quite surprising, with the maximum spotting distances differing only by 10%. This was obtained from a wide class of particle sizes that are located randomly around the fire line, changing density during flight path and so forth, and cannot be interpreted as there are always little things happening that seem to coincide with the target result. Albini model is

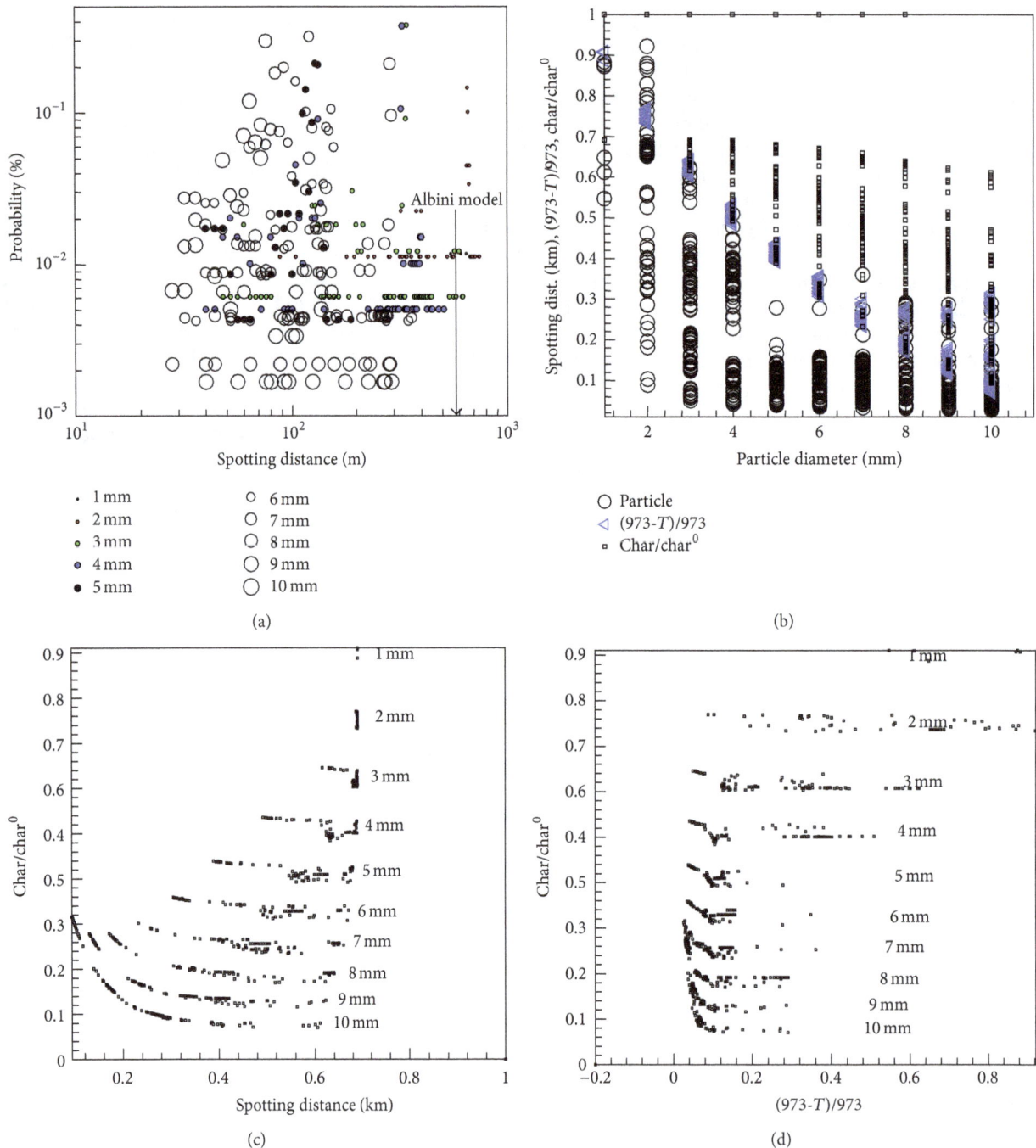

FIGURE 8: Single torching tree results: (a) spotting distance; (b) char, temperature, and spotting distance as a function of particle size; (c) char as a function of particle spotting distance; and (d) particle char as a function of particle temperature.

very popular in forest fire research and it is a simple integral model to predict the maximum spotting distance. The present work shows that for some cases it is in agreement with 3D multiphysics model predictions.

For the strong wind and high intensity, 50000 kW/m, vegetation fire where the predictions show a shorter maximum spotting distance, large differences are present between the two models and this may be attributed to the complex vortical flows formed or to the different particle combustion

model that changes the particle density and alters the particle relaxation time.

The present predictions show also that the particles deposit from the source to the maximum spotting distance, in an inverted exponential function, and most of the particles deposit in the fire vicinity. Some of them were not entrained by the convection column and have a short flight path. In addition to the cases investigated particles up to 10 mm diameter are able to travel several hundred metres. The probability

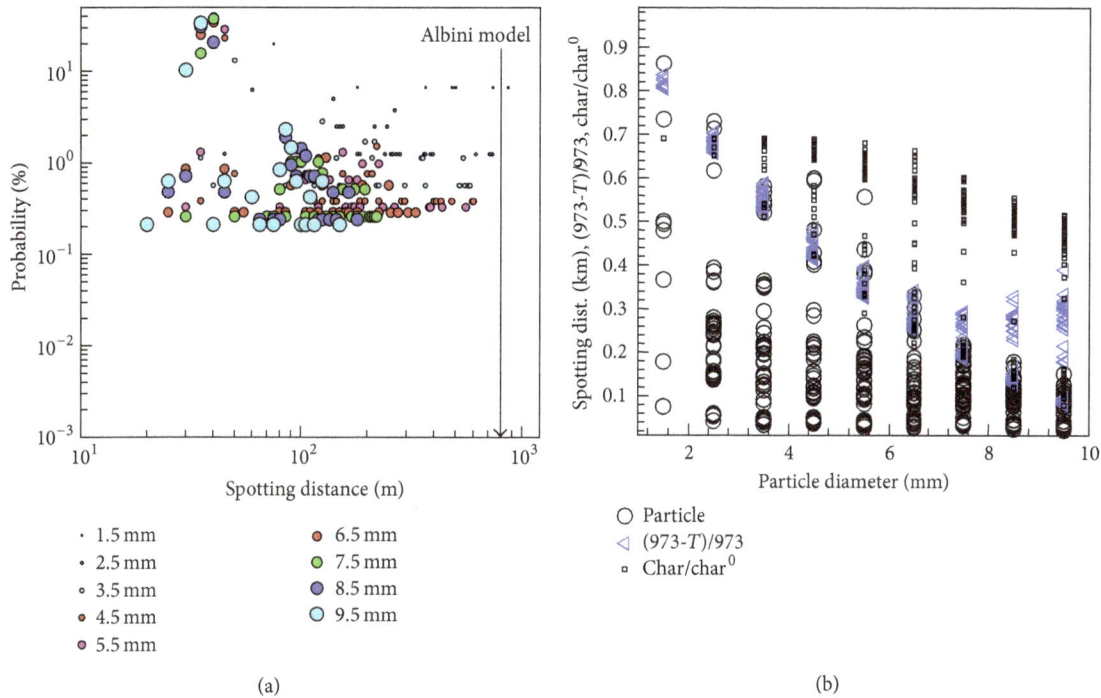

FIGURE 9: Four torching trees: (a) particle spotting distance; (b) char, temperature, and spotting distance as a function of particle size.

to reach longer distances increases for particles in the range 2 to 4 mm, but low probability events may transport a larger particle, 10 mm close to the maximum spotting distance. This is a stochastic phenomenon that should be in the future considered by uncertainty quantification prediction tools.

Significant different assumptions have been made; several of them may be improved and made closer to the reality, namely, the particle combustion model for specific vegetation debris. The unsteady fire structure and modelling may be affected by the specific cases in question, the particle drag or spherical shape, and so forth. Another important influencing factor is the local geophysical and atmospheric conditions that may influence the plume structure. Others may be more easily solved by those with supercomputers access such as grid resolution or improvements in the heterogeneity of the fuel and the structure of mixed fuel types vegetation and terrain topology. The foreseeable continuous increase in speed and memory of the computers favours the development and application of coupled fire-atmosphere LES models because it increases the scope of current empirical or semiempirical operational models.

## Conflict of Interests

The authors declare that there is no conflict of interests regarding the publication of this paper.

## Acknowledgments

This work was supported by Fundação para a Ciência e Tecnologia (FCT, Portugal), through IDMEC, under LAETA, Project UID/EMS/50022/2013. The third author would like to thank the support received by Fundação para a Ciência e Tecnologia (FCT, Portugal) under the project "Extreme—PTDC/EMEMFE/11343/2009." The last author would like to acknowledge the support received during his Ph.D. study from Fundação para a Ciência e Tecnologia (FCT, Portugal) (Grant SFRH/BD/48150/2008 cofinanced by POPH/FSE).

## References

[1] E. K. Noonan-Wright, T. S. Opperman, M. A. Finney et al., "Developing the US wildland fire decision support system," *Journal of Combustion*, vol. 2011, Article ID 168473, 14 pages, 2011.

[2] A. A. Ager, N. M. Vaillant, and M. A. Finney, "Integrating fire behavior models and geospatial analysis for wildland fire risk assessment and fuel management planning," *Journal of Combustion*, vol. 2011, Article ID 572452, 19 pages, 2011.

[3] P.-A. Santoni, A. Sullivan, D. Morvan, and W. E. Mell, "Forest fire research: the latest advances tools for understanding and managing wildland fire," *Journal of Combustion*, vol. 2011, Article ID 418756, 2 pages, 2011.

[4] C. S. Tarifa, P. P. Del Notario, and F. G. Moreno, "On the flight paths and lifetimes of burning particles of wood," *Proceedings of the Combustion Institute*, vol. 10, no. 1, pp. 1021–1037, 1965.

[5] S.-L. Lee and J. M. Hellman, "Firebrand trajectory study using an empirical velocity-dependent burning law," *Combustion and Flame*, vol. 15, no. 3, pp. 265–274, 1970.

[6] F. A. Albini, "Spot fire distance from burning trees—a predictive model," Tech. Rep. INT-56, USDA, Intermountain Forest and Range Experiment Station, 1979.

[7] F. A. Albini, "A model for the wind-blown flame from a line fire," *Combustion and Flame*, vol. 43, pp. 155–174, 1981.

[8] F. A. Albini, "Potential spotting distance from wind-driven surface fires," Research Note INT-309, US Department of Agriculture, Forest Service Intermountain Forest and Range Experiment Station, Ogden, Utah, USA, 1983.

[9] F. A. Albini, "Transport of firebrands by line thermals," *Combustion Science and Technology*, vol. 32, no. 5-6, pp. 277–288, 1983.

[10] F. A. Albini, M. E. Alexander, and M. G. Cruz, "A mathematical model for predicting the maximum potential spotting distance from a crown fire," *International Journal of Wildland Fire*, vol. 21, no. 5, pp. 609–627, 2012.

[11] R. C. Rothermel, "A mathematical model for predicting fire spread in wildland fuels," Research Report, USDA-FS, Ogden, Utah, USA, 1972.

[12] R. E. Burgan and R. C. Rothermel, "BEHAVE: fire behavior prediction and fuel modeling system—fuel subsystem," General Technical Report INT-167, USDA Forest Service, Intermountain Forest and Range Experimental Station, Ogden, Utah, USA, 1984.

[13] P. L. Andrews, "Behave: fire behavior prediction and fuel modeling system—burn subsystem. Part 1," General Technical Report INT-194, USDA Forest Service, Intermountain Forest and Range Experiment Station, Ogden, Utah, USA, 1986.

[14] J. R. Coleman and A. L. Sullivan, "A real-time computer application for the prediction of fire spread across the Australian landscape," *Simulation*, vol. 67, no. 4, pp. 230–240, 1996.

[15] D. X. Viegas, "Forest fire propagation," *Philosophical Transactions of the Royal Society A: Mathematical, Physical and Engineering Sciences*, vol. 356, no. 1748, pp. 2907–2928, 1998.

[16] E. Pastor, L. Zárate, E. Planas, and J. Arnaldos, "Mathematical models and calculation systems for the study of wildland fire behaviour," *Progress in Energy and Combustion Science*, vol. 29, no. 2, pp. 139–153, 2003.

[17] M. A. Finney and S. S. McAllister, "A review of fire interactions and mass fires," *Journal of Combustion*, vol. 2011, Article ID 548328, 14 pages, 2011.

[18] C.-P. Mao, A. C. Fernandez-Pello, and J. A. C. Humphrey, "An investigation of Steady Wall-Ceiling and partial Enclosure fires," *Journal of Heat Transfer*, vol. 106, no. 1, pp. 221–228, 1984.

[19] G. Cox and S. Kuman, "Field modeling of fire in forced ventilated enclosures," *Combustion Science and Technology*, vol. 52, pp. 7–23, 1987.

[20] D. Drysdale, *An Introduction to Fire Dynamics*, Wiley, Chichester, UK, 2nd edition, 1998.

[21] V. K. Schilling, "A parameterization for modelling the meteorological effects of tall forests—a case study of a large clearing," *Boundary-Layer Meteorology*, vol. 55, no. 3, pp. 283–304, 1991.

[22] R. H. Shaw and U. Schumann, "Large-eddy simulation of turbulent flow above and within a forest," *Boundary-Layer Meteorology*, vol. 61, no. 1-2, pp. 47–64, 1992.

[23] G. Gross, *Numerical Simulation of Canopy Flows*, Springer, Berlin, Germany, 1993.

[24] M. G. Inclan, R. Forkel, R. Dlugi, and R. B. Stull, "Application of transilient turbulent theory to study interactions between the atmospheric boundary layer and forest canopies," *Boundary-Layer Meteorology*, vol. 79, no. 4, pp. 315–344, 1996.

[25] A. Wenzel, N. Kalthoff, and V. Horlacher, "On the profiles of wind velocity in the roughness sublayer above a coniferous forest," *Boundary-Layer Meteorology*, vol. 84, no. 2, pp. 219–230, 1997.

[26] M. R. Raupach and A. S. Thom, "Turbulence in and above plant canopies," *Annual Review of Fluid Mechanics*, vol. 13, pp. 97–129, 1981.

[27] D. D. Baldocchi and T. P. Meyers, "Turbulence structure in a deciduous forest," *Boundary-Layer Meteorology*, vol. 43, no. 4, pp. 345–364, 1988.

[28] Z. Li, J. D. Lin, and D. R. Miller, "Air flow over and through a forest edge: a steady-state numerical simulation," *Boundary-Layer Meteorology*, vol. 51, no. 1-2, pp. 179–197, 1990.

[29] U. Svensson and K. Häggkvist, "A two-equation turbulence model for canopy flows," *Journal of Wind Engineering and Industrial Aerodynamics*, vol. 35, no. 1–3, pp. 201–211, 1990.

[30] M. H. Kobayashi, J. C. F. Pereira, and M. B. B. Siqueira, "Numerical study of the turbulent flow over and in a model forest on a 2D hill," *Journal of Wind Engineering and Industrial Aerodynamics*, vol. 53, no. 3, pp. 357–374, 1994.

[31] G. G. Katul, L. Mahrt, D. Poggi, and C. Sanz, "One- and two-equation models for canopy turbulence," *Boundary-Layer Meteorology*, vol. 113, no. 1, pp. 81–109, 2004.

[32] A. S. Lopes, J. M. L. M. Palma, and J. V. Lopes, "Improving a two-equation turbulence model for canopy flows using large-eddy simulation," *Boundary-Layer Meteorology*, vol. 149, no. 2, pp. 231–257, 2013.

[33] J. W. Deardorff, "A three–dimensional numerical investigation of the idealized planetary boundary layer," *Geophysical Fluid Dynamics*, vol. 1, no. 3-4, pp. 377–410, 2008.

[34] P. J. Mason, "Large-eddy simulation: a critical review of the technique," *Quarterly Journal—Royal Meteorological Society*, vol. 120, no. 515, pp. 1–26, 1994.

[35] H. R. Baum, R. G. Rehm, and J. P. Gore, "Transient combustion in a turbulent eddy," in *Proceedings of the 23rd International Symposium on Combustion*, pp. 715–722, The Combustion Institute, Pittsburgh, Pa, USA, 1990.

[36] K. B. McGrattan, H. R. Baum, and R. G. Rehm, "Large Eddy simulations of smoke movement," *Fire Safety Journal*, vol. 30, no. 2, pp. 161–178, 1998.

[37] H. R. Baum and B. J. McCaffrey, "Fire induced flow field—theory and experiment," in *Proceedings of the 2nd International Symposium on Fire Safety Science*, pp. 129–148, Hemisphere, New York, NY, USA, June 1989.

[38] K. B. McGrattan, H. R. Baum, and R. G. Rehm, "Numerical simulation of smoke plumes from large oil fires," *Atmospheric Environment*, vol. 30, no. 24, pp. 4125–4136, 1996.

[39] M. Schatzmann, "An integral model of plume rise," *Atmospheric Environment*, vol. 13, no. 5, pp. 721–731, 1979.

[40] G. A. Davidson, "Gaussian versus top-hat profile assumptions in integral plume models," *Atmospheric Environment*, vol. 20, no. 3, pp. 471–478, 1986.

[41] G. N. Mercer and R. O. Weber, "Plumes above line fires in a cross wind," in *Proceedings of the 2nd International Conference on Forest Fire Research*, D. X. Viegas, Ed., Coimbra, Portugal, November 1994.

[42] J. Trelles, K. B. McGrattan, and H. R. Baum, "Smoke dispersion from multiple fire plumes," *AIAA Journal*, vol. 37, no. 12, pp. 1588–1601, 1999.

[43] H. Schmidt and U. Schumann, "Coherent structure of the convective boundary layer derived from large-eddy simulations," *Journal of Fluid Mechanics*, vol. 200, pp. 511–562, 1989.

[44] X.-Q. Chen and J. C. F. Pereira, "Computational modeling of two-phase flows with particles and droplets," *Trends in Heat, Mass & Momentum Transfer*, vol. 5, pp. 81–99, 1999.

[45] E. Koo, P. J. Pagni, D. R. Weise, and J. P. Woycheese, "Firebrands and spotting ignition in large-scale fires," *International Journal of Wildland Fire*, vol. 19, no. 7, pp. 818–843, 2010.

[46] I. K. Knight, "The design and construction of a vertical wind tunnel for the study of untethered firebrands in flight," *Fire Technology*, vol. 37, no. 1, pp. 87–100, 2001.

[47] S. L. Manzello, J. R. Shields, T. G. Cleary et al., "On the development and characterization of a firebrand generator," *Fire Safety Journal*, vol. 43, no. 4, pp. 258–268, 2008.

[48] M. Almeida, D. X. Viegas, A. I. Miranda, and V. Reva, "Effect of particle orientation and of flow velocity on the combustibility of *Pinus pinaster* and *Eucalyptus globulus* firebrand material," *International Journal of Wildland Fire*, vol. 20, no. 8, pp. 946–962, 2011.

[49] A. Hölzer and M. Sommerfeld, "New simple correlation formula for the drag coefficient of non-spherical particles," *Powder Technology*, vol. 184, no. 3, pp. 361–365, 2008.

[50] S. Suzuki, S. L. Manzello, M. Lage, and G. Laing, "Firebrand generation data obtained from a full-scale structure burn," *International Journal of Wildland Fire*, vol. 21, no. 8, pp. 961–968, 2012.

[51] M. Zastawny, G. Mallouppas, F. Zhao, and B. van Wachem, "Derivation of drag and lift force and torque coefficients for non-spherical particles in flows," *International Journal of Multiphase Flow*, vol. 39, pp. 227–239, 2012.

[52] M. J. Antal Jr., H. L. Friedman, and F. E. Rogers, "Kinetics of cellulose pyrolysis in nitrogen and steam," *Combustion science and technology*, vol. 21, no. 3-4, pp. 141–152, 1980.

[53] J. P. Woycheese and P. J. Pagni, "Combustion models for wooden brands," in *Proceedings of the 3rd International Conference on Fire Research and Engineering (ICFRE3 '99)*, pp. 53–71, Society of Fire Protection Engineers, Chicago, Ill, USA, October 1999.

[54] B. Benkoussas, J.-L. Consalvi, B. Porterie, N. Sardoy, and J.-C. Loraud, "Modelling thermal degradation of woody fuel particles," *International Journal of Thermal Sciences*, vol. 46, no. 4, pp. 319–327, 2007.

[55] N. Sardoy, J.-L. Consalvi, B. Porterie, and A. C. Fernandez-Pello, "Modeling transport and combustion of firebrands from burning trees," *Combustion and Flame*, vol. 150, no. 3, pp. 151–169, 2007.

[56] J. Tomeczek, *Coal Combustion*, Kriger Publishing Company, Malabar, Fla, USA, 1994.

[57] C. Di Blasi, "Modeling chemical and physical processes of wood and biomass pyrolysis," *Progress in Energy and Combustion Science*, vol. 34, no. 1, pp. 47–90, 2008.

[58] A. F. Roberts, "A review of kinetics data for the pyrolysis of wood and related substances," *Combustion and Flame*, vol. 14, no. 2, pp. 261–272, 1970.

[59] A. M. Costa, J. C. F. Pereira, and M. Siqueira, "Numerical prediction of fire spread over vegetation in arbitrary 3D terrain," *Fire and Materials*, vol. 19, no. 6, pp. 265–273, 1995.

[60] T. L. Clark, J. Coen, and D. Latham, "Description of a coupled atmosphere-fire model," *International Journal of Wildland Fire*, vol. 13, no. 1, pp. 49–63, 2004.

[61] S. Bhutia, M. A. Jenkins, and R. Sun, "Comparison of firebrand propagation prediction by a plume model and a coupled-fire/atmosphere large–eddy simulator," *Journal of Advances in Modeling Earth Systems*, vol. 2, article 4, pp. 1–15, 2010.

[62] J. P. Woycheese, P. J. Pagni, and D. Liepman, "Brand Lofting from large-scale fires," *Journal of Fire Protection Engineering*, vol. 10, no. 2, pp. 32–44, 1999.

[63] R. A. Anthenien, S. D. Tse, and A. C. Fernandez-Pello, "On the trajectories of embers initially elevated or lofted by small scale ground fire plumes in high winds," *Fire Safety Journal*, vol. 41, no. 5, pp. 349–363, 2006.

[64] X. Y. Zhou and J. C. F. Pereira, "Multidimensional model for simulating vegetation fire spread using a porous media sub-model," *Fire and Materials*, vol. 24, no. 1, pp. 37–43, 2000.

[65] B. Porterie, D. Morvan, J. C. Loraud, and M. Larini, "Firespread through fuel beds: modeling of wind-aided fires and induced hydrodynamics," *Physics of Fluids*, vol. 12, no. 7, pp. 1762–1782, 2000.

[66] X. Zhou, S. Mahalingam, and D. Weise, "Experimental study and large eddy simulation of effect of terrain slope on marginal burning in shrub fuel beds," *Proceedings of the Combustion Institute*, vol. 31, pp. 2547–2555, 2007.

[67] W. Mell, M. A. Jenkins, J. Gould, and P. Cheney, "A physics-based approach to modelling grassland fires," *International Journal of Wildland Fire*, vol. 16, no. 1, pp. 1–22, 2007.

[68] W. Mell, A. Maranghides, R. McDermott, and S. L. Manzello, "Numerical simulation and experiments of burning douglas fire trees," *Combustion and Flame*, vol. 156, no. 10, pp. 2023–2041, 2009.

[69] S. Kortas, P. Mindykowski, J. L. Consalvi, H. Mhiri, and B. Porterie, "Experimental validation of a numerical model for the transport of firebrands," *Fire Safety Journal*, vol. 44, no. 8, pp. 1095–1102, 2009.

[70] L. A. Oliveira, A. G. Lopes, B. R. Baliga, M. Almeida, and D. X. Viegas, "Numerical prediction of size, mass, temperature and trajectory of cylindrical wind-driven firebrands," *International Journal of Wildland Fire*, vol. 23, no. 5, pp. 698–708, 2014.

[71] E. Koo, R. R. Linn, P. J. Pagni, and C. B. Edminster, "Modelling firebrand transport in wildfires using HIGRAD/FIRETEC," *International Journal of Wildland Fire*, vol. 21, no. 4, pp. 396–417, 2012.

[72] J.-B. Filippi, F. Bosseur, X. Pialat, P.-A. Santoni, S. Strada, and C. Mari, "Simulation of coupled fire/atmosphere interaction with the MesoNH-ForeFire models," *Journal of Combustion*, vol. 2011, Article ID 540390, 13 pages, 2011.

[73] R. C. Rothermel, "How to predict the spread and intensity of forest and range fires," Tech. Rep. INT-143, Intermountain Forest and Range Experiment Station, Ogden, Utah, USA, 1983.

[74] C. B. da Silva and J. C. F. Pereira, "The effect of subgrid-scale models on the vortices computed from large-eddy simulations," *Physics of Fluids*, vol. 16, no. 12, pp. 4506–4534, 2004.

[75] C. B. da Silva and J. C. F. Pereira, "Analysis of the gradient-diffusion hypothesis in large-eddy simulations based on transport equations," *Physics of Fluids*, vol. 19, no. 3, 2007.

[76] H. Schmidt, "Grobstruktur-simulation konvektiver Grenzschichten, Thesis," Report DFVLR-FB 88-30, DLR, University of Munich, Oberpfaffenhofen, Germany, 1988.

[77] J. Smagorinsky, "General circulation experiments with the primitive equations. I. The basic experiment," *Monthly Weather Review*, vol. 91, no. 3, pp. 99–164, 1963.

[78] P. Sagaut, *Large Eddy Simulation for Incompressible Flows*, Springer, Berlin, Germany, 2nd edition, 2002.

[79] B. Porterie, D. Moryan, J. C. Loraud, and M. Larini, "A multi-phase model for predicting line fire propagation," in *Proceedings of the 3rd International Conference on Forest Fire Research, 14th Conference on Fire and Forest Meteorology*, vol. 1, pp. 343–360, Luso, Portugal, 1998.

[80] M. J. Antal, H. L. Friedman, and F. E. Rogers, "Kinetics of cellulose pyrolysis in nitrogen and steam," *Combustion Science and Technology*, vol. 21, no. 3-4, pp. 141–152, 1980.

[81] B. P. Leonard, "A stable and accurate convective modelling procedure based on quadratic upstream interpolation," *Computer*

*Methods in Applied Mechanics and Engineering*, vol. 19, no. 1, pp. 59–98, 1979.

[82] F. Durst, J. C. F. Pereira, and C. Tropea, "The plane symmetric sudden-expansion flow at low Reynolds numbers," *Journal of Fluid Mechanics*, vol. 248, pp. 567–581, 1993.

[83] J. C. F. Pereira and J. M. M. Sousa, "Finite volume calculations of self-sustained oscillations in a grooved channel," *Journal of Computational Physics*, vol. 106, no. 1, pp. 19–29, 1993.

[84] J. P. Magalhães, D. M. S. Albuquerque, J. M. C. Pereira, and J. C. F. Pereira, "Adaptive mesh finite-volume calculation of 2D lid-cavity corner vortices," *Journal of Computational Physics*, vol. 243, pp. 365–381, 2013.

[85] D. M. S. Albuquerque, J. M. C. Pereira, and J. C. F. Pereira, "Residual least squares error estimate for unstructured h-adaptive meshes," *Numerical Heat Transfer, Part B: Fundamentals*, vol. 67, no. 3, pp. 187–210, 2015.

[86] C. H. Chase, *Spotting Distance from Wind-Driven Surface Fires—Extensions of Equations for Pocket Calculations*, Northen Forest Fire Laboratory, Missoula, Mont, USA, 1984.

[87] L. L. Yuan, R. L. Street, and J. H. Ferziger, "Large-eddy simulations of a round jet in crossflow," *Journal of Fluid Mechanics*, vol. 379, pp. 71–104, 1999.

**4**

# Conditional Moment Closure Modelling of a Lifted $H_2/N_2$ Turbulent Jet Flame Using the Presumed Mapping Function Approach

**Ahmad El Sayed[1,2] and Roydon A. Fraser[1]**

[1]*Department of Mechanical and Mechatronics Engineering, University of Waterloo, 200 University Avenue West, Waterloo, ON, Canada N2L 3G1*
[2]*Institut de Combustion Aérothermique Réactivité et Environnement (CNRS), 1C avenue de la Recherche Scientifique, 45071 Orléans Cedex 2, France*

Correspondence should be addressed to Ahmad El Sayed; aselsaye@uwaterloo.ca

Academic Editor: Hong G. Im

A lifted hydrogen/nitrogen turbulent jet flame issuing into a vitiated coflow is investigated using the conditional moment closure (CMC) supplemented by the presumed mapping function (PMF) approach for the modelling of conditional mixing and velocity statistics. Using a prescribed reference field, the PMF approach yields a presumed probability density function (PDF) for the mixture fraction, which is then used in closing the conditional scalar dissipation rate (CSDR) and conditional velocity in a fully consistent manner. These closures are applied to a lifted flame and the findings are compared to previous results obtained using $\beta$-PDF-based closures over a range of coflow temperatures ($T_c$). The PMF results are in line with those of the $\beta$-PDF and compare well to measurements. The transport budgets in mixture fraction and physical spaces and the radical history ahead of the stabilisation height indicate that the stabilisation mechanism is susceptible to $T_c$. As in the previous $\beta$-PDF calculations, autoignition around the "most reactive" mixture fraction remains the controlling mechanism for sufficiently high $T_c$. Departure from the $\beta$-PDF predictions is observed when $T_c$ is decreased as PMF predicts stabilisation by means of premixed flame propagation. This conclusion is based on the observation that lean mixtures are heated by downstream burning mixtures in a preheat zone developing ahead of the stabilization height. The spurious sources, which stem from inconsistent CSDR modelling, are further investigated. The findings reveal that their effect is small but nonnegligible, most notably within the flame zone.

## 1. Introduction

In a previous Conditional Moment Closure (CMC) study [1], the lifted $H_2/N_2$ turbulent jet flame of Cabra et al. [2] was thoroughly investigated using several CMC submodels and chemical kinetic mechanisms over a narrow range of coflow temperatures ($T_c$). For the most part, this work was aimed at the implementation of a fully consistent CMC realisation. Therefore, the consistency of the conditional CMC submodels with the mixture fraction Probability Density Function (PDF) transport equation was emphasised. The commonly used $\beta$-distribution was adopted throughout to presume the PDF. The Conditional Velocity (CV) fluctuations were modelled using the PDF gradient diffusion model of Pope [3].

One important feature of this model is its consistency with the first and second moments of the PDF [4] and with the modelling of the unconditional passive and reactive scalar fluxes [5, 6]. As for the closure of the Conditional Scalar Dissipation Rate (CSDR), the models of Girimaji [7] and Mortensen [8] were considered. Both models are derived by doubly integrating the PDF transport equations and using the same set of boundary conditions. The former is based on the homogeneous form of the equation, while, in the latter, the inhomogeneous terms are retained and PDF gradient modelling is applied to close the CV fluctuations. As such, Mortensen's model provides a fully consistent CSDR closure and degenerates exactly to Girimaji's when the inhomogeneous terms are discarded.

In both CMC realisations, it was found that autoignition is the controlling stabilisation mechanism over the considered $T_c$ range. This conclusion is in full agreement with the findings of Stanković and Merci [9] and in partial agreement with those of Patwardhan et al. [10] and Navarro-Martinez and Kronenburg [11] who report stabilisation via premixed flame propagation as $T_c$ is decreased. Another conclusion drawn in [1] is that Mortensen's fully consistent CSDR model results in delayed ignition and consequently yields larger liftoff heights. Hence, the occurrence of earlier ignition in the realisation employing Girimaji's model is attributed to the spurious (false) sources that arise from the inconsistency of this model with the CMC equations. Therefore, the consistent modelling of the CSDR is influential and ought to be investigated further.

The Mapping Closure (MC) devised by Chen et al. [12] has been frequently employed in the modelling of passive and reactive turbulent scalar mixing. O'Brian and Jiang [13] employ the MC to attain the Amplitude Mapping Closure (AMC) for passive scalar mixing. The closure is widely used in the CMC literature for the modelling of the CSDR [10, 14, 15]. It is the exact equivalent of the counterflow model employed in the framework of the Laminar Flamelet Model (LFM) [16]. Klimenko and Pope [17] employ a generalisation of the MC to formulate the Multiple Mapping Conditioning (MMC) for the modelling of turbulent reacting flows. MMC has close ties with the joint PDF approach [18] and CMC. When turbulent fluctuations are absent, MMC with the mixture fraction chosen as the only major conditioning scalar is equivalent to CMC. When all scalars in composition space are included, MMC is equivalent to the joint PDF approach [19].

In a recent work, Mortensen and Andersson [5] cast the solution of the MC for homogeneous turbulence into a Presumed Mapping Function (PMF) for inhomogeneous turbulent flows. Using a Gaussian reference field, the PMF yields a presumed PDF for the mixture fraction. The resulting PMF-PDF is employed to derive analytical closures for the CV and the CSDR. The CV closure is obtained by inserting the PMF-PDF into the PDF gradient diffusion model of Pope [3]. The CSDR closure is achieved by incorporating the PMF-PDF into the fully consistent, inhomogeneous CSDR expression previously devised by Mortensen [8]. To test the capabilities of the newly proposed approach, Mortensen and Andersson validate the PMF closures against the Direct Numerical Simulation (DNS) of a nonreacting Scalar Mixing Layer (SML). The PMF-PDF yields superior predictions compared to the $\beta$-PDF and the CV and CSDR closures yield a remarkable agreement with the DNS. It is shown that the CV closure is well-behaved at low probabilities compared to the $\beta$-PDF-based gradient diffusion model. The latter is known to result in instabilities in CMC as the CV diverges to infinity at small probabilities [20]. The authors further compare the homogeneous and inhomogeneous versions of the CSDR closure to DNS. They show that the influence of the inhomogeneous modification is small compared to the $\beta$-PDF approach [8]. The study concludes that the PMF approach is well suited for the modelling of mixing statistics in mixture fraction-based presumed PDF combustion models, such as CMC and the LFM.

The closures presented in [5] are concerned with binary mixing; nevertheless, PMF is extensible to the multistream mixing of multiple injections. The derivation and validation of PMF for trinary (three-stream) mixing are addressed in subsequent works. Cha et al. [21] validate the trinary PMF against the DNS of nonreacting Double Scalar Mixing Layers (DSML). The study reveals that trinary PMF is capable of capturing the fine-scale scalar mixing statistics manifesting in the DSML. Mortensen et al. [22] incorporate the trinary PMF into the CMC and stationary LFM of a reacting DSML. The investigated DSML is a representative problem for piloted nonpremixed flames where the fuel and oxidiser streams are separated by a pilot stream. Single-step reversible chemistry is employed in both the CMC and LFM computations. Given the negligible influence of the spurious sources, the CSDR is modelled using the homogeneous version of the trinary PMF closure. A remarkable agreement between both combustion models and DNS is achieved at low-to-moderate extinction levels. Deviations from DNS are reported for higher extinction levels. The discrepancies are attributed to issues specific to the considered combustion models (e.g., the first-order closure for the conditional chemical source in CMC, which is not suitable for the modelling of extinction). In a more recent work, El Sayed et al. [23] assess the trinary PMF in the context of a piloted methane/air flame. The study reveals that the pilot has significant influence on the structure of the flamelets in the near field where mixing is trinary. It is also shown that, compared to the classical flamelet approach that employs the counterflow solution for the modelling of the CSDR, flamelets generated using the trinary PMF can withstand higher strain rates before they extinguish and are more sensitive to transport by means of differential diffusion.

To date, a limited number of attempts have been made to implement PMF in the CMC of well documented laboratory-scale turbulent flames with detailed chemistry. Brizuela and Roudsari [24] use the *homogeneous* trinary PMF approach in the CMC calculations of Sandia flame D [25]. The reported results show that the PMF closures are more accurate than their classical counterparts ($\beta$-distribution for the PDF and the $\beta$-PDF-based AMC and gradient diffusion model for the CSDR and the CV, resp.) in the near field of the flame where the pilot influence is important. The current study is an extension to previous work [1] concerned with the CMC modelling of the lifted $H_2/N_2$ jet flame of Cabra et al. [2]. This flame has been thoroughly investigated using a number of turbulent combustion models such as CMC [9–11], PDF methods [2, 26–28], and the Eddy dissipation concept [2, 29]. To the authors' knowledge, the *inhomogeneous* PMF closures have never been applied to the CMC of laboratory-scale flames. In this paper, the lifted flame is revisited and investigated using CMC supplemented by the PMF approach. The PDF, the CV, and the CSDR are modelled using the *inhomogeneous binary* PMF closures since mixing occurs between two streams. The objectives of this study are to assess the applicability of the PMF-based submodels and to compare the obtained results to previous $\beta$-PDF results. As in [1], a range of coflow temperatures is considered and the stabilisation mechanism is determined by investigating the CMC budgets in mixture fraction and physical spaces and through the analysis of the radical history ahead of the stabilisation height. In addition, the importance of the spurious sources, which are due to the inconsistent modelling of the CSDR, is thoroughly investigated.

## 2. Investigated Flame

The burner of Cabra et al. [2] consists of a central $H_2/N_2$ turbulent jet which issues into a hot coflow. The coflow consists of the combustion products of a lean premixed $H_2$/air flame stabilised on a perforated disk. The disk is surrounded by an exit collar in order to delay the entrainment of ambient air into the coflow. The nozzle exit is placed above the surface of the disk so that the fuel stream exits into the coflow with a uniform composition. Table 1 shows the details of the experimental conditions. The experimental criterion for the determination of the liftoff height ($H_{exp}$) is taken as the first location where the mass fraction of OH reaches 600 ppm. The normalised measured height is $H_{exp}/d = 10$.

## 3. Mathematical Model

*3.1. Conditional Moment Closure.* As in the previous study [1], the turbulence-chemistry interactions are modelled by means of the first-order CMC [30]. For completeness, a brief overview of CMC theory is provided here. In CMC, a reactive scalar $\phi$ is conditionally averaged with respect to the mixture fraction, $\xi$, and its conditional transport equation is solved. The conditional average of $\phi$ is defined as $Q_\phi(\eta, x_i, t) = \langle \phi(x_i, t) \mid \xi(x_i, t) = \eta \rangle$, where $\eta$ is a sample variable of $\xi$, such that $0 \le \eta \le 1$. Using the decomposition approach [30], $\phi$ is expressed as the sum of $Q_\phi$ and a fluctuation $\phi''$ ($\phi = Q_\phi + \phi''$) such that $\langle \phi'' \mid \eta \rangle = 0$. The substitution of this sum into the transport equation of $\phi$ followed by the conditional averaging of the resulting expression leads to the equation of $Q_\phi$. Applying this procedure to the mass fraction of a species $\kappa$ ($Y_\kappa$) and the temperature ($T$) and invoking the primary closure hypothesis [30] yield

$$\frac{\partial Q_\kappa}{\partial t} = \underbrace{-\langle u_i \mid \eta \rangle \frac{\partial Q_\kappa}{\partial x_i}}_{\text{Convection}} \underbrace{- \frac{1}{\bar{\rho}\tilde{P}(\eta)} \frac{\partial}{\partial x_i}\left[\bar{\rho}\langle u_i'' y_\kappa'' \mid \eta \rangle \tilde{P}(\eta)\right]}_{\text{Diffusion in physical space}} + \underbrace{\frac{\langle \chi \mid \eta \rangle}{2} \frac{\partial^2 Q_\kappa}{\partial \eta^2}}_{\text{Diffusion in } \eta\text{-space (micro-mixing)}} + \underbrace{\frac{\langle \dot{\omega}_\kappa \mid \eta \rangle}{\langle \rho \mid \eta \rangle}}_{\text{Chemical source}}, \tag{1}$$

$$\frac{\partial Q_T}{\partial t} = \underbrace{-\langle u_i \mid \eta \rangle \frac{\partial Q_T}{\partial x_i}}_{\text{Convection: } T_{C,x} \text{ and } T_{C,y}}$$

$$\underbrace{- \frac{1}{\bar{\rho}\tilde{P}(\eta)} \frac{\partial}{\partial x_i}\left[\bar{\rho}\langle u_i'' T'' \mid \eta \rangle \tilde{P}(\eta)\right]}_{\text{Diffusion in physical space: } T_{D,x} \text{ and } T_{D,y}} + \underbrace{\frac{\langle \chi \mid \eta \rangle}{2}\left\{\frac{\partial^2 Q_T}{\partial \eta^2} + \frac{1}{\langle c_p \mid \eta \rangle}\left[\frac{\partial \langle c_p \mid \eta \rangle}{\partial \eta} + \sum_{\kappa=1}^{N}\left(\langle c_{p,\kappa} \mid \eta \rangle \frac{\partial Q_\kappa}{\partial \eta}\right)\right]\frac{\partial Q_T}{\partial \eta}\right\}}_{\text{Diffusion in } \eta\text{-space (micro-mixing): } T_{MM}} \tag{2}$$

$$\underbrace{- \frac{\sum_{\kappa=1}^{N}\langle h_\kappa \mid \eta \rangle \langle \dot{\omega}_\kappa \mid \eta \rangle}{\langle \rho \mid \eta \rangle \langle c_p \mid \eta \rangle}}_{\text{Chemical source: } T_{CS}} \underbrace{- \frac{\langle \dot{\omega}_r \mid \eta \rangle}{\langle \rho \mid \eta \rangle \langle c_p \mid \eta \rangle}}_{\text{Radiative Source: } T_{RS}},$$

where $Q_\kappa = \langle Y_\kappa \mid \eta \rangle$, $Q_T = \langle T \mid \eta \rangle$, $\langle u_i \mid \eta \rangle$ is the CV, $\langle \chi \mid \eta \rangle$ is the CSDR, $\langle u_i'' y_\kappa'' \mid \eta \rangle$ and $\langle u_i'' T'' \mid \eta \rangle$ are the conditional species and temperature turbulent fluxes, $\tilde{P}(\eta)$ is the Favre mixture fraction PDF, $\langle \dot{\omega}_\kappa \mid \eta \rangle$ and $\langle h_\kappa \mid \eta \rangle$ are the conditional chemical source and enthalpy of species $\kappa$, $N$ is the number of species, $\langle \dot{\omega}_r \mid \eta \rangle$ is the conditional radiative source, and $\langle \rho \mid \eta \rangle$ and $\langle c_p \mid \eta \rangle$ are the conditional density and specific heat. It is noted that in (1) and (2) the Lewis numbers are set to unity and all diffusivities are assumed to be equal.

The unconditional (Favre) averages of the reactive scalars are obtained by integrating their conditional counterparts weighted by the PDF over mixture fraction space via

$$\tilde{\phi} = \int_0^1 \langle \phi \mid \eta \rangle \tilde{P}(\eta)\, d\eta, \quad \phi = Y_\kappa, T. \tag{3}$$

The quantities $\tilde{P}(\eta)$, $\langle u_i'' y_\kappa'' \mid \eta \rangle$, $\langle u_i'' T'' \mid \eta \rangle$, $\langle u_i \mid \eta \rangle$, $\langle \chi \mid \eta \rangle$, $\langle \dot{\omega}_\kappa \mid \eta \rangle$, and $\langle \dot{\omega}_r \mid \eta \rangle$ appearing in (1)–(3) are unclosed

and require additional modelling. The submodels employed in this study are discussed in the following section.

*3.2. The Presumed Mapping Function Approach.* In CMC, an assumed distribution described by the moments of the mixture fraction is selected directly to presume the PDF. A commonly used distribution for the description of binary mixing statistics is the two-parameter $\beta$-PDF [30]. Although this choice is supported by DNS [7, 31], it lacks a sound physical basis. The recently proposed PMF approach of Mortensen and Andersson [5] employs the MC to offer a less *ad hoc*, naturally evolving alternative for the presumption of the PDF. In PMF, a known reference field is chosen and a mapping function between the true, unknown mixture fraction scalar field and the chosen reference field is established, ultimately leading to a presumed PDF. The PMF-PDF is subsequently utilised to derive fully consistent closures for the CV and CSDR. This section outlines the PMF approach and the resulting closures.

TABLE 1: Conditions of the lifted flame of Cabra et al. [2].

|  | Jet |  | Coflow |
| --- | --- | --- | --- |
| $d$ (mm) | 4.57 | $D$ (mm) | 210 |
| $T_j$ (K) | 305 | $T_c$ (K) | 1045 |
| $U_j$ (m/s) | 107 | $U_c$ (m/s) | 3.5 |
| $X_{H_2}$ | 0.2537 | $X_{H_2}$ | 0.0005 |
| $X_{O_2}$ | 0.0021 | $X_{O_2}$ | 0.1474 |
| $X_{N_2}$ | 0.7427 | $X_{N_2}$ | 0.7534 |
| $X_{H_2O}$ | 0.0015 | $X_{H_2O}$ | 0.0989 |
| $\xi_{st}$ | 0.474 | $\phi$ | 0.25 |

$d$ = jet diameter; $D$ = coflow diameter; $T$ = temperature; $U$ = velocity; $X$ = mole fraction; $\xi$ = mixture fraction; $\phi$ = equivalence ratio. Subscripts: $j$ = jet; $c$ = coflow; st = stoichiometric.

*3.2.1. Probability Density Function.* As in [5], the known reference field is denoted by $\psi$ with sample space variable $\phi$ and consistent with conventional CMC notation and the unknown scalar field is denoted by $\xi$ with sample space variable $\eta$. The reference field is mapped to the scalar field according to

$$X(\psi) = \xi,$$
$$X(\phi) = \eta, \tag{4}$$

where $X$ is a unique mapping function. The PDFs of $\xi$ and $\psi$, $\widetilde{P}(\eta)$, and $r(\phi)$ are related by

$$\widetilde{P}(\eta) = \widetilde{P}(X(\phi)) = r(\phi)\left(\frac{\partial X}{\partial \phi}\right)^{-1}. \tag{5}$$

A zero-mean Gaussian distribution is employed in [5] to prescribe the PDF of $\psi$. Accordingly,

$$r(\phi) = \frac{1}{\sqrt{2\pi}\sigma_r}\exp\left(-\frac{\phi^2}{2\sigma_r^2}\right), \tag{6}$$

where $\sigma_r^2 = 1 - 2\tau$ is the variance. The quantity $\tau$ is a scaled time parameter that is bounded by 0 (no mixing) and 0.5 (complete mixing). Using the cumulative distribution functions of $\widetilde{P}(\eta)$ and $r(\phi)$, the mapping function is [5]

$$X(\phi)$$
$$= \frac{1}{2\sqrt{\pi\tau}}\int_{-\infty}^{+\infty} X(\phi', \tau = 0)\exp\left[-\frac{(\phi'-\phi)^2}{4\tau}\right]d\phi', \tag{7}$$

where $X(\phi', \tau = 0)$ is the initial mapping. For binary mixing, the initial PDF of $\xi$ evolves from the double-delta distribution

$$\widetilde{P}(\eta, \tau = 0) = (1 - \widetilde{\xi})\delta(\eta) + \widetilde{\xi}\delta(\eta - 1), \tag{8}$$

where $\delta$ is the Dirac delta function and $\widetilde{\xi}$ is the mixture fraction mean. The corresponding initial mapping is

$$X(\phi, \tau = 0) = H(\phi - \alpha), \tag{9}$$

where $H$ is the Heaviside step function. The parameter $\alpha$ in (9) is determined from

$$1 - \widetilde{\xi} = \int_{-\infty}^{\alpha} r(\phi, \tau = 0)\, d\phi. \tag{10}$$

Equation (10) has the analytical solution

$$\alpha = \sqrt{2}\,\text{erf}^{-1}\left(1 - 2\widetilde{\xi}\right), \tag{11}$$

where $\text{erf}^{-1}$ is the inverse error function. The substitution of (9) in (7) yields the mapping

$$X(\phi) = \frac{1}{2}\left[1 + \text{erf}\left(\frac{\phi - \alpha}{2\sqrt{\tau}}\right)\right]. \tag{12}$$

Finally, the insertion (12) in (5) leads to the PMF-PDF

$$\widetilde{P}(\eta) = \sqrt{\frac{2\tau}{\sigma_r^2}}\exp\left[E^2(\eta) - \frac{\phi^2}{2\sigma_r^2}\right], \tag{13}$$

where

$$\phi = \alpha + 2\sqrt{\tau}E(\eta),$$
$$E(\eta) = \text{erf}^{-1}(2\eta - 1). \tag{14}$$

The time parameter $\tau$ is obtained by finding the zero of

$$\widetilde{\xi}^2 + \widetilde{\xi''^2} = \int_{-\infty}^{+\infty} X^2(\phi)\, r(\phi)\, d\phi, \tag{15}$$

where $\widetilde{\xi''^2}$ is the mixture fraction variance.

*3.2.2. Conditional Velocity.* The CV fluctuations are modelled using the PDF gradient diffusion model [3, 32, 33], which is the only known model that guarantees consistency with the modelling of the unconditional passive and reactive scalar fluxes. The CV expression takes the form

$$\langle u_i \mid \eta \rangle$$
$$= \widetilde{u}_i - \frac{D_t}{\widetilde{P}(\eta)}\frac{\partial \widetilde{P}(\eta)}{\partial x_i} \tag{16}$$
$$= \widetilde{u}_i - D_t\left\{\frac{\partial \ln[\widetilde{P}(\eta)]}{\partial \widetilde{\xi}}\frac{\partial \widetilde{\xi}}{\partial x_i} + \frac{\partial \ln[\widetilde{P}(\eta)]}{\partial \widetilde{\xi''^2}}\frac{\partial \widetilde{\xi''^2}}{\partial x_i}\right\},$$

where $\widetilde{u}_i$ is the Favre-averaged velocity and $D_t = (C_\mu/Sc_t)(\widetilde{k}^2/\widetilde{\varepsilon})$ is the turbulent diffusivity, $\widetilde{k}$ is the turbulence kinetic energy, and $\widetilde{\varepsilon}$ is the Eddy dissipation. The constant $C_\mu$ is equal to 0.09 and the turbulent Schmidt number $Sc_t$ is set to 0.7 [34]. By inserting (13) in (16), it can be shown that [5]

$$\langle u_i \mid \eta \rangle = \widetilde{u}_i + \frac{D_t}{\sigma_r^2}\left\{\frac{\partial \widetilde{\xi}}{\partial x_i}\frac{d\alpha}{d\widetilde{\xi}}\phi(\eta)\right.$$
$$-\frac{1}{2\tau}\left[\frac{\partial \tau}{\partial \widetilde{\xi}}\frac{\partial \widetilde{\xi}}{\partial x_i} + \frac{\partial \tau}{\partial \widetilde{\xi''^2}}\frac{\partial \widetilde{\xi''^2}}{\partial x_i}\right] \tag{17}$$
$$\left.\cdot\left[1 + \alpha\phi(\eta) - \frac{\phi^2(\eta)}{\sigma_r^2}\right]\right\},$$

where $d\alpha/d\widetilde{\xi} = -\sqrt{2\pi}\exp(\alpha^2/2)$.

*3.2.3. Conditional Scalar Dissipation Rate.* Recently, Mortensen [8] derived a fully consistent closure for $\langle \chi \mid \eta \rangle$. The closure is obtained by the doubly integrating the inhomogeneous mixture fraction PDF transport equation subject to the boundary conditions $\langle \chi \mid \eta = 0 \rangle = 0$ and $\langle \chi \mid \eta = 1 \rangle = 0$. The CV fluctuations, which appear unclosed in this equation, are closed using the PDF gradient diffusion model. The CSDR closure is finally achieved by presuming the PDF with a distribution described by the vector of mixture fraction moments. When mixing is binary, the PDF is described by its first two moments, $\widetilde{\xi}$ and $\widetilde{\xi''^2}$, and the CSDR becomes (it is assumed in (18) that molecular diffusivity is negligible compared to $D_t$)

$$
\langle \chi \mid \eta \rangle = \frac{2}{\widetilde{P}(\eta)} \left\{ -\frac{\partial II(\eta)}{\partial \widetilde{\xi''^2}} S_{\widetilde{\xi''^2}} \right.
$$
$$
+ D_t \left[ \frac{\partial^2 II(\eta)}{\partial \widetilde{\xi''^2} \partial \widetilde{\xi''^2}} \frac{\partial \widetilde{\xi''^2}}{\partial x_i} \frac{\partial \widetilde{\xi''^2}}{\partial x_i} + \frac{\partial^2 II(\eta)}{\partial \widetilde{\xi} \partial \widetilde{\xi}} \frac{\partial \widetilde{\xi}}{\partial x_i} \frac{\partial \widetilde{\xi}}{\partial x_i} \right.
$$
$$
\left. \left. + 2 \frac{\partial^2 II(\eta)}{\partial \widetilde{\xi} \partial \widetilde{\xi''^2}} \frac{\partial \widetilde{\xi}}{\partial x_i} \frac{\partial \widetilde{\xi''^2}}{\partial x_i} \right] \right\}, \tag{18}
$$

where

$$
II(\eta) = \int_0^\eta \int_0^{\eta'} \widetilde{P}(\eta'') \, d\eta'' \, d\eta' \tag{19}
$$

is the integral from 0 to $\eta$ of the cumulative distribution function of $\eta$. The term $S_{\widetilde{\xi''^2}}$ in (18) represents the source term of the mixture fraction variance transport equation. It is given by

$$
S_{\widetilde{\xi''^2}} = 2 D_t \frac{\partial \widetilde{\xi}}{\partial x_i} \frac{\partial \widetilde{\xi}}{\partial x_i} - \widetilde{\chi}, \tag{20}
$$

where $\widetilde{\chi}$ is the Favre-averaged scalar dissipation rate. Using (13) and (19) in (18), Mortensen and Andersson [5] show that CSDR can be expressed analytically as

$$
\langle \chi \mid \eta \rangle = \left\{ \widetilde{\chi} - 2 D_t \left[ \left( \frac{\partial \widetilde{\xi}}{\partial x_i} \right)^2 \left( 2 + \frac{\tau_m^2 \tau_{vv}}{\tau_{vv}^3} - \frac{\tau_{mm}}{\tau_v} \right) \right. \right.
$$
$$
- 2 \frac{d\alpha}{d\widetilde{\xi}} \frac{\partial \widetilde{\xi}}{\partial x_i} \frac{\partial \widetilde{\xi''^2}}{\partial x_i} A(\eta) - \left( \left( \frac{\partial \widetilde{\xi''^2}}{\partial x_i} \right)^2 \tau_v \right.
$$
$$
\left. \left. + 2 \frac{\partial \widetilde{\xi}}{\partial x_i} \frac{\partial \widetilde{\xi''^2}}{\partial x_i} \tau_m + \left( \frac{\partial \widetilde{\xi}}{\partial x_i} \right)^2 \frac{\tau_m^2}{\tau_v} \right) B(\eta) \right] \right\} \frac{\langle \chi \mid \eta \rangle_H}{\widetilde{\chi}}
$$
$$
- 2 D_t \left( \frac{\partial \widetilde{\xi}}{\partial x_i} \right)^2 C(\eta), \tag{21}
$$

where

$$
A(\eta) = \frac{\alpha}{1 - \tau} - \frac{\phi(\eta)}{\sigma_r^2},
$$

$$
B(\eta)
$$
$$
= \frac{\alpha^2}{2(1 - \tau)^2}
$$
$$
- \frac{1}{\sigma_r^2} \left( \frac{\phi(\eta) E(\eta)}{\sqrt{\tau}} + \frac{\phi^2(\eta)}{\sigma_r^2} - \frac{1}{1 + \sigma_r^2} \right),
$$

$$
C(\eta)
$$
$$
= \sqrt{\frac{2}{\pi}} \left( \frac{d\alpha}{d\widetilde{\xi}} + \frac{\tau_m}{\tau} \frac{\phi(\eta)}{\sigma_r^2} \right) \exp\left[ -2E^2(\eta) + \frac{\alpha^2}{2} \right], \tag{22}
$$

$$
\tau_m = \frac{\partial \tau}{\partial \widetilde{\xi}},
$$

$$
\tau_v = \frac{\partial \tau}{\partial \widetilde{\xi''^2}},
$$

$$
\tau_{mm} = \frac{\partial^2 \tau}{\partial \widetilde{\xi}^2},
$$

$$
\tau_{vv} = \frac{\partial^2 \tau}{\partial \widetilde{\xi''^2}^2}.
$$

The term $\langle \chi \mid \eta \rangle_H$ in (21) represents the homogeneous portion of (18); that is, $\langle \chi \mid \eta \rangle_H = 2 \widetilde{\chi} / \widetilde{P}(\eta) \partial II(\eta) / \partial \widetilde{\xi''^2}$. Using (13) and (19), it can be shown that

$$
\langle \chi \mid \eta \rangle_H = \widetilde{\chi} \sqrt{\frac{1 - \tau}{\tau}} \exp\left( -2E^2(\eta) + \frac{\alpha^2}{\sigma_r^2 + 1} \right). \tag{23}
$$

Equation (23) is the exact equivalent of the AMC [13] and the counterflow model [16] as implied by the PMF-PDF.

*3.3. Conditional Fluxes and Sources.* The conditional turbulent fluxes are modelled using the gradient diffusion assumption given by

$$
\langle \phi'' u_i'' \mid \eta \rangle = -D_t \frac{\partial \langle \phi \mid \eta \rangle}{\partial x_i}, \quad \phi = Y_\kappa, T. \tag{24}
$$

This closure does not account for countergradient effects, which are mostly encountered in premixed flames. This approximation may not be suitable when lifted flames are stabilised by premixed flame propagation. A correction such as the one proposed by Richardson and Mastorakos [35] may be necessary. However, following previous CMC investigations of the same flame considered here [1, 9–11], (24) is adopted throughout this study.

The conditional chemical source, $\langle \dot{\omega}_\kappa \mid \eta \rangle$, is modelled using a first-order closure [30]. In this closure, the conditional fluctuations about the conditional averages of the reactive scalars are assumed to be small. Accordingly, $\langle \dot{\omega}_\kappa \mid \eta \rangle$

is modelled as a function of the conditional density, mass fractions, and temperature via

$$\langle \dot{\omega}_\kappa (\rho, T, \mathbf{Y}) \mid \eta \rangle$$

$$\approx \langle \dot{\omega}_\kappa (\langle \rho \mid \eta \rangle, \langle T \mid \eta \rangle, \langle \mathbf{Y} \mid \eta \rangle) \mid \eta \rangle \quad (25)$$

$$= \dot{\omega}_\kappa (\langle \rho \mid \eta \rangle, Q_T, \mathbf{Q}),$$

where $\mathbf{Y} = \{Y_\kappa \mid \kappa = 1, 2, \ldots, N\}$ and $\mathbf{Q} = \langle \mathbf{Y} \mid \eta \rangle$. This closure has been successfully applied in the CMC of lifted flames [1, 9–11, 14].

The conditional radiative source, $\langle \dot{\omega}_r \mid \eta \rangle$, is modelled using the optically thin assumption [36] with $H_2O$ being the predominantly participating species. Previous modelling studies [1, 37] reveal that radiation has a negligible effect on the prediction of the stabilisation height in lifted flames. In the flame under investigation, this is due to the fact that $H_2$ in the fuel stream is highly diluted with $N_2$. The radiative source is nevertheless retained here.

*3.4. Other Considered Submodels.* In addition to the PMF closures presented in Section 3.2, the $\beta$-PDF closures investigated in [1] are also considered in this work for the purpose of comparison with the PMF results. These closures are provided here for completeness. The $\beta$-PDF is given by

$$\tilde{P}(\eta; v, w) = \frac{\eta^{v-1}(1-\eta)^{w-1}}{B(v, w)}, \quad (26)$$

where $v = \gamma \tilde{\xi}$ and $w = \gamma(1 - \tilde{\xi})$ with $\gamma = \tilde{\xi}(1 - \tilde{\xi})/\widetilde{\xi''^2} - 1 \geq 0$ and $B(v, w) = \int_0^1 \eta^{v-1}(1-\eta)^{w-1}d\eta$ is the $\beta$-function.

Two $\beta$-PDF-based CSDR models are considered in [1]. The first is the homogeneous model of Girimaji [7]

$$\langle \chi \mid \eta \rangle = -2\tilde{\chi} \frac{\tilde{\xi}(1-\tilde{\xi})}{\left(\widetilde{\xi''^2}\right)^2} \frac{I(\eta)}{\tilde{P}(\eta)}, \quad (27)$$

where

$$I(\eta) = \int_0^\eta \left\{ \tilde{\xi} \left[ \ln \eta' - \int_0^1 \ln\eta'' \tilde{P}(\eta'') d\eta'' \right] + (1 - \tilde{\xi}) \right.$$

$$\left. \cdot \left[ \ln(1 - \eta') - \int_0^1 \ln(1 - \eta') \tilde{P}(\eta'') d\eta'' \right] \right\} \quad (28)$$

$$\times \tilde{P}(\eta')(\eta - \eta') d\eta'.$$

The second is the inhomogeneous model of Mortensen given by (18) with $\tilde{P}(\eta)$ presumed using the $\beta$-PDF. In this case, it can be shown that (19) simplifies to [8, 38]

$$II(\eta; v, w) = (\eta - \tilde{\xi}) I(\eta; v, w)$$

$$+ \widetilde{\xi''^2} \tilde{P}(\eta; v + 1, w + 1), \quad (29)$$

where $I(\eta; v, w) = \int_0^\eta [\eta^{v-1}(1 - \eta)^{w-1}/B(v, w)]d\eta$ is the incomplete $\beta$-function. When the inhomogeneous terms are

discarded, this version of Mortensen's model reduces exactly to (27). Hence, Girimaji's model is the exact equivalent of the homogeneous version of Mortensen's.

As for the closure of the CV, the PDF gradient diffusion model, (16), is employed with $\tilde{P}(\eta)$ presumed using the $\beta$-PDF.

# 4. Implementation

The implementation details are available in [1]. Briefly, the calculations are performed on a two-dimensional axisymmetric computational domain. A quadratic unstructured mesh is used. The $k - \varepsilon$ model is employed to compute the mean flow quantities. Standard model constants are used except for $C_{\varepsilon 1}$ which is modified from 1.44 to 1.6 in order to improve the spreading rate predictions [39]. The mixing field is obtained by solving transport equations for $\tilde{\xi}$ and $\widetilde{\xi''^2}$. It is important to emphasise that $\tilde{\chi}$ is not calculated algebraically. In practice, this quantity is computed by invoking the assumption of proportionality between the flow time ($\tilde{k}/\tilde{\varepsilon}$) and the time scale of scalar turbulence ($\widetilde{\xi''^2}/\tilde{\chi}$) leading to $\tilde{\chi} = C_\chi \tilde{\varepsilon}/\tilde{k} \widetilde{\xi''^2}$, where the time scale ratio $C_\chi$ is usually set to a constant value of 2. Alternatively, the transport equation of $\tilde{\chi}$ [40] is solved. This approach predicts correctly the asymptotic decay of the time scale ratio from high levels in the near field to a constant value in the neighbourhood of 2 at downstream locations (see Figure 3 in [1]). The SIMPLE algorithm is used for pressure-velocity coupling. Spatial discretisation is performed using the second-order upwind scheme. The inlet and coflow boundary conditions are set following the experimental conditions. Zero-gradient boundary conditions are applied elsewhere.

The CMC equations are discretised using the finite difference method. The first-order derivative of the convective term is discretised using the second-order upwind difference scheme with the kappa flux limiter [41]. The second-order derivatives appearing in the diffusive terms are discretised using the second-order central difference scheme. The mixture fraction grid consists of 80 evenly spaced points. Compared to the flow and mixing calculations, a relatively coarser mesh is used in physical space in order to achieve significant computational savings. This is approach is valid due to the lower spatial dependence of the conditional averages compared to their unconditional counterparts. The ODE solver VODPK [42] is used to solve the system of equations. A three-step fractional method [43] is implemented in order to reduce stiffness of the system. The $H_2$ oxidation chemical kinetics mechanism of Mueller et al. [44] is employed throughout this study.

The details of the numerical implementation of the $\beta$-PDF-based submodels are available in [1] and are not repeated here for brevity. The PMF closures for the PDF (13), the CV (17), and the CSDR (21) are computed using the open-source code *"PMFpack"* [45]. The computation of the closures is straightforward. The major task is to determine $\tau$ and its partial derivatives. $\tau$ is determined by inserting (6) and (12) in (15) and finding the zero of the resulting equation in the interval $[0, 0.5]$. $\tilde{\xi}$ and $\widetilde{\xi''^2}$ are available from the mixing field. Bisection is used to accomplish this task and adaptive quadrature integration is employed to evaluate the integral. The partial derivatives of $\tau$ are determined by means of numerical differentiation.

# 5. Results and Discussion

*5.1. Comparison of PMF and $\beta$-PDF Closures.* In this section, the results of the PMF and $\beta$-PDF closures are presented. The coflow temperature is set to the experimentally reported value of 1045 K [2]. Three CMC realisations are compared. The first two correspond to the realisations investigated in [1]. In both, the PDF is presumed using the $\beta$-distribution, the CV is modelled using the $\beta$-PDF gradient model, and the CSDR is modelled using either Girimaji's mode (27) or Mortensen's model ((18) with $II(\eta)$ calculated using (29)). Hereafter, these realisations will be, respectively, referred to as CMC-$\beta$G and CMC-$\beta$M. The third realisation where the PMF closures are employed ((13) for the PDF, (17) for the CV and (21) for the CSDR) will be referred to as CMC-PMF.

*5.1.1. Results in Physical Space.* The contours of $\widetilde{Y}_{OH}$ ($\times 10^{-3}$) are shown in Figure 1(a). As in the experiments of Cabra et al. [2], the criterion for the determination of the liftoff height is taken as the first location in the flow field where $\widetilde{Y}_{OH}$ reaches 600 ppm. Given the experimental uncertainty of 10% in $\widetilde{Y}_{OH}$ [2], the liftoff height uncertainty is roughly $\pm d$. The liftoff height normalised by $d$ predicted by CMC-PMF is 9.80 compared to 10.50 for CMC-$\beta$G and 10.61 for CMC-$\beta$M. All heights fall within the experimental uncertainty. In comparison with the experimental value $H_{exp}/d = 10$, CMC-PMF results in the smallest relative error. The usage of the PMF closures does not affect the radial location of stabilisation (the distance from the stabilisation point to the centreline normalised by $d$) as CMC-PMF yields 1.47, which is exactly the same value predicted by CMC-$\beta$G and CMC-$\beta$M. The radial profiles of the Favre-averaged temperature and selected species mass fractions are shown in Figures 1(b)–1(f) at the axial locations $x/d = 8, 9, 10, 11, 14$ and 26. The profiles obtained using the three CMC realisations overlap at $x/d = 8$ and 9 and agree well with the experimental data of Cabra et al. [2]. Compared to the results of CMC-PMF, those of CMC-$\beta$G and CMC-$\beta$M are in better agreement with the experiments at $x/d = 10$. The temperature and product mass fractions are overpredicted while the reactant mass fractions are underpredicted in the range $1.1 \leq y/d \leq 1.8$ when the PMF closures are used. As indicated earlier, CMC-PMF predicts a liftoff height of 9.8 compared to 10.50 for CMC-$\beta$G and 10.61 for CMC-$\beta$M. Therefore, at $x/d = 10$, the mixture is not inert and a flame is present, which explains the higher levels of temperature and products and the lower levels of reactants. This effect propagates downstream. As shown at $x/d = 11$, the CMC-PMF results are overall in better agreement with the measurements. The differences between the three CMC realisations diminish at $x/d = 14$, with CMC-$\beta$G and CMC-$\beta$M being slightly closer to the experiments. Further downstream at $x/d = 26$ the results are almost identical.

*5.1.2. Results in Mixture Fraction Space*

*Comparison of the Reactive Scalars.* Figures 2(a) and 2(b) show the axial evolution of the conditional temperature, $Q_T$, and the conditional OH mass fraction, $Q_{OH}$, around the stabilisation locations indicated in Figure 1(a). In all three realisations, as $x/d$ increases, the peaks of $Q_T$ and $Q_{OH}$ shift from lean mixtures towards stoichiometry and their amplitudes increase dramatically from the inert state. Compared to CMC-$\beta$G and CMC-$\beta$M, CMC-PMF yields relatively higher $Q_T$ and $Q_{OH}$ levels and therefore leads to the smallest liftoff height. The profiles of CMC-$\beta$G and CMC-$\beta$M are very similar, which explains why the liftoff heights resulting from these realisations are nearly identical (see Figure 1(a)). It is important to note that $Q_T$ and $Q_{OH}$ evolve differently in the CMC-PMF realisation. In CMC-$\beta$G and CMC-$\beta$M, $Q_T$ and $Q_{OH}$ start to increase around the "most reactive" mixture fraction ($\eta_{mr} = 0.0543$ [1]), as shown at $x/d = 8.86$, and their peaks remain around $\eta_{mr}$ up to $x/d = 9.42$ before shifting slowly to less lean mixtures at $x/d = 9.69$ and eventually closer towards stoichiometry at $x/d = 11.09$. This behaviour indicates the occurrence of spontaneous ignition (autoignition) ahead of the stabilisation height. In CMC-PMF, although $Q_T$ and $Q_{OH}$ reach their maxima at $\eta_{mr}$ at $x/d = 8.86$, it is obvious that their peaks shift more aggressively towards stoichiometry as $x/d$ increases. This shows that the flame is stabilised by a different mechanism other than autoignition. As it will be shown in Section 5.3, a preheat zone exists ahead of the stabilisation height where lean mixtures are preheated by downstream burning mixtures as in premixed flame propagation.

*Comparison of the Presumed PDFs.* The evolution of the PDF is displayed in Figure 2(c). Both presumed PDFs peak at the same location in mixture fraction space. The PMF-PDF is slightly narrower than the $\beta$-PDF and presents a relatively higher peak. Although the differences between the two are small, the mildest changes in the shape of the PDF can have a nonnegligible impact on the CV and CSDR distributions and can affect the unconditional averages of the different reactive scalars (3).

*Comparison of the CSDR Models.* Figure 2(d) shows the evolution of the CSDR. Girimaji's homogeneous model results in almost symmetric profiles. The near-symmetric shape is due to the fact that $\widetilde{\xi} \approx 0.474$ at the stoichiometric isoline. This value is very close to 0.5 at which the model is symmetric. The model can however yield skewed profiles when $\widetilde{\xi} \neq 0.5$ [46]. On the other hand, Mortensen's inhomogeneous models (based on the $\beta$-PDF and the PMF-PDF) yield skewed profiles. The effect of inhomogeneity manifests away from the mean of the mixture fraction ($\eta \geq \widetilde{\xi}$) as the PDF decays to zero (see Figure 2(c)). Mortensen's models are in general less dissipative as they produce substantially lower CSDR levels for $\eta \geq \widetilde{\xi}$, most notably around stoichiometry and in rich mixtures. It is clear that the presumed form of the PDF in Mortensen's model affects the shape of the CSDR. Although the trends of the $\beta$-PDF and the PMF-PDF versions of the model are similar, significant difference can be clearly

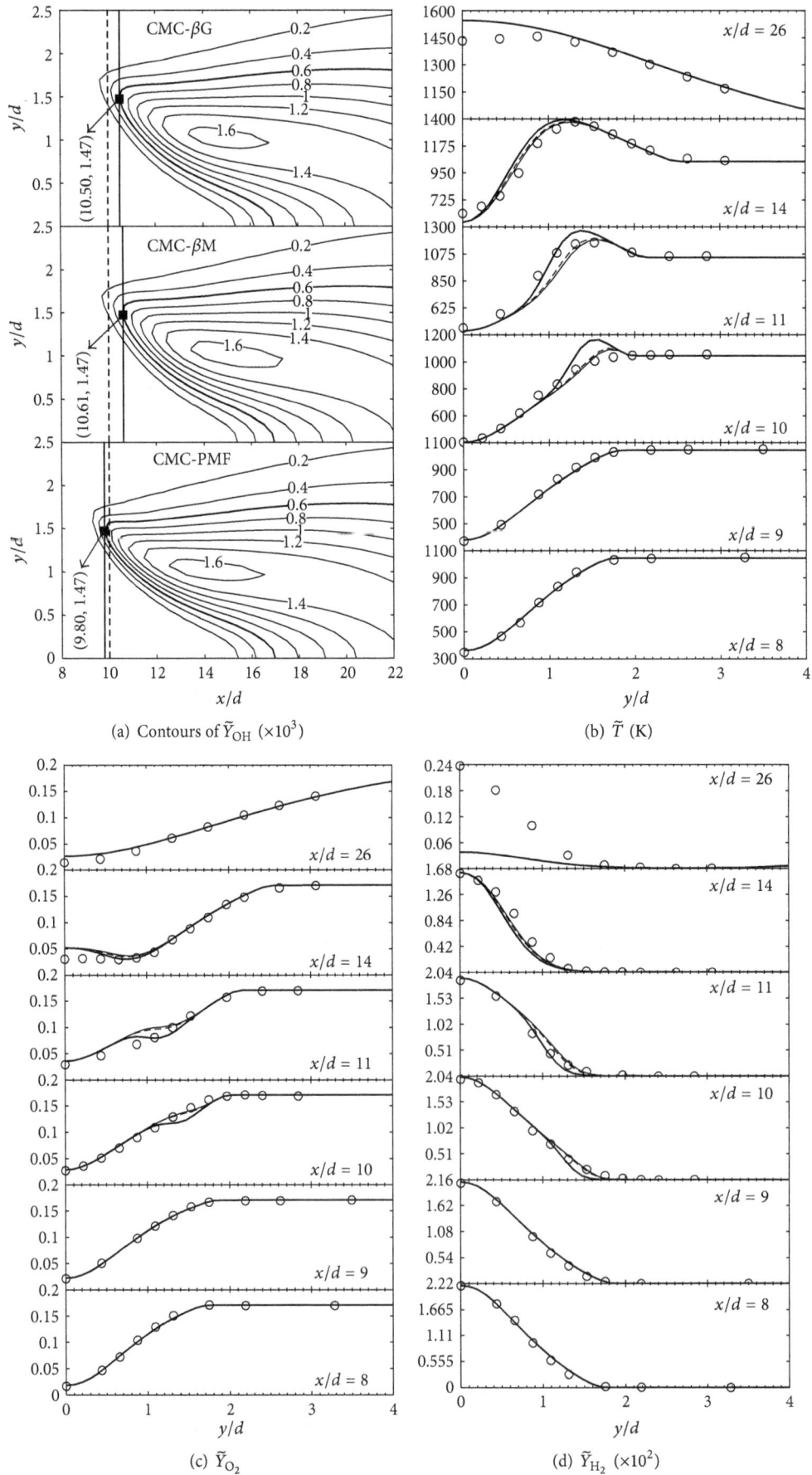

(a) Contours of $\widetilde{Y}_{OH}$ $(\times 10^3)$

(b) $\widetilde{T}$ (K)

(c) $\widetilde{Y}_{O_2}$

(d) $\widetilde{Y}_{H_2}$ $(\times 10^2)$

FIGURE 1: Continued.

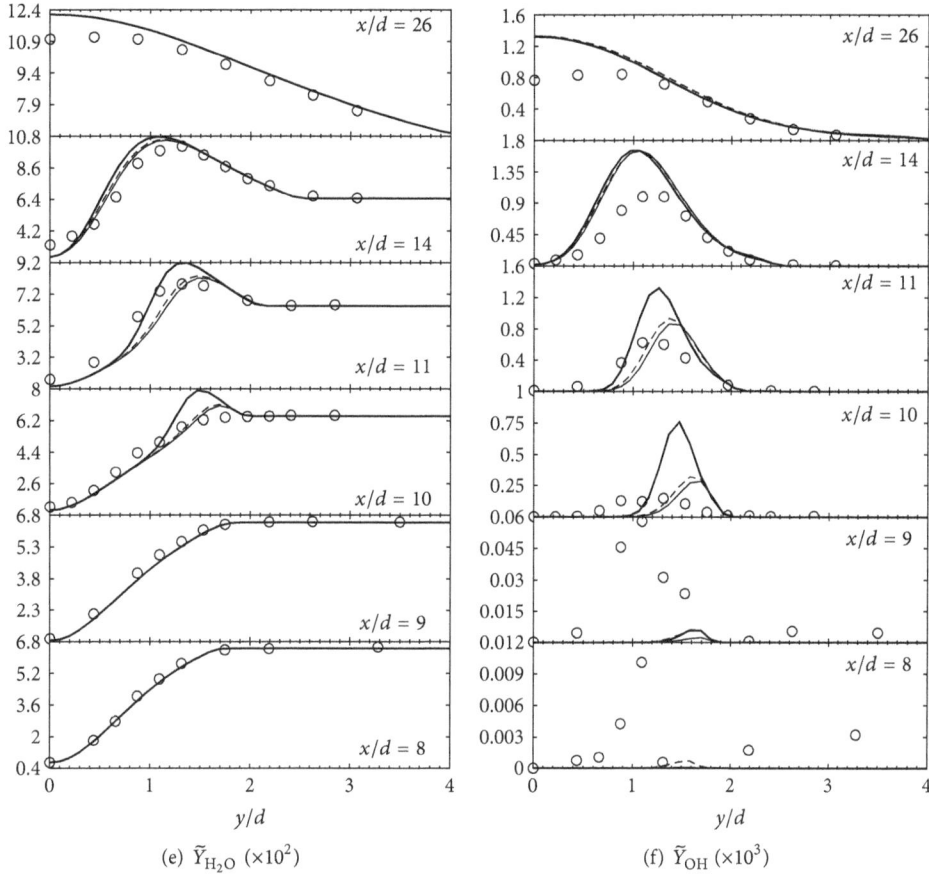

(e) $\widetilde{Y}_{H_2O}$ ($\times 10^2$)    (f) $\widetilde{Y}_{OH}$ ($\times 10^3$)

FIGURE 1: Results of the three CMC realisations for $T_c = 1045$: (a) contours of $\widetilde{Y}_{OH}$ ($\times 10^3$) and radial profiles of (b) $\widetilde{T}$, (c) $\widetilde{Y}_{O_2}$, (d) $\widetilde{Y}_{H_2}$ ($\times 10^2$), (e) $\widetilde{Y}_{H_2O}$ ($\times 10^2$), and (f) $\widetilde{Y}_{OH}$ ($\times 10^3$). In (a) the thick contours correspond to 600 ppm. The solid and dashed vertical lines correspond to the normalised numerical and experimental liftoff heights, respectively. In (b)–(f), dashed lines, CMC-$\beta$G; thin solid lines, CMC-$\beta$M; thick solid lines, CMC-PMF; symbols, experimental data [2].

seen around stoichiometry and in rich mixture, with the PMF-PDF version being less dissipative in these regions. This observation illustrates the strong influence of the presumed form of the PDF on the modelling of the CSDR.

*Comparison of the CV Models.* The axial and radial CV components, $\langle u \mid \eta \rangle$ and $\langle v \mid \eta \rangle$, are displayed in Figures 2(e) and 2(f), respectively. The $\beta$-PDF and PMF-PDF gradient diffusion models yield very similar results within two-to-three standard deviations of the mixture fraction mean. As in the CSDR profiles, the differences between the two closures become substantial away from the mean. The $\beta$-PDF gradient model tends to $\pm\infty$ as the PDF approaches zero. The PMF-PDF closure is generally better behaved over the whole mixture fraction space as it does not overshoot significantly at low probabilities. This behaviour demonstrates again the large influence of the presumed PDF. By inspecting Figure 2(e), although the trends of $\langle u \mid \eta \rangle$ are quite different, the results of the $\beta$-PDF and PMF-PDF closures are of the same order of magnitude and do not differ much from $\tilde{u}$. This is attributed to the fact that the axial velocity fluctuations are small, which is in turn due to the small magnitude of $\partial \ln[\widetilde{P}(\eta)]/\partial x$ over the whole range of $\eta$. On the other hand, the magnitude of

$\partial \ln[\widetilde{P}(\eta)]/\partial y$ can vary substantially, depending on the radial variations of the PDF at the point of interest. As shown in Figure 2(f), in the range where the PDF is finite ($0 < \eta \lesssim \widetilde{\xi}_{st}$ in Figure 2(c)), the magnitude of $\partial \ln[\widetilde{P}(\eta)]/\partial y$ is small regardless of whether the $\beta$-PDF or the PMF-PDF is used. However, at low probabilities ($\eta \gtrsim \widetilde{\xi}_{st}$), the two PDFs yield substantially different velocities. When the $\beta$-PDF is employed, $\partial \ln[\widetilde{P}(\eta)]/\partial y$ is one order of magnitude larger than the PMF-PDF fluctuations. In the absence of experimental measurements, it is difficult to judge which closure is more accurate. Nevertheless, the fact that the PMF closure does no overshoot at low probabilities is desirable for numerical stability.

*Comparison to Conditional Measurements.* The conditional profiles of the temperature and species mass fractions obtained from the three CMC realisations are shown in Figure 3. The calculation of the conditional data from the experimental scatter is described in [1]. The numerical results are reported at the axial locations $x/d$ = 9, 10, 11, 14 and 26 near the stoichiometric isocontour. As shown, the CMC-PMF results are generally in better agreement with the experimental data compared to CMC-$\beta$G and CMC-$\beta$M.

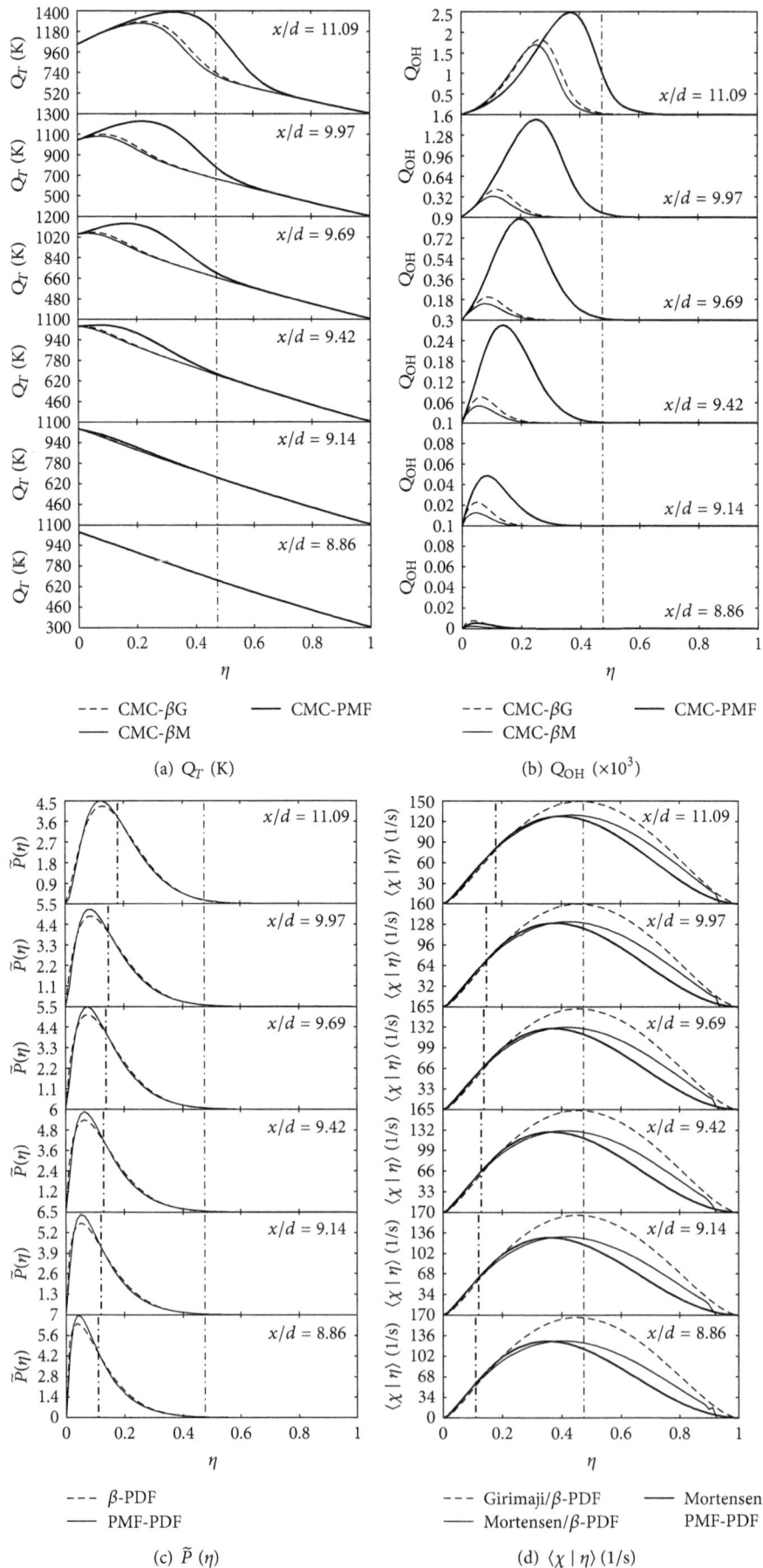

(a) $Q_T$ (K)

(b) $Q_{OH}$ ($\times 10^3$)

(c) $\tilde{P}(\eta)$

(d) $\langle \chi \mid \eta \rangle$ (1/s)

FIGURE 2: Continued.

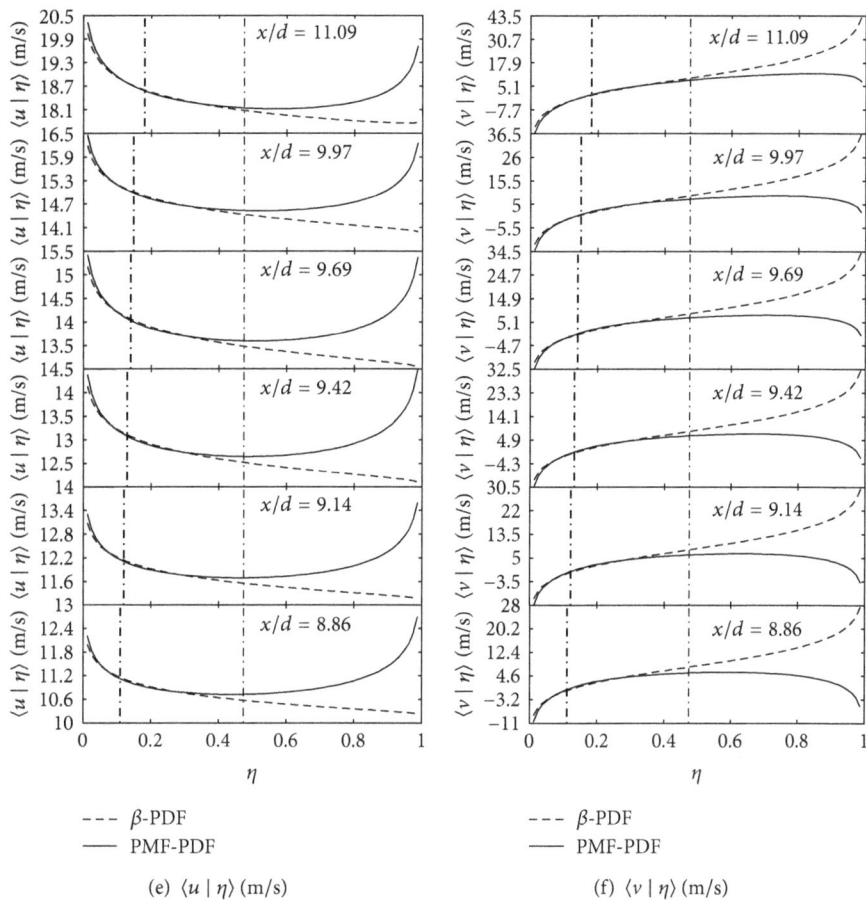

FIGURE 2: Axial evolution of (a) $Q_T$ (K), (b) $Q_{OH}$, (c) $\tilde{P}(\eta)$, (d) $\langle \chi \mid \eta \rangle$ (1/s), (e) $\langle u \mid \eta \rangle$ (m/s), and (f) $\langle v \mid \eta \rangle$ (m/s) at $y/d = 1.47$ (the radial location of stabilisation). The thin vertical dash-dotted line indicates the location of the stoichiometric mixture fraction ($\eta = \tilde{\xi}_{st}$). The thick vertical dash-dotted line in (c)–(f) indicates the location of the mixture fraction mean ($\eta = \tilde{\xi}$). The coflow temperature is 1045 K.

The improved predictions are mostly notable in lean mixture at $x/d = 10$ and 11 and in rich mixtures at $x/d = 14$. The results show that PMF is a reliable and accurate approach for the modelling of the unclosed terms in the CMC equations.

*5.1.3. Effect of the PMF and β-PDF Closures.* Having identified the differences between the β-PDF and the PMF-PDF closures, it becomes obvious why the results of CMC-PMF differ from those of CMC-βG and CMC-βM (Figures 2(a) and 2(b)). In [1], the differences between the results of CMC-βG and CMC-βM were solely attributed to the distinct CSDR levels at the "most reactive" mixture fraction, simply because the same presumed PDF (β-PDF) and CV model (the β-PDF gradient diffusion model) were employed in both realisations. It is obvious from Figure 2(d) that Mortensen's model based on the PMF-PDF results in comparable CSDR levels at $\eta_{mr}$ and produces profiles similar in shape and magnitude to those of the same model based on the β-PDF. Despite this, the predictions of the CMC-PMF realisation show departure from the CMC-βM results as shown in Figures 2(a) and 2(b). Thus, it becomes clear that the modelling of the CV also plays an important role in the modelling of this flame. Therefore, arguments based exclusively on the grounds of

intensity of micromixing at $\eta_{mr}$ are not sufficient (though necessary) in order explain the variability in the results. To illustrate this, the axial profiles of the CSDR models employed in the three realisation are plotted in Figure 4 at $\eta_{mr}$ ($\langle \chi \mid \eta_{mr} \rangle$). Although the flame is stabilised by autoignition in CMC-βM [1] and by premixed flame propagation CMC-PMF (see Section 5.3), $\langle \chi \mid \eta_{mr} \rangle_{CMC-PMF}$ (thick solid lines) is lower than $\langle \chi \mid \eta_{mr} \rangle_{CMC-βM}$ (thin solid line). Hence, although the lower CSDR levels at $\eta_{mr}$ favour the occurrence of autoignition, stabilisation is achieved by a different mechanism. For this reason, any analysis based exclusively on the CSDR is insufficient. Returning to Figure 2, the lower CSDR levels around stoichiometry in the CMC-PMF realisation lead to a decrease in the leakage of fuel and oxidiser from stoichiometric mixtures towards lean and rich mixtures. Therefore, the oxidation of the fuel becomes more intense in the vicinity of the stoichiometric mixture fraction. This behaviour, accompanied by the heating of lean mixtures in the preheat zone, promotes the early formation of a flame. Further, the smaller magnitude of $\langle v \mid \eta \rangle$ around stoichiometry and in rich mixtures results in a long residence time, which leads to increased chemical activity and hence higher $Q_T$ and $Q_{OH}$ levels.

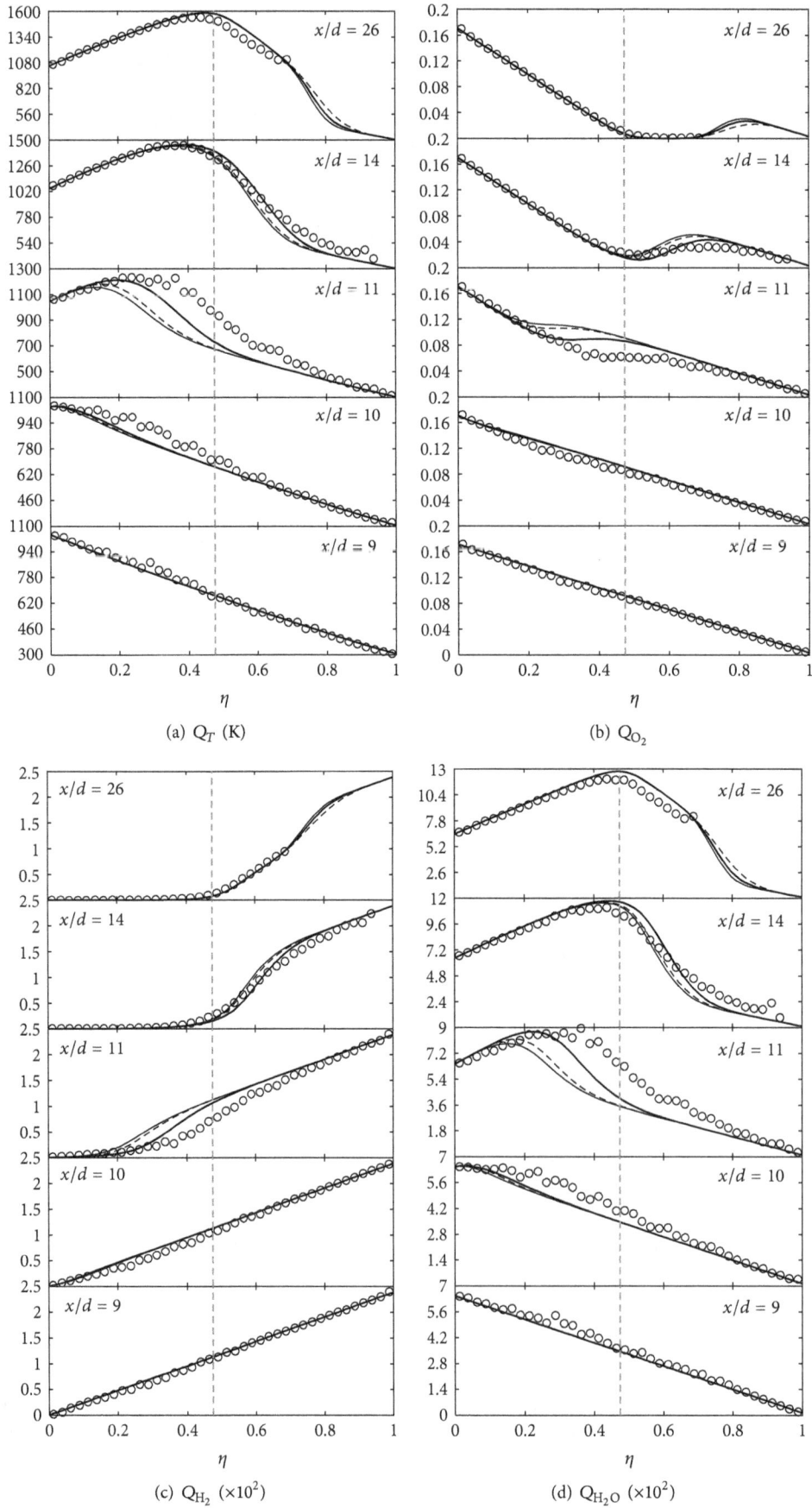

(a) $Q_T$ (K)

(b) $Q_{O_2}$

(c) $Q_{H_2}$ ($\times 10^2$)

(d) $Q_{H_2O}$ ($\times 10^2$)

FIGURE 3: Continued.

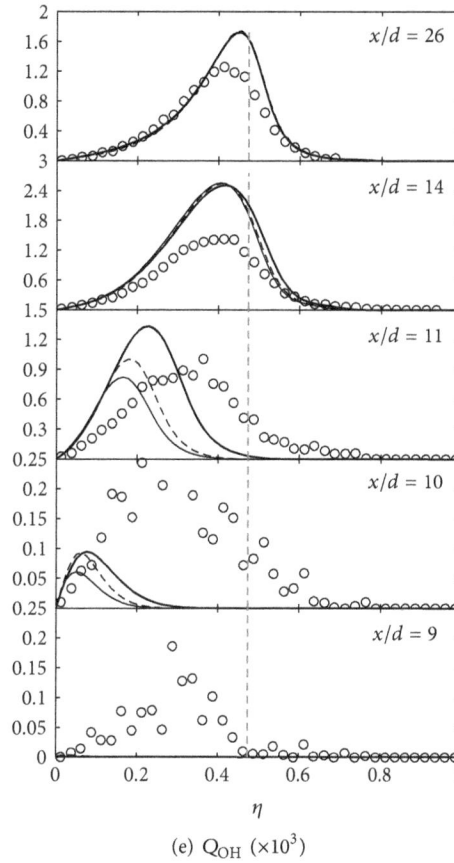

$$(e) \; Q_{OH} \; (\times 10^3)$$

FIGURE 3: Conditional profiles near the stoichiometric isocontour: (a) $Q_T$ (K), (b) $Q_{O_2}$, (c) $Q_{H_2}$ ($\times 10^2$), (d) $Q_{H_2O}$ ($\times 10^2$), and (e) $Q_{OH}$ ($\times 10^3$). Dashed lines, CMC-$\beta$G; thin solid lines, CMC-$\beta$M; thick solid lines, CMC-PMF; symbols, experimental data [2]. The vertical grey dashed line indicates the location of the stoichiometric mixture fraction.

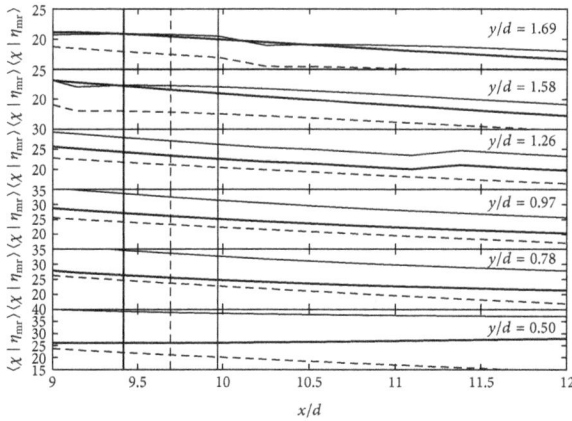

FIGURE 4: Axial profiles of $\langle \chi \mid \eta_{mr} \rangle$ at several radial locations for $T_c = 1045$ K: dashed lines, Girimaji's model ($\langle \chi \mid \eta_{mr} \rangle_{CMC-\beta G}$); thin solid lines, Mortensen's model based on the $\beta$-PDF ($\langle \chi \mid \eta_{mr} \rangle_{CMC-\beta M}$); thick solid lines, Mortensen's model based on the PMF-PDF ($\langle \chi \mid \eta_{mr} \rangle_{CMC-PMF}$). The vertical lines designate the ignition locations obtained with each CSDR model. The pattern of each vertical line follows that of the corresponding CSDR.

*5.2. Sensitivity to the Coflow Temperature.* Several CMC [1, 9–11] and PDF [27, 28] calculations show that the flame

under investigation is very sensitive to $T_c$. Cabra et al. [2] report an experimental uncertainty of 3% in the temperature measurements. As in [1], small perturbations ($\sim \pm 1.435\%$) are applied to $T_c$ in order to assess flame response when the PMF closures are employed. Figure 5 shows the radial profiles of the Favre-averaged temperature and species mass fractions at the axial locations $x/d = 8, 9, 10, 11, 14$ and 26 for $T_c = 1030$ and 1060 K. The results of all three realisations are displayed. When $T_c$ is decreased to 1030 K (black lines), the profiles of CMC-PMF are in close agreement with those of CMC-$\beta$G. This trend differs from the $T_c = 1045$ K case (Figure 1), where the profiles of CMC-$\beta$G and CMC-$\beta$M show closer agreement. In comparison to the experimental data, the predictions of CMC-PMF and CMC-$\beta$G are in general superior to those CMC-$\beta$M, particularly at $x/d = 11$ and 14. Overall, the results of all three realisations remain in reasonable agreement with the experiments. When $T_c$ is increased to 1060 K (grey lines), the profiles of CMC-PMF are in close agreement with those of CMC-$\beta$M. Again, this trend differs from those observed in the $T_c = 1030$ and 1045 K cases. The experimental measurements are grossly mispredicted at all axial locations as a result of the occurrence of early ignition (discussed in Section 5.3), which is in turn due to the higher coflow temperature.

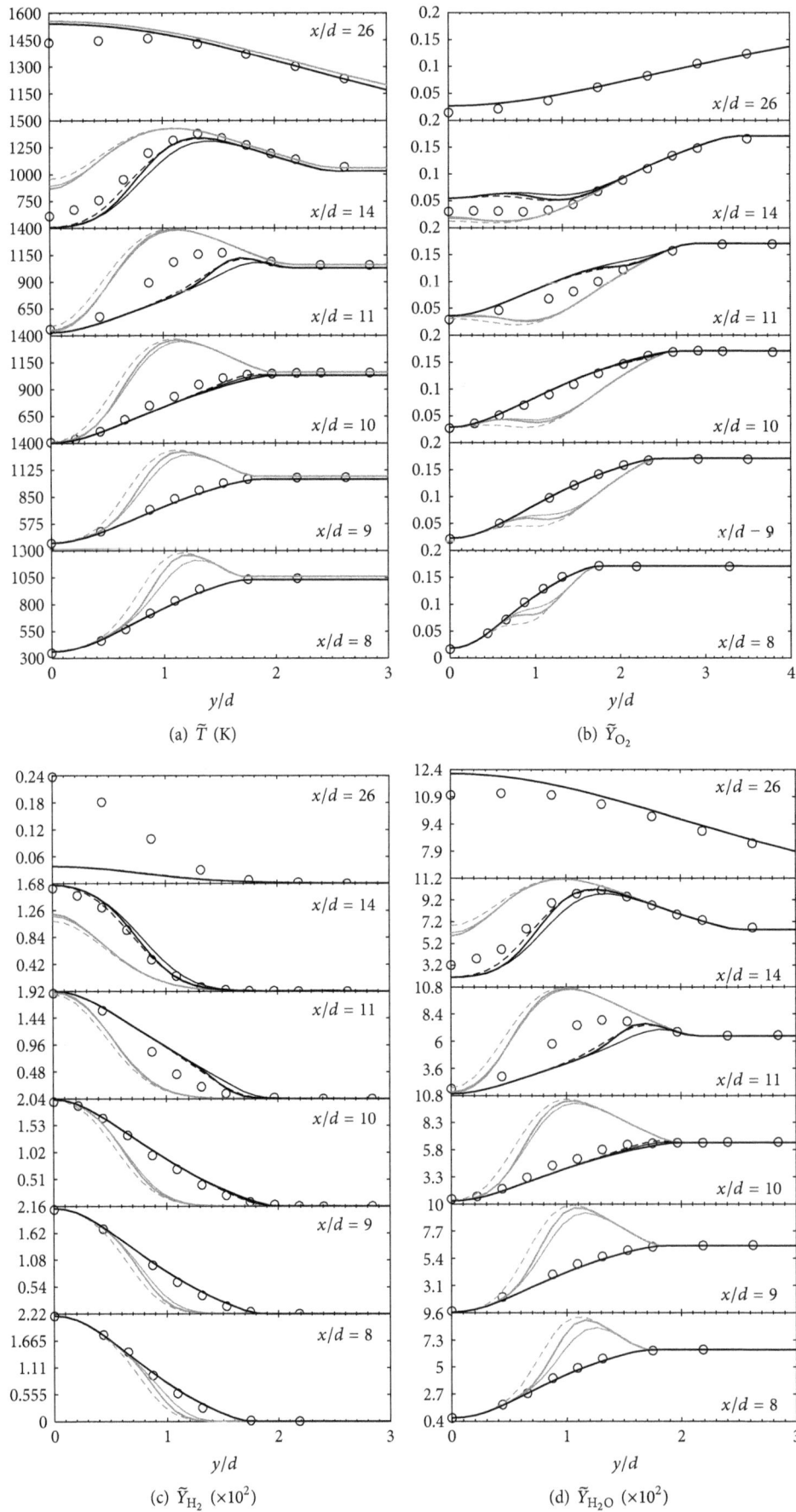

(a) $\tilde{T}$ (K)

(b) $\tilde{Y}_{O_2}$

(c) $\tilde{Y}_{H_2}$ ($\times 10^2$)

(d) $\tilde{Y}_{H_2O}$ ($\times 10^2$)

FIGURE 5: Continued.

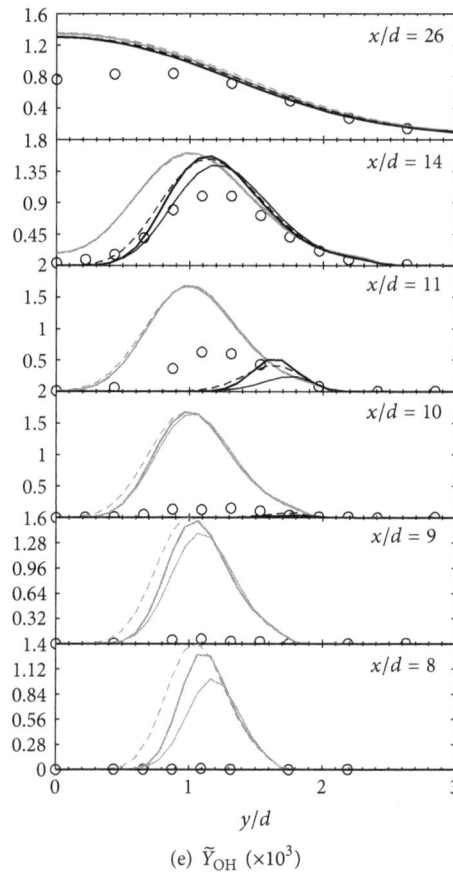

(e) $\tilde{Y}_{OH}$ ($\times 10^3$)

FIGURE 5: Radial profiles of (a) $\tilde{T}$, (b) $\tilde{Y}_{O_2}$, (c) $\tilde{Y}_{H_2}$ ($\times 10^2$), (d) $\tilde{Y}_{H_2O}$ ($\times 10^2$), and (e) $\tilde{Y}_{OH}$ ($\times 10^3$) obtained from the three CMC realisations for $T_c$ = 1030 K (black lines) and 1060 K (grey lines). Lines: dashed, CMC-$\beta$G; thin solid, CMC-$\beta$M; thick solid, CMC-PMF; symbols, experimental data [2].

The normalised stabilisation coordinates obtained using the three realisations are displayed in Table 2. PMF-CMC yields the smallest liftoff height in the $T_c$ = 1030 K case, followed by CMC-$\beta$G and then CMC-$\beta$M. Quantitatively, the CMC-PMF prediction is the closest to $H_{exp}/d$. The predicted radial stabilisation location is the same in all realisations (1.58) and is slightly larger than the one calculated at $T_c$ = 1045 K (1.47). Therefore, the flame base becomes wider as $T_c$ is decreased. When $T_c$ = 1060 K, the liftoff height predicted in CMC-PMF falls between the heights calculated in CMC-$\beta$G and CMC-$\beta$M. The radial stabilisation locations obtained in CMC-PMF and CMC-$\beta$M are the same (1.26), while the predicted CMC-$\beta$G value is slightly smaller (1.16). As such, it can be seen that the flame base becomes narrower when $T_c$ is increased.

5.3. Stabilisation Mechanism. Several stabilisation mechanisms have been proposed in the literature of lifted flames. Some are extinction due to excessive straining [47], premixed flame propagation [48, 49], and autoignition [50]. In the flame under investigation, the latter two theories are more plausible due to the presence of the vitiated coflow [50]. It was found in [1] to be autoignition-stabilised over a range of coflow temperatures (1030–1060 K) when the $\beta$-PDF closures

TABLE 2: Locations of the stabilisation points obtained using the three CMC realisations with the coflow temperatures 1030 K and 1060 K.

| Realisation | $T_c$ = 1030 K | | $T_c$ = 1030 K | |
|---|---|---|---|---|
| | $x/d$ | $y/d$ | $x/d$ | $y/d$ |
| CMC-$\beta$G | 11.48 | 1.58 | 6.61 | 1.16 |
| CMC-$\beta$M | 11.91 | 1.58 | 7.39 | 1.26 |
| CMC-PMF | 11.22 | 1.58 | 6.97 | 1.26 |

are used (CMC-$\beta$G and CMC-$\beta$M). The distinction between stabilisation by autoignition and premixed flame propagation was achieved by means of the numerical indicators developed by Gordon et al. [28], which involve the analyses of the transport budgets and the history of radical build-up ahead of the stabilisation height. In this section, these indicators are invoked in order to determine whether the usage of the PMF approach changes the previous conclusions.

5.3.1. Budgets in Mixture Fraction Space. Figure 6 shows the transport budgets of the steady-state conditional temperature equation (the right-hand side terms of (2)) obtained using the CMC-PMF realisation for $T_c$ = 1030, 1045 and

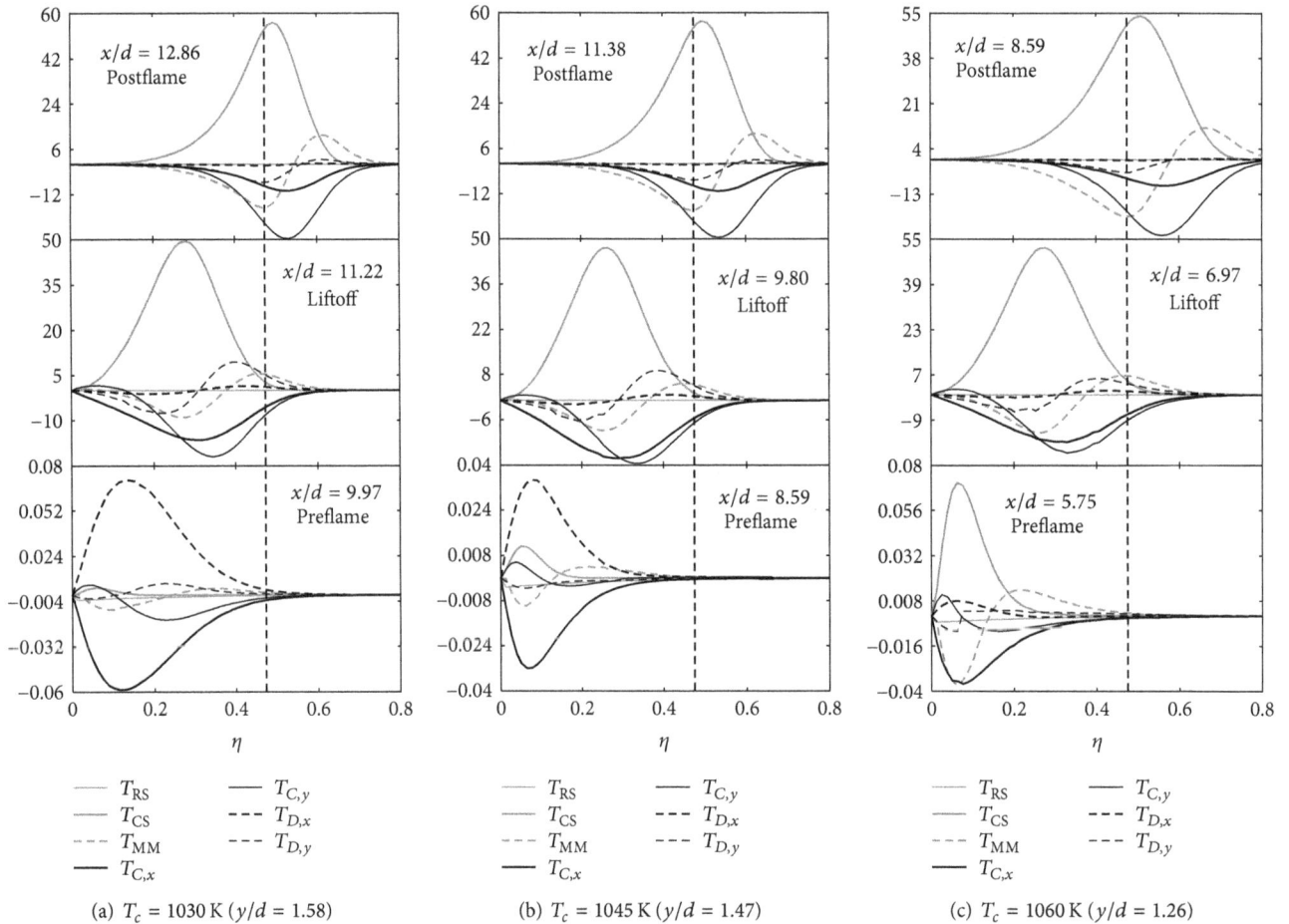

FIGURE 6: Transport budget of the steady-state $Q_T$ equation (r.h.s. terms of (2)) obtained using the CMC-PMF realisation for different coflow temperatures: (a) $T_c = 1030$ K ($y/d = 1.58$), (b) $T_c = 1045$ K ($y/d = 1.47$), and (c) $T_c = 1060$ K ($y/d = 1.26$). $T_{RS}$: radiative source; $T_{CS}$: chemical source; $T_{MM}$: micromixing; $T_{C,x}$: axial convection; $T_{C,y}$: radial convection; $T_{D,x}$: axial diffusion; $T_{D,y}$: radial diffusion. The vertical dashed line corresponds to the location of the stoichiometric mixture fraction. All terms are scaled down by a factor of $10^5$ and the units are K/s.

1060 K. For each $T_c$, the budgets are reported at three axial locations around the predicted stabilisation height covering the preflame, liftoff, and postflame regions. In the CMC-$\beta$G and CMC-$\beta$M realisations considered in [1], a balance in lean mixtures between the chemical source, $T_{CS}$, the axial convection term, $T_{C,x}$, and micromixing, $T_{MM}$, was found in the pre-flame region for all $T_c$. The axial and radial diffusion terms, $T_{D,x}$ and $T_{D,y}$, and the radial convection term, $T_{C,y}$, were found to have little contribution to the overall budget. This balance leads to the conclusion that the mixtures ignite spontaneously, and therefore the flame was deemed to be stabilised by autoignition. As shown in Figures 6(a) and 6(b), a different balance manifests in the preflame regions for $T_c = 1030$ and 1045 K (bottom panes) as $T_{C,x}$ is essentially counterbalanced by $T_{D,x}$. The contributions of the remaining terms are smaller in comparison, most notably that of $T_{CS}$. This balance suggests the presence of a preheat zone as in premixed flames. It is important to note that $T_{D,x}$ and $T_{C,x}$ do not peak at stoichiometry, but rather in lean mixtures. Therefore, as postulated by Patwardhan et al. [10], lean mixtures are preheated by downstream burning mixture. As such,

a premixed flame front propagates upstream and anchors the base of the lifted flame in lean mixtures. It is worth noting that when $T_c = 1045$ K, $T_{CS}$ yields a larger contribution compared to the $T_c = 1030$ K case. This observation indicates that there is a weak competition from autoignition. However, $T_{CS}$ is not sufficiently large to change the nature of the stabilisation mechanism. When $T_c$ is increased to 1060 K, $T_{CS}$ peaks at $\eta_{mr}$ and the $T_{CS}$-$T_{C,x}$-$T_{MM}$ balance recurs in the preflame zone as displayed in Figure 6(c) (bottom pane), which indicates the occurrence of autoignition as in the CMC-$\beta$G and CMC-$\beta$M realisations [1]. At liftoff (middle panes in Figures 6(a)–6(c)), the peak of $T_{CS}$ shifts to $\eta \approx 0.27$ and its amplitude increases dramatically. In the cases where $T_c = 1030$ and 1045 K, $T_{CS}$ is balanced by $T_{C,x}$ and $T_{C,y}$. The terms $T_{MM}$ and $T_{D,y}$ are more important than in upstream locations and $T_{D,x}$ is negligible. A similar balance is observed when $T_c = 1060$ K, except that $T_{MM}$ is more prevalent. Compared to the CMC-$\beta$G and CMC-$\beta$M realisations [1], when $T_c = 1060$ K the roles of $T_{MM}$ and $T_{C,y}$ vary significantly due to the different shapes of the CSDR and CV predicted by the PMF closures. $T_{MM}$ has a weaker effect because of the smaller CSDR levels for $\eta \gtrsim \tilde{\xi}$

(not shown). Conversely, $T_{C,y}$ is more influential due to the larger residence time caused by the smaller magnitude of $\langle v \mid \eta \rangle$ over the same range of $\eta$ (not shown). Further downstream in the postflame regions (top panes in Figures 6(a)–6(c)), $T_{CS}$ peaks around the stoichiometric mixture fraction and it is essentially balanced by $T_{MM}$ and $T_{C,y}$. $T_{C,x}$ remains important but its role diminishes from upstream locations, and $T_{D,x}$ is virtually zero. In comparison to the CMC-$\beta$G and CMC-$\beta$M in [1], for all $T_c$ values, $T_{C,y}$ acts as the major heat sink around stoichiometry due to the larger residence time. Beyond the postflame locations indicated in Figure 6, $T_{C,x}$ and $T_{C,y}$ diminish gradually and the flame budgets approach the structure of a nonpremixed flame, which is largely characterised by a $T_{CS}$-$T_{MM}$ balance.

*5.3.2. Budgets in Physical Space.* As in [1], the controlling stabilisation mechanism can be determined as well via the analysis the Integrated Transport (IT) budget of the conditional temperature in physical space. To compute the individual contributions, each term on the right-hand side of (2) is weighted by the appropriate presumed PDF and then integrated over the mixture fractions space. The axial profiles of the resulting contributions are shown in Figure 7 for all combinations of CMC realisations and coflow temperatures. The indicated $y/d$ values correspond to the radial locations of stabilisation obtained in each realisation. As shown in the top and middle panes of Figures 7(a)–7(c), for all coflow temperatures in the CMC-$\beta$G and CMC-$\beta$M realisations $IT_{CS}$ is balanced by $IT_{C,x}$ and $IT_{MM}$ in the preflame regions. The remaining terms have little contribution to the overall balance. However, they become more important right ahead of the stabilisation height. The $IT_{CS}$-$IT_{C,x}$-$IT_{MM}$ balance indicates that the mixture ignites spontaneously, and therefore the flame is autoignition-stabilised for all $T_c$ values. The nature of the controlling stabilisation mechanism in the CMC-PMF realisation depends on $T_c$. For $T_c$ = 1030 (bottom pane in Figure 7(a)), there is a clear balance between $IT_{D,x}$ and $IT_{C,x}$ up to $x/d \approx 10.25$ and $IT_{CS}$ is small. This structure indicates the presence of a preheat zone, which is indicative of stabilisation by means of premixed flame propagation. Beyond this location and prior to liftoff, $IT_{CS}$ increases rapidly. The terms $IT_{MM}$, $IT_{C,y}$, and $IT_{D,y}$ become more important and $IT_{D,x}$ has the smallest contribution to the overall budget. Similar trends are observed for $T_c$ = 1045 K (bottom pane in Figure 7(b)). However, in this case $IT_{CS}$ is more significant in the preheat zone, which indicates that there is competition from the autoignition mechanism. Nevertheless, $IT_{CS}$ is not sufficiently large in order for autoignition to happen. Increasing $T_c$ further to 1060 K, the budgets in the preflame region (bottom pane in Figure 7(c)) are very similar to those of the CMC-$\beta$G and CMC-$\beta$M. $IT_{CS}$ is primarily balanced by $IT_{C,x}$ and hence autoignition takes place. In the vicinity of the stabilisation height, the budgets of all realisations show similar structures as $IT_{CS}$ is counterbalanced by $IT_{D,y}$, $IT_{C,x}$, $IT_{C,y}$, and $IT_{MM}$. However, the relative importance of $IT_{C,y}$ and $IT_{MM}$ with respect to $IT_{CS}$ and $IT_{C,x}$ varies significantly in the CMC-PMF realisations. Compared to CMC-$\beta$G and CMC-$\beta$M, $IT_{C,y}$ is larger in magnitude due to the lower radial CV component (longer residence time) whereas $IT_{MM}$

is smaller because of the lower CSDR levels. Beyond the stabilisation heights, the structure of a nonpremixed flame is gradually approached as $IT_{CS}$ is primarily balanced $IT_{MM}$ with smaller contributions from the remaining terms.

*5.3.3. Radical History ahead of the Stabilisation Height.* The analyses of the transport budgets of the temperature in mixture fraction and physical spaces reveal that the nature of the stabilisation mechanism becomes sensitive to the coflow temperature when the PMF approach is employed. Further analysis of the history of radical build-up ahead of the stabilisation height can provide more insight into this matter. Figure 8 shows the axial profiles of the normalised Favre-averaged temperature $((\widetilde{T} - \widetilde{T}_{min})/(\widetilde{T}_{max} - \widetilde{T}_{min}))$ and mass fractions $(\widetilde{Y}/\widetilde{Y}_{max})$ of H, O, OH, and HO$_2$ for all combinations of CMC realisations and coflow temperatures. The subscripts "min" and "max" denote the minimum and maximum values of the reactive scalars at the axial locations of stabilisation. For all coflow temperatures, the CMC-$\beta$G and CMC-$\beta$M realisations (top and middle panes in Figure 8) show that HO$_2$ builds up rapidly prior to the runaway of H, O and OH. Therefore, HO$_2$ acts as a precursor to the production of H, O, and OH as in autoignition scenarios [28]. The evolution of radicals in the CMC-PMF realisation (bottom panes in Figure 8) shows similar trends, which is at first glance indicative of the occurrence of autoignition. As shown by Gordon et al. [28], in premixed flame propagation radicals build up simultaneously in the preheat zone. It is obvious here that HO$_2$ builds up prior to the remaining radicals. However, for $T_c$ = 1030 and 1045 K, there is a notable decrease in the axial distance between the runaway location of HO$_2$ and those of the remaining radicals (see the filled circles) and between the peak of HO$_2$ and the stabilisation height (distance between dash-dotted and dashed vertical lines). Hence, although radicals do not start building up at the same point, there is a clear tendency towards simultaneous radical production ahead of the stabilisation height. It is also important to note that when lifted flames are stabilised by means of premixed flame propagation, liftoff takes place at locations where the local mixture fraction mean, and $\widetilde{\xi}$ is equal to the stoichiometric mixture fraction, $\xi_{st}$ [51]. This is not the case here since the flame stabilises at $\widetilde{\xi}$ = 0.183 when $T_c$ = 1030 K and $\widetilde{\xi}$ = 0.139 when $T_c$ = 1045 K. These $\widetilde{\xi}$ values are well below $\xi_{st}$ = 0.474. Still, it is clear that $\widetilde{\xi}$ approaches $\widetilde{\xi}_{st}$ as $T_c$ is decreased. A further decrease in $T_c$ is therefore expected to lead to a more simultaneous radical build-up.

*5.4. Spurious Sources.* Theoretically, if the closed CMC equations are multiplied by the PDF and then integrated over the mixture fraction space, the unconditional equations should be fully recovered [30]. When an inconsistent CSDR model (e.g., a homogeneous model or an inhomogeneous model employing an inconsistent CV closure), the process outline above would result in additional spurious (false) source terms. The analysis of the spurious sources is a valuable tool that enables the identification of the flaws of inconsistent

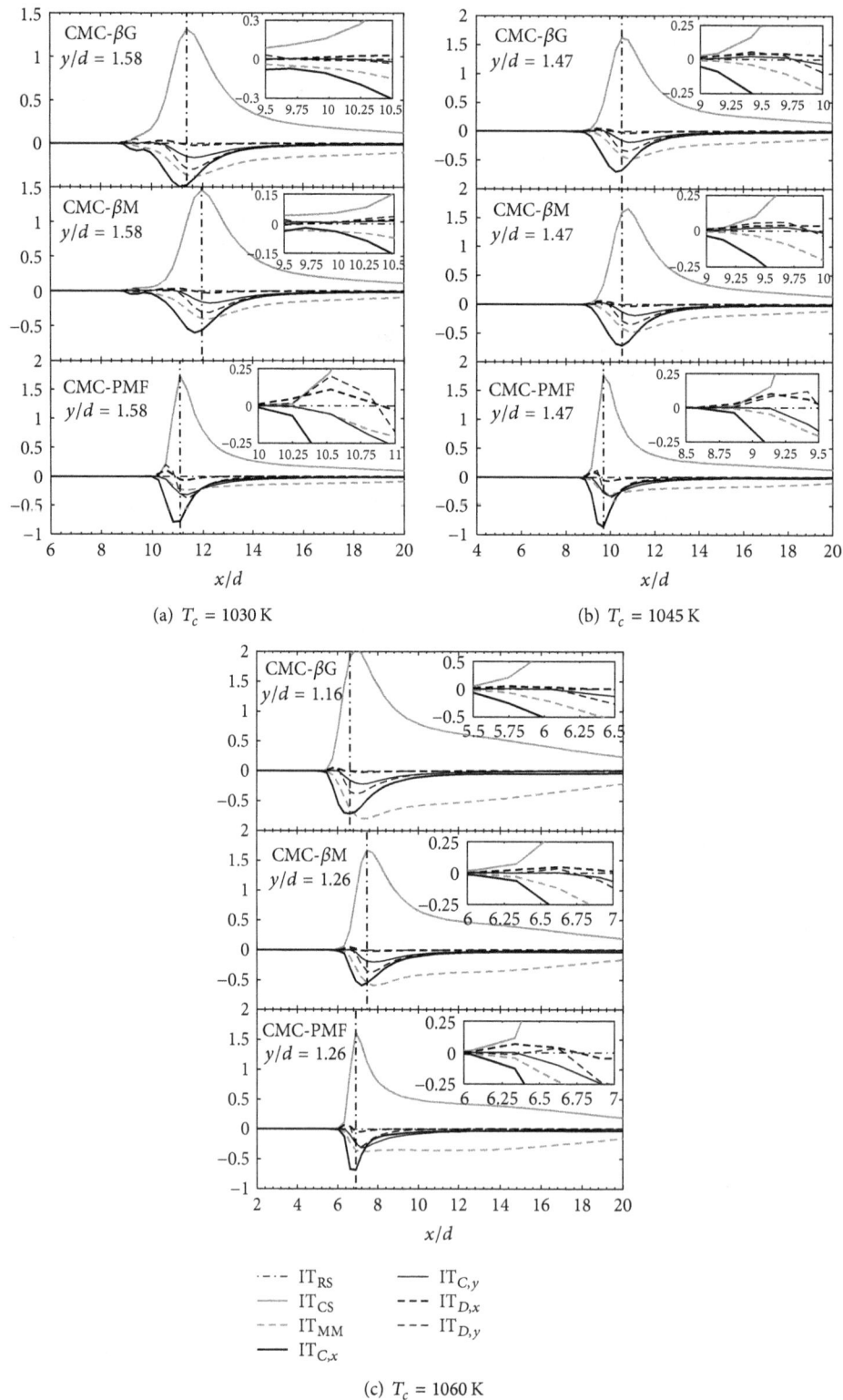

(a) $T_c = 1030\,\text{K}$

(b) $T_c = 1045\,\text{K}$

(c) $T_c = 1060\,\text{K}$

FIGURE 7: Axial profiles of the Integrated Transport (IT) budget of the steady-state $Q_T$ equation (integrated r.h.s. terms of (2)) obtained using the three CMC realisations: (a) $T_c = 1030\,\text{K}$, (b) $T_c = 1045\,\text{K}$, and (c) $T_c = 1060\,\text{K}$. The bottom, middle, and top panes correspond to CMC-$\beta$G, CMC-$\beta$M, and CMC-PMF, respectively. The indicated $y/d$ values correspond to the radial locations of stabilisation. The vertical dash-dotted lines correspond to the stabilisation heights. $\text{IT}_{RS}$: radiative source; $\text{IT}_{CS}$: chemical source; $\text{IT}_{MM}$: micromixing; $\text{IT}_{C,x}$: axial convection; $\text{IT}_{C,y}$: radial convection; $\text{IT}_{D,x}$: axial diffusion; $\text{IT}_{D,y}$: radial diffusion. All terms are scaled down by a factor of $10^6$ and the units are K/s.

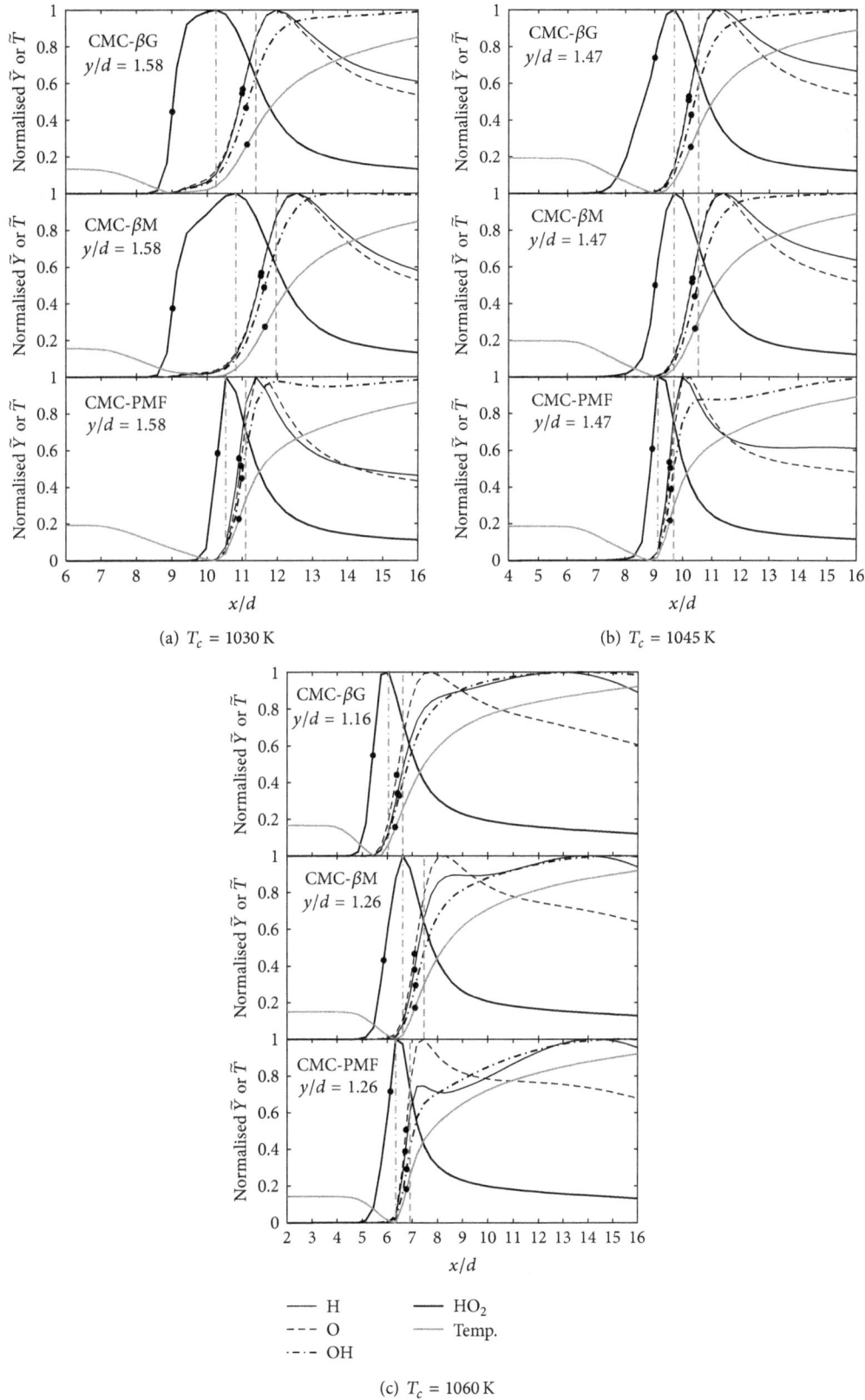

FIGURE 8: Axial profiles of the normalised Favre-averaged temperature and mass fractions of H, O, OH, and HO$_2$ obtained using the three CMC realisations: (a) $T_c$ = 1030 K, (b) $T_c$ = 1045 K, and (c) $T_c$ = 1060 K. The bottom, middle, and top panes correspond to CMC-$\beta$G, CMC-$\beta$M, and CMC-PMF, respectively. The indicated $y/d$ values correspond to the radial locations of stabilisation. The vertical dash-dotted and dashed lines correspond to the axial locations of maximum HO$_2$ and liftoff height, respectively. The circles indicate the locations of the maximum slopes.

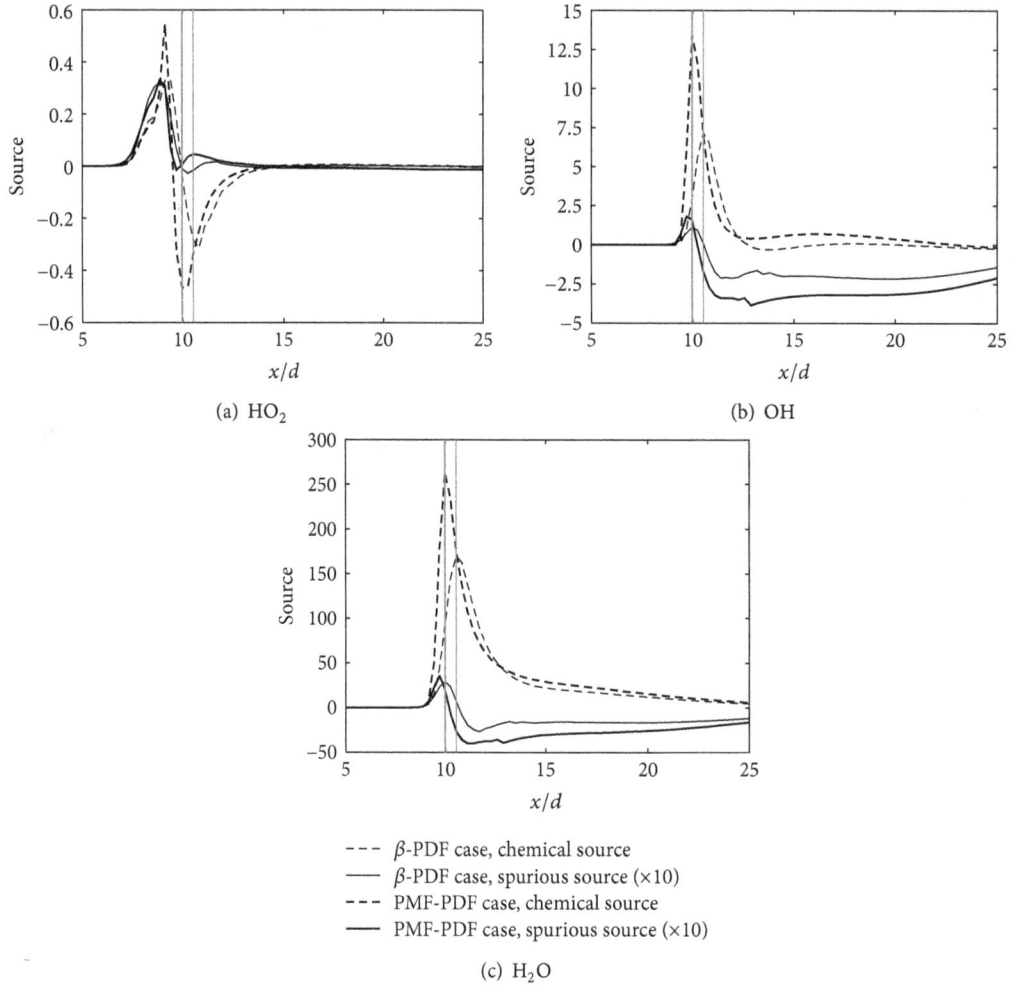

FIGURE 9: Axial profiles of the chemical and spurious sources at $y/d = 1.47$ (radial locations of stabilisation) for $T_c = 1045$ K: (a) HO$_2$, (b) OH, and (c) H$_2$O. All spurious sources are multiplied by 10 and the units are s$^{-1}$. The vertical grey lines indicate the axial locations of the stabilisation heights: thin, $\beta$-PDF; thick, PMF-PDF.

CMC implementations. The spurious source associated with a species $\kappa$ is calculated as [30, 52]

$$\widetilde{S}_\kappa = \frac{1}{2} \int_0^1 [\langle \chi \mid \eta \rangle_i - \langle \chi \mid \eta \rangle_c] \frac{\partial^2 Q_{\kappa,i}}{\partial \eta^2} \widetilde{P}(\eta)\, d\eta, \qquad (30)$$

where $\langle \chi \mid \eta \rangle_i$ and $\langle \chi \mid \eta \rangle_c$ are, respectively, the inconsistent and consistent CSDR models (the same PDF is used in both) and $Q_{\kappa,i}$ is the conditional mass fraction of species $\kappa$ obtained in the inconsistent realisation. The consistent inhomogeneous CSDR model of Mortensen [8], (18), is benchmarked against its inconsistent homogeneous version with $\widetilde{P}(\eta)$ modelled using both the $\beta$- and PMF-PDFs. As such, the difference $[\langle \chi \mid \eta \rangle_i - \langle \chi \mid \eta \rangle_c]$ in (30) is exactly the negative of the inhomogeneous contribution of the model. The details of the two cases are as follows:

(i) $\beta$-PDF case is as follows:

    (a) $\langle \chi \mid \eta \rangle_i$: (27) (the equivalent of the homogeneous version of (18));

    (b) $\langle \chi \mid \eta \rangle_c$: (18) with $II(\eta)$ obtained from (29).

(ii) PMF-PDF case is as follows:

    (a) $\langle \chi \mid \eta \rangle_i$: (23);
    (b) $\langle \chi \mid \eta \rangle_c$: (21).

Figure 9 compares the axial evolution of the integrated conditional chemical sources (IT$_{CS}$) and spurious sources of HO$_2$, OH, and H$_2$O for $T_c = 1045$ K. The profiles are plotted at the radial locations of stabilisation ($y/d = 1.47$ for the $\beta$-PDF case and $y/d = 1.37$ for the PMF-PDF case). As displayed in Figure 9(a), $\widetilde{S}_{HO_2}$ is negligible in both cases before the runaway of HO$_2$. Ahead of the stabilisation height, $\widetilde{S}_{HO_2}$ acts as a source. It increases as the production of HO$_2$ proceeds and peaks in the vicinity of the maximum of IT$_{CS,HO_2}$; then decreases as soon as the consumption of HO$_2$ begins, reaching a local minimum right ahead of the base of the flame. As the consumption of HO$_2$ continues, $\widetilde{S}_{HO_2}$ increases again and peaks downstream of the stabilisation height before it decays gradually inside the flame zone. Overall, the magnitude of $\widetilde{S}_{HO_2}$ is more important ahead of the stabilisation height. Figure 9(b) displays the axial

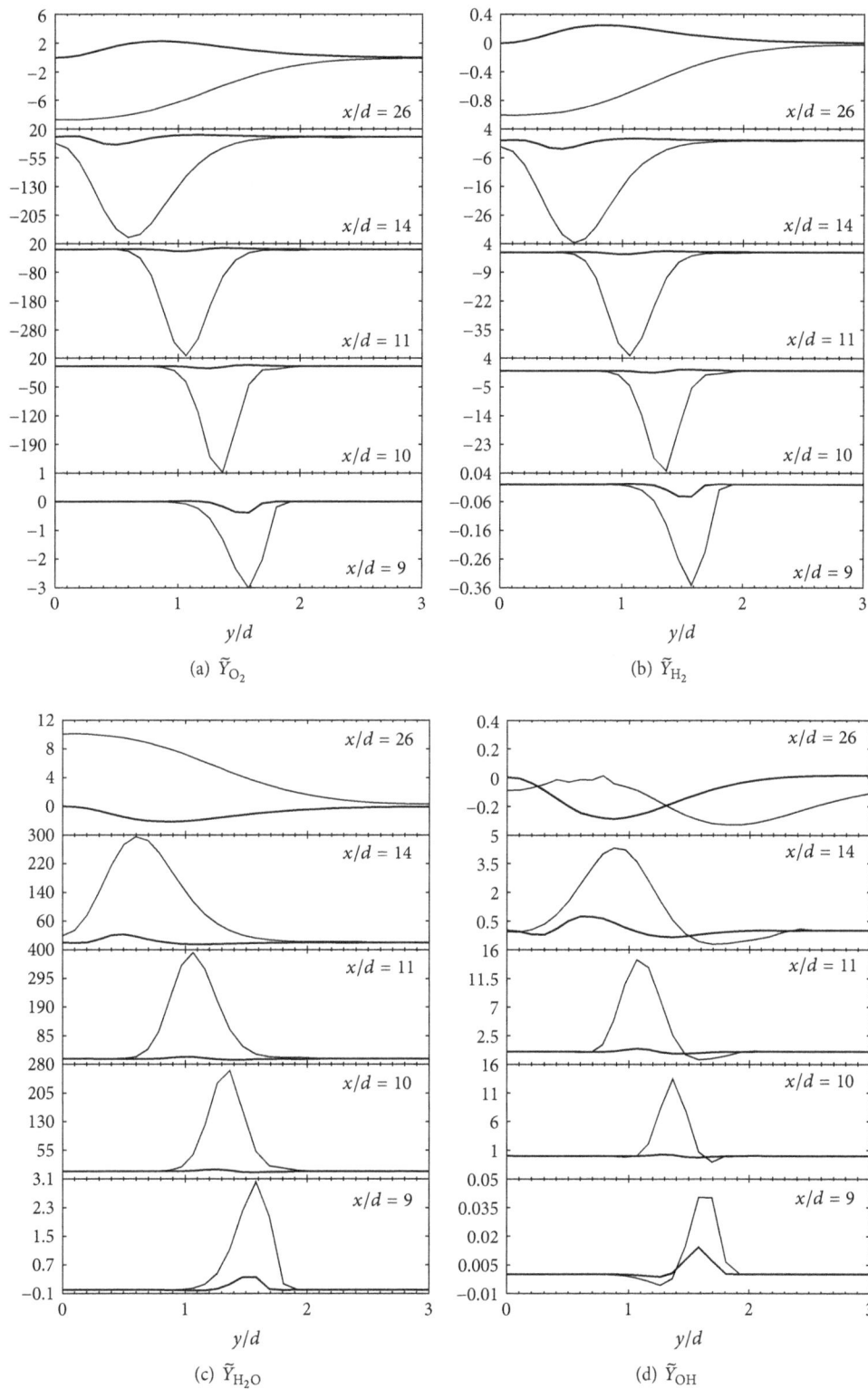

FIGURE 10: Radial profiles of the chemical and spurious sources obtained in the PMF-PDF case for $T_c = 1045$ K: (a) $O_2$, (b) $H_2$, (c) $H_2O$, and (d) OH. Thin lines, chemical sources; thick lines, spurious sources. The units are $s^{-1}$.

variation of $\widetilde{S}_{OH}$. Similar to what is observed in Figure 9(a), $\widetilde{S}_{OH}$ in Figure 9(b) acts as a source ahead of the stabilisation height. Its magnitude is negligible before the runaway of OH then increases as OH is produced and peaks in the vicinity of the maximum of $\mathrm{IT}_{CS,OH}$. $\widetilde{S}_{OH}$ decreases sharply at the base of the flame before increasing at a much slower rate in the flame zone (its magnitude decreases gradually with increasing $x/d$) and continues to act as a sink. As opposed to $\widetilde{S}_{HO_2}$, $\widetilde{S}_{OH}$ is more important in the flame zone compared to the region located ahead of the stabilisation height. The trends observed in the axial variation of $\widetilde{S}_{H_2O}$ (Figure 9(c)) are similar to those of $\widetilde{S}_{OH}$.

The comparison of the results of the $\beta$-PDF and PMF-PDF cases ahead of the stabilisation heights in Figures 9(a)–9(c) LOH reveals that the spurious sources in the PMF-PDF case yield higher maxima, however, prevailing over narrower axial bands. In the flame zone, the magnitude of the spurious sources obtained in the $\beta$-PDF case is smaller at all axial locations.

Overall, it can be concluded that the errors arising from the inconsistent modelling of the CSDR are small but nonnegligible. To ascertain this conclusion, the radial profiles of the chemical and spurious sources of $O_2$, $H_2$, $H_2O$, and OH are plotted in Figure 10. Only the PMF-PDF case is shown here. The results of the $\beta$-PDF case are very similar (see [1]). It is evident that the effect of the spurious sources is nonnegligible at all axial locations, particularly within the flame zone at $x/d = 14$ and 26. The ratios of spurious-to-chemical sources at the axial locations considered in Figure 10 (not shown) indicate that the spurious contribution increases with $x/d$, which is expected to take place as the scalar dissipation rate decays.

The findings presented in the current section show that inconsistent CMC implementations may influence the results as the additional spurious terms stemming from inconsistencies in the modelling of the CSDR can behave as significant sources or sinks.

## 6. Conclusions

A previously investigated lifted $H_2/N_2$ jet flame was revisited in order to assess the applicability of the PMF approach in the context of CMC. The findings were compared to previous results obtained using $\beta$-PDF-based closures over a range of coflow temperatures ($T_c$). In view of the current results, the following conclusions are drawn:

(i) The PMF-PDF is in general narrower than the $\beta$-PDF and presents a higher peak. The shape of the PDF has a large influence on the modelling of the CSDR and the CV.

(ii) Girimaji's CSDR model results in nearly symmetric profiles near the stoichiometric isoline whereas the Models of Mortensen (based on the PMF- and $\beta$-PDFs) yield skewed profiles. The latter models are less dissipative away from the mean of the mixture fraction.

(iii) The gradient diffusion CV model is well-behaved over the whole mixture fraction space when the PMF-PDF is employed as it does not overshoot at low probabilities. The $\beta$-PDF-based model tends to infinity as the PDF approaches zero.

(iv) The liftoff heights predicted in all realisations fall within experimental uncertainty. In comparison to the $\beta$-PDF-based calculations at $T_c = 1045\,\mathrm{K}$, the PMF approach yields improved results in physical and mixture fraction spaces and a more accurate liftoff height. The flame remains very sensitive to small perturbations in $T_c$.

(v) As opposed to the $\beta$-PDF closures, the transport budgets in mixture fraction and physical spaces and the radical history ahead of the stabilisation height reveal that the nature of the stabilisation mechanism is sensitive to $T_c$ when the PMF closures are used. For sufficiently high $T_c$ (e.g., $T_c = 1060\,\mathrm{K}$), the mixture ignites spontaneously at the "most reactive" mixture fraction, and the flame is stabilised by autoignition. As $T_c$ is decreased (e.g., to $T_c = 1045$ and $1030\,\mathrm{K}$), lean mixtures are preheated by a premixed flame front propagating upstream, and, therefore, the flame is stabilised by means of premixed flame propagation.

(vi) The effect of the spurious sources resulting from the inconsistent modelling of the CSDR is small in general but nonnegligible, in particular within the flame zone.

PMF is a consistent and reliable approach for the modelling of the conditional mixing and velocity statistics in CMC. Further application to other combustion phenomena is necessary to fully assess its applicability.

## Disclosure

This work had been done at the University of Waterloo and the second address is the current address of the corresponding author Ahmad El Sayed.

## Conflict of Interests

The authors declare that there is no conflict of interests regarding the publication of this paper.

## Acknowledgment

This project is funded by the Natural Sciences and Engineering Research Council of Canada.

## References

[1] A. El Sayed and R. A. Fraser, "Consistent Conditional Moment Closure Modelling of a lifted turbulent jet flame using the presumed $\beta$-PDF approach," *Journal of Combustion*, vol. 2014, Article ID 507459, 25 pages, 2014.

[2] R. Cabra, T. Myhrvold, J. Y. Chen, R. W. Dibble, A. N. Karpetis, and R. S. Barlow, "Simultaneous laser Raman-Rayleigh-lif measurements and numerical modeling results of a lifted turbulent $H_2/N_2$ jet flame in a vitiated coflow," *Proceedings of the Combustion Institute*, vol. 29, no. 2, pp. 1881–1888, 2002.

[3] S. B. Pope, "The probability approach to the modelling of turbulent reacting flows," *Combustion and Flame*, vol. 27, no. 3, pp. 299–312, 1976.

[4] M. J. Cleary, *CMC modelling of enclosure fires [Ph.D. thesis]*, University of Sydney, Sydney, Australia, 2004.

[5] M. Mortensen and B. Andersson, "Presumed mapping functions for eulerian modelling of turbulent mixing," *Flow, Turbulence and Combustion*, vol. 76, no. 2, pp. 199–219, 2006.

[6] M. Mortensen and S. M. de Bruyn Kops, "Conditional velocity statistics in the double scalar mixing layer—a mapping closure approach," *Combustion Theory and Modelling*, vol. 12, no. 5, pp. 929–941, 2008.

[7] S. S. Girimaji, "On the modeling of scalar diffusion in isotropic turbulence," *Physics of Fluids A*, vol. 4, no. 11, pp. 2529–2537, 1992.

[8] M. Mortensen, "Consistent modeling of scalar mixing for presumed, multiple parameter probability density functions," *Physics of Fluids*, vol. 17, no. 1, Article ID 018106, 2005.

[9] I. Stanković and B. Merci, "LES-CMC simulations of a turbulent lifted hydrogen flame in vitiated co-flow," *Thermal Science*, vol. 17, no. 3, pp. 763–772, 2013.

[10] S. S. Patwardhan, K. N. Lakshmisha, and B. N. Raghunandan, "CMC simulations of lifted turbulent jet flame in a vitiated coflow," *Proceedings of the Combustion Institute*, vol. 32, no. 2, pp. 1705–1712, 2009.

[11] S. Navarro-Martinez and A. Kronenburg, "Flame stabilization mechanisms in lifted flames," *Flow, Turbulence and Combustion*, vol. 87, no. 2-3, pp. 377–406, 2011.

[12] H. Chen, S. Chen, and R. H. Kraichnan, "Probability distribution of a stochastically advected scalar field," *Physical Review Letters*, vol. 63, no. 24, pp. 2657–2660, 1989.

[13] E. E. O'Brien and T.-L. Jiang, "The conditional dissipation rate of an initially binary scalar in homogeneous turbulence," *Physics of Fluids A*, vol. 3, no. 12, pp. 3121–3123, 1991.

[14] I. S. Kim, E. Mastorakos, and B. Merci, "Simulations of turbulent lifted jet flames with two-dimensional conditional moment closure," *Proceedings of the Combustion Institute*, vol. 30, no. 1, pp. 911–918, 2005.

[15] A. El Sayed, A. Milford, and C. B. Devaud, "Modelling of autoignition for methane-based fuel blends using conditional moment closure," *Proceedings of the Combustion Institute*, vol. 32, no. 1, pp. 1621–1628, 2009.

[16] N. Peters, "Laminar diffusion flamelet models in non-premixed turbulent combustion," *Progress in Energy and Combustion Science*, vol. 10, no. 3, pp. 319–339, 1984.

[17] A. Y. Klimenko and S. B. Pope, "The modeling of turbulent reactive flows based on multiple mapping conditioning," *Physics of Fluids*, vol. 15, no. 7, pp. 1907–1925, 2003.

[18] S. B. Pope, "PDF methods for turbulent reactive flows," *Progress in Energy and Combustion Science*, vol. 11, no. 2, pp. 119–192, 1985.

[19] M. J. Cleary and A. Kronenburg, "'Hybrid' multiple mapping conditioning on passive and reactive scalars," *Combustion and Flame*, vol. 151, no. 4, pp. 623–638, 2007.

[20] S. M. de Bruyn Kops and M. Mortensen, "Conditional mixing statistics in a self-similar scalar mixing layer," *Physics of Fluids*, vol. 17, no. 9, Article ID 095107, 11 pages, 2005.

[21] C. M. Cha, S. M. de Bruyn Kops, and M. Mortensen, "Direct numerical simulations of the double scalar mixing layer. Part I: passive scalar mixing and dissipation," *Physics of Fluids*, vol. 18, no. 6, Article ID 067106, 2006.

[22] M. Mortensen, S. M. de Bruyn Kops, and C. M. Cha, "Direct numerical simulations of the double scalar mixing layer. Part II: reactive scalars," *Combustion and Flame*, vol. 149, no. 4, pp. 392–408, 2007.

[23] A. El Sayed, M. Mortensen, and J. Z. Wen, "Assessment of the presumed mapping function approach for the stationary laminar flamelet modelling of reacting double scalar mixing layers," *Combustion Theory and Modelling*, vol. 18, no. 4-5, pp. 552–581, 2014.

[24] E. A. Brizuela and M. Z. Roudsari, "Comparison of RANS/CMC modeling of flame D with conventional and with presumed mapping function statistics," *Combustion Theory and Modelling*, vol. 15, no. 5, pp. 671–690, 2011.

[25] R. S. Barlow and J. H. Frank, "Effects of turbulence on species mass fractions in methane/air jet flames," *Symposium (International) on Combustion*, vol. 27, no. 1, pp. 1087–1095, 1998.

[26] A. R. Masri, R. Cao, S. B. Pope, and G. M. Goldin, "PDF calculations of turbulent lifted flames of $H_2/N_2$ fuel issuing into a vitiated co-flow," *Combustion Theory and Modelling*, vol. 8, no. 1, pp. 1–22, 2004.

[27] R. R. Cao, S. B. Pope, and A. R. Masri, "Turbulent lifted flames in a vitiated coflow investigated using joint PDF calculations," *Combustion and Flame*, vol. 142, no. 4, pp. 438–453, 2005.

[28] R. L. Gordon, A. R. Masri, S. B. Pope, and G. M. Goldin, "A numerical study of auto-ignition in turbulent lifted flames issuing into a vitiated co-flow," *Combustion Theory and Modelling*, vol. 11, no. 3, pp. 351–376, 2007.

[29] T. Myhrvold, I. S. Ertesvåg, I. R. Gran, R. Cabra, and J.-Y. Chen, "A numerical investigation of a lifted $H_2/N_2$ turbulent jet flame in a vitiated coflow," *Combustion Science and Technology*, vol. 178, no. 6, pp. 1001–1030, 2006.

[30] A. Y. Klimenko and R. W. Bilger, "Conditional moment closure for turbulent combustion," *Progress in Energy and Combustion Science*, vol. 25, no. 6, pp. 595–687, 1999.

[31] C. K. Madnia, S. H. Frankel, and P. Givi, "Reactant conversion in homogeneous turbulence: mathematical modeling, computational validations, and practical applications," *Theoretical and Computational Fluid Dynamics*, vol. 4, no. 2, pp. 79–93, 1992.

[32] N. Swaminathan and R. W. Bilger, "Analyses of conditional moment closure for turbulent premixed flames," *Combustion Theory and Modelling*, vol. 5, no. 2, pp. 241–260, 2001.

[33] N. Swaminathan and R. W. Bilger, "Scalar dissipation, diffusion and dilatation in turbulent $H_2$-air premixed flames with complex chemistry," *Combustion Theory and Modelling*, vol. 5, no. 3, pp. 429–446, 2001.

[34] W. P. Jones and J. H. Whitelaw, "Calculation methods for reacting turbulent flows: a review," *Combustion and Flame*, vol. 48, pp. 1–26, 1982.

[35] E. S. Richardson and E. Mastorakos, "Simulations of non-premixed edge-flame propagation in turbulent non-premixed jets," in *Proceedings of the 3rd European Combustion Meeting (ECM '07)*, Chania, Greece, April 2007.

[36] R. S. Barlow, A. N. Karpetis, J. H. Frank, and J.-Y. Chen, "Scalar profiles and NO formation in laminar opposed-flow partially premixed methane/air flames," *Combustion and Flame*, vol. 127, no. 3, pp. 2102–2118, 2001.

[37] C. Y. Ma, T. Mahmud, M. Fairweather, E. Hampartsoumian, and P. H. Gaskell, "Prediction of lifted, non-premixed turbulent flames using a mixedness-reactedness flamelet model with radiation heat loss," *Combustion and Flame*, vol. 128, no. 1-2, pp. 60–73, 2002.

[38] M. Mortensen, "Implementation of a conditional moment closure for mixing sensitive reactions," *Chemical Engineering Science*, vol. 59, no. 24, pp. 5709–5723, 2004.

[39] S. B. Pope, "An explanation of the turbulent round-jet/plane-jet anomaly," *AIAA Journal*, vol. 16, no. 3, pp. 279–281, 1978.

[40] W. P. Jones and P. Musonge, "Closure of the Reynolds stress and scalar flux equations," *The Physics of Fluids*, vol. 31, no. 12, pp. 3589–3604, 1988.

[41] B. Koren, "A robust upwind discretization method for advection, diffusion and source terms," in *Numerical Methods for Advection-Diffusion Problems*, C. B. Vreugdenhil and B. Koren, Eds., vol. 45 of *Notes on Numerical Fluid Mechanics*, pp. 117–138, Vieweg, Braunschweig, Germany, 1993.

[42] G. D. Byrne, "Pragmatic experiments with Krylov methods in the stiff ODE setting," in *Computational Ordinary Differential Equations*, J. Cash and I. Gladwell, Eds., pp. 323–356, Oxford University Press, Oxford, UK, 1992.

[43] G. Strang, "On the construction and comparison of difference schemes," *SIAM Journal on Numerical Analysis*, vol. 5, no. 3, pp. 506–517, 1968.

[44] M. A. Mueller, T. J. Kim, R. A. Yetter, and F. L. Dryer, "Flow reactor studies and kinetic modeling of the H2/O2 reaction," *International Journal of Chemical Kinetics*, vol. 31, no. 2, pp. 113–125, 1999.

[45] M. Mortensen, PMFpack, https://github.com/mikaem/PMFpack.

[46] N. Swaminathan and R. W. Bilger, "Assessment of combustion submodels for turbulent nonpremixed hydrocarbon flames," *Combustion and Flame*, vol. 116, no. 4, pp. 519–545, 1999.

[47] N. Peters and F. A. Williams, "Lift-off characteristics of turbulent jet diffusion flames," *AIAA Journal*, vol. 21, no. 3, pp. 423–429, 1983.

[48] S. Ruan, N. Swaminathan, and O. Darbyshire, "Modelling of turbulent lifted jet flames using flamelets: *a priori* assessment and *a posteriori* validation," *Combustion Theory and Modelling*, vol. 18, no. 2, pp. 295–329, 2014.

[49] Z. Chen, S. Ruan, and N. Swaminathan, "Simulation of turbulent lifted methane jet flames: effects of air-dilution and transient flame propagation," *Combustion and Flame*, vol. 162, no. 3, pp. 703–716, 2015.

[50] R. L. Gordon, A. R. Masri, S. B. Pope, and G. M. Goldin, "Transport budgets in turbulent lifted flames of methane autoigniting in a vitiated co-flow," *Combustion and Flame*, vol. 151, no. 3, pp. 495–511, 2007.

[51] L. Vanquickenborne and A. van Tiggelen, "The stabilization mechanism of lifted diffusion flames," *Combustion and Flame*, vol. 10, no. 1, pp. 59–69, 1966.

[52] K. Tsai and R. O. Fox, "Modeling multiple reactive scalar mixing with the generalized IEM model," *Physics of Fluids*, vol. 7, no. 11, pp. 2820–2830, 1995.

# 5

# Experimental Study of Constant Volume Sulfur Dust Explosions

**Joseph Kalman,**[1,2] **Nick G. Glumac,**[1] **and Herman Krier**[1]

[1]*University of Illinois at Urbana-Champaign, 1206 W. Green Street, Urbana, IL 61801, USA*
[2]*Naval Air Warfare Center Weapons Division, 1 Administration Circle, China Lake, CA 93555, USA*

Correspondence should be addressed to Joseph Kalman; kalmanjm@gmail.com

Academic Editor: Hong G. Im

Dust flames have been studied for decades because of their importance in industrial safety and accident prevention. Recently, dust flames have become a promising candidate to counter biological warfare. Sulfur in particular is one of the elements that is of interest, but sulfur dust flames are not well understood. Flame temperature and flame speed were measured for sulfur flames with particle concentrations of 280 and 560 g/m³ and oxygen concentration between 10% and 42% by volume. The flame temperature increased with oxygen concentration from approximately 900 K for the 10% oxygen cases to temperatures exceeding 2000 K under oxygen enriched conditions. The temperature was also observed to increase slightly with particle concentration. The flame speed was observed to increase from approximately 10 cm/s with 10% oxygen to 57 and 81 cm/s with 42% oxygen for the 280 and 560 g/m³ cases, respectively. A scaling analysis determined that flames burning in 21% and 42% oxygen are diffusion limited. Finally, it was determined that pressure-time data may likely be used to measure flame speed in constant volume dust explosions.

## 1. Introduction

Sulfur dust cloud combustion is a potential candidate to counter biological weapons. Sulfur dust has been used as a pesticide [1]. However, the more intriguing aspect of sulfur dust clouds is that they produce sulfur oxides which are chemical precursors to sulfuric acid [1]. It is well-known that sulfuric acid is extremely corrosive and dangerous to living organisms. The concept is that burning sulfur clouds will produce sulfur oxides. In the presence of water, sulfuric acid can be formed. It is thought that the sulfuric acid created, coupled with the elevated temperatures and ultraviolet radiation produced, will kill the spores. Other strong acids have been shown to have sporicidal tendencies [2].

In recent years, compositions have been studied for biodefeat applications. Mechanical alloys of aluminum and iodine [3, 4] as well as aluminum-iodine pentoxide thermites have been studied [5] and shown to be effective in killing biological spores, at least in part, due to the release of iodine. Other mechanical alloys of titanium and boron have also been investigated for this purpose [3, 6, 7].

Fundamentally, sulfur dust flames are unique. Sulfur is one of two elemental dusts whose combustion products are gaseous at standard conditions (298 K, 1 atm) with carbon being the other. Unlike carbon, the melting and boiling points of sulfur are much lower than the adiabatic flame temperature of a sulfur-air flame. This combination of factors provides conditions for a dust flame with a potentially strong gas-phase component.

To date, very few studies have investigated sulfur dust clouds. The limited work on sulfur dust flames by Proust [8] is among the only publications to do so; however, that work has provided few details into the combustion mechanism. Therefore, the primary goal for the current work is to measure fundamental aspects of sulfur dust cloud combustion in terms of fundamental quantities (e.g., flame speed) and to gain insight into the physical and chemical mechanisms involved. Moreover, the use of pressure-time data from constant volume dust explosions to determine flame speed is investigated.

## 2. Experimental Methods

The current study uses a 31 L cube chamber to maximize optical access. Five of the sides (including the door) have acrylic windows with circular viewing areas of 6.7 in diameter. Each window is clamped onto the side of the chamber with a size 6 pipe flange. Gas, vacuum, and pressure transducer ports are located on the top and sides of the chamber. A piezoresistive Kulite pressure transducer (XTM-190-250G) is used, and the signals are conditioned by an Endevco PR Conditioner model 106. The bottom (sixth side) has five 1 in diameter ports. One port is placed in the center with the other four being 3 inches away from the center, each in a different direction (i.e., left, right, front, and back). A single off-center port is used for wire feedthrough for the ignition source.

The port in the center of the chamber has a nozzle with forty, 0.889 mm diameter holes (number 65 drill bit) at a 45 degree angle. The two-piece particle injector is mounted underneath the center port of the chamber. The first piece is attached to the chamber through 1/4 in-20 screws. It contains a port on the side that attaches to a compressed air line and 1/4 in stainless steel tube which extends to the center and is bent 90 degrees downward. The powder is placed in an aluminum holder with a conical bottom (from a 1 inch drill bit) that attaches to the second piece of the injector. The centered 1/4 inch tube elbow directs the air burst downward into the powder holder. The pressurized burst rebounds off the bottom of the powder holder and carries the particles upwards through the nozzle into the chamber. The chamber is sealed by o-rings on the windows, injector, and door. Additional descriptions of the chamber are provided in [9].

Sulfur powder (−325 mesh) from Alfa Aesar was used. Particle size analysis was conducted using a Jeol 6060LV scanning electron microscope (SEM). The particles had an average diameter of 22.4 $\mu$m with a Sauter mean diameter of 30.4 $\mu$m. An anticaking agent, Aerosil 200 (Evonik), was used to improve the dispersion characteristics of sulfur. The average diameter of the Aerosil 200 powder was 12 nm according to the manufacturer. A size distribution was not measured as the particles were too small to be resolved by the SEM. The sulfur was mixed with 1% of Aerosil 200 by mass in a low energy tumbler for 3 hours. The addition of the anticaking agent was observed to have a noticeable effect on the flowability and dispersal of sulfur. Further details on the choice of anticaking agent are given in [9].

A known mass of the sulfur mixture was placed within the injector. The mass of powder remaining in the injector after the test was measured to determine the actual amount of powder injected. Alligator clips on the ignition posts held the igniter in the center of the chamber. A 4 J igniter (pyrogen covered bridgewire, Estes) was used to initiate combustion. The charge was ignited by discharging a 1 $\mu$F capacitor at 4000 V from an RISI fireset (Model FS-43).

Prior to the test, the chamber was put under vacuum and filled to slightly below atmospheric pressure (2 in Hg) so that the gas used for injection brought the total pressure to 1 atmosphere. A 100 psig burst of air was used to inject the powder. The injection lasted a total of one second under constant pressure. Ignition occurred 400 ms after injection had ended to allow for a uniform cloud to form and turbulence to dissipate.

The determination of this ignition time was made by analysis of two-dimensional particle concentration measurements. These laser extinction measurements provided quantitative information on how the particle concentration developed in time and space. The pixel intensities from images taken prior to powder injection were averaged and used as the incident intensity, $I_0$, in Beers Law:

$$\frac{I}{I_0} = \exp \frac{-3Q_{ext}LC}{2\rho D_{32}}. \qquad (1)$$

The intensity of each pixel in subsequent images allowed for the particle concentration, $C$, to be determined since the extinction efficiency ($Q_{ext}$) from Mie theory, Sauter mean diameter, $D_{32}$, path length, $L$ (355 mm), and density of sulfur, $\rho$ were all known. The arithmetic mean and standard deviation of the particle concentration for all pixels were calculated to provide a statistical analysis of the uniformity of the cloud. The ignition time was taken as the point in time where the spatially averaged concentration approached the expected value (based upon the amount of powder injected) and the standard deviation approached a minimum or decreased.

This method also provided visual evidence of the turbulence dissipating. Although the turbulence was never directly measured, it is believed that this quantitative measurement of the particle concentration yields sufficient insight into the injection process. Moreover, all of the experimental conditions in this work have the same injection and ignition conditions (e.g., ignition delay and injection back pressure). Comparison of the results from each conditions should see minimal effects from the turbulence. The reader is directed to [9] for additional details and discussion on these measurements.

*2.1. Diagnostics.* Flame speed measurements were made with ionization probes, similar to the work of Nair and Gupta [10], but were applied to dust flames instead of the gaseous flames that they studied. Two of the off-center 1 in diameter holes were used to feedthrough wires for the ionization probes where the wires were terminated by an R-type thermocouple connector (Omega). Two 0.01 in diameter tungsten wires were fed through a two-bore, 1/8 in outer diameter aluminum oxide tube. Approximately 1/4 inch of each wire extended outside of the ceramic tube. The other end of each wire was attached to the complementary thermocouple connector. The placement of the ionization probes are shown in Figure 1 where the relative location of the two ionization probe combs were chosen to provide information of the symmetry of the flame. The distance between the probes in each comb was 15.5 ± 0.76 mm.

Signals were recorded using a pair of Picoscope 4424 oscilloscopes, which collected 200,000 samples over a 2-second period. A pair of ionization probe traces from a sulfur flame is shown in Figure 3. The time when the first voltage drop reached a minimum for each trace was taken as the time when the flame front reached the probe. Laser shadowgraph confirmed this arrival qualitatively, as shown in Figure 2. The red curve indicates the location of the flame front. This feature was observed to reach the ionization probe at the same time (i.e., within the temporal resolution of the camera) the voltage trace spiked from the ionization probe (Figure 3).

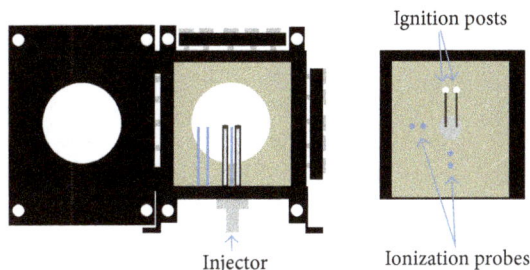

FIGURE 1: Schematic of the flame speed experimental setup.

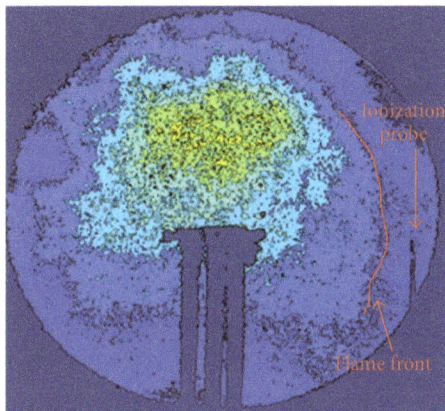

FIGURE 2: Qualitative shadowgraph measurement used to verify the time-of-arrival of the flame to the ionization probe. Image was taken prior to the flame reaching the probe.

The shadowgraph measurement also provided qualitative information on the flame shape. Although portions of the front are smooth, the flame front is not symmetric. The cause of this appearance is likely a combination of remaining turbulence and natural convection.

The $\Delta t$ shown in Figure 3 was used to calculate the flame speed. An uncertainty of 20% of the nominal value was associated with each measurement which is mostly due to the time-of-arrival measurement. The density corrections were calculated using Cantera [11] and the SOx mechanism from the University of Leeds [12].

Temperature measurements were made using thermocouples and pyrometry. Thermocouples (50 $\mu$m R-type, Omega part P13R-002) were covered in a thin coating of an aluminum oxide spray paint (ZYP Coating, A aerosol) to minimize catalytic effects. Thermocouples were attached to the connectors within the dust explosion chamber in the same manner as the previously described ionization probes. The thermocouple extension wire was connected to an Omega data acquasition (DAQ) system (Omega part OMB-DAQ-3005). Temperature was sampled every 100 $\mu$s for the first 2 seconds of each test. The DAQ system was triggered by a TTL pulse generated by the delay generator. The measurements were corrected for radiative and conductive losses.

Pyrometry measures the temperature of the condensed phases by comparing the thermal background to Planck's equation (with the emissivity included). A three-color

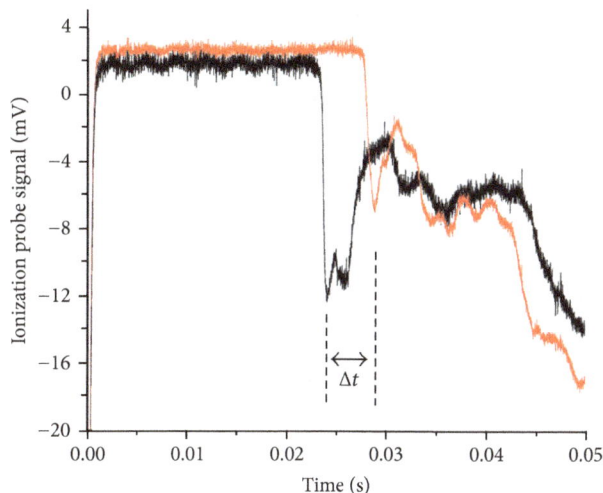

FIGURE 3: Representative voltage traces from the ionization probes from a sulfur flame.

pyrometer was used to obtain time-resolved temperature information by monitoring the emission at 700, 825, and 900 nm. Hamamatsu R928 photomultiplier tubes (PMT) were used for the 700 and 825 nm channels, while an R636-10 PMT was used for the 900 nm channel. Light was collected into a trifurcated fiber optic cable where each of the three branches went to a different PMT. A Stanford Research System (SRS) 300 MHz quad preamplifier (SRS model SR445) conditioned each signal before being recorded by the Picoscope.

A fiber optic-coupled Ocean Optics Jaz spectrometer was also used for pyrometry measurements. This spectrometer records spectra from 200 to 870 nm. However, due to spectral features from SO, $SO_2$, and $S_2$ from the ultraviolet into the visible region of the spectrum, only the thermal emission in the range of 600 to 850 nm was used to determine the condensed phase temperature. The intensity calibration for both devices was conducted with a tungsten lamp (Ocean Optics LS Cal 1) with a known spectral intensity for the spectral regions studied.

## 3. Results and Discussion

Measurements of pressure rise, temperature, and flame speed were made for 6 different conditions. Two different concentrations of the sulfur/anticaking agent mixture were used (280 and 560 g/m$^3$) and three different concentrations of oxygen (10, 21, and 42% by volume). The remaining balance of the gas was nitrogen. The stoichiometric conditions were 280 g/m$^3$ in 21% oxygen and 560 g/m$^3$ in 42% oxygen. The choice of these conditions was based upon consideration of the application (i.e., burning in air) and being able to isolate the effects of particle concentration and stoichiometry. Difficulties establishing flames at lower particle concentrations prevented the use of those concentrations. The tendency of sulfur to agglomerate, especially at the higher sulfur concentrations, yielded an upper limit on the particle concentrations used.

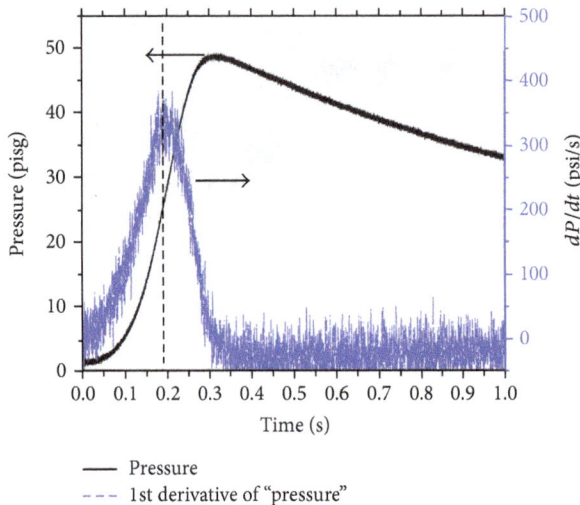

FIGURE 4: Sample pressure-time data from a sulfur explosion.

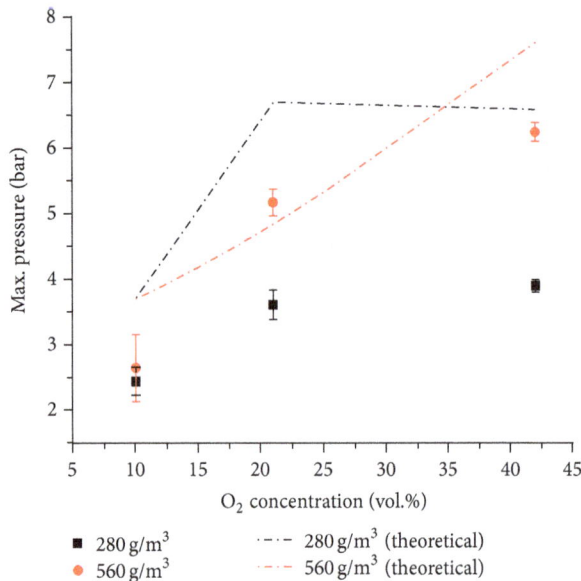

FIGURE 5: Maximum pressure rise (absolute pressure) for sulfur explosions within a 31 L chamber.

The problem of agglomeration is addressed in greater detail in [9].

### 3.1. Pressure Data.

A representative pressure-time curve and its first temporal derivative are shown in Figure 4. The maximum pressure rises are shown in Figure 5 with the theoretical maximum pressure rise and were determined from NASA's CEA program [13] under constant volume. It was observed that the pressure rise increased with particle concentration and increasing oxygen concentration. The data is consistent with work by Cashdollar [14] after scaling their pressure data to account for the different chamber volumes between the facilities used in each respective study.

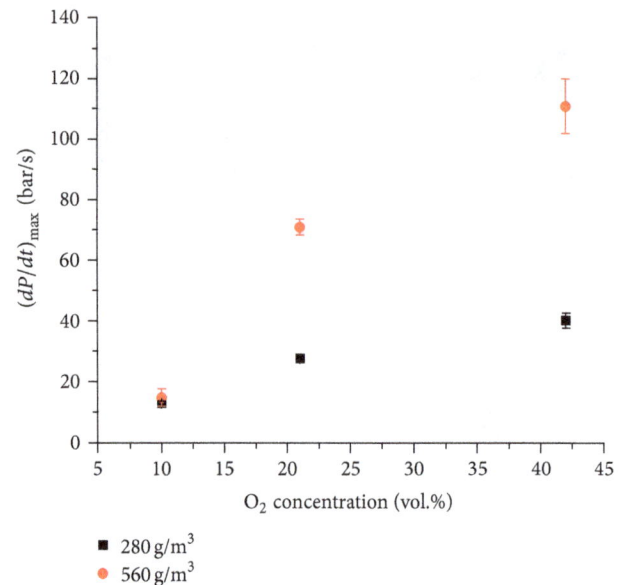

FIGURE 6: Maximum rate of pressure rise (absolute pressure) within the 31 L chamber from sulfur explosions.

The 560 g/m$^3$ condition within a 21% oxygen environment resulted in a pressure rise greater than the maximum pressure. This result may be due to particle settling from increased agglomerations during the experiment. An increase in agglomerations at the higher particle loadings were reported in [9]. If additional particles fall out of the suspension, it will bring the equivalence ratio closer to stoichiometric for these conditions. The maximum pressure will increase and approach the value seen for the lower particle loading. A similar effect of particle settling should then be expected for the 10% and 42% oxygen conditions. The difference in the maximum pressure rise for 10% O$_2$ is negligible because the maximum theoretical pressure rises are almost identical. Figure 5 shows that the maximum pressure (dashed lines) in the oxygen enriched case would be lower if the particle settling increased. This decrease is consistent with what was observed experimentally. For all of these conditions, it is challenging to quantify how much the pressure should change theoretically because the actual mass of settled particles is unknown.

The maximum rate of pressure rise, Figure 6, was seen to increase monotonically with oxygen concentration. For oxygen concentrations above 10% by volume, the rate of pressure rise is greater for higher particle concentration. The rate of pressure rise indicates how quickly the heat is being release. The importance of this quantity will be seen in (2), where the rate of maximum pressure rise is directly proportional to the flame speed.

### 3.2. Validity of Pressure-Time Data for Flame Speed Measurements.

Pressure-time data may also be used to determine flame speed. Dahoe and de Goey [15] analyzed constant volume explosions (not necessarily dust explosions) from a thermodynamic standpoint to relate pressure-time data to laminar flame speed. This equation is shown as

$$\frac{dP}{dt} = \frac{3}{R}\left(\frac{dx}{dP}\right)^{-1}\left[1-\left(\frac{P_i}{P}\right)^{1/\gamma}(1-x)\right]^{2/3}\left(\frac{P}{P_i}\right)^{1/\gamma}S_L, \quad (2)$$

where

$$x(P) = \frac{P - P_i}{P_e - P_i} \quad (3)$$

and $P$ is the instantaneous pressure, $P_i$ and $P_e$ are the initial and maximum pressures, respectively, and $R$ is the spherical equivalent radius taken to be the radius of sphere with the same volume as the chamber used in the current study. The ratio of the specific heats, $\gamma$, was assumed to be constant (and approximately equal to 1.4 here because of the use of air), and $S_L$ is the laminar flame speed. The analysis Dahoe and de Goey [15] conducted on this method included the additional assumption of a linear dependence on pressure for the mass burnt fraction, $x(P)$ (3). The work by Luijten et al. [16] used a multizone approach to develop a more rigorous definition for $x$, although it is not shown here because of its length.

The derivation of (2) was based on ideal assumptions that in practice may not be true for dust explosions. They are as follows.

(1) The chamber is well-insulated and assumed to be filled with reactants that are perfectly mixed and stagnant.

(2) The mixture is ignited from the center and the flame front produced is spherical, infinitely thin, and not wrinkled.

(3) The flame front breaks the chamber into two regions, the burnt and unburnt gases, where the mixtures within each zone are uniform (e.g., composition and temperature).

(4) Since the chamber has a fixed volume and heat cannot escape due to the well-insulated walls, the pressure rises from the heat addition. Moreover, the hot combustion products within the spherical flame will expand, thus compressing the unburnt gas isentropically.

Two concentrations of the sulfur mixture were tested to determine if the use of pressure-time data to measure flame speed is appropriate for dust explosions. A near stoichiometric concentration of $264 \pm 34\,g/m^3$ and a fuel-rich condition of $498 \pm 53\,g/m^3$ were ignited in air by a pyrotechnic igniter. Energy release by the igniter did not have a significant influence on the measurement [9]. The uncertainty in the concentration is due to the distribution of the measured powder mass injected.

The two combs of ionization probes discussed above were used to directly measure the flame propagation speed and compared to those calculated from (2) and (3). The calculated flame speed based upon the pressure-time data shown in Figure 4 is displayed in Figure 7.

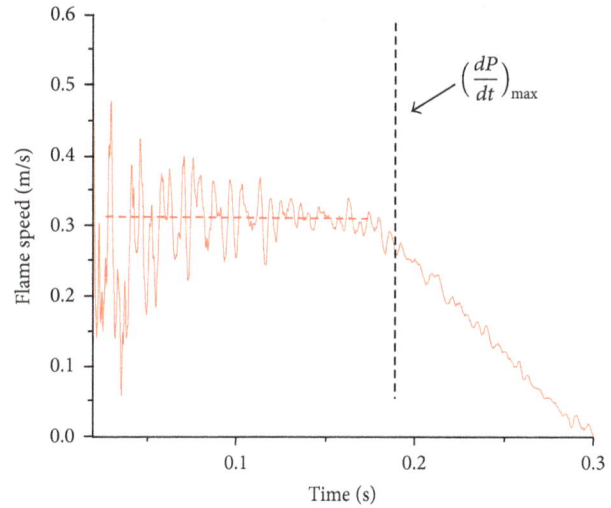

FIGURE 7: Calculate flame speed from pressure-time data.

The calculated laminar flame speed initially oscillates significantly due to a low signal-to-noise ratio from the pressure transducer signal. These oscillations dampen as the pressure rises. Despite the oscillatory nature, each calculated curve was observed to oscillate around a constant value until the time where $dP/dt$ is a maximum. The time of the maximum rate of pressure rise is indicated by the vertical lines in Figures 4 and 7. The laminar flame speed from the pressure data was taken as the average speed from 20 ms after ignition to the time where $(dP/dt)_{max}$ was reached. The time of 20 ms was used to limit the influence of the low signal-to-noise ratio of the pressure data very close to the instant of ignition. The decrease in laminar flame speed after $(dP/dt)_{max}$ was also observed by Santhanam et al. [17]. The reason for the decrease in flame speed is very likely due to heat losses. As the flame approaches the walls of the vessel, additional heat is lost to those walls which are near room temperature. The rate of pressure rise, which is related to the rate of heat release, decreases because some of that energy is absorbed by the chamber walls. The rate of pressure rise is proportional to flame speed (see (2)) so that a decrease in flame speed is observed. An analysis on the primary mode of heat transfer from the flame is discussed in the following section.

*3.2.1. Heat Loss Analysis.* The decrease in flame speed after $(dP/dt)_{max}$ was believed to be due to heat losses. With the increased amount of thermal radiation from dust flames, it is necessary to determine the importance of the various modes of heat transfer, specifically conduction and radiation.

The amount of energy lost to the chamber walls is difficult to quantify because of the complexity of the problem. However, the relative importance of conduction versus radiative losses may be analyzed qualitatively. The ratio of conductive to radiative losses, (4), was approximated for different flame temperatures and location (i.e., distance between the flame and the wall) where $k$ is the thermal conductivity, $\epsilon$ is the emissivity, $\sigma$ is the Stefan-Boltzmann constant, $\Delta x$ is

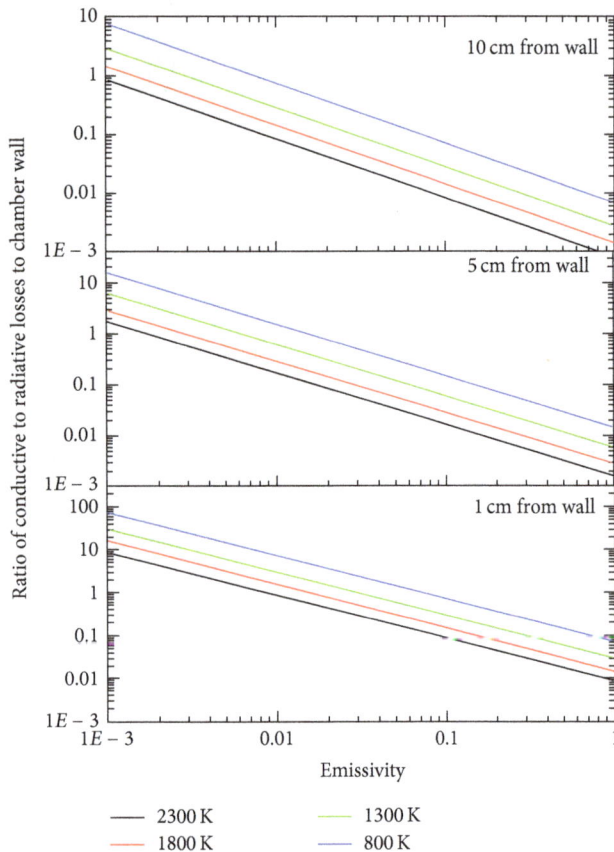

FIGURE 8: Calculated (first-order approximation) ratio of the conductive and radiative heat losses at multiple positions, at flame temperatures, and for $F = 1$.

expected since thermal radiation is dependent on those two parameters. For most of the conditions in this calculation set, radiation is either dominant or comparable to conduction, as illustrated in Figure 8. However, when the actual surface area of the particles compared to the flame front is considered, the value of $F$ decreases from unity to approximately 0.35 and 0.038 when the flame is 1 and 10 mm thick, respectively. This factor will drive the ratio from (4) towards conduction. Therefore, if the flame thickness is in fact on the order of several millimeters or larger, it is unlikely that observed decrease in calculated flame speed is *dominated* by radiative losses unless the emissivity, temperature, and flame diameter (i.e., close to the wall) are large. Although further information is needed to determine the flame thickness accurately, this finding is still significant. Heat losses from sulfur dust flames may not be dominated by radiation which is contrary to what has been concluded about other dust flames (e.g., Al) where the product is also solid [17].

*3.2.2. Discussion.* The measured flame speeds from all of the test for both measurement techniques are shown in Figure 9. A large amount of data scatter is seen for both concentrations. It is believed that the range of flame speeds observed is at least in part due to the turbulence that remained in the system after the injection process. The flame speeds measured in each direction for a single experiment were observed to vary more than 50% in some cases. Moreover, there were instances where the signals from two probes in a single pair would indicate the flame had arrived at the same time. This observation is likely due to the flame approaching the ionization probe pair from a side-on approach rather than head-on as would be expected for a spherical flame. This point is further supported by the asymmetry observed by the highly irregular front seen by shadowgraph measurements [9].

The calculated laminar flame speeds from the pressure data and the ionization probes are plotted in Figure 9. Reasonable agreement is seen between the two measurement techniques. The flame speeds reported by Proust [8] were approximately 20% lower than the stoichiometric flame speed measured in the current work. This result suggests that it is *plausible* to use pressure-time data to estimate laminar flame speed. The higher flame speeds measured here are likely in part due to increased turbulence here, although it is difficult to prove without any doubt, because the level of turbulence was not measured in either study. Natural convection potentially also played a role. The burning sulfur particles produce hot gases, which of course will rise. The upward draft will then distort the flame front potentially increasing the surface area. Evidence of the effect of natural convection can be observed in Figure 2 where the flame has clearly propagated upwards more than downwards.

Therefore, it is somewhat inappropriate to use this technique to determine the *laminar* flame speed because of how both turbulence and natural convection effect the flame shape. It better represents the flame speed of an equivalent spherical flame with the same mass consumption rate. With that being said, the pressure-time data does provide a measure of a flame speed based upon the degree

the distance between the flame and the wall, and $T$ is the temperature at the location indicated by its subscript. The variable $F$ is the ratio of the surface area of the particles within the flame and flame front surface area. Since the emissivity of the powder is unknown, the ratio of the heat losses was calculated as a function of emissivity. Figure 8 shows the results from these calculations. Consider

$$\frac{Q_{cond}}{Q_{rad}} \approx \frac{k}{F\epsilon\sigma} \frac{(T_{flame} - T_{wall})}{\Delta L} \frac{1}{T_{flame}^4 - T_{wall}^4}. \quad (4)$$

The calculated ratios displayed in Figure 8 assumed the ratio of the particle surface area to that of the flame was unity. This assumption is not realistic since the particles only constitute a fraction of the flame front for a given flame thickness. The value of $F$ can be determined using geometric considerations, an average particle diameter, and a flame thickness. Santhanam et al. [17] calculated the flame thickness of Al dust flames by multiplying the flame speed by the burntime of an individual Al particle. This same estimation is more challenging for sulfur dust since the burn time of an individual particle is unknown. The effect of flame thickness will be discussed.

High temperatures and large values of the emissivity favor radiative dominated heat losses when $F = 1$, which is to be

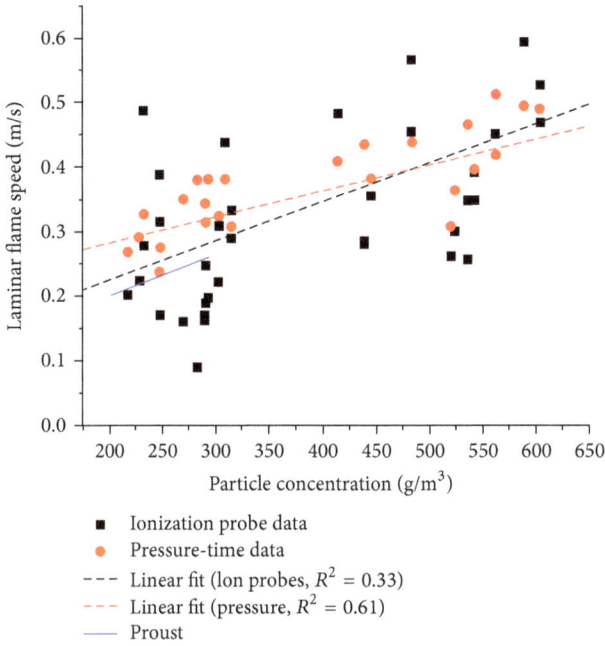

FIGURE 9: Comparison of the flame speed determined by the ionization probes, pressure-time data, and work from Proust [8].

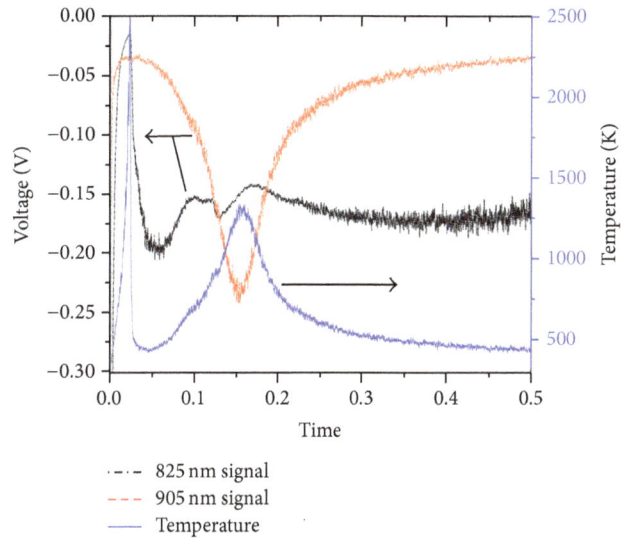

FIGURE 10: Representative filtered signals and calculated temperature for the PMT pyrometer.

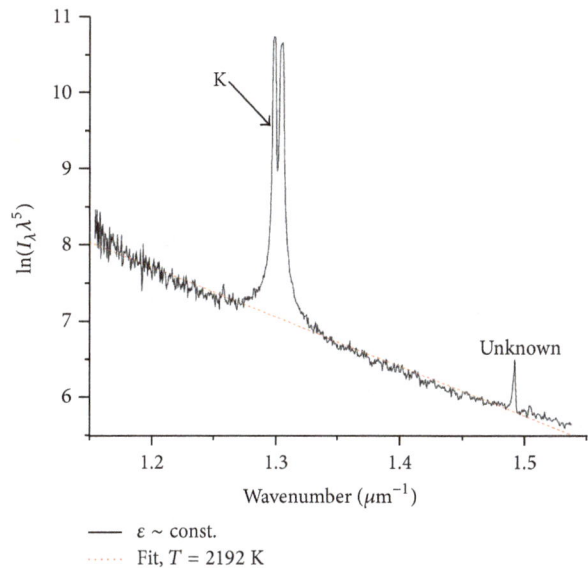

FIGURE 11: Fit of the thermal background from the Jaz spectrometer to determine temperature.

of turbulence. This result should still allow comparisons of flames in different conditions to be compared as long as the degree of turbulence is kept constant (i.e., injection parameters, ignition delay, and energy).

### 3.3. Flame Temperature.

The temperature of the sulfur flames were measured by thermocouple and pyrometry. Both measurements were used to determine the peak temperature. The thermocouple provided the peak temperature locally, while pyrometry measurements indicated the peak temperature within the field of view of the pyrometer. The spatially integrated pyrometry signal is always biased towards the hottest regions due to the strong temperature dependence on the intensity of thermal radiation. Pyrometry data was collected with the 3-color PMT pyrometer and the Jaz Ocean Optics spectrometer. It should be noted that only the 825 nm and 905 nm signals were used because the 700 nm channel did not provide a sufficient signal level. The integration time for all data collected by the spectrometer was 10 ms.

Representative traces of the time-resolved data provided by the PMT pyrometer are displayed in Figure 10 with the calculated temperature using the gray body approximation. The early peak seen in Figure 10 was due to the emission from the pyrotechnic igniter. It was observed that the maximum temperature from the sulfur explosion occurred near the time that the rate of pressure rise was maximum. The peak temperature was recorded for each experiment.

Figure 11 shows a spectrum taken from the Jaz spectrometer. A gray body was believed to be the most appropriate assumption for the spectral emissivity because the $\lambda^{-1}$ or $\lambda^{-2}$ may not be appropriate for all materials [18] and the large optical depths produced by the dust cloud [19]. The measured

peak temperatures from pyrometry and the thermocouples are shown in Figure 12. Only thermocouple data was obtained from the 10% oxygen tests, because the pyrometry signals were very weak and much lower than the noise level. The PMT pyrometer was used only for the 21% because of a spectral interference near 900 nm when the oxygen level was increased to 42% by volume. The thermal background from the Jaz spectrometer was fit for each of the tests for the oxygen-enriched conditions. As such, no time-resolved data were obtained for these conditions.

The dashed lines represent the maximum flame temperatures calculated under equilibrium conditions. The temperature was adjusted by considering the heat absorbed

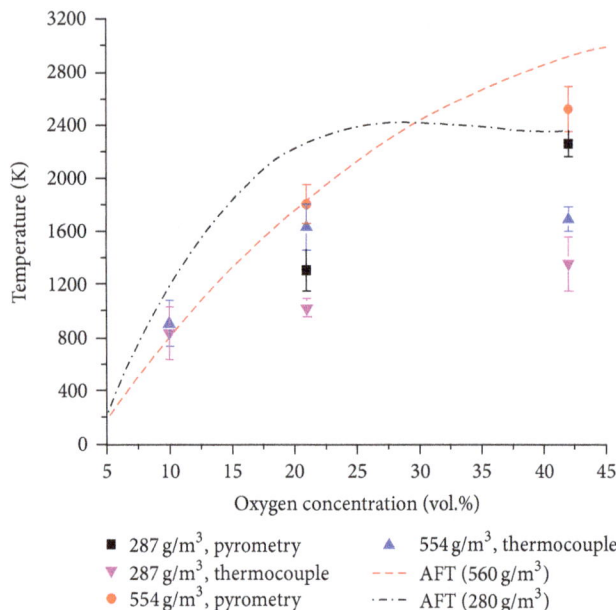

FIGURE 12: Temperature measurements of sulfur dust explosions.

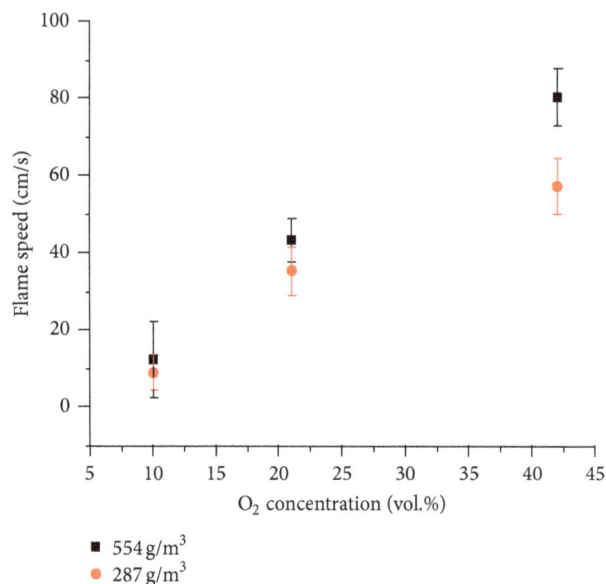

FIGURE 13: Flame speed measurements by analysis of pressure-time data for sulfur dust explosions.

by the anticaking agent. This change in temperature was minimal (typically less than 10 K). It should be noted that this adjustment is a first-order approximation since it did not include the equilibrium composition to be recalculated at the final temperature.

It was observed that the peak temperature steadily increased for the lower particle concentration from approximately 800 K to about 1300 K for the thermocouple measurements (corrected for radiative losses) as the oxygen content was increased from 10% to 42%. The increase in temperature from 21% to 42% oxygen from the pyrometry measurements was much greater as the maximum temperature exceeded 2000 K. This large difference is likely due to the fact that pyrometry measurements are biased towards the highest temperatures within the field-of-view while the thermocouples measured the local temperature.

A large temperature difference between the pyrometry and thermocouple measurements was also seen for the 21% and 42% oxygen tests at the higher particle loading. The temperature was observed to increase with particle loading for all oxygen concentrations. The average peak temperature for the 10% oxygen case was the lowest of the conditions for the 560 g/m$^3$ tests at 909 K.

The increase in temperature as oxygen concentration increased was likely due to the amount of energy liberated in each condition. The pressure data indicated that the maximum pressure rise (i.e., heat release), Figure 5, increased with oxygen concentation.

*3.4. Flame Speed.* Since the pressure-time data was determined to be an accurate way of measuring flame speed within the constant volume explosion chamber, it was decided to use that diagnostic to determine how particle and oxygen concentrations affected the flame speed. The same approach

that was used to determine flame speed, as discussed above, was applied to the other conditions. The data for all of the particle and oxygen concentrations are summarized in Figure 13.

The individual data points for the 21% oxygen case were displayed in the previous section (see Figure 9). The averages at that oxygen concentration also includes additional data points taken after the completion of that portion of the work. A monotonic increase in flame speed was observed with an increase in oxygen concentration. A similar trend is shown in Figure 6 for the $(dP/dt)_{max}$ data. Since the flame speed was calculated from the pressure data (and $S_L$ scales with $dP/dt$), it is expected to see a similar dependence. For a given oxygen concentration, the higher particle loading condition (i.e., higher equivalence ratio) has a greater flame speed.

## 4. Combustion Mechanism Discussion

Gas-phase emission from sulfur oxides and $S_2$, as reported in [9], suggests that sulfur has a gas-phase component, but the extent of the gas-phase chemistry is unknown. However, the data do not explicitly suggest anything about the limiting process(es) during combustion, namely, if sulfur burns in a kinetically or diffusion limited manner.

*4.1. Damköhler Number Analysis.* Calculation of the key Damköhler number will aid in this discussion. This Damköhler number is the ratio of the chemical and diffusion time scales. A chemical time scale was determined by simulating an adiabatic, constant volume, perfectly-stirred reactor (PSR) with the $SO_x$ mechanism from University of Leeds [12]. The ratios of sulfur, oxygen, and nitrogen that were present in the chamber after injection under atmospheric pressure were used as the initial conditions.

TABLE 1: Summary of the normalized Damköhler numbers.

|  | 10% $O_2$ | 21% $O_2$ | 42% $O_2$ |
|---|---|---|---|
| 280 g/m$^3$ | 0.12 | 1.00 | 0.48 |
| 560 g/m$^3$ | 0.80 | 1.00 | 7.00 |

TABLE 2: Flame speed scaling in the diffusion limit.

|  | $V_{10\%}/V_{21\%}$ | $V_{21\%}/V_{21\%}$ | $V_{42\%}/V_{21\%}$ |
|---|---|---|---|
| 280 g/m$^3$ | 0.64 | 1.00 | 1.73 |
| 560 g/m$^3$ | 0.52 | 1.00 | 1.43 |

TABLE 3: Flame speed scaling in the kinetic limit.

|  | $V_{10\%}/V_{21\%}$ | $V_{21\%}/V_{21\%}$ | $V_{42\%}/V_{21\%}$ |
|---|---|---|---|
| 280 g/m$^3$ | 0.89 | 1.00 | 1.15 |
| 560 g/m$^3$ | 0.85 | 1.00 | 1.09 |

TABLE 4: Measured flame speed ratios.

|  | $V_{10\%}/V_{21\%}$ | $V_{21\%}/V_{21\%}$ | $V_{42\%}/V_{21\%}$ |
|---|---|---|---|
| 280 g/m$^3$ | 0.25 | 1.00 | 1.62 |
| 560 g/m$^3$ | 0.29 | 1.00 | 1.86 |

An initial temperature of 1300 K was needed to start the reaction. The time step was set to 1 microsecond. The chemical time scales were taken as the time it takes for the temperature to go from 10% to 90% of the total temperature change. The temperature change was the difference from 1300 K to the adiabatic flame temperature.

The diffusion time scale is inversely proportional to the diffusion coefficient, $D$, which was calculated from kinetic theory [20]. The binary diffusion coefficient of oxygen through nitrogen was calculated using the measured peak temperatures. The diffusion coefficient scales as $T^{1.65}$ and $MW^{-0.5}$. Because of this scaling, even if the temperature were to increase by 1700 K (bringing the temperature up to about 3100 K, the adiabatic flame temperature, for the hottest measured condition), the diffusion time scale would only decrease by about a factor of about 2.8. Similarly, if the diffusion of two other species besides $O_2$ and $N_2$ were considered (i.e., SO, $SO_2$, and $S_2$), the diffusion time scale would change by less than a factor of 2 (increase or decrease). Therefore, the diffusion coefficient will not change significantly (i.e., an order of magnitude) for mixtures of sulfur compounds, oxygen, and nitrogen. The pressure was taken as 1 atm for this calculation. The reasoning for this assumption was that at the time when the flame front passed the thermocouples and ionization probes, the pressure within the chamber had not significantly risen.

The Damköhler numbers calculated are relative to one another (i.e., not absolute) to eliminate concerns on the appropriate area to use to nondimensionalize the value and since the maximum pressure does not change substantially (i.e., less than an order of magnitude). The Da numbers shown in Table 1 are normalized to the value at 21% $O_2$ at each respective particle concentration so that individual values have no meaning, but the relative change is important.

A significant increase (i.e., by an order of magnitude) in Da is seen as the oxygen concentration increased from 10% to 21% for the lower particle concentration. A similar increase is observed for the 560 g/m$^3$ concentration as the oxygen content rises from 21% to 42%. This increase suggests that as the oxygen concentration increases, diffusion becomes more important which at first glance seems counterintuitive. Considering that the oxygen was not the only quantity that changed, this result is justified. The temperature also increased and that affects the chemistry more significantly than diffusion (i.e., exponentially versus $T^{1.65}$). The exponent on temperature (1.65) includes the temperature dependence from the collision integral [21]. The more intriguing aspect of this observation is that for both of the particle loadings, diffusion becomes more important as the flame transitions from fuel-rich conditions to stoichiometric. Stoichiometry

for the 280 g/m$^3$ and 560 g/m$^3$ concentrations is when there is 21% and 42% oxygen by volume, respectively.

*4.2. Flame Speed Scaling Analysis.* The above discussion only provided insight into how the flames in each condition burned relative to one another. Analysis of the flame speed and temperature can be used to further specify the combustion mechanism and any limiting phenomena.

The scaling of the flame speed should be dependent on the burning mechanism. Landau and Lifshitz [22] state that the flame speed for a thermally driven combustion wave scales as $(\alpha/\tau_{comb})^{0.5}$, where $\tau_{comb}$ is a combustion time scale. Goroshin et al. [23] argued that for a diffusion limited flame, $\tau_{comb}$ scales inversely proportional to the diffusion coefficient of the gas. If the flame is diffusion limited, the flame speed should scale as $(\alpha D)^{0.5}$. Goroshin et al. [23] contended that if this thermally driven flame is kinetically limited, the difference in mass diffusivity of the gas mixtures (i.e., in each condition) should not play a role, thus the flame speed should scale with $(\alpha)^{0.5}$. This logic was used for the current work. The theoretical scaling for the diffusion and kinetically limited flames are shown in Tables 2 and 3, respectively. The velocity ratios from the data of the current work is displayed in Table 4.

It is observed that the ratio of experimentally measured velocities from the 21% and 42% oxygen cases (for both particle concentrations) are much closer to the ratio predicted by the diffusion limited theory. This result, in conjuction with the previous discussion, suggests that oxygen enriched sulfur dust flames burn in the diffusion limit.

Intuitively, it would be expected that an oxygen enriched flame would burn in the kinetic limit since a higher concentration of oxygen is closer to the particle surface and potentially significantly reducing the diffusion time scale. However, the oxygen concentration is only one aspect of these flames. The temperature measurements show that the flame burns hotter as more oxygen was added to the system. The diffusion time scale does decrease with temperature ($T^{1.65}$), but it is not affected as much as the kinetics, which scale exponentially with temperature. The increase in temperature causes the chemistry to occur much faster, resulting in the diffusion process to be the limiting step.

Decreasing the oxygen concentration to 10% does not fit the scaling ratios predicted by kinetically or diffusion limited flame. The experiments with higher oxygen concentrations produced very consistent data whereas the 10% oxygen concentrations had a larger spread. The larger spread is not well represented by the data shown here. On multiple occasions, the 10% oxygen tests did not ignite (for both powder concentrations). It is possible that perhaps a flammability unit was approached by these oxygen depleted conditons.

*4.3. Group Combustion Regime.* Finally one must determine if the dust particles burn independently of each other. So a second condition that must be determined is the droplet spacing. The droplet combustion analysis assumes that there are no interactions with other particles [24]. The spray combustion community uses a group combustion number to estimate if the particles burn individually or together within a larger group flame. A group combustion number is defined by Glassman and Yetter [25] as follows:

$$G = 3\left(1 + 0.276 \mathrm{Re}^{0.5} \mathrm{Sc}^{0.5} \mathrm{Le} N^{2/3}\right) \frac{R}{S}, \qquad (5)$$

where Sc is the Schmidt number (ratio of momentum and mass diffusivities), Le is the Lewis number (ratio of thermal and mass diffusivities), $N$ is the number of particles, $R$ is the particle radius, and $S$ is the average particle spacing [25]. The ratio of $R$ and $S$ is equal to the cube root of the quotient of the particle mass loading density (i.e., $g/m^3$) and the particle density [26] and is on the order of 0.1 for the conditions tested in this work. The number of particles within the chamber for a given test is off the order of $10^8$-$10^9$. Both Sc and Le will be on the order of 0.1 to 1. The Reynolds number is unknown because the velocity was not measured but it is likely orders of magnitude larger than $10^{-8}$-$10^{-9}$. Therefore, using (5), $G$ will be much greater than $10^{-2}$. A group number of less than $10^{-2}$ is specified for individually burning particles to occur [25].

## 5. Conclusions

Constant volume sulfur dust explosions were investigated. Measurements of flame speed using ionization probes showed reasonable agreement to the calculated speed from the pressure time data. Although there was agreement, it is inappropriate to call the quantity laminar flame speed because of the turbulence in the system. Flame speed was observed to range from 9 cm/s with 10% oxygen for both particle concetrations studied to as high as 80 cm/s in 42% oxygen for 554 $g/m^3$ of sulfur. The flame speed also increased with particle concentration which may be attributed to a shorter interparticle distance at higher concentrations. The temperature was measured to vary from approximately 800 K for both particle concentrations to approximately 2200 and 2600 for 287 $g/m^3$ and 554 $g/m^3$, respectively. Flame speed and temperature were not observed to be a function solely on equivalence ratio but rather dependent on the concentrations of sulfur and oxygen. Further analysis concluded that

diffusion became the limiting process as the stoichiometry transitioned from fuel-rich to stoichiometric.

## Conflict of Interests

The authors declare that there is no conflict of interests regarding the publication of this paper.

## Acknowledgments

This work was funded by DTRA Grant HDTRA1-11-1-0014 under project manager Dr. Suhithi Peiris. The authors would like to thank undergraduate students Sasank Vemulapati and Chris Murzyn for their assistance. This work was carried out in part in the Frederick Seitz Materials Research Laboratory Central Facilities at the University of Illinois at Urbana, Champaign.

## References

[1] J. O. Nriagu, Ed., *Sulfur in the Environment, Part I: The Atmospheric Cycle*, John Wiley and Sons, 1978.

[2] B. Setlow, C. A. Loshon, P. C. Genest, A. E. Cowan, C. Setlow, and P. Setlow, "Mechanisms of killing spores of *Bacillus subtilis* by acid, alkali and ethanol," *Journal of Applied Microbiology*, vol. 92, no. 2, pp. 362–375, 2002.

[3] D. Allen, *Optical combustion measurements of novel energetic material in a heterogeneous shock tube [M.S. thesis]*, University of Illinois, Mechanical Science and Engineering, Champaign, Ill, USA, 2012.

[4] S. Zhang, M. Schoenitz, and E. L. Dreizin, "Iodine release, oxidation, and ignition of mechanically alloyed Al-I composites," *The Journal of Physical Chemistry C*, vol. 114, no. 46, pp. 19653–19659, 2010.

[5] B. R. Clark and M. L. Pantoya, "The aluminium and iodine pentoxide reaction for the destruction of spore forming bacteria," *Physical Chemistry Chemical Physics*, vol. 12, no. 39, pp. 12653–12657, 2010.

[6] S. A. Grinshpun, A. Adhikari, M. Yermakov et al., "Inactivation of aerosolized *Bacillus atrophaeus* (BG) endospores and MS2 viruses by combustion of reactive materials," *Environmental Science & Technology*, vol. 46, no. 13, pp. 7334–7341, 2012.

[7] M. Clemenson, *Explosive initiation of various forms of the ti/2b energetic system [M.S. thesis]*, University of Illinois at Urbana-Champaign, Mechanical Science and Engineering, 2012.

[8] C. Proust, "Flame propagation and combustion in some dust-air mixtures," *Journal of Loss Prevention in the Process Industries*, vol. 19, no. 1, pp. 89–100, 2006.

[9] J. Kalman, *Experimental investigation of constant volume sulfur dust explosions [Ph.D. thesis]*, University of Illinois at Urbana-Champaign, 2014.

[10] M. R. S. Nair and M. C. Gupta, "Burning velocity measurement by bomb method," *Combustion and Flame*, vol. 22, no. 2, pp. 219–221, 1974.

[11] D. Goodwin, *Cantera: An Object-Oriented Software Toolkit for Chemical Kinetics, Thermodynamics, and Transport Processes*, Caltech, Pasadena, Calif, USA, 2009.

[12] T. Ziehn and A. S. Tomlin, "A global sensitivity study of sulfur chemistry in a premixed methane flame model using HDMR,"

*International Journal of Chemical Kinetics*, vol. 40, no. 11, pp. 742–753, 2008.

[13] S. Gordon and B. J. McBride, "Computer program for calculation of complex chemical equilibrium compositions and applications," NASA Reference Publication 1311, NASA, Washington, DC, USA, 1996.

[14] K. L. Cashdollar, "Flammability of metals and other elemental dust clouds," *Process Safety Progress*, vol. 13, no. 3, pp. 139–145, 1994.

[15] A. E. Dahoe and L. P. H. de Goey, "On the determination of the laminar burning velocity from closed vessel gas explosions," *Journal of Loss Prevention in the Process Industries*, vol. 16, no. 6, pp. 457–478, 2003.

[16] C. C. M. Luijten, E. Doosje, and L. P. H. de Goey, "Accurate analytical models for fractional pressure rise in constant volume combustion," *International Journal of Thermal Sciences*, vol. 48, no. 6, pp. 1213–1222, 2009.

[17] P. R. Santhanam, V. K. Hoffmann, M. A. Trunov, and E. L. Dreizin, "Characteristics of aluminum combustion obtained from constant-volume explosion experiments," *Combustion Science and Technology*, vol. 182, no. 7, pp. 904–921, 2010.

[18] J. Kalman, N. Glumac, and H. Krier, "High temperature metal oxide spectral emissivities for pyrometry applications," *AIAA Journal of Thermophysics and Heat Transfer*. Submitted.

[19] J. Kalman, D. Allen, N. Glumac, and H. Krier, "Optical depth effects on aluminum oxide spectral emissivity," *Journal of Thermophysics and Heat Transfer*, vol. 29, no. 1, pp. 74–82, 2015.

[20] N. M. Laurendeau, *Statistical Thermodynamics: Fundamentals and Applications*, Cambridge University Press, New York, NY, USA, 2010.

[21] L. Monchick and E. A. Mason, "Transport properties of polar gases," *The Journal of Chemical Physics*, vol. 35, no. 5, pp. 1676–1697, 1961.

[22] L. Landau and E. Lifshitz, *Fluid Mechanics*, Pergamon Press, 1959.

[23] S. Goroshin, F.-D. Tang, A. J. Higgins, and J. H. Lee, "Laminar dust flames in a reduced-gravity environment," *Acta Astronautica*, vol. 68, no. 7, pp. 656–666, 2011.

[24] S. R. Turns, *An Introduction to Combustion: Concepts and Applications*, McGraw-Hill Series in Mechanical Engineering, McGraw-Hill Education, 2000, http://books.google.com.eg/books?id=sqVIPgAACAAJ&redir_esc=y.

[25] I. Glassman and R. Yetter, *Combustion*, Academic Press, 2008.

[26] R. K. Eckhoff, "Chapter 4—propagation of flames in dust clouds," in *Dust Explosions in the Process Industries*, pp. 251–384, Gulf Professional Publishing, Burlington, Vt, USA, 3rd edition, 2003.

# 6

# N$_2$O and NO Emissions from CFBC Cofiring Dried Sewage Sludge, Wet Sewage Sludge with Coal and PE

**Zhiwei Li and Hongzhou He**

*Cleaning Combustion and Energy Utilization Research Center of Fujian Province, Jimei University, 9 Shigu Road, Xiamen 361021, China*

Correspondence should be addressed to Hongzhou He; hhe99@126.com

Academic Editor: Constantine D. Rakopoulos

Experiments on cofiring dried sewage sludge, wet sewage sludge with coal and polyethylene (PE) were carried out on a pilot scale 0.15MWt circulating fluidized bed combustion (CFBC) plant, and the influence of furnace temperatures, cofiring rates on N$_2$O and NO emissions was investigated. Temperature is an effective parameter influencing N$_2$O emission, and higher temperature leads to significant N$_2$O reduction and decrease of conversion ratio of fuel-N to N$_2$O. Increasing in cofiring rates leads to higher nitrogen content in the mixed fuel, which could result in higher NO and N$_2$O emissions from combustion. With more sewage sludge addition, higher NO but lower N$_2$O emissions are observed. N$_2$O emission from cofiring wet sewage sludge with coal is higher than that from cofiring dried sewage sludge with coal and PE, and fuel-N conversion ratio to N$_2$O and NO is much higher in cofiring wet sewage sludge with coal than that in cofiring dried sewage sludge with coal and PE.

## 1. Introduction

Vast quantities of sewage sludge are produced as by-product of wastewater treatment in recent years, and the production is expected to rise significantly [1]. Meanwhile, huge quantities of plastic waste are produced due to the ever-increased consumption of plastic all over the world. So much sewage sludge and plastic waste disposal have posed a very serious environmental challenge, and much attention has been paid on this problem.

Volatile contents of sewage sludge as dry ash-free basis are generally higher than 80% [2–4], with heating value similar to that of brown coal, which indicates that sewage sludge could be incinerated with the advantage of thermal recycling and volume reduction [1, 4, 5]. But the low heating value of the sewage sludge is usually very low, because of the high water content in it. Therefore, supplementary fuel is required to ensure stable combustion and burnout. Coal and plastic waste have much higher heating value compared with sewage sludge and is suitable for cofiring with sewage sludge to provide supplementary energy [4, 6]. Circulating fluidized bed combustion (CFBC), as an established technique, is feasible

to deal with these waste materials, with the advantages of burning a wide variety of solid fuels and low emissions [7].

However, CFBC emits much N$_2$O because of its low combustion temperature [8]. In addition, nitrogen content in the sewage sludge is generally in the range of 4–8%, much higher than that in the coal [9, 10], and fuel-N is the main source of NO and N$_2$O emissions at low combustion temperatures [11]. As a result, there is a high potential for N$_2$O and NO emissions with sludge combustion in CFBC. It is reported [4] that N$_2$O emissions were in the range of 300–700 mg/m$^3$ (std., dry basis) with CFBC cofiring wet sewage sludge with coal. Many intensive researches have been carried out on N$_2$O and NO emission characteristics from sludge combustion, and detailed information can be found in the work of Werther and Ogada work [1]. But the work undertaken was mainly based on the cofiring of coal with sludge; the knowledge about cofiring sludge with plastic waste is limited. In this work, cofiring of dried sewage sludge and wet sewage sludge with PE on CFBC was conducted. To compare N$_2$O and NO emissions from cofiring sewage sludge with PE, cofiring sewage sludge with coal was also studied.

## 2. Experiments

*2.1. CFBC Pilot Plant.* The investigations were conducted in a pilot-scale CFBC plant. The experimental setup is shown in Figure 1. It includes combustion system, pressure and temperature measuring system, flue-gas sampling, and analysis system.

The combustion system was composed of a furnace with an inside diameter of 300 mm and height of 6000 mm (from the air distributor to the top), a cyclone, a vertical leg, and a U-valve. The refractory-lined bottom of the furnace was 800 mm in height. The other parts of the furnace, the cyclone, the vertical leg, and the U-valve were all made of high temperature alloy covered with heat insulation material on the outside. Forty kilograms of quartz sand with the size of 0.5–1.0 mm was fed into the furnace as bed material during startup. The thermal input to the furnace with sludge and PE or coal was about $0.15 MW_{th}$, corresponding to a fuel-feeding rate of 20–80 kg/h, depending on different cofiring rates. The fuel was fed into the furnace through screw feeder, 890 mm above the air distributor. The primary air was fed into the furnace through the air distributor in the range of 100–250 $m^3$/h, and no secondary air was supplied. Compressed air at atmosphere temperature was used as fluidizing air of the U-valve, in the range of 8–15 $m^3$/h.

During cofiring dried sewage sludge with coal and PE, the vertical temperature profiles in the furnace were adjusted through regulating the depth of a water tube inserted into the furnace from the top of the furnace. The water tube was 4000 mm in length and 51 mm in outside diameter. But during cofiring wet sewage sludge with coal, the water tube in the furnace was removed to reduce heat loss. During cofiring dried sewage sludge with coal and PE, cofiring wet sewage sludge with coal, the gas concentration of $H_2O$, $SO_2$, CO, $CO_2$, NO, and $N_2O$ was continuously analyzed with an on-line gas analyzer (GasMet DX-3000), and the gas-sampling nozzle was at the exit of the cyclone. Oxygen concentration in flue gas was analyzed with a Zr-type probe. Main operation parameters were recorded in a computer at the interval of 15 seconds. Gas analysis data were stored at an interval of 60 seconds.

*2.2. Characteristics of the Sewage Sludge, Coal, and PE.* Polyethylene (PE) powder, with average diameter between 2 mm and 3 mm and low heat value of 46.30 MJ/kg, was used to simulate plastic waste. The analysis data of PE, dried sewage sludge, wet sewage sludge, and coal is presented in Table 1. The properties of wet sewage sludge in Table 1 are the average of 4 samples of wet sewage sludge, as in Table 2. The water content in dried sewage sludge (as received basis) is as low as 7.4%, and low heating value is about 12.8 MJ/kg. The water content in wet sewage sludge is about 76.3% (as received basis), and the low heating value is only about 0.64 MJ/kg. The dried sewage sludge is in the form of powder with sizes between 0.1 mm and 1.7 mm. The ash compositions of dried sewage sludge and coal are shown in Table 3. Attention should be paid to the fact that CaO content in the dried sewage sludge is as high as 10%. In some tests, wet sewage sludge

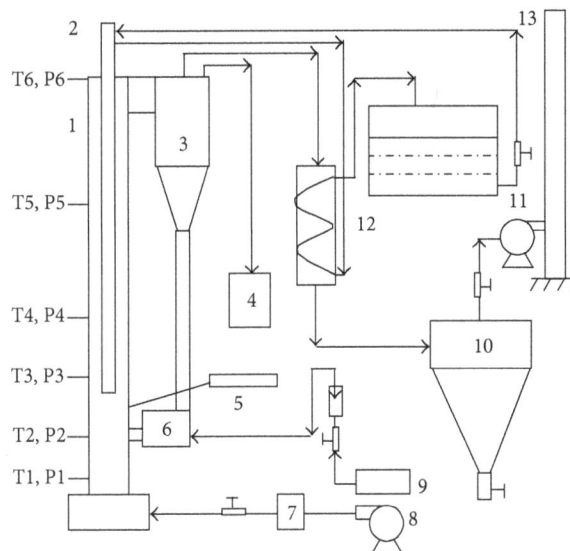

FIGURE 1: Test facilities of the CFBC pilot: 1, furnace; 2, water tube; 3, cyclone; 4, gas analyzer; 5, fuel screw feeder; 6, U-valve; 7, primary air meter; 8, forced draft fan; 9, air compressor; 10, bag filter; 11, induced draft fan; 12, economizer; 13, stack; T1–T6, thermocouples; P1–P6, pressure taps.

was used, and in other tests, dried sewage sludge with water injection through a nozzle at the point 150 mm above the fuel-feed point during the experiments. Dried sewage sludge was mixed with PE or coal before its addition into the furnace, and the wet sewage sludge was added to the furnace at the top of the furnace, through a pump.

*2.3. Experiment Conditions.* In the cofiring of dried sewage sludge with coal and PE, the furnace temperatures were in the range of 964–1184 K, and the cofiring rates were in the range of 50%–100%. The water in the simulated wet sewage sludge was 30%, by injection of water. The water content is defined as the total water of the added and that in the sludge divided by the mass of sludge and water. Cofiring rate is defined as the heat input with the sewage sludge divided by the total heat input with sewage sludge and PE or coal. In the cofiring of wet sewage with coal, the furnace temperature was 1187 K. Because the low heating value of wet sewage sludge is too low, cofiring rate is not used in cofiring wet sewage sludge with coal. In the cofiring of wet sewage sludge and coal, the water content in the sludge is 82% (through addition of water to the original wet sewage sludge), the sludge flow rate is 47.2 kg/h, and coal flow rate is 23.1 kg/h. All gas emissions presented in this paper are normalized to dry gas with an oxygen concentration of 6%, at 273 K and 101.3 kPa. In this paper, the oxygen concentration in the flue gas was controlled in the range of 3.3% to 3.6% in all tests.

## 3. Results and Discussion

*3.1. Influence of Furnace Temperatures on $N_2O$ and NO Emissions.* The furnace temperature is an effective parameter influencing $N_2O$ emission. Raising the furnace temperature

TABLE 1: Properties of sewage sludge and coal.

| | Coal | Dried sewage sludge | Wet sewage sludge | PE |
|---|---|---|---|---|
| Ultimate analysis (wt%, dry ash free basis) | | | | |
| Carbon | 83.21 | 54.96 | 53.44 | 85.71 |
| Hydrogen | 4.55 | 6.84 | 7.75 | 14.29 |
| Oxygen | 9.48 | 26.81 | 29.63 | |
| Sulfur | 1.52 | 3.24 | 3.01 | |
| Nitrogen | 0.84 | 8.15 | 6.17 | |
| Chlorine | 0.04 | 0.48 | 0.48 | |
| Ca/(S + 0.5Cl) molar ratio | 0.19 | 1.01 | 1.02 | |
| LHV (MJ/kg) | 25.87 | 12.81 | 0.64 | 46.34 |
| Proximate analysis (wt%, as received basis) | | | | |
| Moisture | 4.54 | 7.40 | 76.28 | — |
| Ash | 15.10 | 35.16 | 11.70 | — |
| Volatile | 24.60 | 47.40 | 11.67 | — |
| Fixed carbon | 55.76 | 10.04 | 0.36 | — |

TABLE 2: Properties of wet sewage sludge.

| Sample | 1 | 2 | 3 | 4 | Average |
|---|---|---|---|---|---|
| Ultimate analysis (wt%, dry ash free basis) | | | | | |
| Carbon | 58.9 | 50.4 | 52 | 52.4 | 53.4 |
| Hydrogen | 8.47 | 7.23 | 7.72 | 7.56 | 7.55 |
| Oxygen | 21.6 | 32.8 | 31.9 | 32.2 | 29.6 |
| Sulfur | 3.42 | 3.67 | 2.61 | 2.35 | 3.01 |
| Nitrogen | 7.53 | 5.89 | 5.74 | 5.52 | 6.17 |
| Proximate analysis (wt%, as received basis) | | | | | |
| Moisture | 72.9 | 78.8 | 77.4 | 76 | 76.3 |
| Ash | 15 | 10.1 | 11 | 10.7 | 11.7 |
| Volatile | 11.7 | 10.7 | 11.2 | 13 | 11.7 |
| Fixed carbon | 0.41 | 0.38 | 0.37 | 0.26 | 0.36 |
| LHV (MJ/kg) | 0.96 | 0.26 | 0.48 | 0.85 | 0.64 |

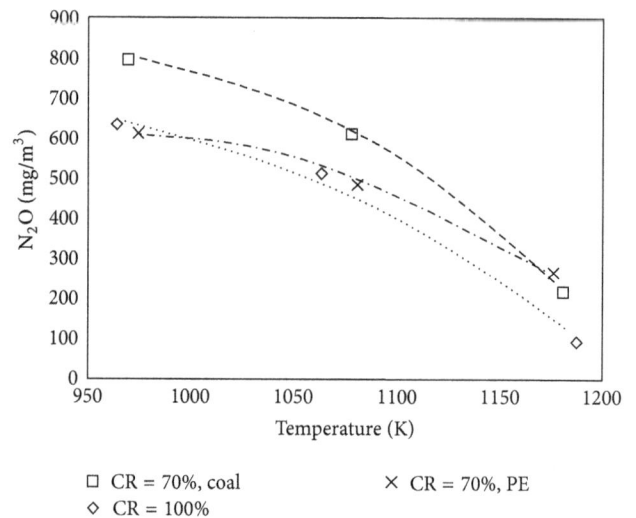

□ CR = 70%, coal    × CR = 70%, PE
◇ CR = 100%

FIGURE 2: Influence of furnace temperature on $N_2O$ emission.

TABLE 3: Composition of the ash of coal and sewage sludge.

| Items | Coal (wt%) | Sewage sludge (wt%) |
|---|---|---|
| $SiO_2$ | 58.72 | 36.33 |
| $Al_2O_3$ | 21.26 | 21.08 |
| $Fe_2O_3$ | 11.22 | 9.17 |
| CaO | 2.7 | 10 |
| MgO | 0.54 | 3.41 |
| $SO_3$ | 1.25 | 0.87 |
| $TiO_2$ | 0.84 | 3.49 |
| $K_2O$ | 2.05 | 6.85 |
| $Na_2O$ | 0.12 | 2.01 |
| $P_2O_5$ | 0.22 | 2.08 |

leads to significant $N_2O$ reduction during the combustion of sludge and coal [9, 12, 13], and $N_2O$ concentrations decreased from 560 mg/m$^3$ to less than 110 mg/m$^3$ as the freeboard temperature increased from 1117 K to 1149 K during wet sludge incineration. The same trend in this study was also found, as in Figure 2. As the furnace temperature increases from about 973 K to about 1173 K, $N_2O$ emission decreases from 612 mg/m$^3$ to 266 mg/m$^3$ for the cofiring rate of 100% and decreases from 794 mg/m$^3$ to 218 mg/m$^3$ for the cofiring rates of 70% with coal, from 674 mg/m$^3$ to 94 mg/m$^3$ for the cofiring rate of 70% with PE. The influence of furnace temperature on the conversion ratio of fuel-N to $N_2O$ is shown in Figure 3, and it is clear that an increase in furnace temperature leads to the reduction of conversion ratio of fuel-N to $N_2O$.

Influence of the furnace temperatures on NO emission is shown in Figure 4. For the test runs with cofiring rate of 100%, an increase in furnace temperature results in a decrease of NO emission and of a conversion ratio of fuel-N to NO, as shown in Figure 5. But for the test runs of cofiring sewage sludge

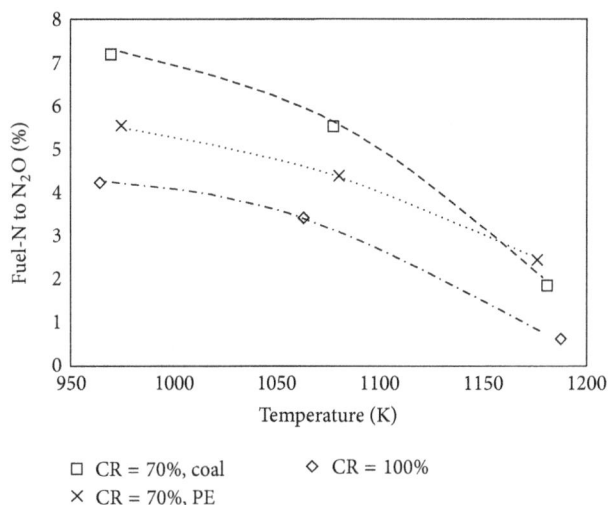

FIGURE 3: Influence of furnace temperature on conversion ratio of fuel-N to $N_2O$.

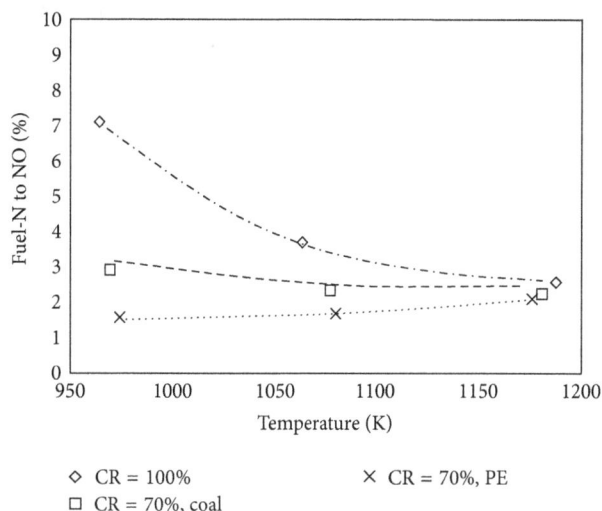

FIGURE 5: Influence of furnace temperatures on conversion ratio of fuel-N to NO.

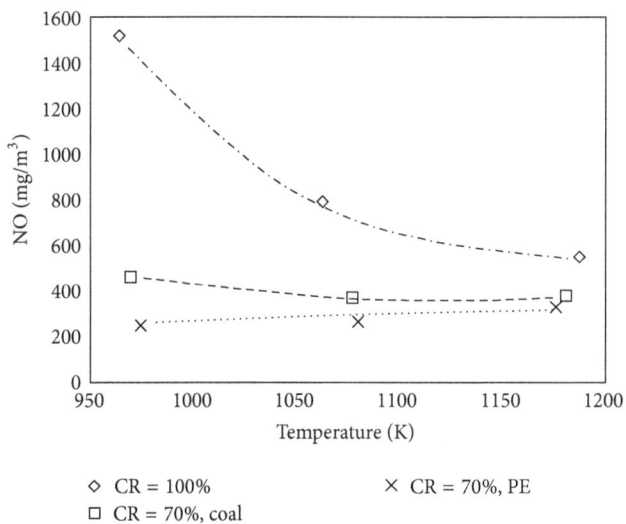

FIGURE 4: Influence of furnace temperatures on NO emission.

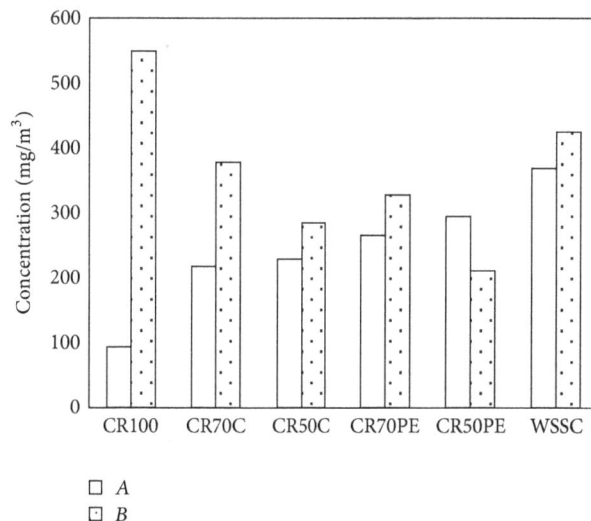

FIGURE 6: Influence of cofiring rates on $N_2O$ emission. A: $N_2O$; B: NO.

with coal and PE, furnace temperature is not an important parameter influencing NO emissions.

*3.2. Influence of Cofiring Rates on $N_2O$ and NO Emissions.* Nitrogen content in the mixed fuel increases with the rising of cofiring rates, which presents a higher potential for $N_2O$ and NO emissions [4, 11], and the influence of cofiring rate on $N_2O$ and NO emissions was studied, and the test conditions were in Table 4, and the furnace temperature was in the range of 1180 K and 1190 K.

As in Figure 6, when the cofiring rates increase from 50% to 100%, NO emission increases from about 211 mg/m$^3$ to 551 mg/m$^3$ with cofiring of PE and increases from 266 mg/m$^3$ to 551 mg/m$^3$ with cofiring of coal, and similar results from cofiring sludge with pulverized coal were also reported [1]. But with the increase of cofiring rates from 50% to 100%,

$N_2O$ emission decreases from 295 mg/m$^3$ to 93 mg/m$^3$ for cofiring of PE and decreases from 229 mg/m$^3$ to 93 mg/m$^3$ for cofiring of coal, but this observation is different from the work of Philippek and Werther [4], in which significant $N_2O$ emission increasing was observed with more sludge addition to coal firing boiler. For cofiring of sewage sludge with coal and PE, increase in cofiring rates leads to lower conversion ratio of fuel-N to $N_2O$, but there is not an apparent tendency about influence of cofiring rate on conversion ratio of fuel-N to NO. As in Figure 6, $N_2O$ emission from cofiring wet sewage sludge with coal is 370 mg/m$^3$, much higher than that from cofiring dried sewage sludge with coal and PE, at the range of 93–295 mg/m$^3$. As in Figure 7, conversion ratio of fuel-N to $N_2O$ and NO in cofiring wet sewage sludge with coal is 14% and 11%, respectively, but in cofiring dried sewage

TABLE 4: The co-firing rate test conditions.

| Test number | Co-firing rate (%) | Coal (%) | PE (%) | Dried sewage sludge (%) | Wet sewage sludge (%) |
|---|---|---|---|---|---|
| CR100 | 100 | 0 | 0 | 100 | |
| CR70C | 70 | 30 | 0 | 70 | |
| CR50C | 50 | 50 | 0 | 50 | |
| CR70PE | 70 | 0 | 30 | 70 | |
| CR50PE | 50 | 0 | 50 | 50 | |
| WSSC | 0 | 50 (mass) | 0 | 0 | 50 (mass) |

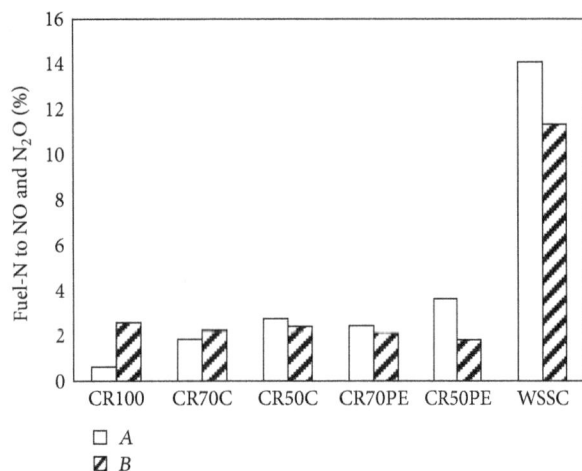

FIGURE 7: Influence of cofiring rate on NO emission. A: $N_2O$; B: NO; CR100: CR no. 100%; CR70C: CR no. 70%, coal; CR50C: CR no. 50%, coal; CR70PE: CR no. 70%, PE; CR50PE: CR no. 50%, PE; WSSC: wet sewage sludge and coal.

sludge, conversion ratio of fuel-N to $N_2O$ and NO is much lower, only 0.6%–3.6% and 1.8%–2.6%, respectively.

In the process of cofiring sewage sludge with PE and coal, CaO plays an important role in influencing $N_2O$ and NO emissions. Because of the high Ca content in the sludge ash and the high ash content in the sludge, CaO concentration in the furnace is higher when the cofiring rates increase, and CaO can catalyze the reactions of NO formation and $N_2O$ reduction [4, 14].

As in Figure 7, increase of cofiring rates leads to an increase in fuel nitrogen conversion ratios to NO and a decrease in ratios of fuel nitrogen conversion to $N_2O$. The catalytic reactions enhance the NO emission, as shown in Figure 7, and are the dominant factor leading to the decrease in $N_2O$ emission.

## 4. Conclusion

Through cofiring of dried sewage sludge, wet sewage sludge with coal and PE on the pilot scale 0.15MW$_{th}$ CFBC plant, the following conclusions were deduced:

(1) the furnace temperature is an effective parameter influencing $N_2O$ emission, and high temperature helps to enhance $N_2O$ decomposition;

(2) there is a strong dependence of $N_2O$ and NO emissions on cofiring rates, and rise of cofiring rates leads to lower $N_2O$ emission but higher NO emission. And increase in cofiring rates also results in the reduction of conversion ratio of fuel-N to $N_2O$;

(3) $N_2O$ emission from cofiring wet sewage sludge is much higher than that from cofiring dried sewage sludge with coal and PE, and fuel-N conversion ratio to $N_2O$ and NO in cofiring wet sewage sludge with coal is much higher than that in cofiring dried sewage sludge with coal and PE.

## References

[1] J. Werther and T. Ogada, "Sewage sludge combustion," *Progress in Energy and Combustion Science*, vol. 25, no. 1, pp. 55–116, 1999.

[2] J. Latva-Somppi, E. I. Kauppinen, T. Valmari et al., "The ash formation during co-combustion of wood and sludge in industrial fluidized bed boilers," *Fuel Processing Technology*, vol. 54, no. 1–3, pp. 79–94, 1998.

[3] A. van der Drift, J. van Doorn, and J. W. Vermeulen, "Ten residual biomass fuels for circulating fluidized-bed gasification," *Biomass and Bioenergy*, vol. 20, no. 1, pp. 45–56, 2001.

[4] C. Philippek and J. Werther, "Co-combustion of wet sewage sludge in a coal-fired circulating fluidised-bed combustor," *Journal of the Institute of Energy*, vol. 70, no. 485, pp. 141–150, 1997.

[5] M. Saito, K. Amagai, G. Ogiwara, and M. Arai, "Combustion characteristics of waste material containing high moisture," *Fuel*, vol. 80, no. 9, pp. 1201–1209, 2001.

[6] P. E. Campbell, S. McCahey, B. C. Williams, and M. L. Beekes, "Coal and plastic waste in a PF boiler," *Energy Policy*, vol. 28, no. 4, pp. 223–229, 2000.

[7] M. Y. Tsai, K. T. Wu, C. C. Huang, and H. T. Lee, "Co-firing of paper mill sludge and coal in an industrial circulating fluidized bed boiler," *Waste Management*, vol. 22, no. 4, pp. 439–442, 2002.

[8] K. Svoboda, D. Baxter, and J. Martinec, "Nitrous oxide emissions from waste incineration," *Chemical Papers*, vol. 60, no. 1, pp. 78–90, 2006.

[9] M. Sänger, J. Werther, and T. Ogada, "NO$_x$ and $N_2O$ emission characteristics from fluidized bed combustion of semi-dried municipal sewage sludge," *Fuel*, vol. 80, no. 2, pp. 167–177, 2001.

[10] F. Winter, C. Wartha, and H. Hofbauer, "NO and $N_2O$ formation during the combustion of wood, straw, malt waste and peat," *Bioresource Technology*, vol. 70, no. 1, pp. 39–49, 1999.

[11] H. Liu and B. M. Gibbs, "The influence of calcined limestone on NO$_x$ and $N_2O$ emissions from char combustion in fluidized bed combustors," *Fuel*, vol. 80, no. 9, pp. 1211–1215, 2001.

[12] H. Liu and B. M. Gibbs, "Modelling of NO and $N_2O$ emissions from biomass-fired circulating fluidized bed combustors," *Fuel*, vol. 81, no. 3, pp. 271–280, 2002.

[13] J. Werther, T. Ogada, and C. Philippek, "$N_2O$ emissions from fluidised bed combustion of sewage sludges," *Journal of the Institute of Energy*, vol. 68, pp. 93–101, 1995.

[14] J. I. Hayashi, T. Hirama, R. Okawa et al., "Kinetic relationship between $NO/N_2O$ reduction and $O_2$ consumption during flue-gas recycling coal combustion in a bubbling fluidized-bed," *Fuel*, vol. 81, no. 9, pp. 1179–1188, 2002.

# 7

# Combustion of Biogas Released from Palm Oil Mill Effluent and the Effects of Hydrogen Enrichment on the Characteristics of the Biogas Flame

Seyed Ehsan Hosseini, Ghobad Bagheri, Mostafa Khaleghi, and Mazlan Abdul Wahid

*High Speed Reacting Flow Laboratory, Faculty of Mechanical Engineering, Universiti Teknologi Malaysia, 81310 Skudai, Johor, Malaysia*

Correspondence should be addressed to Seyed Ehsan Hosseini; seyed.ehsan.hosseini@gmail.com

Academic Editor: Kalyan Annamalai

Biogas released from palm oil mill effluent (POME) could be a source of air pollution, which has illustrated negative effects on the global warming. To protect the environment from toxic emissions and use the energy of POME biogas, POME is conducted to the closed digestion systems and released biogas is captured. Since POME biogas upgrading is a complicated process, it is not economical and thus new combustion techniques should be examined. In this paper, POME biogas (40% $CO_2$ and 60% $CH_4$) has been utilized as a fuel in a lab-scale furnace. A computational approach by standard $k$-$\varepsilon$ combustion and turbulence model is applied. Hydrogen is added to the biogas components and the impacts of hydrogen enrichment on the temperature distribution, flame stability, and pollutant formation are studied. The results confirm that adding hydrogen to the POME biogas content could improve low calorific value (LCV) of biogas and increases the stability of the POME biogas flame. Indeed, the biogas flame length rises and distribution of the temperature within the chamber is uniform when hydrogen is added to the POME biogas composition. Compared to the pure biogas combustion, thermal $NO_x$ formation increases in hydrogen-enriched POME biogas combustion due to the enhancement of the furnace temperature.

## 1. Introduction

The increasingly strict regulations on pollution formation are pushing the energy and environmental research communities to find cleaner fuel and more efficient combustion technologies. Fossil fuel production is slow and taking many years therefore, the natural reserves of fossil fuels are rapidly exhausting. Many investigations have been carried out to find renewable fuels to replace these transient fossil fuels. Hence, biomass was found to have great potential to be applied in current combustion systems and biofuel could be one of the most important alternative fuels in the future energy mix of the world [1]. Meanwhile, biogas released from anaerobic digestion of biomass and organic wastes could be a source of energy for heat and power generation purposes. By capturing biogas from waste materials, not only an acceptable source of energy is provided but also the environment is protected from greenhouse gas emissions [2]. Unlike other alternative

fuels, biogas is not limited geographically and biogas simple production process is one of the most important power points of this fuel [3]. Palm oil as one of the most famous biofuel resources has been developed widely in South East Asian countries like Indonesia, Malaysia, and Thailand and tropical countries in Africa and South America. Palm oil, with approximately 28% total annual production, is known as the biggest vegetable oil in the world [4, 5]. However, sustainability of palm oil-based biodiesel production is under question due to POME generation. Huge amount of biogas generated from POME is released to the environment per annum which leads the world to global warming. The low calorific value (LCV) of biogas is the main barrier of biogas utilization development [6, 7]. Therefore, biogas should be upgraded to remove impurities such as $CO_2$ and $H_2S$ [8]. Since biogas upgrading is not economical, pure biogas was applied in flameless combustion technology successfully [9]. Today, with biogas utilization development in heat and power generation, comprehensive knowledge about various biogas combustion techniques is needed to select efficient energy conversion by biogas combustion [10]. On the other hand, hydrogen ($H_2$) as a clean fuel with low carbon dioxide ($CO_2$), carbon monooxide (CO), sulfur oxide ($SO_x$), and unburned hydrocarbon (UHC) emissions has great potential as a major fuel in the future [11]. However, $H_2$ is not freely found in nature and when it is used as a fuel, due to its high flammability and the diffusivity of $H_2$, some concerns about safety exist in storage and transport which leads to high explosion risk [12]. Before the 1990s, since most of the researches about $H_2$ enriched combustion were about liquid fuel enrichment, they were not applicable, because atomization and mixing of the species as the mechanical processes play a crucial role in the combustion of $H_2$ enriched liquid fuels. However, the effects of $H_2$ addition to gaseous fuel combustion are more radical because the combustion characteristics of such flames depend heavily on the properties of the gaseous fuels and conditions of the flame. Therefore, $H_2$ enrichment of different gaseous fuels like methane ($CH_4$), propane ($C_3H_8$), and natural gas (NG) had been developed during the last decade. Ignition of such fuel mixtures under their lean flammability limits ensures the fuel saving targets under the radical development method [13]. The raised temperature due to $H_2$ combustion as well as quick reaction rate with $O_2$ can justify this physical phenomenon. Although hydrogen-enriched gaseous fuel combustion has been developed experimentally and numerically, the effects of hydrogen enrichment on biogas conventional combustion have not been taken into consideration seriously. Since biogas utilization has become one of the valuable energy sources in the world, the effects of hydrogen enrichment of biogas on the conventional flame stability and pollutant formation are investigated in this paper.

*1.1. Biogas Composition.* Biogas is a flammable renewable gas formed in the anaerobic digestion (AD) of biomass which needs a relatively short formation time. The process of biogas generation and the type of feedstock play important roles in the biogas ingredients mixture [14]. Biogas consists of noncombustible $CO_2$, combustible $CH_4$ with low amounts of hydrogen sulfide ($H_2S$), water vapor ($H_2O$), carbon monoxide (CO), ammonia ($NH_3$), hydrogen ($H_2$), nitrogen ($N_2$), oxygen ($O_2$), dust, and occasionally siloxanes [3]. The most important biogas resources in the world are municipal solid waste (MSW) [15], domestic garbage landfills and old waste deposits [16], palm oil mill effluent [17], sewage sludge [18], cattle ranching and manure fermentation [19], coal mining [20], and agricultural products, rice paddies [21]. The average calorific value of biogas is 21.5 MJ/m$^3$ which is low in comparison with the calorific value of natural gas at 36 MJ/m$^3$. Based on the feedstock, $CH_4$ forms about 40–80% of the composition of biogas. Since the lower heating value of $CH_4$ is around 34,300 kj/m$^3$ at the standard temperature and pressure, the lower heating value of biogas is about 13,720–27,440 kJ/m$^3$. The physical characteristics of biogas are usually modeled by $CO_2$ and $CH_4$ because more than 98% of biogas is a combination of these two gases.

*1.2. Hydrogen-Enriched Gaseous Fuel Combustion Modelling.* Arrhenius reaction rate of hydrogen-enriched fuel increases due to the growth of the temperature; consequently, the rate of $O_2$ consumption in the lean mixture increases. In the simulation of hydrogen-enriched gaseous fuel combustion, fast chemistry models are superior due to their low computational cost. Indeed, the conserved scalar model as a subcategory of some other models such as flame sheet, laminar flamelet, and conserved scalar models with equilibrium chemistry is defined based on the relationship of the flame thermochemical characteristics as a function of the mixture fraction [22]. Ilbas et al. [23] applied the conserved scalar model to simulate a nonpremixed turbulent combustion of hydrogen-enriched methane. Similar flame was simulated with the eddy dissipation concept (EDC) by Frassoldati et al. [24] and Mardani and Tabejamaat [25] and an unacceptable accuracy was found for mass fraction of minor species like O and OH which conducts to wrong expectation for NO formation behavior. It was found that, for modelling the combustion of hydrogen-enriched methane, the steady laminar flamelet method has better performance in terms of minor species prediction [26]. Probability density function (PDF) was calculated for mass fraction of minor and major species, mixture fraction, and temperature by Suo [27]. It was reported that, compared to the equilibrium model, better results within the reaction zone were gained by flamelet model. The simulation of a nonpremixed hydrogen-enriched methane bluff-body flame with respect to laminar flamelet model, equilibrium chemistry, flame sheet model, and constrained equilibrium chemistry which was done by Hossain and Malalasekera [28] proves that just major species could be predictable by flame sheet model. Although acceptable results for $H_2O$ mass fraction and combustion temperature could be achieved by the flame sheet model, overprediction of $CO_2$ is one of the weak points of this model. On the other hand, the reliable results were recorded for temperature, $CO_2$, and $H_2O$ mass fraction by using the constrained equilibrium model; however, the predicted levels of CO and OH mass fraction were not accurate. Similarly prediction of temperature and species mass fraction were reported poor when the equilibrium model was applied. It was concluded that accurate results for temperature and major and minor species mass fraction could be achieved by laminar flamelet model. Hossain and Malalasekera [29, 30] modeled a similar flame applying flamelet model and pointed out that reasonable results could be achieved at upstream locations when a coupled radiation/flamelet model is utilized [31].

*1.3. Chemical Reaction Mechanisms in Hydrogen-Enriched Gaseous Fuel Combustion.* Various researchers applied different chemical reaction mechanisms in their simulation and it was found that simulation of chemical reaction mechanism plays an important role in the performance of the laminar flamelet model. For instance in the combustion simulation carried out by Ilbas et al. [23] only seven species were employed and Suo [27] applied GRI mechanism with 18 species. GRI2.11 was used by Ravikanti et al. [26] and it was found the simulated results are in good agreement with the results of a reduced DRM-22 mechanism [25]. In the combustion simulation done by Frassoldati et al. [24], 600 reactions were implemented when 48 species were considered.

*1.4. Hydrogen-Enriched Gaseous Fuel Turbulence Modelling.* A great variety of models are introduced for modeling turbulence, chemical reaction, and interaction of these two; yet there is no universal suitable model for all turbulent combustion applications achieved. Each model shows several advantages and disadvantages and poses better performance only in specific applications. Reynolds averaged Navier-Stokes modeling (RANS), large eddy simulation (LES), and direct numerical simulation (DNS) are the main methods that could be applied for numerical investigation of non-premixed combustion. In turbulence modelling, despite the Reynolds Stress Model (RSM) yielding proper results for prediction of high strain rate flows and streamline curvature, its function is under question due to different results issued by various researchers. The inability of RSM to predict flow field in the simulation of swirling flame was reported by Meier et al. [32]. The failure of RSM in swirling flow modelling with a processing vortex core (PVC) in the case of local velocity gradients capturing prediction was reported by Erdal and Shirazi [33]. Moreover, the convergence problems of RSM simulation have been reported by some researchers [34, 35]. Besides, some reports indicate that $k$-$\varepsilon$ mode has great capacity to model various combustion systems [36]. Dally et al. [37] modelled a bluff body flame with standard modified $k$-$\varepsilon$ and RSM flame sheet model applied with a beta probability density function and claimed that both standard $k$-$\varepsilon$ and RSM are not capable of predicting the flow field with acceptable accuracy. To improve projecting the flow field, a fine-tuning $k$-$\varepsilon$ model constant ($C\varepsilon1$ from 1.44 to 1.6) was assumed. Based on this assumption, Kim and Huh [38] simulated a bluff body methane/hydrogen flame using a conditional moment closure combustion model to predict NO formation in a turbulent condition. It was concluded that, although the local results of velocity fields and the variations of the mixture faction were not reliable, the modified $k$-$\varepsilon$ model is eligible to predict the overall mixture fraction fields and velocity. Indeed, acceptable accuracy was reported by [23–25] when fine-tuning of the standard $k$-$\varepsilon$ model with EDC combustion was simulated. Also, the reliability of both RSM and fine-tuned $k$-$\varepsilon$ model is enhanced when they applied with laminar flamelet model [26]. The performance of large eddy simulation (LES) was compared to fine-tuned $k$-$\varepsilon$ model by Suo [27] when the Smagorinsky-Lilly and RNG/$k$-$\varepsilon$ models were applied as the subgrid models. It was found that the results of both LES and modified $k$-$\varepsilon$ model are in good agreement with experimental records.

## 2. Methodology

The objective of this project is mainly to study the impacts of hydrogen addition on the biogas conventional flame stability and pollutant formation. In experimental step, biogas (40%

$CO_2$ and 60% $CH_4$) is injected to the combustor as fuel. In numerical modeling, biogas flow field and related chemical reactions are simulated. After model validation, hydrogen is added to the biogas and the impacts of such change in fuel composition on the temperature distribution in the furnace, flame stability, and pollutant formation are investigated in three cases (Case 1: 0% $H_2$, 40% $CO_2$, 60% $CH_4$, Case 2: 5% $H_2$, 40% $CO_2$, 55% $CH_4$, and Case 3: 10% $H_2$, 40% $CO_2$, 50% $CH_4$).

*2.1. Experimental Setup.* The diameter and the length of the chamber are 264 mm and 600 mm, respectively, made of carbon steel. The inside diameter of the furnace is 150 mm after installation of refractory. Five holes are located at the top of the furnace in a specific distance from burner to record temperature and pollutants by $K$-type thermocouples and gas analyzer, respectively. The diameter of fuel inlet jet is 5 mm surrounded by holes with 5 mm diameter for air inlet. A spark ignition system is used to ignite the reactants. Some flow meters are applied to check the flow of air, biogas, hydrogen, and the hydrogen-enriched biogas line. At the first step, the combustion system is run by biogas and for the second and third steps hydrogen is added to th biogas components 5% and 10%, respectively. Figure 1 displays the experimental setup.

*2.2. Numerical Solution.* Three-dimensional (3D) simulation is done by ANSYS 14 using ANSYS Modeler to design the chamber and ANSYS Meshing to mesh the combustor [39]. Mesh refinement could be effective to improve the convergence rate and scalar properties; thus grid resolution for smooth flow representation could be ensured. Due to symmetry, only one-eighth of the furnace is simulated. The grid consists of 7769 nodes and 33798 elements.

*2.2.1. Boundary Conditions.* In each cell, it has been assumed that all the properties have an average value at the center of the cell. The second order upwind scheme is set to calculate the temperature, velocity, and pressure through the governing equations with coupled algorithm. A mass-weighted averaging method is set to interpolate the velocity values of cell-center to face values. The temperature of fuel and air inlet and pressure outlet is 300 K and $1.013 \times 10^5$ Pa, respectively, and free stream turbulence is set at 5%. The residual energy equation should drop below $10^{-6}$ and for all other variables it is set at $10^{-3}$ to ensure the convergence of the solution. The swirl velocity of components is neglected in this steady-state CFD simulation and stoichiometric equivalence ratios ($\Phi = 1$) are taken into consideration. The density of biogas includes 40% $CO_2$ and 60% $CH_4$ in the standard temperature and pressure is considered 1.2146 g/l. Table 1 displays various fuel conditions.

*2.2.2. Grid Independent Check.* The results of numerical solution with 7769 nodes and 33798 elements are in good agreement with the experimental records. The grid independent of the simulation could be tested by changing

TABLE 1: Fuel data in various cases.

| Case | $CO_2$ | $CH_4$ | $H_2$ | Fuel density (kg/m$^3$) | $V_{air}$ (m/s) | $V_{fuel}$ (m/s) | $\dot{m}_{air}$ (kg/s) | $\dot{m}_{fuel}$ (kg/s) |
|------|--------|--------|-------|-------------------------|-----------------|------------------|------------------------|-------------------------|
| H0 | 40 | 60 | 0 | 1.2146 | 30 | 22.64 | 0.005544 | 0.00054 |
| H5 | 40 | 55 | 5 | 1.1834 | 30 | 24.78 | 0.005544 | 0.000576 |
| H10 | 40 | 50 | 10 | 1.1521 | 30 | 27.27 | 0.005544 | 0.000617 |

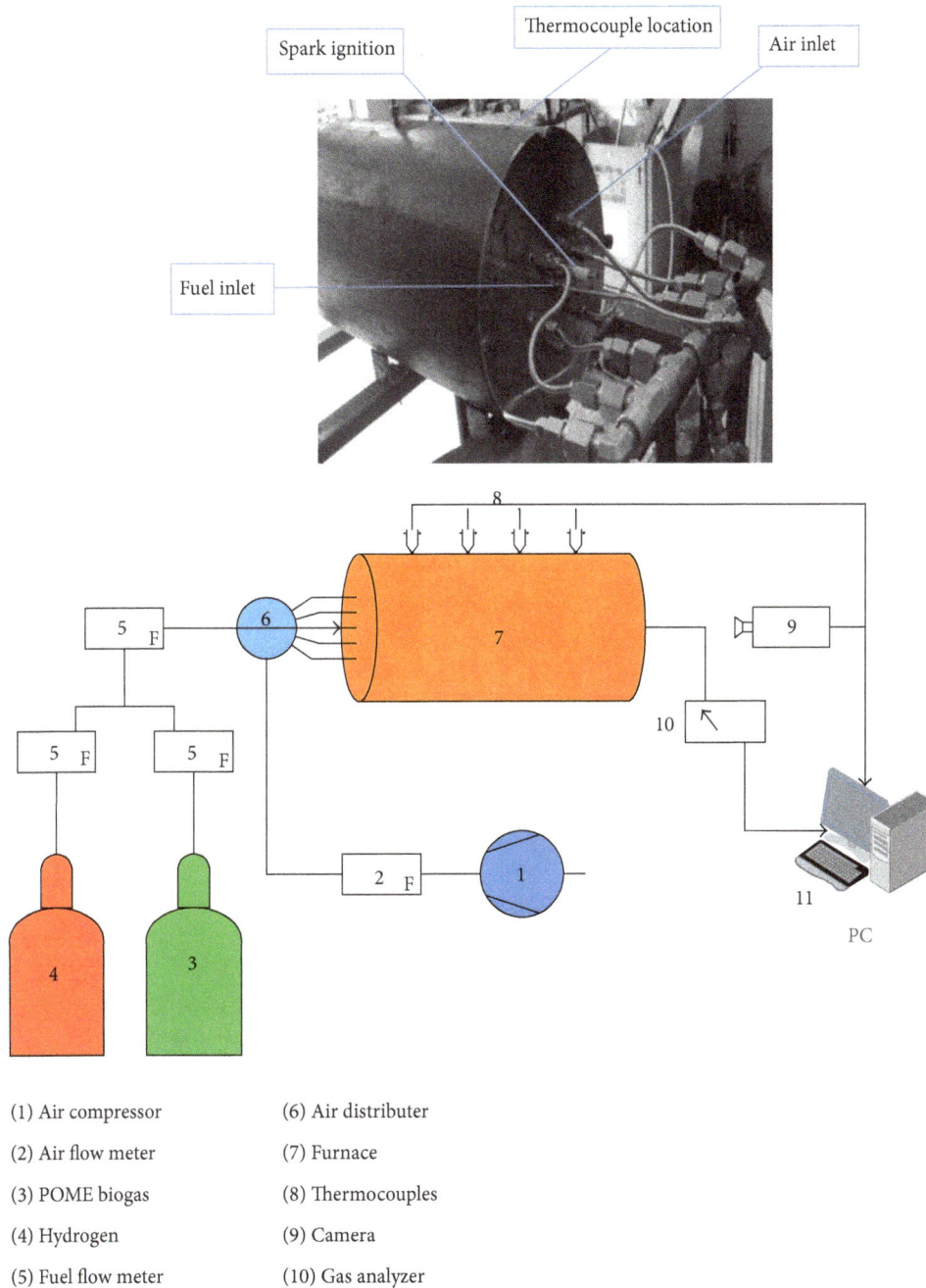

(1) Air compressor
(2) Air flow meter
(3) POME biogas
(4) Hydrogen
(5) Fuel flow meter
(6) Air distributer
(7) Furnace
(8) Thermocouples
(9) Camera
(10) Gas analyzer

FIGURE 1: Experimental setup.

the number of meshes to the finer meshes. In this simulation the number of elements was adopted 33798, 68453, and 100356. These adoptions were motivated by the fact that the most significant conformity to the experimental measurements could be achieved when the mesh is so fine. However, meaningful changes were not observed in the results and the grid independent of solution is confirmed by this test. Figure 2 compares the radial temperature profile of experimental results and simulation data related to biogas conventional combustion. This figure reveals that the captured data from

X = 100 mm

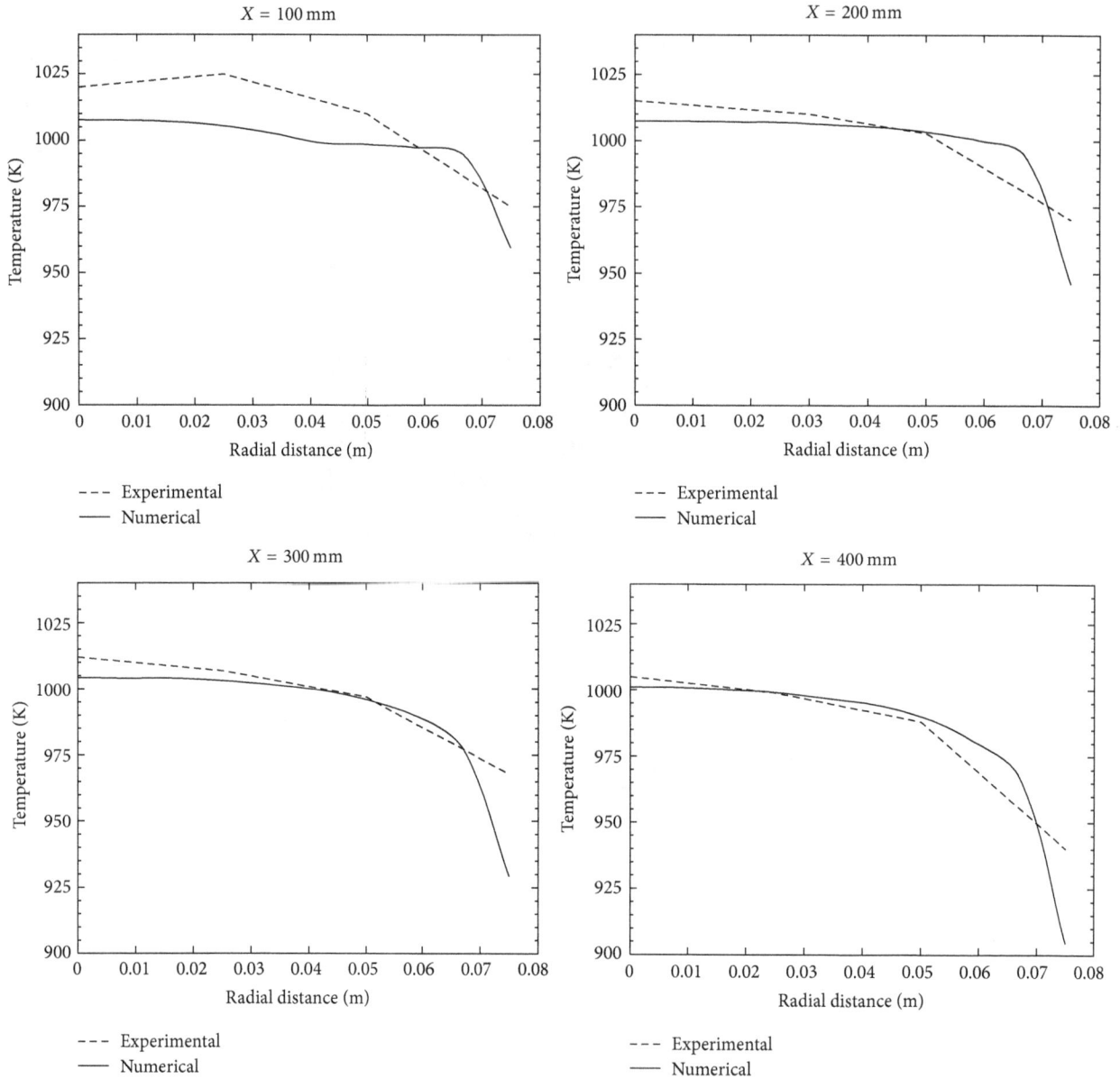

X = 200 mm

--- Experimental
— Numerical

--- Experimental
— Numerical

X = 300 mm

X = 400 mm

--- Experimental
— Numerical

--- Experimental
— Numerical

FIGURE 2: The radial temperature profile of experimental results and simulation data.

experimental and simulated study have the same trend in terms of the temperature profile except at $X = 100$ mm.

### 2.2.3. Turbulence Modelling.

The standard $k$-$\varepsilon$ model is employed to model conventional flame with the laminar flamelet combustion model. In the standard $k$-$\varepsilon$ model, $k$ and $\varepsilon$ are modelled with following transport equations:

$$\frac{\partial}{\partial t}(\rho k) + \frac{\partial}{\partial x_i}(\rho k u_i) = \frac{\partial}{\partial x_j}\left[\left(\mu + \frac{\mu_t}{\sigma_k}\right)\frac{\partial k}{\partial x_j}\right] + P_k + P_b - \rho\varepsilon - Y_M + S_k,$$

$$\frac{\partial}{\partial t}(\rho\varepsilon) + \frac{\partial}{\partial x_i}(\rho\varepsilon u_i) = \frac{\partial}{\partial x_j}\left[\left(\mu + \frac{\mu_t}{\sigma_\varepsilon}\right)\frac{\partial k}{\partial x_j}\right] + C_{1\varepsilon}\frac{\varepsilon}{k}(P_k + C_{3\varepsilon}P_b) - C_{2\varepsilon\rho}\frac{\varepsilon^2}{k} + S_\varepsilon,$$

(1)

where $\mu_t$, $P_k$, and $Y_M$ represent turbulent viscosity, turbulence kinetic energy production, and the contribution of the fluctuating dilatation in compressible turbulence to the overall dissipation rate, respectively, which are described by

$$\mu_t = \rho C_\mu \frac{k^2}{\epsilon},$$

$$P_k = -\rho \overline{u_i' u_j'} \frac{\partial u_j}{\partial x_i},$$

$$Y_M = 2\rho\varepsilon M_t^2. \tag{2}$$

*2.2.4. Combustion Modelling.* The nonpremixed modelling approach is usually applied to the simulation of turbulent diffusion flames with fast chemistry. This model offers many benefits over the eddy dissipation (ED) formulation and allows radical species prediction, dissociation impacts, and precise turbulence-chemistry coupling. The steady laminar flamelet method models a turbulent flame as an ensemble of discrete, steady laminar flames. It is assumed that the individual flamelets have the same structure as laminar flames in simple configurations and are achieved by calculation or experiments. Using detailed chemical mechanisms, FLUENT can calculate laminar opposed-flow diffusion flamelets for nonpremixed combustion. The laminar flamelets are then set in a turbulent flame using statistical PDF methods. The sensible chemical kinetic impacts can be incorporated into turbulent flames by using the laminar flamelet approach. The chemistry can be preprocessed and tabulated, offering enormous computational savings. A set of flamelet profiles in a flamelet library is built in terms of independence scalar dissipation rate and immediate mixture fraction. It means that

$$P(Z, X) = P(Z) P(X), \tag{3}$$

where the mean value of the scalar dissipation rate is calculated by

$$X = C_x \frac{\widetilde{\varepsilon}}{k} Z''^2. \tag{4}$$

$C_x$ is equal to 2.5.

The flamelet library is then integrated with a probability density function (PDF) to compute the average scalar properties:

$$\widetilde{\emptyset} = \int_0^\infty \int_0^1 \emptyset(Z, x) P(Z, x) \, dZ \, dx. \tag{5}$$

Mixture fraction could be estimated based on Drake formula [40]. Since GRI3.0 mechanism includes 325 reactions and 53 species, it is superior to GRI 2.11 in terms of up-to-date kinetics and accuracy; thus for developing the flamelet library, GRI3.0 mechanism was employed. Indeed, kinetics related to prompt $NO_x$ calculation have been improved in this revision [41].

*2.2.5. Radiation Modelling.* Prediction of radiative heat transfer is a crucial factor in the simulation of turbulent combustion. Notable discrepancies between numerical predictions and experimental results in terms of pollutant formation and combustion characteristics could emerge if an accurate radiative heat transfer method is not applied. Since $NO_x$ formation is sensitive to the trend of furnace temperature,

overprediction of $NO_x$ formation takes place if the radiative heat loss is not considered. For prediction of radiation, discrete ordinates (DO) radiation model is employed in this simulation because of its reasonable computational cost. DO is widely used in such similar computational investigation with no significant error. The related formula and more details of this radiative model can be found in [42, 43].

*2.2.6. $NO_x$ Formation Modelling.* In conventional combustion regime, $NO_x$ formation reduction plays an important role to control acid rain, smog, ozone depletion, and greenhouse effects. $NO_x$ is usually formed in the presence of nitrogen and oxygen within a locally high temperature region. Thermal $NO_x$, prompt $NO_x$, $N_2O$ intermediate mechanism, and fuel-bound nitrogen are mentioned as the main regimes for $NO_x$ formation in the combustion process. At extremely high temperatures within the combustion chamber, $N_2$ and $O_2$ can react through chemical mechanisms that are named Zeldovich formulation. The rate of thermal $NO_x$ formation increases quickly with increasing temperature. Prompt or Fenimore $NO_x$ formation occurs in fuel rich conditions and it was found that the prompt $NO_x$ formation increases near equivalence ratio of 1.4 [44]. $N_2O$ intermediate mechanism takes place in fuel lean, elevated temperature, and low pressures. $NO_x$ formation by $N_2O$ mechanism was proposed because of lower flame temperature in the combustion process [45]. Fuel-bound $NO_x$ formation mechanism is related to the presence of nitrogen species in the molecular structure of the fuel. Due to the characteristics of biogas conventional combustion, thermal $NO_x$ and prompt $NO_x$ are considered in the simulation. The transport equation for NO species formation is

$$\frac{\partial}{\partial t}\left(\rho Y_{NO}\right) + \nabla \cdot \left(\rho \vec{v} Y_{NO}\right) = \nabla \cdot \left(\rho D \nabla Y_{NO}\right) + S_{NO}. \tag{6}$$

$S_{NO}$ includes thermal $NO_x$ which should be determined by Zeldovich equations and prompt $NO_x$. Zeldovich equations could be written as [46]

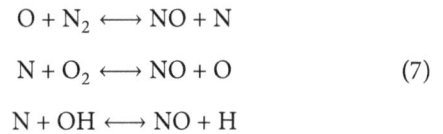

$$O + N_2 \longleftrightarrow NO + N$$

$$N + O_2 \longleftrightarrow NO + O \tag{7}$$

$$N + OH \longleftrightarrow NO + H$$

The reaction rate constants of reactions are selected from [47] and partial equilibrium approach is considered to estimate the concentration of OH and O radicals. The calculation of prompt $NO_x$ formation is done from global model presented in [48]:

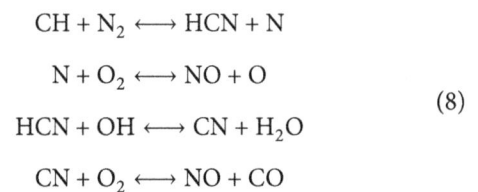

$$CH + N_2 \longleftrightarrow HCN + N$$

$$N + O_2 \longleftrightarrow NO + O$$

$$HCN + OH \longleftrightarrow CN + H_2O \tag{8}$$

$$CN + O_2 \longleftrightarrow NO + CO$$

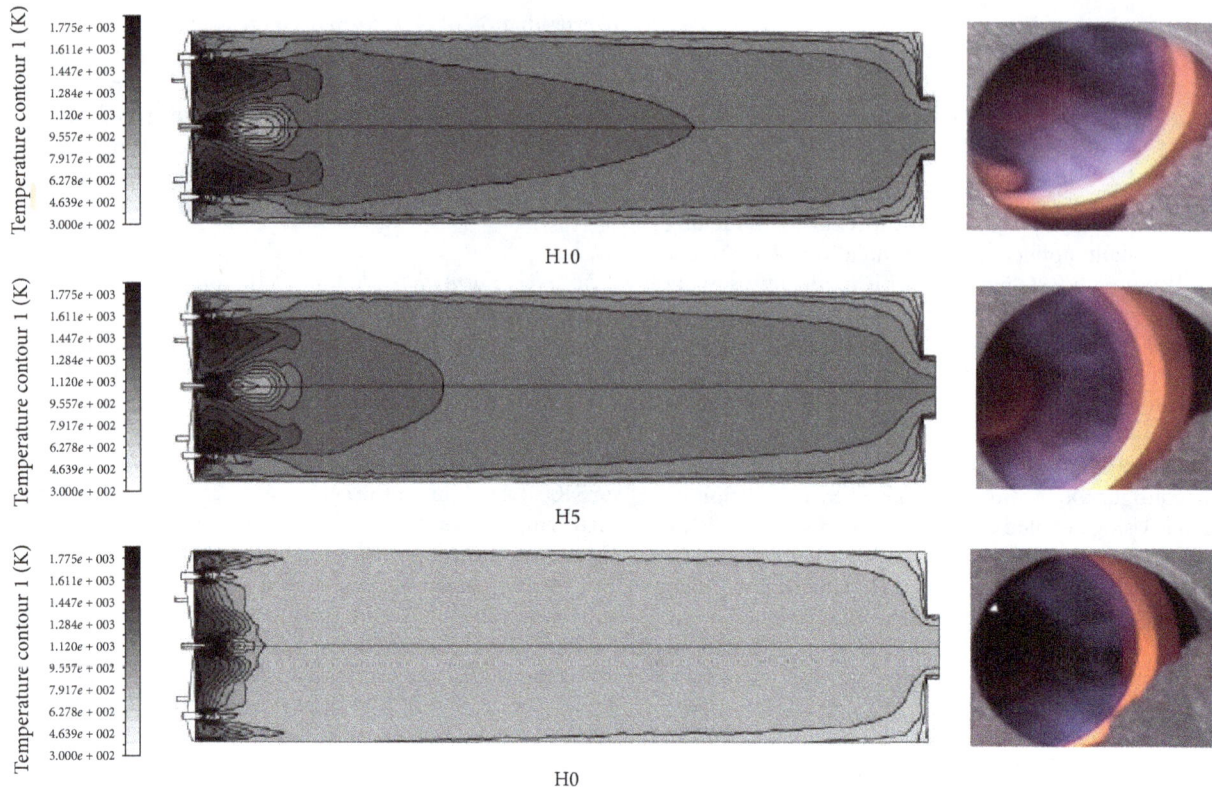

FIGURE 3: Temperature distribution in the chamber and the pictures of the flames.

To consider the interaction of $NO_x$ formation and turbulence, a PDF of temperature is applied to compute a time average rate of $NO_x$ constitution when generation rate of thermal $NO_x$ and prompt $NO_x$ is calculated. The computed time averaged $NO_x$ results are applied in (6).

## 3. Results and Discussion

The velocity of air jet is kept constant at 30 m/s in all cases. When hydrogen is added to the fuel the density of the flow reduces; thus the mass flow rate of the mixture should increase to get stoichiometric conditions. As hydrogen is added to the biogas ingredients by 5%, the structure of biogas flame changes noticeably and the peak of temperature increases. Furthermore, hydrogen addition to the bogus combination causes some changes in the pattern of the flame. The high temperature of the flame shrinks and moves slightly further away from the furnace axis. Indeed, a small hot region is formed at the flame tip and when the concentration of hydrogen raised to 10%, this region becomes bigger. However, the flame temperature is not changed significantly when the percentage of hydrogen intensifies to 10%. Indeed, the length of flame increases when hydrogen is added to the biogas mixture and when further growth in hydrogen percentage is happening, the length of the flame increases. Figure 3 demonstrates the contour of temperature inside the chamber in the three cases.

It can be interpreted that added hydrogen changes the density of biogas mixture and thus the flow rate of mixture increases; thus the flammability of the biogas increases because of hydrogen addition (the flammability limit of H0, H5, and H10 is 9, 12, and 15 (percent by volume), respectively). To further hydrogen content (10% hydrogen), the flammability of biogas intensifies due to lower density of mixture and higher flow field. Figure 4 displays the numerical results of hydrogen addition effects on the axial temperature and radial temperature profile (at $x = 65$ mm) of the biogas flame. From Figure 4(b), it can be seen that the biogas flame becomes narrow when the percentage of hydrogen content of biogas increases. The flame thickness reduction can be attributed to the enhancement of the mixing. In the simulation of nonpremixed combustion, mixture fraction indicates the mixing rate. While the mixture fraction increases, it can be construed that mixing does not occur properly. Figure 5 displays the mixture fraction in axial and radial directions. When the hydrogen content of biogas increases, the radial spreading rate of mixture fraction decreases which is a sign of mixing growth.

The axial profile of $H_2$, $CH_4$, and $O_2$ mass fraction in hydrogen-enriched biogas combustion is presented in Figure 6.

The contour of $NO_x$ formation and the effects of hydrogen enrichment on $NO_x$ formation of biogas combustion are presented in Figures 7 and 8, respectively. $NO_x$ formation significantly increases when hydrogen is added to

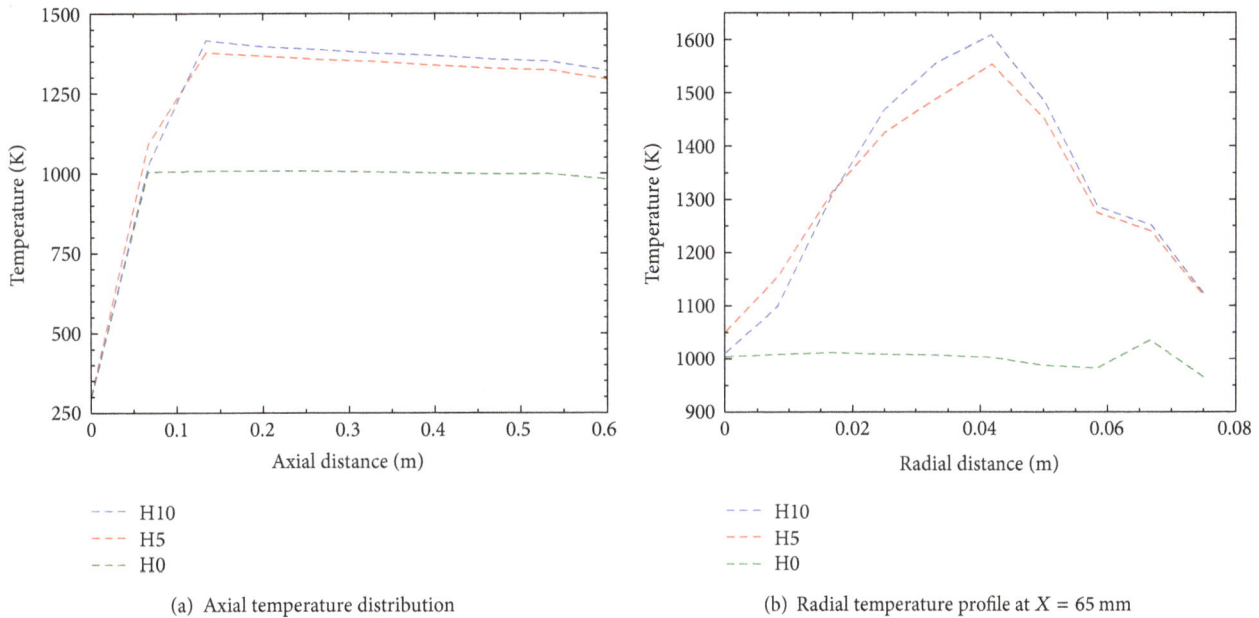

(a) Axial temperature distribution

(b) Radial temperature profile at $X = 65\,mm$

FIGURE 4: Effect of biogas hydrogen enrichment on (a) axial temperature distribution and (b) radial temperature profile.

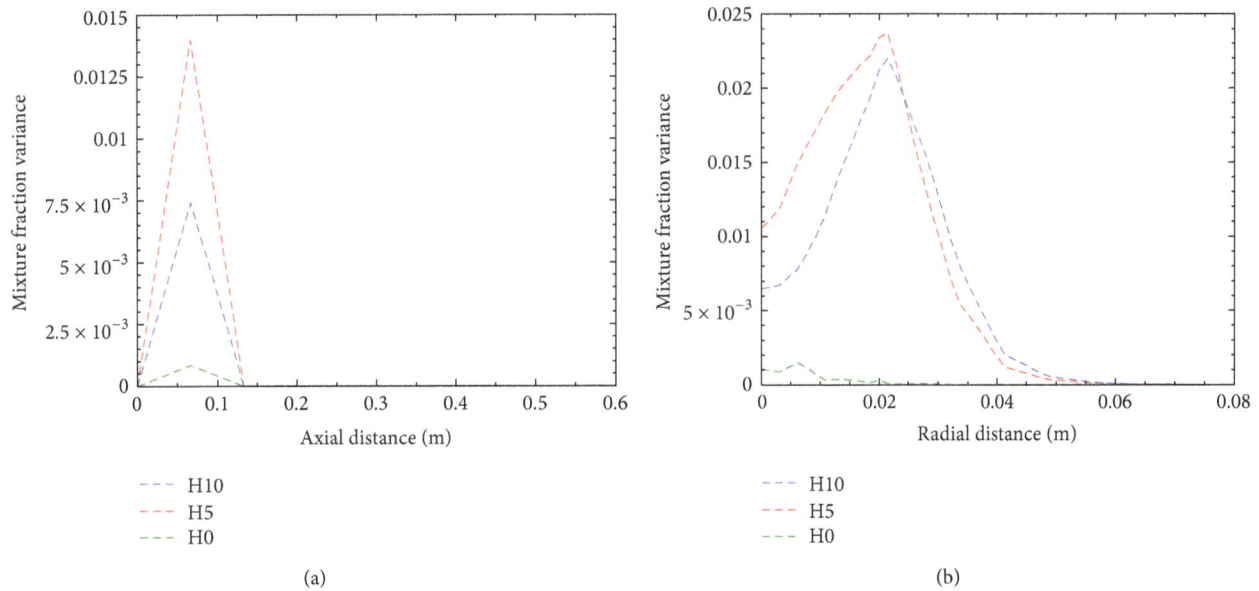

(a)

(b)

FIGURE 5: Effect of hydrogen enrichment on the (a) axial profile of mixture fraction and (b) radial profile of mixture fraction in biogas combustion.

the biogas content. Although the contribution of prompt $NO_x$ decreases in the total generated $NO_x$ (because of the carbon content reduction in the fuel mixture), thermal $NO_x$ is increased dramatically due to considerable growth in the flame temperature. It means that hydrogen in the biogas stream contributes to the $NO_x$ formation only through an increase in temperature and consequently via thermal $NO_x$. The maximum temperature in cases H0, H5, and H10 was recorded 1860 K, 1790 K, and 1710 K, respectively. Indeed,

from numerical simulation (Figure 3) it can be seen that hot spots are developed in case H10. Because of that, $NO_x$ formation region changes with the same pattern that peak temperature distribution region changes.

It was mentioned that Figure 4 indicates that the case H10 has higher temperature from others; however, Figure 7 demonstrates that the $NO_x$ concentration of H10 is lower than case H5. The increasing of hydrogen percentage from 5% to 10% led the combustion to be more complete. Therefore,

(a)

(b)

(c)

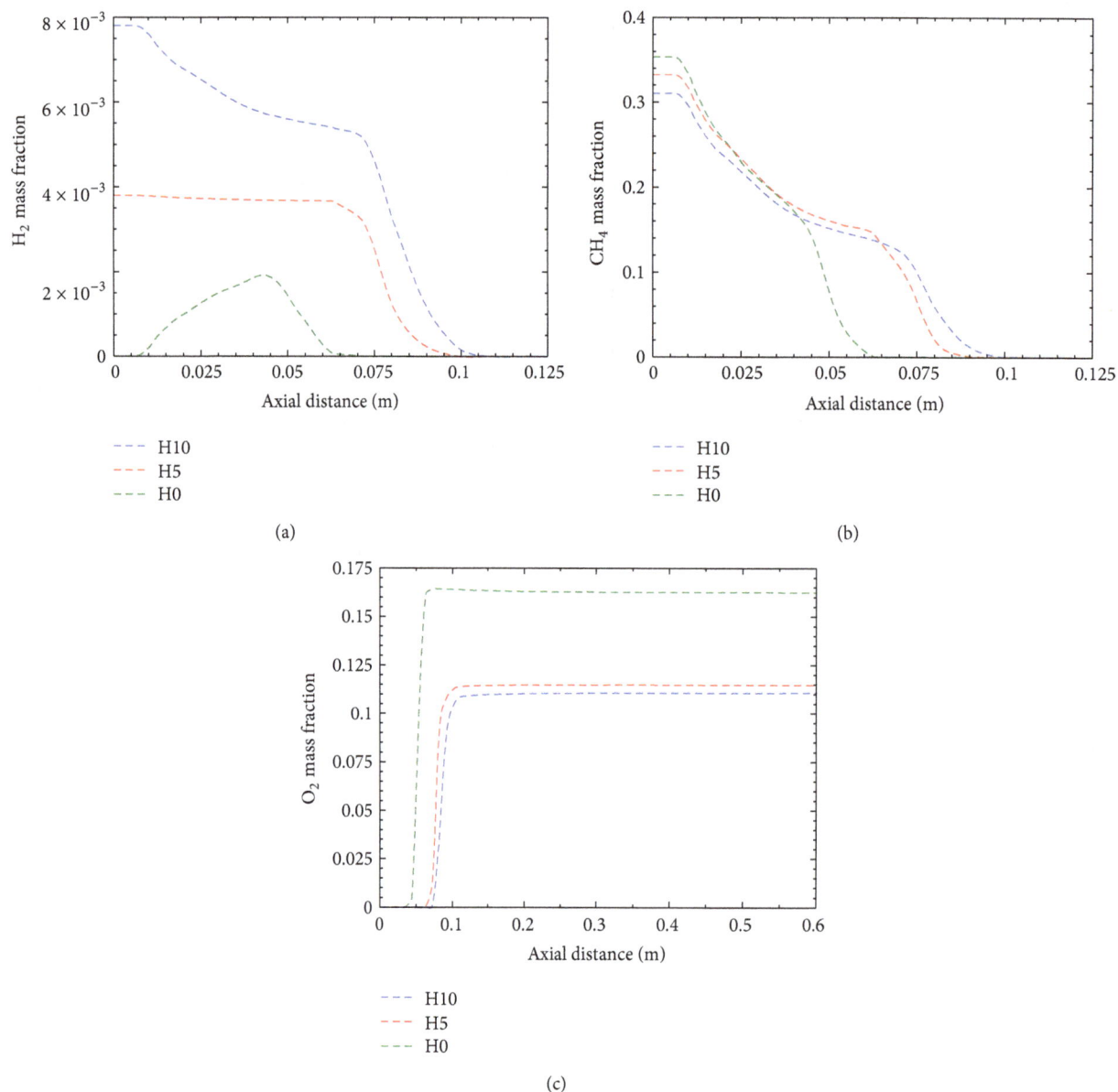

FIGURE 6: The axial profile of $H_2$, $CH_4$, and $O_2$ mass fraction in hydrogen-enriched biogas combustion.

according to Figure 6, the concentration of oxygen ($O_2$) is less than H5. It could be argued that lower rate of $NO_x$ production in H10 is related to oxygen ($O_2$) shortage, even though it has higher temperature.

## 4. Conclusion

Although POME biogas could be an acceptable source of energy, LCV of biogas is the most important barrier of this renewable and sustainable fuel development. Since POME biogas upgrading is not economic, new methods should be taken into consideration to improve the combustion of biogas. Since the percentage of hydrogen in the POME biogas components could be controlled by some chemical strategies,

the effects of hydrogen enrichment on biogas conventional coflow flame were investigated. Combustion characteristics and flame stability of pure biogas (H0: 40% $CO_2$ and 60% $CH_4$) and hydrogen-enriched biogas (H5: 40% $CO_2$, 55% $CH_4$, 5% $H_2$ and H10: 40% $CO_2$, 50% $CH_4$, 10% $H_2$) were studied. It was found that adding hydrogen to the biogas content could improve LCV of biogas and consequently the stability of biogas flame increases. Also, the distribution of temperature becomes uniform when hydrogen is added to the biogas. Indeed, the length of the biogas flame is stretched when hydrogen is introduced to the fuel mixture. The simulated results show that the mixing process of fuel and air improves rapidly in the presence of hydrogen. When the concentration of $H_2$ is increased, the density of the inlet fuel

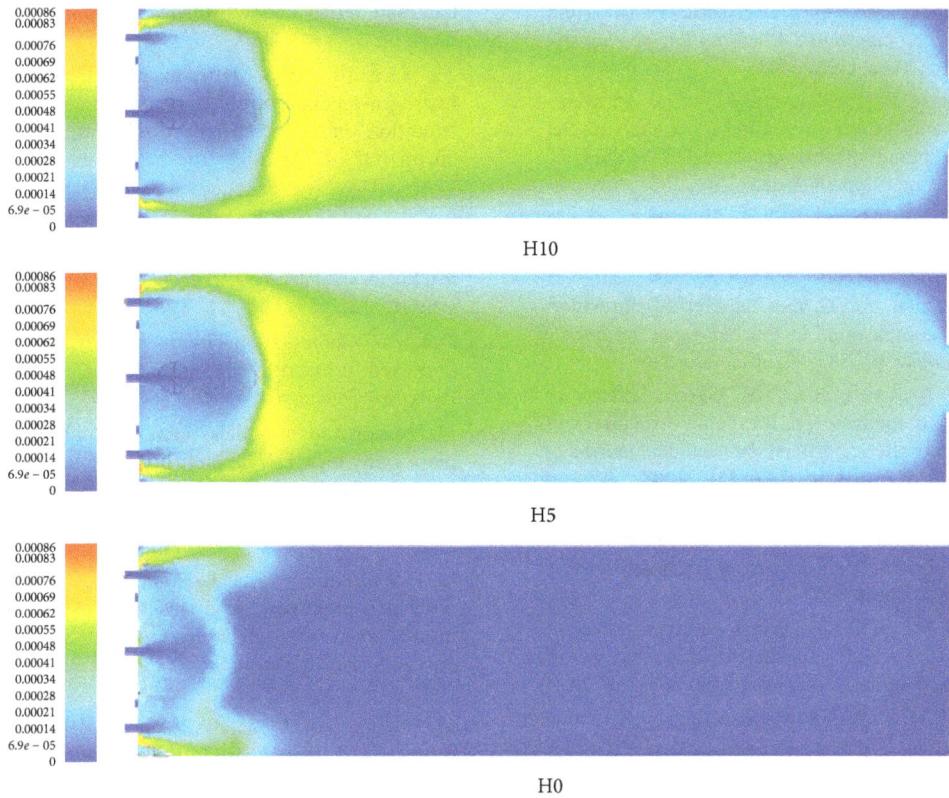

FIGURE 7: $NO_x$ formation contour.

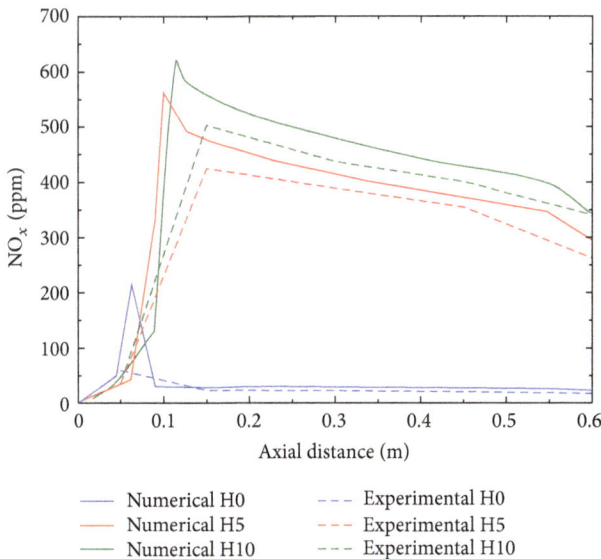

FIGURE 8: The axial profile of NO mass fraction in hydrogen-enriched biogas combustion.

mixture and thus the flame structure is changed. Compared to the pure biogas combustion, thermal $NO_x$ formation increases in hydrogen-enriched biogas combustion due to temperature enhancement.

## Conflict of Interests

The authors declare that there is no conflict of interests regarding the publication of this paper.

## References

[1] S. E. Hosseini, M. A. Wahid, S. Salehirad, and M. M. Seis, "Evaluation of palm oil combustion characteristics by using the Chemical Equilibrium with Application (CEA) software," *Applied Mechanics and Materials*, vol. 388, pp. 268–272, 2013.

[2] S. E. Hosseini, M. A. Wahid, and N. Aghili, "The scenario of greenhouse gases reduction in Malaysia," *Renewable and Sustainable Energy Reviews*, vol. 28, pp. 400–409, 2013.

[3] G. Taleghani and A. S. Kia, "Technical-economical analysis of the Saveh biogas power plant," *Renewable Energy*, vol. 30, no. 3, pp. 441–446, 2005.

[4] S. B. Hansen, S. I. Olsen, and Z. Ujang, "Greenhouse gas reductions through enhanced use of residues in the life cycle of Malaysian palm oil derived biodiesel," *Bioresource Technology*, vol. 104, pp. 358–366, 2012.

[5] H. Stichnothe and F. Schuchardt, "Life cycle assessment of two palm oil production systems," *Biomass and Bioenergy*, vol. 35, no. 9, pp. 3976–3984, 2011.

[6] S. E. Hosseini and M. A. Wahid, "Biogas utilization: experimental investigation on biogas flameless combustion in lab-scale furnace," *Energy Conversion and Management*, vol. 74, pp. 426–432, 2013.

[7] S. Jahangirian and A. Engeda, "Biogas combustion and chemical kinetics for gas turbine applications," in *Proceedings of the*

*ASME 2008 International Mechanical Engineering Congress and Exposition, Volume 3: Combustion Science and Engineering*, pp. 13–22, ASME, Boston, Mass, USA, 2008.

[8] T. Patterson, S. Esteves, R. Dinsdale, and A. Guwy, "An evaluation of the policy and techno-economic factors affecting the potential for biogas upgrading for transport fuel use in the UK," *Energy Policy*, vol. 39, no. 3, pp. 1806–1816, 2011.

[9] S. E. Hosseini, M. A. Wahid, and A. A. A. Abuelnuor, "Biogas flameless combustion: a review," *Applied Mechanics and Materials*, vol. 388, pp. 273–279, 2013.

[10] B. Yan, X. Bai, G. Chen, and C. Liu, "Numerical simulation of turbulent biogas combustion," in *Proceedings of the Energy Sustainability Conference*, pp. 885–896, usa, June 2007.

[11] H. J. Burbano, A. A. Amell, and J. M. García, "Effects of hydrogen addition to methane on the flame structure and CO emissions in atmospheric burners," *International Journal of Hydrogen Energy*, vol. 33, no. 13, pp. 3410–3415, 2008.

[12] F. Markert, S. K. Nielsen, J. L. Paulsen, and V. Andersen, "Safety aspects of future infrastructure scenarios with hydrogen refuelling stations," *International Journal of Hydrogen Energy*, vol. 32, no. 13, pp. 2227–2234, 2007.

[13] G. Hu, S. Zhang, Q. F. Li et al., "Experimental investigation on the effects of hydrogen addition on thermal characteristics of methane/air premixed flames," *Fuel*, vol. 115, pp. 232–240, 2014.

[14] P. Gupta, R. S. Singh, A. Sachan, A. S. Vidyarthi, and A. Gupta, "Study on biogas production by anaerobic digestion of garden-waste," *Fuel*, vol. 95, pp. 495–498, 2012.

[15] M. L. de Souza-Santos and K. Ceribeli, "Technical evaluation of a power generation process consuming municipal solid waste," *Fuel*, vol. 108, pp. 578–585, 2013.

[16] R. M. Barros, G. L. T. Filho, and T. R. da Silva, "The electric energy potential of landfill biogas in Brazil," *Energy Policy*, vol. 65, pp. 150–164, 2014.

[17] M. J. Chin, P. E. Poh, B. T. Tey, E. S. Chan, and K. L. Chin, "Biogas from palm oil mill effluent (POME): opportunities and challenges from Malaysia's perspective," *Renewable and Sustainable Energy Reviews*, vol. 26, pp. 717–726, 2013.

[18] W. Yuan and T. J. Bandosz, "Removal of hydrogen sulfide from biogas on sludge-derived adsorbents," *Fuel*, vol. 86, no. 17-18, pp. 2736–2746, 2007.

[19] D. Raha, P. Mahanta, and M. L. Clarke, "The implementation of decentralised biogas plants in Assam, NE India: the impact and effectiveness of the National Biogas and Manure Management Programme," *Energy Policy*, vol. 68, pp. 80–91, 2014.

[20] D. Luo and Y. Dai, "Economic evaluation of coalbed methane production in China," *Energy Policy*, vol. 37, no. 10, pp. 3883–3889, 2009.

[21] G. K. Kafle and S. H. Kim, "Effects of chemical compositions and ensiling on the biogas productivity and degradation rates of agricultural and food processing by-products," *Bioresource Technology*, vol. 142, pp. 553–561, 2013.

[22] W. P. Jones and J. H. Whitelaw, "Calculation methods for reacting turbulent flows: a review," *Combustion and Flame*, vol. 48, pp. 1–26, 1982.

[23] M. Ilbas, I. Yilmaz, and Y. Kaplan, "Investigations of hydrogen and hydrogen-hydrocarbon composite fuel combustion and NOx emission characteristics in a model combustor," *International Journal of Hydrogen Energy*, vol. 30, no. 10, pp. 1139–1147, 2005.

[24] A. Frassoldati, P. Sharma, A. Cuoci, T. Faravelli, and E. Ranzi, "Kinetic and fluid dynamics modeling of methane/hydrogen jet flames in diluted coflow," *Applied Thermal Engineering*, vol. 30, no. 4, pp. 376–383, 2010.

[25] A. Mardani and S. Tabejamaat, "Effect of hydrogen on hydrogen-methane turbulent non-premixed flame under MILD condition," *International Journal of Hydrogen Energy*, vol. 35, no. 20, pp. 11324–11331, 2010.

[26] M. Ravikanti, M. Hossain, and W. Malalasekera, "Laminar flamelet model prediction of $NO_x$ formation in a turbulent bluff-body combustor," *Proceedings of the Institution of Mechanical Engineers. Part A: Journal of Power and Energy*, vol. 223, no. 1, pp. 41–54, 2009.

[27] D. Y. Suo, "Computational modeling of hydrogen enriched non-premixed turbulent methane air flames," in *Proceedings of the European Combustion Meeting (ECM '05)*, pp. 1–6, 2005.

[28] M. Hossain and W. Malalasekera, "A combustion model sensitivity study for CH4/H2 bluff-body stabilized flame," *Proceedings of the Institution of Mechanical Engineers, Part C: Journal of Mechanical Engineering Science*, vol. 221, no. 11, pp. 1377–1390, 2007.

[29] M. Hossain and W. Malalasekera, "Numerical study of bluff-body non-premixed flame structures using laminar flamelet model," *Proceedings of the Institution of Mechanical Engineers Part A: Journal of Power and Energy*, vol. 219, no. 5, pp. 361–370, 2005.

[30] M. Hossain and W. Malalasekera, "Modelling of a bluff stabilized CH4/H2 flame based on a laminar flamelet model with emphasis on NO prediction," *Proceedings of the Institution of Mechanical Engineers, Part A: Journal of Power and Energy*, vol. 217, no. 2, pp. 201–210, 2003.

[31] M. Hossain, J. C. Jones, and W. Malalasekera, "Modelling of a bluff-body nonpremixed flame using a coupled radiation/flamelet combustion model," *Flow, Turbulence and Combustion*, vol. 67, no. 3, pp. 217–234, 2002.

[32] W. Meier, O. Keck, B. Noll, O. Kunz, and W. Stricker, "Investigations in the TECFLAM swirling diffusion flame: laser Raman measurements and CFD calculations," *Applied Physics B: Lasers and Optics*, vol. 71, no. 5, pp. 725–731, 2000.

[33] F. M. Erdal and S. A. Shirazi, "Local velocity measurements and computational fluid dynamics (CFD) simulations of swirling flow in a cylindrical cyclone separator," *Journal of Energy Resources Technology*, vol. 126, no. 4, pp. 326–333, 2004.

[34] S.-Y. Liu, Y. Zhang, and B.-G. Wang, "Cyclone separator three-dimensional turbulent flow-field simulation using the Reynolds stress model," *Transaction of Beijing Institute of Technology*, vol. 25, no. 5, pp. 377–383, 2005.

[35] A. M. Jawarneh and G. H. Vatistas, "Reynolds stress model in the prediction of confined turbulent swirling flows," *Transactions of the ASME, Journal of Fluids Engineering*, vol. 128, no. 6, pp. 1377–1382, 2006.

[36] M. R. Halder and S. K. Som, "Numerical and experimental study on cylindrical swirl atomizers," *Atomization and Sprays*, vol. 16, no. 2, pp. 223–236, 2006.

[37] B. B. Dally, D. F. Fletcher, and A. R. Masri, "Flow and mixing fields of turbulent bluff-body jets and flames," *Combustion Theory and Modelling*, vol. 2, no. 2, pp. 193–219, 1998.

[38] S. H. Kim and K. Y. Huh, "Use of the conditional moment closure model to predict NO formation in a turbulent $CH_4/H_2$ flame over a bluff-body," *Combustion and Flame*, vol. 130, no. 1-2, pp. 94–111, 2002.

[39] *Fluent A. 12.0 Theory Guide*, Ansys, 2009.

[40] M. C. Drake and R. J. Blint, "Relative importance of nitric oxide formation mechanisms in laminar opposed-flow diffusion flames," *Combustion and Flame*, vol. 83, no. 1-2, pp. 185–203, 1991.

[41] B. Rohani and K. M. Saqr, "Effects of hydrogen addition on the structure and pollutant emissions of a turbulent unconfined swirling flame," *International Communications in Heat and Mass Transfer*, vol. 39, no. 5, pp. 681–688, 2012.

[42] G. D. Raithby and E. H. Chui, "Finite-volume method for predicting a radiant heat transfer in enclosures with participating media," *Journal of Heat Transfer*, vol. 112, no. 2, pp. 415–423, 1990.

[43] J. Y. Murthy and S. R. Mathur, "Finite volume method for radiative heat transfer using unstructured meshes," *Journal of Thermophysics and Heat Transfer*, vol. 12, no. 3, pp. 313–321, 1998.

[44] L. Pillier, A. El Bakali, X. Mercier et al., "Influence of $C_2$ and $C_3$ compounds of natural gas on NO formation: an experimental study based on LIF/CRDS coupling," *Proceedings of the Combustion Institute*, vol. 30, pp. 1183–1191, 2005.

[45] P. C. Malte and D. T. Pratt, "Measurement of atomic oxygen and nitrogen oxides in jet-stirred combustion," *Symposium (International) on Combustion*, vol. 15, no. 1, pp. 1061–1070, 1975.

[46] C. T. Bowman, "Kinetics of pollutant formation and destruction in combustion," *Progress in Energy and Combustion Science*, vol. 1, no. 1, pp. 33–45, 1975.

[47] R. K. Hanson and S. Salimian, *Survey of Rate Constants in the N/H/O System*, Springer, New York, NY, USA, 1984.

[48] G. G. de Soete, "Overall reaction rates of NO and $N_2$ formation from fuel nitrogen," *Symposium (International) on Combustion*, vol. 15, no. 1, pp. 1093–1102, 1975.

# 8

# Physical and Combustion Characteristics of Briquettes Made from Water Hyacinth and Phytoplankton Scum as Binder

**R. M. Davies[1] and O. A. Davies[2]**

[1] Department of Agricultural and Environmental Engineering, Niger Delta University, PMB 071, Yenagoa, Bayelsa State, Nigeria
[2] Department of Fisheries and Aquatic Environment, Rivers State University of Science and Technology, PMB 5080, Port Harcourt, Rivers State, Nigeria

Correspondence should be addressed to R. M. Davies; rotimidavies@yahoo.com

Academic Editor: Essam Eldin Khalil

The study investigated the potential of water hyacinths and phytoplankton scum, an aquatic weed, as binder for production of fuel briquettes. It also evaluated some physical and combustion characteristics. The water hyacinths were manually harvested, cleaned, sun-dried, and milled to particle sizes distribution ranging from <0.25 to 4.75 mm using hammer mill. The water hyacinth grinds and binder (phytoplankton scum) at 10% ($B_1$), 20% ($B_2$), 30% ($B_3$), 40% ($B_4$), and 50% ($B_5$) by weight of each feedstock were fed into a steel cylindrical die of dimension 14.3 cm height and 4.7 cm diameter and compressed by hydraulic press at pressure 20 MPa with dwell time of 45 seconds. Data were analysed using analysis of variance and descriptive statistics. Initial bulk density of uncompressed mixture of water hyacinth and phytoplankton scum at different binder levels varied between 113.86 ± 3.75 ($B_1$) and 156.93 ± 4.82 kg/m³ ($B_5$). Compressed and relaxed densities of water hyacinth briquettes at different binder proportions showed significant difference $P < 0.05$. Durability of the briquettes improved with increased binder proportion. Phytoplankton scum improved the mechanical handling characteristics of the briquettes. It could be concluded that production of water hyacinth briquettes is feasible, cheaper, and environmentally friendly and that they compete favourably with other agricultural products.

## 1. Introduction

The Niger Delta of Nigeria is characterized by extensive network of rivers and creeks which discharge their waters into the Atlantic Ocean. As a result, fishing is the major occupation of its inhabitants [1]. One of the most invasive and prolific aquatic weeds that devastate lakes, canals, rivers, and ponds in the Niger Delta is water hyacinth (*Eichhornia crassipes*). This aquatic weed blooms heavily in the Niger Delta due to favourable climatic condition [2]. In Niger Delta, the average weight or volume of fuelwood per day (16.45 kg or 7.5 m³) exceeds the Food and Agriculture Organization (FAO) average allowance (0.46 m³) [1].

The major source of energy in the rural community is fuelwood as other sources of energy are either not available or grossly inadequate. The demand for fuelwood is expected to rise to about 213.4 × 10³ metric tonnes, while the supply will decrease to about 28.4 × 10³ metric tonnes by the year 2030 [3]. Increasing pressure on forest resources for energy has led to what is called "other energy crisis of wood fuel" [4]. This has led to environmental degradation, deforestation, and misuse of soil forests and water resources. The uncontrolled level of cutting of wood for firewood and charcoal for combustion and for other domestic and industrial uses is now a serious problem in Nigeria. Total annual consumption of wood in Nigeria is estimated about 50–55 million cubic meters of which 90% is firewood, while estimated shortfall of fuelwood in the northern part of the country is about 5–8 million cubic meters [5]. The annual deforestation of the wood lands in the northern part of Nigeria runs to about 92,000 hectare a year. The fuelwood extraction rate in the country is estimated to be about 3.85 times the rate of regrowth or afforestation.

Water hyacinth is an aquatic weed that grows at an extremely rapid pace and its production is about 2 tonnes of biomass per acre and population doubles every 5–15 days [6]. The harvest frequency for aquatic plants tends to be in the order of days, whereas the frequency for trees and crops is in the order of years and months. The abundance, availability, low cost, and rapid growth of water hyacinths make them an ideal candidate for biofuel, particularly in the developing countries [7]. The objectives of this study are to investigate the potential of water hyacinths and phytoplankton, an aquatic weed, as organic binder for production of fuel briquettes and also determine some physical and combustion characteristics of the briquettes.

## 2. Materials and Methods

This study involved collection of samples in Port Harcourt, Niger Delta located between latitudes 40 2$''$ and 60 2$''$ north of the equator and longitudes 50 1$''$ and 70 2$''$ east of the Greenwich meridian [2]. Water hyacinth was harvested manually from the earthy fish ponds using drag net. Phytoplankton scum was harvested from concrete fish pond using scoop net and subsequently sun-dried. It was ground to fine particle size using plate mill and later sieved to particle size 0.075 mm with Tyler sieve. Water hyacinth sample was cleaned to be devoid of foreign matters (stone, dust, and plant materials) prior drying. It was sun-dried and finally milled to particle sizes distribution ranging from <0.25 to 4.75 mm using hammer mill. The particle size distribution was achieved by using particle size analysis equipment consisting of sieve shaker and Tyler sieves of various diameter or particles size openings. The percentages of binder used in the mixture were 10, 20, 30, 40, and 50% of residue weight. The agitating process was done in a mixer to enhance proper blending prior compaction. A steel cylindrical die of dimension 14.3 cm height and 4.7 cm diameter was used for this study. The die was freely filled with known amount of weight (charge) of each sample mixture and positioned in the hydraulic powered press machine for compression into briquettes.

The piston was actuated through hydraulic pump at the speed of 30 mm/min of piston movement to compress the sample. Compacted pressure was 20 MPa. A known pressure was applied at a time to the material in the die and allowed to stay for 45 seconds (dwell time) before released and the briquette formed was extruded. A stopwatch was used for the purpose of timing. Prior to the release of applied pressure to the maximum depth of piston movement was measured for the purpose of calculating the volume displacement to enable the determination of compressive density of the briquette. Each briquette was replicated three times according to the level of process variables. The moisture content of the ground material before and after compaction was determined using ASABE [8] standard. Bulk density of the loose materials was determined according to ASABE [8] standard.

Compressed density was determined according to Bamgboye and Bolufawi [9] and Olorunnisola [10]. A steel cylindrical die of 14.3 by 4.7 cm was filled with 50 g of each sample mixture and was hydraulically compressed. A known pressure was applied through hydraulic at a time to the material in the die and allowed for 45 secs (dwell time) before released. The pressure was monitored through dial gauge installed

on the machine. A stopwatch was used for timing. After compression, the height of briquette was measured and the volume was calculated. The briquette density was calculated by dividing the average mass of the briquette by its volume. The height and diameter of the briquette were consistently measured until they were stable. The stable height and diameter were used to calculate the new volume of the briquette since the charge was known initially.

Relaxed density and relaxation ratio were calculated as the ratio of compressed density to relaxed density according to Bamgboye and Bolufawi [9] and Olorunnisola [10]:

$$\text{relaxation ratio} = \frac{\text{compressed density}}{\text{relaxed density}}. \quad (1)$$

Percentage of water resistance capacity of dry briquette (10.3% wet basis) when immersed in cylindrical glass container containing distilled water at $29 \pm 2°C$ for 120 seconds was investigated. Relative change in weight of the briquette was measured. Percentage water gain was calculated using the following relationship:

$$\%\text{water gained by briquette} = \frac{M_2 - M_1}{M_1}, \quad (2)$$

where $M_1$ is the Initial weight of briquette before immersion and $M_2$ is the Final weight of briquette after immersion.

The equation becomes

$$\text{water resistance capacity}\%$$
$$= 100\% - \%\text{water absorbed.} \quad (3)$$

See [11]. Briquettes shattering index (durability index) was determined according to ASTM D440-86 [12] of drop shatter developed for coal using the following equation:

$$\text{shattering index}$$
$$= \frac{\text{weight of briquettes retained on the screen after dropping}}{\text{weight of briquettes before dropping}}. \quad (4)$$

The static coefficient of friction of briquettes was investigated with respect to four test surfaces, namely, galvanised steel, rubber, plywood sheet, and aluminium sheet. A fiber box of 75 mm length, 50 mm width, and 30 mm height without base or lid was filled with sample and placed on an adjustable tilting plate, faced with test surface. The sample container was raised slightly (10 mm) so as not to touch the surface. The inclination of the test surface was increased gradually with a screw device until the box just started to slide down and the angle of tilt was measured from a graduated scale. For each replicate, the briquettes in the container were emptied and refilled with new briquettes. The static coefficient of friction was calculated as described by Mohsenin [13].

Briquette burning rate was determined according to the method used by Onuegbu et al. [14]. The insulator, Bunsen burner, tripod stand, and wire gauze were arranged on the balance and their weight was recorded. Briquette sample of known weight was placed on wire gauze and the burner ignited. This was positioned on top of a mass balance monitored to record instantaneous measurements of the mass every 10 seconds throughout the combustion process using a stopwatch, until the briquettes were completely burnt and constant weight was obtained. The weight loss at specific time was computed from the expression

$$\text{burning rate} = \frac{\text{total weight of the burnt briquette}}{\text{total time taken}}. \quad (5)$$

Calorific value of the sample was determined using Gallenkamp Ballistic Bomb Calorimeter according to ASTM E711-87 (2004).

Ignition time was determined according to Onuegbu et al., [14]. Each briquette was ignited by placing a Bunsen burner on a platform 4 cm directly beneath. Bunsen burner was used to ensure that the whole of the bottom surface of the briquette was ignited simultaneously after adjusting it to blue flame. Caution was taken to avoid flame spread in the transverse directions. The burner was left in until the briquette was well ignited and had entered into its steady state burn phase.

Thermal fuel efficiency of the energy was calculated according to Oladeji [15]

$$\text{TFE} = \frac{M_w C_p \ (T_b - T_o) + M_c L}{M_f E_f} \times 100\%. \quad (6)$$

The numerator gives the net heat supplied to the water while the denominator gives the net heat liberated by the fuel, where TFE is the thermal fuel efficiency of the energy, $M_w$ is the mass of water in the pot (kg), $C_p$ is the specific heat of water (kJ/kgK), $T_o$ is the initial temperature of water (K), $T_b$ is the boiling temperature of the water (K), $M_c$ is the mass of water evaporated (kg), $L$ is the latent heat of evaporation (kg), $M_f$ is the mass of fuel burnt (kg), $E_f$ is the calorific value of the fuel (kJ/kg).

## 3. Results and Discussion

The obtained values for initial density of uncompressed mixture of water hyacinth at different binder levels varied from 113.86 to 156.93 kg/m$^3$ (Figure 1). The initial bulk density increase with increase in binder proportion. This signified a desirable development for the densification process. The obtained result is much lower than the corresponding values of uncompressed bulk densities of the residue materials as reported by Oladeji [16] were 95.33 and 98.00 kg/m$^3$ for corncob from white and yellow maize. The initial bulk density of uncompressed rice husk and corncob and water hyacinth of 138 kg/m$^3$, 155 kg/m$^3$, and 40 kg/m$^3$ as reported [17, 18]. These values are higher than the minimum value of 40 kg/m$^3$ recommended for wooden materials [19, 20]. Importance of these results indicated the actualization of volume reduction of the raw material which provides a technological benefit. Density is an important parameter, which characterizes the briquetting process. If the density is higher, the energy/volume ratio is higher too. Hence, high density products are

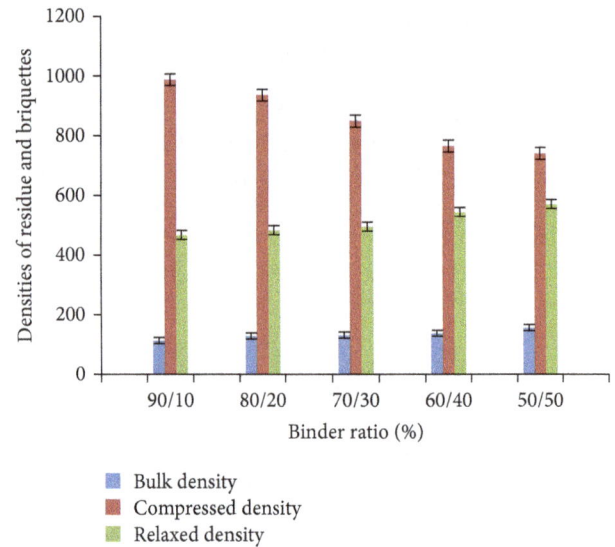

FIGURE 1: Densities of water hyacinth residue and briquettes.

desirable in terms of transportation, storage, and handling and are more cost effective than the natural state.

*3.1. Compressed Density of Briquettes and Binder Proportions.* Compressed densities of briquettes at the different binder proportions are presented in Figure 1. The recorded values showed an increase in binder (10–30%) with decreased compressive density (739.61 ($B_5$) to 987.65 kg/m$^3$ ($B_1$)). The decline observed in compressed density with increased binder inclusion could be attributed to the binder occupying pores in-between the particles of water hyacinth. The recorded values of compressed density were higher than those of the initial bulk density (113.86 to 156.93 kg/m$^3$) of the uncompressed mixture of water hyacinth and binder. It is clearly shown that compressed density is inversely proportional to binder proportions. This trend was in disagreement with the values reported for production of fuel briquettes from waste paper and coconut husk admixture ranging from 8.1 to 11.2 kg/m$^3$ at different binder levels [10]. The effect of binder proportion on compressed density was studied and it was observed that the difference in binder type and blending ratio had significant effect on the compressed density of the briquettes ($P < 0.05$) [21]. These values are higher than the initial densities of the uncompressed mixture of corncob from white maize 151 to 235 kg/m$^3$ and 145 to 225 kg/m$^3$ for corncob from yellow maize. The values of compressed densities obtained are more than the minimum recommended value of 600 kg/m$^3$ and for efficient transportation and safe storage [20]. The equation representing the relationship between bulk density of mixture of water hyacinth and phytoplankton residue ($B_d$), compressed density ($C_d$), and relaxed density ($R_d$) of water hyacinth briquettes with binder ratio ($B_r$) and their coefficient of determination ($R^2$) are presented as follows. The relationship existing between initial bulk density ($B_d$) and binder ratio ($B_r$) appears to be linear and a strongly positive correlation. Compressed and relaxed

TABLE 1: Feedstock particle size distribution for production of briquettes.

| Sieve size (mm) | Percentage of material retained on the sieve (%) |
|---|---|
| 4.75 | 1 |
| 3.0 | 3 |
| 2.0 | 4 |
| 1.0 | 17 |
| 0.5 | 32 |
| 0.25 | 28 |
| <0.25 | 15 |

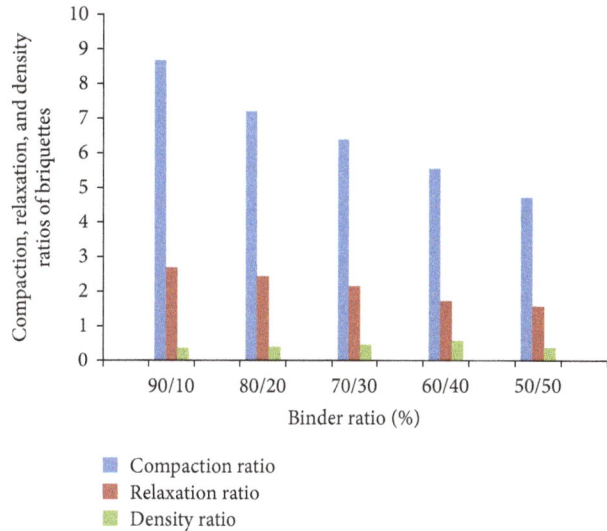

FIGURE 2: Effect of binder on compaction, relaxation, and density ratios of briquettes.

densities of the briquettes also showed linear relationship with very high coefficient of determination:

$$B_d = 4.7105B_r + 105.92, \qquad R^2 = 0.9189,$$
$$C_d = 33.28B_r + 1054.8, \qquad R^2 = 0.9409, \qquad (7)$$
$$R_d = 13.495B_r + 430.84, \qquad R^2 = 0.9764.$$

The interaction between relaxed density and binder levels varied from $466.97 \pm 7.91$ kg/m$^3$ ($B_1$) to $571.24 \pm 10.37$ kg/m$^3$ ($B_5$) (Figure 1). The relaxed density increased with the increasing binder proportion. It could be inferred that the optimum amount of binder required for densification was 50% ($B_5$). At this level of binder, the produced briquettes have the required strength to withstand handling, transportation, and storage. Conversely, the corresponding report revealed that the binder types and blending ratio had no significant influence ($P > 0.05$) on compressed density [21]. The binder (phytoplankton scum) used competed favourably with more than 50 organic and inorganic binders that have been reported for densification. A similar trend was reported on the relationship between relaxed density and binder proportions [22, 23]. Those studies reported increased the relaxed density with the increase in binder proportion for the production of sawdust and palm oil sludge briquettes. Increase in relaxed density with increased binder proportion was equally observed for production of some briquettes from sawdust, rice husk, peanut shell, coconut fibre, and palm fibre [24] (Table 1).

The effect of binder on the compaction ratio ranged from 4.713 ($B_1$) to 8.684 ($B_1$) for all the five binder proportions utilized (Figure 2). This is an indication that the volume displacement is high. This is good for packaging, storage and transportation and above all, it is an indication of good quality briquettes. This showed that void spaces are expelled at higher binder ratio. There was more resistance to compression as the binder ratio increased.

The values of compaction ratio obtained in this study compare and compete favourably with other biomass residues. Compaction ratio of 3.80 for briquetting of rice husk was observed [17, 18]. Compaction ratios of 3.5 and 4.2 were reported for densification of groundnut and melon shells, respectively [25]. Compaction ratio varied from 3.194 to 9.730 for briquettes from Guinea corn (*Sorghum Bi-color*) residue [9]. The compaction ratio of briquettes produced from white corncob increased with increasing binder ratio [16].

The relaxation of the briquettes varied from $1.569 \pm 0.12$ ($B_1$) to $2.691 \pm 0.07$ ($B_5$) for the five studied binder levels (Figure 2). The difference in the relaxation ratio of briquette at the different binder proportions was significant ($P < 0.001$). The obtained range of relaxation ratio in this study is still within the reported range of 1.8 to 2.5 and 1.65 to 1.8 [10, 26]. Relaxation ratio values were 1.11 and 1.32 for briquettes produced from charcoal and Arabic gum, respectively, but briquettes made from charcoal and cassava starch had relaxation ratio values of 1.17 and 1.34 [21]. The obtained values of relaxation ratio signified that briquettes of low relaxation ratio exhibited low elastic property and more stability while briquettes of high relaxation ratio exhibited high tendency of elastic property and less stability. A similar observation was made for briquettes produced from hay material which had relaxation ratio of 1.68 to 1.8 [26]. The lower values ratio indicates a more stable briquette, while higher value indicates high tendency towards relaxation, that is, less stable briquette.

The density ratio of the briquettes ranged from $0.371 \pm 0.02$ ($B_1$) to $0.580 \pm 0.07$ ($B_5$) for the five studied binder levels (Figure 2). The obtained range of relaxation ratio in this study is within the reported range of 0.173 to 0.497 [9].

The result of water resistance property of the briquettes varied from $52 \pm 2.42\%$ ($B_1$) to $97.1 \pm 3.39\%$ ($B_5$) for the five studied binder levels (Figure 3). It was observed that the briquette produced from binder (50%) had good hygroscopic properties as compared to the briquettes from the other four combinations. The briquette from $B_5$ exhibits the least water absorption characteristic. This is an indication that hygroscopic property of briquettes at different binder proportions showed a decrease in water absorption capacity with increased quantity of utilized binder. The percentages of water resistance penetration of carbonized cashew shell, rice husk, and grass briquettes were investigated when immersed in water at 27°C for 30 seconds. It was observed that the briquetted fuel from carbonized cashew shell had low percentage

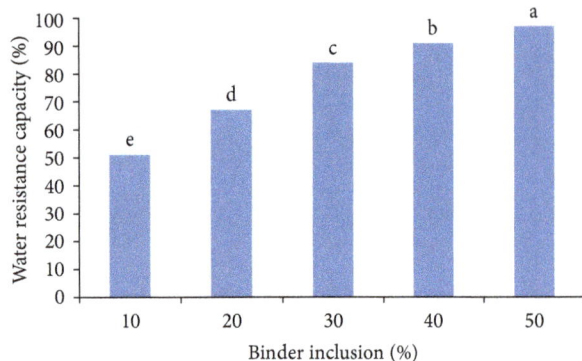

FIGURE 3: Effect of binder on water resistance of briquette. Means of the same letter are significantly different ($P < 0.05$).

TABLE 2: Mechanical handling characteristic of briquettes.

| Binder ratio | Shattering index | Crushing strength (N) |
|---|---|---|
| 90 : 10 | $0.59 \pm 0.01$ | $104.00 \pm 3.86$ |
| 80 : 20 | $0.74 \pm 0.03$ | $139.60 \pm 4.17$ |
| 70 : 30 | $0.89 \pm 0.04$ | $164.00 \pm 3.08$ |
| 60 : 40 | $0.91 \pm 0.02$ | $222.30 \pm 5.26$ |
| 50 : 50 | $0.98 \pm 0.03$ | $263.50 \pm 5.76$ |

of water resistance penetration of less than 10% as compared to the briquetted fuel from carbonized rice husk and grass that had percentage of water resistance penetration of about 35 and 45%. The briquetted fuel from carbonized cashew shell required minimum energy for production and low water absorption properties [11].

The effect of binder proportion on the shattering index of the briquettes was conducted as shown in Table 2. The mean shattering index ranged between $0.59 \pm 0.01$ ($B_1$) and $0.98 \pm 0.03$ ($B_5$) and variation of the values was significant ($P < 0.001$). It could be inferred that the amounts of binder used have significant influence on the durability rating of the briquettes ($P < 0.05$). The mean values of shattering index for binders $B_1$ ($0.59 \pm 0.01$) and $B_2$ ($0.74 \pm 0.03$) were low and showed significant difference ($P < 0.05$); thus, they might not be suitable for briquettes production.

Meanwhile, the mean values of shattering index of $B_4$ ($0.91 \pm 0.01$) and $B_5$ ($0.98 \pm 0.01$) fall within the acceptable range of DIN51731 [27] and Kaliyan and Morey [19] for production briquettes. This implies that $B_5$ is the optimum binder level requirement to produce durable, reliable, and stable briquettes that stand mechanical handling and transportation, with economical feasibility and environmental friendliness. It discovered that increase in binder proportion and types of binder have a significant effect on the durability rating of the briquettes [21, 22].

The effect of types of binders and quantity on the durability of briquettes was reported [28]. It was observed that adding 10–25% (by weight) of molasses or sodium silicate or a mixture of 50% molasses and 50% sodium silicate with rice straw produced briquettes with 40–80% durability at a particle size of 0.15 mm and forming pressure of 29.4 MPa [19]. It was also found that the higher the amount of binder inclusion, the higher the briquette durability rating. Addition (by weight) of any of the following six binders did not improve the alfalfa pellet durability over the control: 4% bentonite, 1.5% Perma-Pel (lignosulfonate), 1.5% Lignosite 458, 4% of neutralized liquid Lignosite, 4% of liquid molasses, and 40% of ground barley grain [19].

The interaction between crushing strength and binder levels varied from $104.00 \pm 3.86$ N ($B_1$) to $263.50 \pm 5.76$ N ($B_5$) (Table 2). The load required to rupture briquettes at different

binder ratios was significantly different ($P < 0.05$). The crushing strength increased with increasing binder proportion. This is an indication that phytoplankton as binder has a good binding power that competed favourably with binders from other biomasses. It could be inferred that the optimum amount of binder required to produce high quality briquettes is 50% ($B_5$). At this level of binder, the produced briquettes have the required strength to withstand handling, transportation and storage.

The coefficient of static friction of the water hyacinth briquette ranged from 0.35 ($B_5$) to 0.47 ($B_1$) for galvanized steel as shown in Table 3. This is an indication that at higher binder ratio, the briquette becomes more pliable and smoother due to the glossy nature of phytoplankton. The briquettes on aluminium and plywood sheet had the lowest and highest values for static coefficient of friction at different binder ratio, respectively. The lowest static coefficient of friction was recorded for aluminium sheet, 0.31 ($B_5$). The highest static coefficient of friction corresponds to plywood sheet, 0.56 ($B_1$). The coefficient of static friction of water hyacinth briquette on plywood sheet at different binder ratios were statistically different ($P < 0.05$).

The obtained values of thermal fuel efficiency of water hyacinth briquettes are shown in Table 4. The results showed that increased binder subsequently increasing the thermal fuel efficiency of briquettes from $19.67 \pm 0.23$ ($B_1$) to $31.73 \pm 0.93$% ($B_5$). The result of analysis variance showed that there was a significant difference among the obtained values ($P < 0.05$). Binder $B_4$ could be regarded as the optimum binder level required to produce briquettes of acceptable thermal fuel efficiency and low smoke as compared to firewood briquettes. The consequent of selecting any other binder level higher than binder $B_4$ amounted to energy and economic losses. In addition, it could be inferred that any increase in binder proportion beyond $B_4$ has no significant influence on the fuel efficiency of the briquettes. However, briquettes with binder levels lower than $B_4$ might not be acceptable.

The effect of binder on ignition time of the briquettes varied from $73.54 \pm 3.37$ sec ($B_1$) to $123.42 \pm 3.47$ sec ($B_5$) as shown in Table 4. The obtained trend of the ignition time indicated that ignition time increased with increasing binder proportion. The recorded lowest ignition time ($73.54 \pm 3.37$ sec) recorded for $B_1$ could be attributed to high porosity exhibited between inter- and intraparticles which enable easy percolation of oxygen and outflow of combustion briquettes due to low bonding force. The values were significantly different at all levels of binder ($P < 0.05$). Ignition time for 100% coal briquette sample took 286 sec to ignite [15].

TABLE 3: Coefficient of static friction.

| Binder ratio | Galvanised steel | Rubber | Plywood sheet | Aluminium sheet |
|---|---|---|---|---|
| 90 : 10 | 0.47 (±0.02) | 0.48 (±0.03) | 0.56 (±0.02) | 0.43 (±0.01) |
| 80 : 20 | 0.46 (±0.01) | 0.44 (±0.02) | 0.51 (±0.03) | 0.37 (±0.01) |
| 70 : 30 | 0.41 (±0.05) | 0.44 (±0.03) | 0.49 (±0.01) | 0.38 (±0.03) |
| 60 : 40 | 0.37 (±0.03) | 0.38 (±0.01) | 0.45 (±0.04) | 0.33 (±0.05) |
| 50 : 50 | 0.35 (±0.04) | 0.34 (±0.02) | 0.43 (±0.03) | 0.31 (±0.03) |

TABLE 4: Combustion characteristics of water hyacinth briquettes and binder proportions.

| Combustion parameters | Binder ratio | | | | |
|---|---|---|---|---|---|
| | 10% | 20% | 30% | 40% | 50% |
| Thermal fuel efficiency (%) | 19.67 ± 0.23d | 21.82 ± 0.35c | 23.67 ± 0.21b | 31.24 ± 0.48a | 31.73 ± 0.93a |
| Calorific value (Kcal/kg) | 3563 ± 76.94e | 3791 ± 83.15d | 3864 ± 41.03c | 4195 ± 32.96b | 4281 ± 90.78a |
| Ignition time (min.) | 73.54 ± 3.37e | 88.27 ± 1.23d | 93.54 ± 3.82c | 114.37 ± 4.12b | 123.42 ± 3.47a |
| Burning rate (g/min) | 2.25 ± 0.01a | 2.01 ± 0.03b | 1.89 ± 0.04c | 1.71 ± 0.02d | 1.63 ± 0.02e |

Means with same letter along the column are not significantly different ($P > 0.05$).

The effect of binder on the burning rate was studied. From Table 4, burning rate of water briquettes significantly varied between $1.63 \pm 0.02 \, g^{-1} \, min$ ($B_5$) and $2.25 \pm 0.01 \, g^{-1} \, min$ ($B_1$) ($P < 0.05$). The obtained burning rate values of the briquettes decreased with increasing binder proportion. The implication of this observation is that more fuel might be required for cooking with briquettes produced from $B_1$ than from $B_5$.

The calorific values of briquettes produced from mixture of water hyacinth and binder at different levels are presented in Table 4. The calorific values of the briquettes ranged between $3563 \pm 76.94$ kcal/kg ($B_1$) and $4281 \pm 90.38$ kcal/kg ($B_5$). This showed that phytoplankton scum as binder improved the calorific value of water hyacinth from 3190 kcal/kg (sample 100 : 0). The recorded values of calorific values were significant at the different binder levels ($P < 0.05$). Adegoke [29] reported an improvement in calorific value of briquettes of palm kernel shell mixed with sawdust from 19.91 MJ/kg (4755.4 kcal/kg) MJ/kg to 20.54 MJ/kg (4905.9 kcal/kg). The calorific value of the briquettes is within the acceptable range for commercial briquette (>4179.8 kcal/kg) DIN 51731 [27]. It was observed that briquettes produced from binder ratio $B_1$ to $B_3$ are not be suitable for the production of commercial briquettes.

## 4. Conclusion

The optimum binder level required to produce the briquettes with the highest durability strength is 50% binder ratio. The best shatter and durability indices showed that they have good shock and impact resistance and are good for handling and transportation. They also have good density ratio. Therefore, combination of water hyacinth and phytoplankton scum is very suitable for briquette production for domestic and industrial uses. The physical and mechanical handling characteristics of water hyacinth briquettes compete favourably with other biomass briquettes. Binder $B_4$ could be regarded as the optimum binder level required to produce briquettes of acceptable thermal fuel efficiency and low smoke as compared to firewood briquettes. Water hyacinth only without

binder might not satisfy the minimum calorific value. Utilization of phytoplankton scum, an aquatic weed, as organic binder exhibits a good binding characteristic.

## Conflict of Interests

The authors declared that there is no conflict of interests.

## References

[1] O. A. Osi, *Survey of fish processing machinery in Bayelsa State [B.Sc. Thesis]*, Niger Delta University, Nigeria, 2008.

[2] C. C. Tawari, *Effectiveness of agricultural agencies in fisheries management and production in the Niger Delta, Nigeria [Ph.D. thesis]*, Rivers State University of Science and Technology, Port Harcourt, Nigeria, 2006.

[3] A. O. Adegbulugbe, "Energy-environmental issites in Nigeria," *International Journal of Global Energy Issues*, vol. 6, no. 12, pp. 7–18, 1994.

[4] J.-F. K. Akinbami, "Renewable energy resources and technologies in Nigeria: present situation, future prospects and policy framework," *Mitigation and Adaptation Strategies for Global Change*, vol. 6, no. 2, pp. 155–181, 2001.

[5] Nigeria Environmental Action Team (NEST), "Nigeria Threatened Environment. A Natural Profile 'Atmosphere'," NEST, Ibadan, Nigeria.Pp. 116–117, 2001.

[6] M. A. Olal, M. N. Muchilwa, and P. L. Woomer, "Water Hyacinth, Utilizations and the use of waste material for Handicraft production in Kenya," in *Micro and Small Enterprises and Natural Resource Use*, D. L. M. Nightingale, Ed., pp. 119–127, Micro-Enterprises Support Programme UNRP, Nairobi, Kenya, 2001.

[7] D. Sophie, "A fast-Growing Plant Becomes mod furniture," in *Connecticut Cottage Gardens*, Cuttoges and Gardens, Norwalk, Conn, USA, 2006.

[8] American Society of Agricultural and Biological Engineering (ASABE), "Cubes, pellet and crumbles definitions and methods for determining density, durability and moisture content," ASAE DEC96, St. Joseph, Mich, USA, 2003.

[9] A. Bamgboye and S. Bolufawi, "Physical characteristics of briquettes from Guinea corn (sorghum bi-color) residue," *Agricultural Engineering International*, article 1364, 2008.

[10] A. O. Olorunnisola, "Production of fuel briquettes from waste paper and coconut husk admixtures," *Agricultural Engineering International*, vol. IX, article EE 06 066, 2007.

[11] S. H. Sengar, A. G. Mohodl, Y. P. Khandetod, S. S. Patil, and A. D. Chendake, "Performance of briquetting machine for briquette fuel," *International Journal of Energy Engineering*, vol. 2, no. 1, pp. 28–34, 2012.

[12] American Society for Testing and Materials (ASTM. D440-86), "Standard test method of drop shatter test for coal," in *Annual Book of ASTM Standards*, vol. 05, pp. 188–191, West Conshohocken, Pa, USA, 1998.

[13] N. N. Mohsenin, *Physical Properties of Plant and Animal Materials*, Gordon and Breach Press, New York, NY, USA, 1986.

[14] T. U. Onuegbu, U. E. Ekpunobi, I. M. Ogbu, M. O. Ekeoma, and F. O. Obumselu, "Comparative studies of ignition time and water boiling test of coal and biomass briquettes blend," *International Journal of Research & Reviews in Applied Sciences*, vol. 7, pp. 153–159, 2012.

[15] J. T. Oladeji, *The effects of some processing parameters on physical and combustion characteristics of corncob briquettes [Ph.D. thesis]*, Department of Mechanical Engineering, Ladoke Akintola University of Technology, Ogbomoso, Nigeria, 2011.

[16] J. T. Oladeji, "A comparative study of effects of some processing parameters on densification characteristics of briquettes produced from two species of corncob," *The Pacific Journal of Science and Technology*, vol. 13, no. 1, pp. 182–192, 2012.

[17] J. T. Oladeji, "Pyrolytic conversion of sawdust and rice husk to medium grade fuel," in *Proceedings of the Conference of the Nigerian Institute of Industrial Engineers (NIIE '10)*, pp. 81–86, Ibadan, Nigeria, April 2010.

[18] R. M. Davies and U. S. Mohammed, "Moisture-dependent engineering properties water hyacinth parts," *Singapore Journal of Scientific Research*, vol. 1, no. 3, pp. 253–263, 2011.

[19] N. Kaliyan and R. Morey, "Densification characteristics of corn stover and switchgrass," in *Proceedings of the ASABE Annual International Meeting*, paper 066174, St. Joseph, Mich, USA, 2006.

[20] S. Mani, L. G. Tabil, and S. Sokhansanj, "Specific energy requirement for compacting corn stover," *Bioresource Technology*, vol. 97, no. 12, pp. 1420–1426, 2006.

[21] O. A. Sotannde, A. O. Oluyege, and G. B. Abah, "Physical and combustion properties of charcoal briquettes from neem wood residues," *International Agrophysics*, vol. 24, no. 2, pp. 189–194, 2010.

[22] O. A. Ajayi and C. T. Lawal, "Hygroscopic and combustion characteristics of sawdust briquettes with palm oil sludge as binder," *Journal of Agricultural Engineering and Technology*, vol. 5, pp. 29–36, 1997.

[23] W. H. Engelleitner, "Binders: how they work and how to select one," *Powder and Bulk Engineering*, vol. 15, no. 2, pp. 31–37, 2001.

[24] O. C. Chin and K. M. Siddiqui, "Characteristics of some biomass briquettes prepared under modest die pressures," *Biomass and Bioenergy*, vol. 18, no. 3, pp. 223–228, 2000.

[25] J. T. Oladeji, C. C. Enweremadu, and E. O. Olafimihan, "Conversion of agricultural wastes into biomass briquettes," *International Journal of Applied Agricultural and Apiculture Research*, vol. 5, no. 2, pp. 116–123, 2009.

[26] M. J. O'Dogherty, "A review of the mechanical behaviour of straw when compressed to high densities," *Journal of Agricultural Engineering Research*, vol. 44, no. C, pp. 241–265, 1989.

[27] Deutsches Institut für Normunge, "Testing on solid fuels compresses untreated wood-requirements and testing," V. DIN, 51731, 1996.

[28] J. P. Singh, T. C. Thakur, S. Sharma, and R. K. Srivastava, "Effect of manner of stacking on changes in nutritional value of treated baled paddy straw by dripping technique," *Agricultural Mechanization in Asia, Africa and Latin America*, vol. 42, no. 4, pp. 84–87, 2011.

[29] C. O. Adegoke, "Preliminary investigation of sawdust as high grade solid fuel," *Journal of Renewal Energy*, vol. 1-2, pp. 102–107, 1999.

# 9

# A Comparison of the Characteristics of Planar and Axisymmetric Bluff-Body Combustors Operated under Stratified Inlet Mixture Conditions

**G. Paterakis, K. Souflas, E. Dogkas, and P. Koutmos**

*Laboratory of Applied Thermodynamics, Department of Mechanical Engineering and Aeronautics, University of Patras, 26504 Patras, Greece*

Correspondence should be addressed to P. Koutmos; koutmos@mech.upatras.gr

Academic Editor: Constantine D. Rakopoulos

The work presents comparisons of the flame stabilization characteristics of axisymmetric disk and 2D slender bluff-body burner configurations, operating with inlet mixture stratification, under ultralean conditions. A double cavity propane air premixer formed along three concentric disks, supplied with a radial equivalence ratio gradient the afterbody disk recirculation, where the first flame configuration is stabilized. Planar fuel injection along the center plane of the *leading face* of a slender square cylinder against the approach cross-flow results in a stratified flame configuration stabilized alongside the wake formation region in the second setup. Measurements of velocities, temperatures, OH* and CH* chemiluminescence, local extinction criteria, and large-eddy simulations are employed to examine a range of ultralean and close to extinction flame conditions. The variations of the reacting front disposition within these diverse reacting wake topologies, the effect of the successive suppression of heat release on the near flame region characteristics, and the reemergence of large-scale vortical activity on approach to lean blowoff (LBO) are investigated. The cross-correlation of the performance of these two popular flame holders that are at the opposite ends of current applications might offer helpful insights into more effective control measures for expanding the operational margin of a wider range of stabilization configurations.

## 1. Introduction

The requirements in the design of fuel flexible, low emissions and versatile transportation and power generation systems are placing heavy demands on the development and performance of current combustion devices [1, 2]. As a result the pursuit of new combustion concepts or modes is ongoing and a significant part of the overall design effort is devoted to maintain satisfactory flame stability and efficient emission targets [2, 3].

Over the years lean premixed combustion has gained popularity and has become widely exploited as it addresses some of the above issues over a range of applications [2–4]. However, the anticipated operation of future combustors under increased loads and mixing and burning rates may result in complications such as sensitivity to mixing, low reaction and heat release rates, extinctions, and instabilities [3–5].

Partially premixing and stratifying the reactive mixture is becoming an increasingly widespread strategy for burner design in the effort to expand the stability margin of the lean fully premixed concept and at the same time meet these combined requirements [2, 4–6]. The effective control of spatially varying local fuel-air ratio distributions within the combustor offers significant operational flexibility. Leaner regions increase efficiency and reduce pollutant formation, and richer mixtures improve ignition performance and allow for better control of instabilities. However nonuniformities in the local fuel-air ratio can also lead to spatially varying combustion performance with implications that have to be addressed in relation to the local turbulent chemistry, the stabilization method, and its interaction with the stability margin (lean blow-out (LBO)) of the particular flame configuration [4–7].

Evidently due to the complexity of the above described phenomena, better knowledge of the *details* of the turbulent transport and reaction processes over a range of operating conditions from lean to ultra-lean and approaching blowoff (LBO) is important, if a full exploitation or even extension of the operational margin is to be realised (e.g., [4, 6–8]). A range of systematic experiments have been performed to study the structure of such flame configurations (e.g., [4–8]) along with a number of supporting simulations that employ various levels of complexity for the turbulence-chemistry treatment and usually involve adaptive local gridding within a time-dependent procedure (e.g., [9]). Phenomenological analyses (e.g., [2, 6, 8]) and global correlations (e.g., [3, 10]) are also utilized to complement the overall development effort and relieve the burden and cost involved in the multiscalar approaches.

These experimental and computational works have so far produced detailed descriptions of the ultra-lean behaviour in traditional stabilizers such as fully premixed axisymmetric (e.g., [3, 4, 7, 8]) and slender two-dimensional (e.g., [6, 11]) bluff-body arrangements. However, stratified bluff-body flames operated at ultra-lean and near blow off conditions have not been investigated as extensively. Moreover the growing combination of burner and mixture placement setups in lean combustors has increased current interest for a more complete description of such flame types. Therefore comparative examinations of the performance of the axisymmetric versus the slender two-dimensional bluff-body stabilization configuration under stratified conditions are warranted; these could provide useful information regarding the more effective control, exploitation, or even extension, for example, through appropriate placement of pilot injection, of the operational margin under both setups (e.g., [2, 4, 6, 11, 12]).

This work then compares the ultra-lean performance and topology characteristics of propane flames stabilized in an axisymmetric, double cavity fuel-air premixer/disk arrangement [13] and a counterinjected 2D slender square cylinder [11] under stratified conditions. The above geometries were chosen because these comprise two of the most popular stabilizers, at the opposite end of the field of practical applications, and comparisons of their limiting performance could provide useful directions in the design process for a wide range of configurations. Identification of the individual characteristics such as, for example, flame and recirculation lengths, flame front topologies, turbulence levels, vortical activity, and under successively weakening stabilization within the same geometry arrangement may help determine suitable methodologies to extend stability in a fashion tailored not only for the particular flame geometries but also for a range of intermediate variants, upgrades, or retrofits [2, 4, 12]. Furthermore both of these setups have been systematically investigated under *fully premixed* conditions and their operation under *stratified* conditions could then be directly contrasted against such a range of well documented data.

Measurements of turbulent velocities, temperatures, chemiluminescence imaging of $OH^*$ and $CH^*$, flame front visualization, and gas analysis provided information for lean, ultra-lean, and close to blowoff flames. Supporting large-eddy simulations were also undertaken employing the thickened flame model [14, 15] and a nine-step mechanism for propane combustion [11, 13] to complement the experimental effort

and assist in the more complete interpretation of the optical measurements.

The combined methodology helped to elucidate some of the parametric characteristics of these diverse flame topologies and improve understanding regarding the differences and similarities between these two stabilizers when operated with a stratified inlet mixture as opposed to a fully premixed operation. In addition, the local extinction behavior was examined by using the combination of three, measured or computed, parameters, namely, a local Karlovitz number, a local temperature threshold, and the ratio of the $OH^*/CH^*$ chemiluminescence signature. The credibility of this parameterization is appraised, where possible, by using both measurements and simulations.

## 2. Flame Stabilization Configurations

The two burner setups are shown in Figures 1(a), 1(b), 1(c), 1(d), 1(e), and 1(f). In the first configuration stratified flames, established by staged fuel-air premixing in a double-cavity arrangement (Figure 1(c)) formed along three concentric disks and stabilized in the vortex region of the afterbody A ($D_b = 0.025$) under the additional action of a coflowing swirl stream (Figure 1(b)) are investigated. The combustion tunnel and the studied burner are similar to those reported in [13] and have been optimized through a series of systematic tests [16]. Fuel is supplied through the central shaft ($D_p = 0.010$) that is running along the central air supply tube ($D_c = 0.052$) into the internal hollow of the active bluff-body disk B (Figure 1(c)). Then it is injected through an annular 1 mm slot into the primary fuel air-mixing cavity and is mixed with the central air supply. This maintains an afterbody equivalence ratio gradient between a $\Phi_{min} \approx 0.35$ to $0.20$ and a $\Phi_{max} \approx 0.98$ to $0.5$, depending on flame condition, as measured with flame ionization detection from local gas samples at the afterbody exit (Figure 1(d)). Limiting conditions approaching blowoff were obtained by reducing the fuel level which then decreased the peak $\Phi$ at the primary zone inlet. In all cases a regulated stratified reactive mixture profile is attained across the most favourable position for leading edge flame front stabilization at the disk burner rim (Figure 1(d)). The simulated cavity flows for a lean flame at $S = 1.0$ are shown in Figure 1(d) to illustrate the overall flow patterns obtained within and at the exit of the premixer system; these remain largely unaltered between lean and ultra-lean operation. The global $\Phi$, based on the total mass flows of fuel and central air together with the range of investigated conditions and parameters, is shown in Table 1. The blockage ratio (BR) in the afterbody annular jet supplying the primary zone is BR $= (D_b/D_c)^2 = 0.23$ [13].

The second setup involves the study of partially premixed and stratified flames established by planar injection (discrete jets of small aspect ratio) of propane along the center plane of the *leading face* of a slender square cylinder. Propane is supplied through the cylinder ends, issued through 135 holes of 1 mm diameter spaced at 0.5 mm apart on the symmetry plane over a width of 200 mm along the upstream face of the square, and reaction is stabilized within and adjacent to the downstream wake formation region (Figures 1(e) and

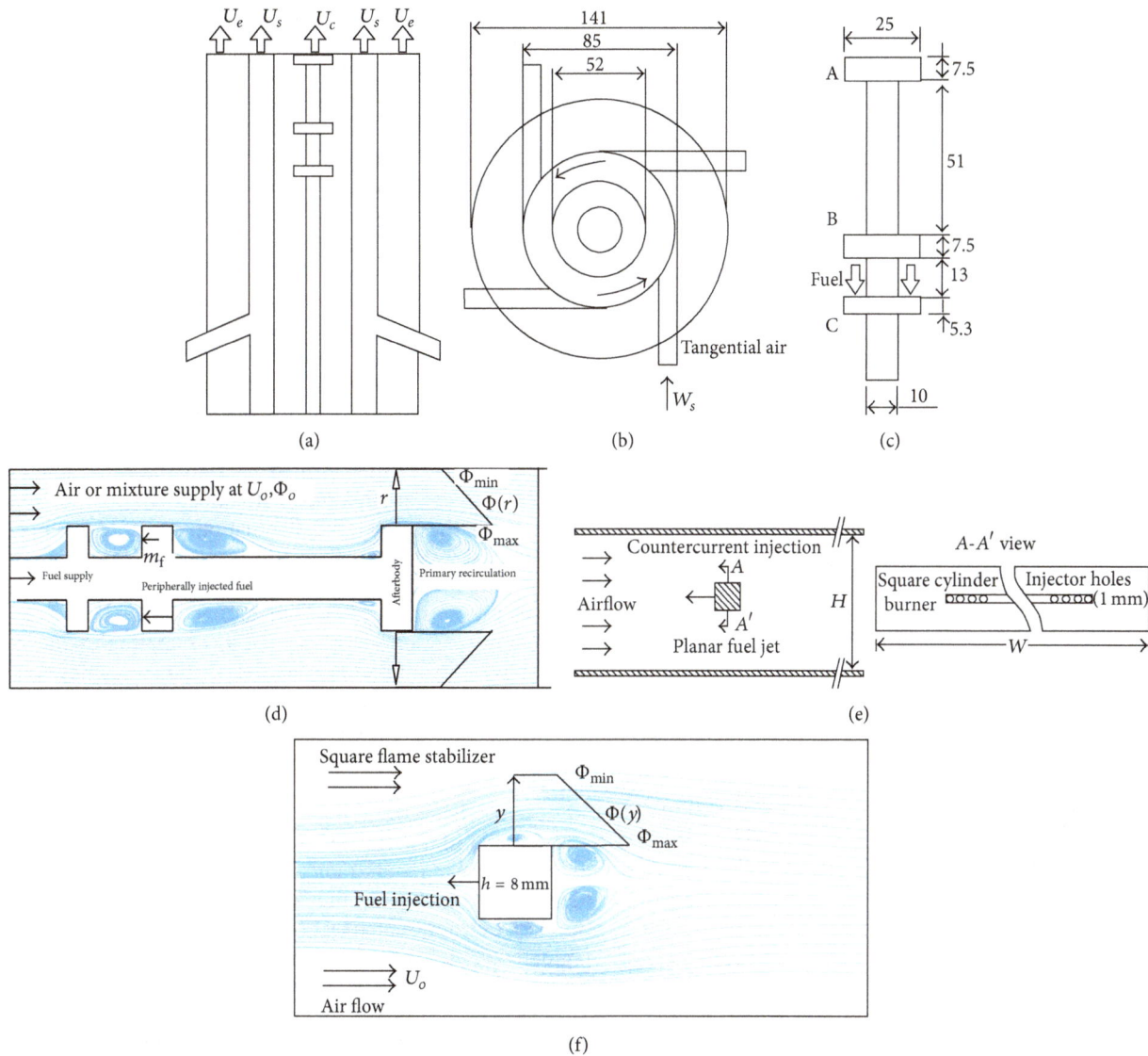

FIGURE 1: Axisymmetric disk ((a), (b), (c), and (d)) and slender square ((e), (f)) burner arrangements, including respective flow patterns, fuel-air placements, and mixture stratifications at inlet to the stabilization regions.

1(f)). The level of stratification across the square flanks is controlled by the velocity ratio between the fuel injection and the approach cross-flow with peak levels ranging from 1.6 to 0.6. The blockage ratio of the square burner in the two-dimensional tunnel was BR = $h/H$ = 0.10 [11]. The Reynolds numbers of the cold flow were maintained at 7980 and 5800 for the disk and the square, respectively.

A high swirl case was chosen for study in the disk configuration. The geometric swirl number, $S$, ($S$ = 1.00 for this investigation), used for the quantitative representation of swirl intensity, was expressed here as the ratio of integrated (bulk) tangential to primary axial air velocities ($W_s/U_s$), measured through LDV in the coflow swirl stream. Square burner flames formed by counterinjection with a low fuel-to-approach-air velocity ratio operated toward the lean LBO limit were chosen for the second case. Under both cases strong recirculation regions assist in flame stabilization under high stretch rates leading to a spread in the disposition of

the flame front alongside the wake flanks. The wake development in the square is affected by periodically asymmetric vortex shedding with strength that drastically varies between isothermal and low or high fuel injection reacting conditions [6, 11, 12, 17]. On the other hand the axisymmetric disk exhibits low frequency vortex rollup from the rim periphery, which is again stronger under isothermal conditions [13, 18]. The *interaction of the locally evolving extinction process with the underlying turbulent wake dynamics* presents several differences between the above two flameholders (e.g., [4, 6–8, 12, 13, 17]). This facilitates a useful evaluation of their different behaviors under ultra-lean operation and allows for an appraisal of the adopted modelling methodology.

All fuel flows were regulated by Bronkhorst MV-304/306 *High-Tech* mass flow controllers. Necessary supply valves, the ignite, and a flame-out detection device were used and were all controlled electronically by a central unit.

TABLE 1: Operating conditions and parameters.

| Case | $\delta$ (%) | $U_{Fuel}$ | $T_{max}$ (K) | $L_f/D_b$ | $L_R/D_b$ | $\Phi_{Global}$ | Ka |
|---|---|---|---|---|---|---|---|
| Disk | | | | | | | |
| Lean | 50 | 1.43 | 1860 | 5.6 | 1.43 | 0.147 | 6 |
| Ultralean | 18 | 1.12 | 1732 | 5.2 | 1.36 | 0.115 | 18.5 |
| Near LBO | 6 | 1 | 1508 | 4.4 | 1.28 | 0.102 | 47 |
| Square | | | | | | | |
| Lean | 94 | 1.24 | 2050 | 8.5 | 5.5 | 0.031 | 15 |
| Ultralean | 65 | 1.05 | 1850 | 3.5 | 3.5 | 0.025 | 35 |
| Near LBO | 14 | 0.73 | 1600 | 1.4 | 1.45 | 0.018 | 72 |

(1) $\delta$ is percent deviation from LBO, $\delta = (m_{Fuel} - m_{Fuel,LBO})/m_{fuel,LBO}$ (%), ($m_{fuel,LBO}$, fuel flow at lean blowoff). $U_{FBO} = 0.95$ m/s for disk and 0.64 m/s for square.

(2) $T_{max}$: maximum measured wake temperature.

(3) $L_f$: visible flame length; $L_R$: measured recirculation length.

(4) Central air supply velocity for disk, $U_c = 4.87$ m/s and $Re_{Db} = 7980$ for all cases, coflow velocities in annular swirl stream at $S = 1.0$, $U_s = W_s = 10.5$ m/s, surrounding shielding stream velocity, $U_e = 10$ m/s. Approach airflow for square, $U_\infty = 6.7$ m/s and $Re_H = 5800$ for all cases.

(5) $\Phi_{Global}$, global equivalence ratio, based on mass flows of injected fuel and, (a) air supply in the central pipe for the disk, or (b) approach airflow over the tunnel height for the square.

(6) $T_{max}$ location for the disk was at $(x/D_b, r/D_b) = (0.4, 0.45)$, while for the square this was at $(x/h, y/h) = (0.7$ to $1.0, 0.9)$.

Flame nomenclature in both cases is denoted on the basis of the proximity to LBO, defined by $\delta = (m_{Fuel} - m_{Fuel,LBO})/m_{Fuel,LBO}$ (%), where $m_{Fuel,LBO}$ is the fuel flow at lean blow-out. The global $\Phi$ based on the total mass flows of fuel and approach air together with the range of investigated conditions and parameters is included in Table 1.

## 3. Experimental Methods

The burners have been previously investigated under isothermal operation to establish the mixing topologies that sustain the reacting fields presented below [11, 13, 16, 19]. Mean and turbulent velocities, temperatures, and OH* and CH* chemiluminescence (CL) images were obtained using a two-component LDV system, thin digitally compensated thermocouples, and a chemiluminescence imaging system (LaVision Flame Master consisting of a CCD camera, an image intensifier IRO unit, camera optical filter at $307 \pm 10$ nm and achromat lens for OH imaging).

Transverse profiles of the time-averaged mean and turbulent streamwise and cross-stream velocities and statistics were obtained with a two-component 2 Watt Argon-Ion laser, fiber optics linked to TSI transmitting and receiving optics and described in detail in [19]. The fuel and air flows were seeded with dried magnesium oxide powder (approximately $5\,\mu$m nominal diameter before agglomeration) dispersed by two purpose-built cyclone separators. In order to minimize bias errors due to unequal particle densities in the fuel and air streams, the two flows were seeded separately, with rates of particles in proportion to the flow rates in each one. Extreme deviations were determined by seeding only one of the two streams. The filtered Doppler signals were processed by a TSI frequency counter (1980B). Mean and statistical values were postprocessed from 20480 data weighed by the time between particles to correct for velocity bias. Velocity spectra were obtained by sampling at 4 kHz and performing a fast fourier transform (FFT) on 20 sets of 1024 values [11, 19].

Temperatures were measured with Pt-Pt/10%Rh uncoated beaded S-type 25 to $75\,\mu$m thermocouples. A twin bore ceramic cladding of $130 \times 0.9$ mm with 0.2 mm capillary holes, oriented along the flow, supported the stem. The T/C output was interfaced to a DaqTemp 7A Omega card. The thermocouple acts as a low-pass filter with a response of the order of 50 Hertz for, for example, $100\,\mu$m wires that depend on local parameters (position, temperature, junction geometry and material, gas velocity, conductivity, heat capacity and density, and flame structure). Here, using an FFT algorithm, the signal's frequency spectrum was estimated, the amplitude and the phase were corrected, and the signal was transformed back to the time domain according to the methodology described in [20], so that about 60% of the temperature fluctuations spectrum could be calculated [11, 19, 20]. No correction was applied for radiation, but for propane flames and uncoated wires, the systematic error in the mean can be up to 10% at 1900 K [19]. Mean and rms temperatures were derived at a sampling frequency of 400 Hz. Maximum uncertainties in the velocities were less than 8.5% in the mean and less than 15% in the rms. Uncertainties in the estimation of the thermocouple time constant are rather unimportant for the mean temperature, but according to the present compensation procedure may affect the variance by between 25 and 35%.

Chemiluminescence measurements of the electronically excited OH* and CH* radicals are frequently employed to identify the topology of the flame front, since, for example, the electronically excited OH* exists mainly in the flame front region (e.g., [4–7, 12]). Image acquisition and data reduction were performed using the Davis 8.0 software from LaVision. The background images taken under the same integration time and gain were subtracted from the originals. Averages were produced from 300 instantaneous images recorded at 16 to 60 frames per second depending on the binning and the interrogation window. The signal-to-noise ratio of the instantaneous images was better than $8:1$, and the exposure time was 2.1 ms under a single frame mode operation. The maximum CCD chip resolution was $1626 \times 1236$ pixels. Intensifier gate times of the order of $100\,\mu$s were used for the measurements with a gain up to about 85%. Since the CCD is combined with an intensifier, the effective exposure is determined by the image intensifier gate, which typically is of the order of several nsecs, whilst the image collection also depends on the attainable CCD recording rate. The camera exposure is a temporal envelope that integrates the phosphor emission of the intensifier whilst maintaining an exposure time greater than the sum of the delay and the gate times.

A direct qualitative comparison between the time-mean simulated OH mole fraction fields with the experimentally measured mean OH* chemiluminescence fields is not immediately possible. This is because the experimental measurements are effectively a 2D image of projected line of sight measurements from a cylindrically symmetric process (in the mean), whilst the numerical results are a 2D axisymmetric slice through the axis of symmetry of the cylindrically symmetric process. To compare the two different datasets, a transformation process is required to be applied to one or both of the datasets. As the time-average flame is axisymmetric, the location of the reaction zone can therefore be observed from converting the ensemble average chemiluminescence image (obtained from a set of 300 images) with an Abel transform. Three-point Abel deconvolution formulations given by Dasch [21] were used to extract two-dimensional information from the ensemble average chemiluminescence images, as the three-point Abel inversion algorithm was previously found to be more accurate and efficient when compared to onion peeling, filtered backprojection, and two-point Abel deconvolution methods, as demonstrated in [7, 21].

A quartz microprobe was used to obtain samples at the afterbody annular exit, and flame ionization detection (employing a Signal 3010 MINIFID device) was employed to measure the concentration of the hydrocarbon at this position and allow the evaluation of the equivalence ratio levels with an uncertainty of the order of 4%. Global exhaust emissions could also be measured by extracting flue gases and measuring bulk species ($NO_x$, CO, $CO_2$, $O_2$, and $C_xH_y$) concentrations with a Kane-May KM9106 Quintox flue gas analyzer further downstream of the burner [11, 13].

## 4. Simulation Model Formulation

*4.1. Aerodynamic Model.* The flames were simulated with the ANSYS Fluent 14.0 (ANSYS Inc.) software [22]. It offered mesh adaption flexibility near the fuel injectors and facilitated the exploitation of reduced chemistry within the context of an adequate modeling scheme for the turbulent reactions. Within the Large Eddy Simulation procedure employed here, the flow variables, $F$, can be decomposed into resolvable $\widetilde{F}$ and subgrid, $F'$, scale quantities using a Favre-weighed filter, $\widetilde{F} = \overline{\rho F}/\overline{\rho}$. The equations describing the resolvable flow quantities are, for example, [9, 13, 14, 22].

*Continuity, Momentum is*

$$\frac{\overline{\partial \rho}}{\partial t} + \frac{\partial \left( \overline{\rho} \widetilde{u}_i \right)}{\partial x_i} = 0, \tag{1}$$

$$\frac{\partial \overline{\rho} \widetilde{u}_i}{\partial t} + \frac{\partial \left( \overline{\rho} \widetilde{u}_i \widetilde{u}_j \right)}{\partial x_j} = -\frac{\partial \overline{p}}{\partial x_i} + \frac{\partial}{\partial x_j} \left[ \mu \left( \frac{\partial \widetilde{u}_i}{\partial x_j} + \frac{\partial \widetilde{u}_j}{\partial x_i} \right) \right]$$
$$- \frac{\partial \tau_{ij}^a}{\partial x_j} + \left( \Delta \overline{\rho}_\infty \right) g_i \tag{2}$$

*and Scalars are*

$$\frac{\partial \overline{\rho} \widetilde{Y}_k}{\partial t} + \frac{\partial \left( \overline{\rho} \, \widetilde{u}_j \widetilde{Y}_k \right)}{\partial x_j} = \frac{\partial}{\partial x_l} \left( \overline{\rho} D_k \frac{\partial \widetilde{Y}_k}{\partial x_l} \right)$$
$$- \frac{\partial J_l}{\partial x_l} + \overline{\rho} \widetilde{\dot{\omega}}_{Y_k}, \quad k = 1, N. \tag{3}$$

The solution of an energy equation completes the above system ($\rho, \mu, T, u_i$, and $Y_k$ are the density, viscosity, temperature, velocity and species composition of the gas, and $i =$

1, 2, 3 in a Cartesian system $(x, y, z)$). $\tau_{ij}^a = \tau_{ij} - 1/3\tau_{ij}\delta_{ij}$ is the anisotropic part of the subgrid stress tensor, $\tau_{ij}$, with the isotropic part of the viscous and subgrid stress being adsorbed into the pressure. The subgrid stresses are modeled as $\tau_{ij}^a = -2\mu_{sgs}\widetilde{S}_{ij}$, where $\mu_{sgs} = \overline{\rho}(C_s\overline{\Delta})^2(2\widetilde{S}_{ij}\widetilde{S}_{ij})^{1/2}$, $\widetilde{S}_{ij}$ is the resolvable strain tensor and $\Delta$ is the filter width related to the local mesh spacing, $\Delta = \sqrt[3]{\Delta x_i \Delta y_i \Delta z_i}$. A gradient assumption is used for the subgrid scalar flux, $J_k = \overline{\rho}(\widetilde{u_l Y}_k - \widetilde{u}_l \widetilde{Y}_k) = -(\mu_{sgs}/Sc_{sgs})(\partial \widetilde{Y}_k/\partial x_l)$, where $Sc_{sgs}$ is the subgrid Schmidt number. Turbulent transport is modelled with the dynamic Smagorinsky model (e.g., [9, 11, 22]) with bounded $C_s$ ($0 < C_s < 0.23$). Pressure-velocity coupling was handled via a time-dependent SIMPLE algorithm (e.g., [9, 22]). A second-order accurate time discretization was used, and the time step was chosen to maintain a maximum Courant number between 0.4 and 0.6. Rms values were computed from the resolved scale quantities, including the contributions from the subgrid scale stresses. The simple P-1 radiation model available in the software (ANSYS Inc.) was adopted in the simulations (although no correction for radiation was applied to the temperature measurements; its use in the simulations allows some flexibility in future processing of the present data).

The employed commercial software has frequently been used to investigate with success premixed flame stabilization in complex bluff-body configurations under stable (e.g., [23]) and approaching blowoff [24] conditions in a fashion similar to the present study. One should nevertheless exercise great caution in interpreting the results obtained regarding the behaviour and development of the flame front at the final stages toward LBO; centers of excellence in combustion studies worldwide are currently trying to develop such a capability and capture this complex phenomenon with sufficient accuracy [23]. Notwithstanding the above comments, here every effort was made to improve the basic capability of the software through use (a) of the higher order differencing schemes available within the software for LES, (b) of a well-established turbulence-chemistry closure such as the thickened flame model, (c) of the extended nine-step chemical scheme for propane chemistry, and (d) through careful mesh refinement in the main reaction zone. With these improvements the performance of the software was deemed satisfactory in providing useful information at ultralean and near LBO conditions, and this was tested where possible against the present measurements.

*4.2. Combustion Model.* To represent the mixed regime combustion that may result from the various fuel settings the thickened flame model (TFM) (e.g., [14, 15, 22, 23]) was employed. This formulation was coupled to a nine-step reduced scheme for propane (i.e., Rxn 1: $C_3H_8 + O_2 + OH + H \rightarrow 3CO + 5H_2$, Rxn 2: $CO + OH \leftrightarrow CO_2 + H$, Rxn 3: $3H_2 + O_2 \leftrightarrow 2H_2O + 2H$, Rxn 4: $2H + M \rightarrow H_2$, Rxn 5: $H_2 + O_2 \leftrightarrow 2OH$, Rxn 6: $C_3H_8 + O + OH + H \rightarrow C_2H_2 + CO + H_2O + 3H_2$, Rxn 7: $C_2H_2 + O_2 \leftrightarrow 2CO + H_2$, Rxn 8: $2O + M \leftrightarrow O_2$, Rxn 9: $O_2 + H \leftrightarrow O + OH$). This represents an extension of the mechanism described in [11, 13, 16], now including the radical OH to facilitate more direct comparisons with the measured chemiluminescent $OH^*$ and has been tuned to reproduce the flame speed over a section of the lean $\Phi$ range as described in [11]. All the species are explicitly solved on the computational grid, and it is likely that intermediate radicals with shorter time scales might not be properly resolved within the context of the employed refinement. In the present low Reynolds number laboratory flames, the exploitation of the above combination of turbulence models and chemistry scheme can be considered attractive, in providing a more detailed and extensive description of the developing flame front properties while also allowing for a more direct comparison with optical experimental results.

The averaged reaction rate source terms were computed with the ISAT algorithm (e.g., [7, 19, 22]). Previous measurements [11, 13] have suggested that the lean flames lie in the thin reaction regime. The present ultra-lean flames are closer to the broken reaction zone boundary with Karlovitz numbers [1] ($Ka = \tau_{ch}/\tau_k$, where $\tau_{ch}$ and $\tau_k$ are the chemical and Kolmogorov timescales) of about 80 and local grid-based Karlovitz numbers (e.g., [9]) of higher values.

*4.3. Local Extinction Monitoring Parameters.* An effort was also made to examine the local extinction behaviour at near LBO conditions by considering the combination of three, measured or computed, parameters. First, the parameter $\Lambda = Ka/Ka_{ex}$, computed from the transient simulations, was chosen, where $Ka$ is the local Karlovitz number and $Ka_{ex}$ is the Karlovitz stretch factor for symmetric counterflow laminar flame extinction being a function of the local equivalence ratio with $\Lambda > 1$ signifying local extinction (e.g. [2, 10]).

Second, the computed statistical distribution of the temperature difference $\Delta T_c$ between the local instantaneous temperature, $T_{inst}$, and a chosen threshold temperature, $T_{limit}$, $\Delta T_c = T_{inst} - T_{limit}$ was evaluated, and local extinction is considered when $\Delta T_c < 0$; the usefulness of a threshold temperature value has been demonstrated from the extensive experimental investigations on $C_3H_8$ flame extinction of reference [3]. For conditions close to the present experiment, values of $T_{limit}$ of around 1400 K have been reported for swirl, grid, and porous plate burners [3], while a similar range of limiting extinction temperatures was measured and computed in laminar counterflow setups [2]. Present tests in the simulations suggested a value between 1325 and 1350 K.

Third, the trends in the variations of the ratio of the measured time-mean $OH^*/CH^*$ chemiluminescence signature, evaluated at selected locations close to the reacting flame front stabilization region, within the recirculation zone, at $x/D_b = 0.44$ for the disk and at $x/h = 1.6$ for the square, were examined for possible correlation with the proximity to blowoff. Although local extinction events are undoubtedly instantaneous phenomena, the proximity to LBO has routinely been assessed, for preliminary as well as practical studies, through time-averaged criteria, for example, Damkholer numbers, and so forth. Therefore the exploitation of the average $OH^*$ to $CH^*$ ratio cannot be excluded as a useful criterion. Furthermore, due to the complexity of the phenomena involved within time scales of a few milliseconds, apart from fast recording instrumentation, customarily more

than one parameter has been examined to determine the flame condition, something that in most cases is and should be technical practice. Here the possibility of using the above parameters, in combination, as likely local markers for delineating reactive from nonreactive regions, in the present stratified flames with inlet mixture nonuniformities that place them neither in the nonpremixed nor in the fully premixed regime and the likelihood of extrapolating this behaviour to determine the proximity to LBO are tentatively explored.

*4.4. Computational Details.* Particular emphasis was given to the mesh distributions near the low aspect ratio injectors with respective dimensions, ($1 : 25 : 52$ mm) → (annular injection slot : afterbody : central pipe) for the disk and ($1 : 8 : 80$ mm) → (injection hole : square cylinder height : tunnel height) for the square (see Figure 1). The mesh for the disk extended over a 360° sector, while for the square the resolved span was chosen to include seven injection holes plus one half spacing on each side, over a width of 1.75 h, with symmetry boundary conditions at the lateral boundaries. The complete premixer cavities section was included in the mesh for the disk and the computations started well upstream of the complex premixer/burner system to incorporate the effects of the mean and turbulent inflow conditions measured at this position. Similarly the computations for the square started $7D_b$ upstream of the burner face, where measured inlet velocity and turbulence profiles were available. Hybrid meshes were utilised, unstructured near the compound regions of the disk or the square injectors and the burner boundaries, and structured in other locations with expansion to the outlet, which was placed $15D_b$ downstream of the bluff bodies. Typical meshes of 1.45 and 1.75 Mcells were employed for the disk and the square, respectively, in the basic runs. No-slip boundary conditions were used near the walls with the node close to the burner surface placed at $\Delta y/D_b = 0.00175$, while the law of the wall was applied elsewhere. Inlet conditions were taken from measurements, while a static pressure boundary condition was applied at the outlet and all outflowing quantities were extrapolated from the interior node. The LES quality was checked by calculating the resolved fraction of the turbulence energy, $R_k = k_{res}/(k_{res} + k_{sgs})$, with values up to 95% within the main reaction zones. Run times on 48 3.0 GHz processors (4 XEON 5660) and 24 3.8 GHz processors (4 i7) run in parallel were about 1.7 CPU hrs for the simulation of one through flow time ($\tau = D_b/U_c = 5 \times 10^{-3}$ s) for the disk, while more computational time was necessary for the square due to the denser meshes utilized to resolve the discrete, small aspect ratio, series of injector holes. Time series of various quantities were collected for, for example, the disk over about $225\tau$ and relevant statistics were postprocessed from this sample.

## 5. Results and Discussion

A description of the isothermal flow and the stable flame characteristics under operation away from the blow off limit region has been presented in references [11, 13, 16, 19] for the two burner configurations, respectively. The present work

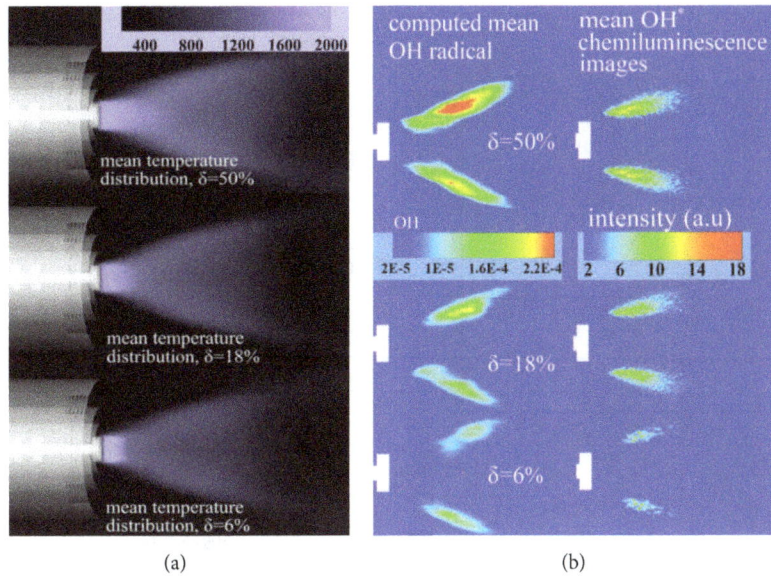

FIGURE 2: Simulated time-mean disk burner wake characteristics, (a) temperature, (b) OH mole fraction radical and OH* chemiluminescence distributions for various ultra-lean flame settings.

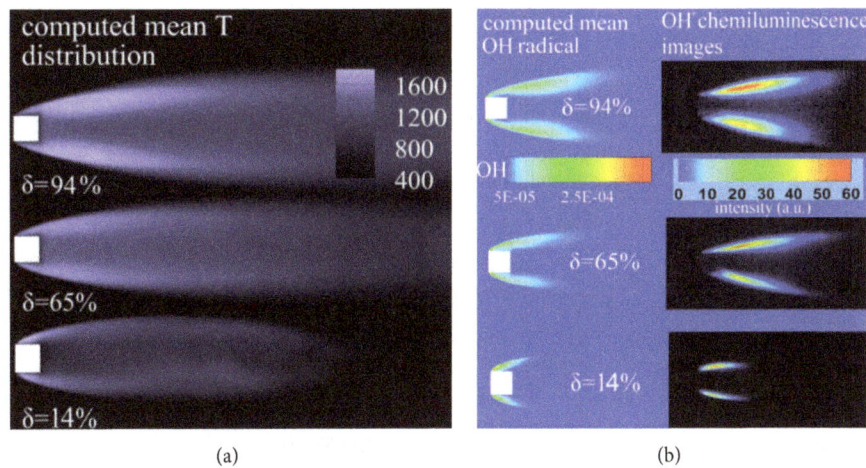

FIGURE 3: Simulated time-mean square burner wake characteristics, (a) temperature, (b) OH mole fraction radical and OH* chemilumines-cence distributions for various ultra-lean flame settings.

focuses on the comparative examination of the topologies obtained in the disk flame operated under a high swirl level ($S = 1$) and the square burner counterinjected with a low fuel-to-approach-air velocity ratio at ultra-lean conditions located near the LBO margin. The data presented for the two configurations are obtained for flame settings with successively reduced fuel flows, that is, at reduced $\delta$ levels (as indicated in Table 1).

An overall picture of the two reacting wake flows emerges in the simulated time-averaged patterns displayed in Figures 2 and 3, respectively, in the form of mean temperature and OH radical mole fraction contours together with the corresponding distributions of measured OH* chemilumi-nescence. A peak temperature region is initially situated within and adjacent to the afterbody disk recirculation and extends up to $2.25D_b$, with a conical growth of the hot wake

at the fuel lean condition ($\delta$ = 50%, Figure 2(a)). At the leaner setting, the flame retains its conical shape, but the peak temperature envelope suffers a noticeable reduction with a moderate narrowing of the downstream wake spread. Closer to blowoff (at $\delta$ = 6%) the peak temperature region now recedes considerably closing in on the axis with the peak temperatures withdrawn toward the rim shear layers. The computed time-averaged OH mole fraction contour plot in Figure 2(b) also suggests that the leading edge of the toroidal flame front gradually weakens and detaches from the disk rim, while its trailing edge moderately retracts. The accompanying time-averaged OH* chemiluminescence images in Figure 2(b) (Abel transformed images) corroborate to some extent the flame detachment but suggest a thinner reacting front with a much weaker activity particularly close to the disk axis. These plots imply that the present

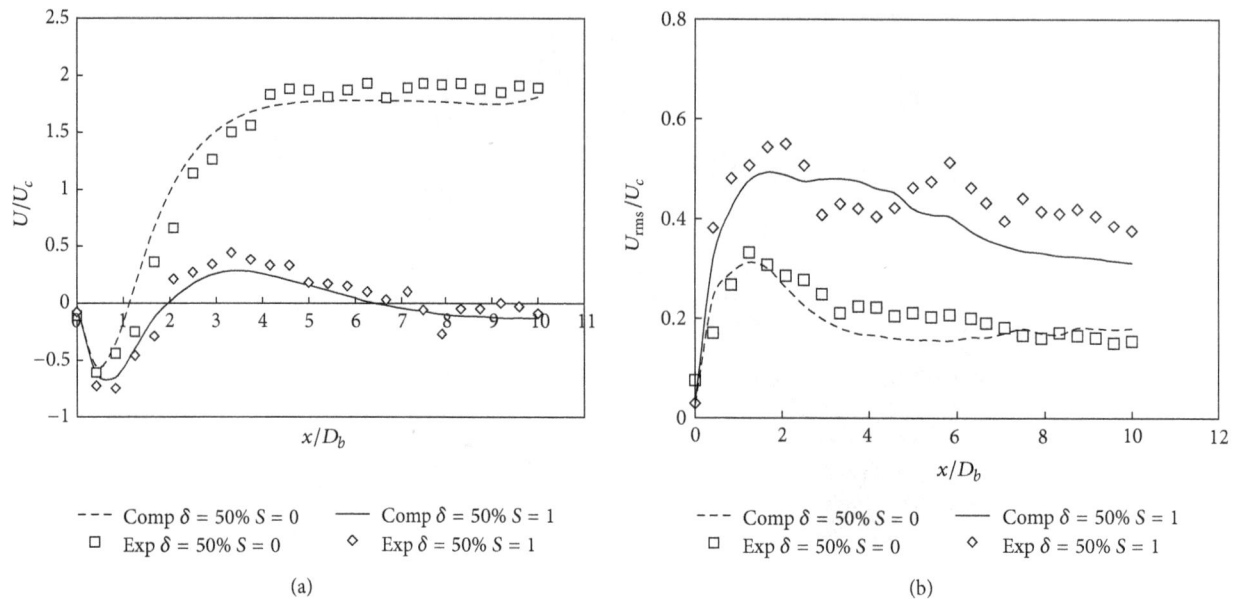

FIGURE 4: Comparisons between simulated and measured time-mean streamwise velocity (a) and turbulence intensity (b) distributions along the near wake of the disk burner.

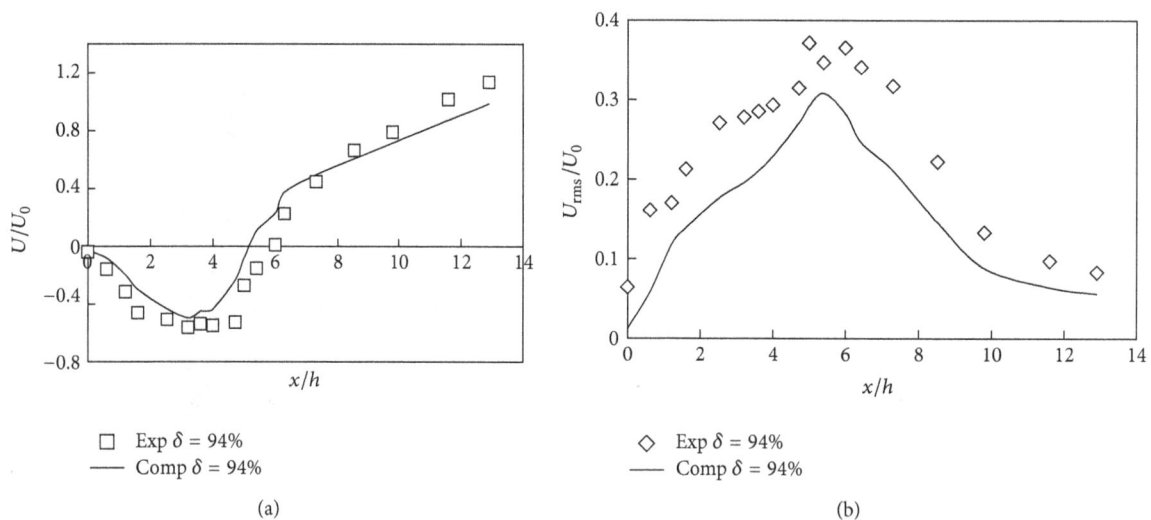

FIGURE 5: Comparisons between simulated and measured time-mean streamwise velocity (a) and turbulence intensity (b) distributions for the near wake of the square burner.

partially premixed disk flames behave somewhat differently than the fully premixed axisymmetric configurations burning methane that were visualized in [7] to drastically close in on the axis at the limiting lean operation; the disk flame disposition with the present leaner outward gradient inlet stratification follows closely the behaviour reported in [5] for similar stratification conditions.

Time-mean temperature contours for the square cylinder reacting wakes are presented in Figure 3(a) for medium, low, and limiting fuel injection velocities. The counterinjected configuration allows for partial premixing prior to ignition and stabilization around the cylinder flanks thus resulting in temperature shear layers that enclose the recirculation zone. Gradual fuel reduction here seems to lead to an attendant

shortening of these flanking flame fronts, as illustrated by the computed time-mean OH radical contours and the included OH* images in Figure 3(b). It should be noted that fuel injection *directly* into the vortex region would result in an altogether different reacting configuration of the *nonpremixed* type [11, 19].

The development of the near wake recirculation regions in terms of the measured and computed time-mean streamwise velocity and turbulence intensity distributions is shown in Figures 4 and 5 for the disk and the square burners, respectively, operating at lean conditions (Table 1). Good agreement is maintained in the axial velocity profile comparisons (Figure 4(a)) at a range of conditions spanning the plane wake ($S = 0$) with lower turbulence levels and the

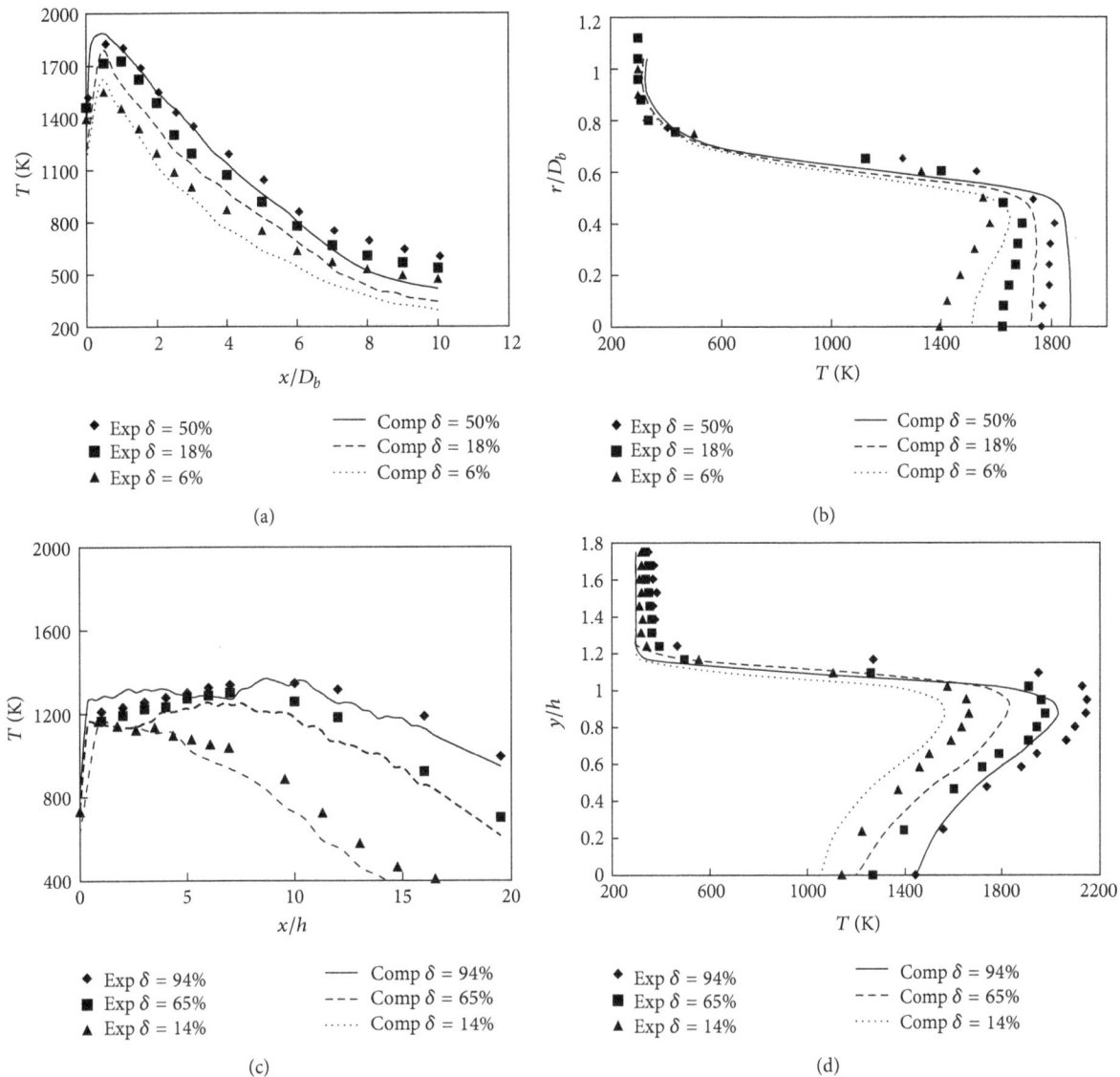

FIGURE 6: Comparisons between simulated and measured time-mean streamwise and cross-stream temperature distributions for disk ((a), (b)) and square ((c), (d)) burners, respectively. Cross-stream temperature distributions were taken at $x/D_b = 0.4$ for the disk and at $x/h = 0.6$ for the square.

higher swirl case ($S = 1$) with elevated turbulence levels (Figure 4(b)). In the disk wake, the recirculation lengths were found to vary from $1.43D_b$ for the lean ($\delta = 50\%$) to $1.28D_b$ for the weaker ($\delta = 6\%$) flame, respectively, gradually approaching the cold vortex zone length, and this trend was followed well by the simulations. On the other hand in the square burner the recirculation region was significantly elongated, in both measurements and simulations, by up to 5.5 times with respect to the cold wake, due to heat release and this effect is depicted by the extensive backflow region in the time-mean streamwise velocity profile comparisons shown in Figure 5(a). Maximum axial turbulence intensities are obtained for the disk at the higher swirl level of $S = 1$ (Figure 4(b)) and are of similar magnitude to the intensities attained in the square burner flame (Figure 5(b)) justifying the choice of the higher swirl disk case for direct comparisons with the square burner configuration.

Time-mean temperature distributions for the disk and the square are shown in Figures 6(a), 6(b), 6(c), and 6(d) in the form of center-line (-plane) profiles along the wakes and radial (cross-stream) traverses taken close to the stabilizers, across the recirculation zone, respectively. The measured axial temperature variations (Figure 6(a)) are reproduced satisfactorily by the combined methodology in the near axisymmetric wake up to $x/D_b = 4$, while discrepancies can be identified very close to the disk wall and in the downstream region where a faster wake development is computed. The lack of a more complete heat transfer modelling near the wall (apart from the use of the simple radiation model), the deficiencies of the TFM model in the vicinity of the wall due to a combination of inadequate mesh resolution, and likely shortcomings in the performance of the chemistry scheme there or the increased thermocouple measurement errors closer to the wall may be in part responsible for this discrepancy.

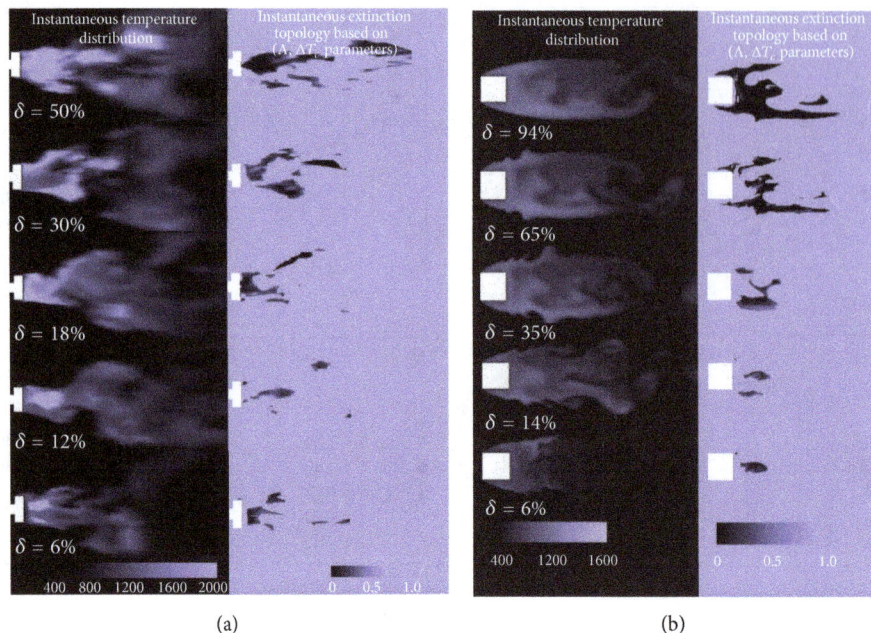

FIGURE 7: Snapshots of temperature and extinction topologies for a range of simulated flame settings as LBO is approached, (a) disk and (b) square burner.

The cross-stream traverses in Figure 6(b) (obtained at $x/D_b = 0.4$) are consistent with the overall distributions shown in Figure 2(a) and suggest that a wider flame spread is obtained in the computed profiles. The peak $T$, located in line with the disk rim vortex where flame stabilization occurs, suffers a gradual drop from 1860 K to 1700 K and then to 1500 K with the leaner profile producing a more discernible off-axis temperature peak in accordance with the overall topology of Figure 2(a). These distributions could offer some guidance as far as the possible methodologies, for example, pilot injection placement that can be adopted to mitigate the evolution toward LBO in the stabilization region close to the burner rim.

The influence of varying the local equivalence ratio stratification on the simulated and measured time-mean centerplane and cross-stream (taken at $x/h = 0.6$) temperature distributions for the square burner and for the fuel levels already seen in Figure 3(a) is depicted in the plots of Figures 6(c) and 6(d). Significantly lower temperatures, within a much longer recirculation zone, are now exhibited in this topology, and this trend is maintained for all successive fuel reductions. Peak values are found at the flanks of the developing wake of the square center-body, at $y/h = 0.9$, in line with a more spreadout flame front stabilization disposition due to the displacing effect of the flanking recirculations on the square sides (Figures 3(a) and 3(b)). In fully premixed slender bluff-body stabilization setups studied in, for example, [6, 12], the flame was distinctly drawn within the recirculation and interacted with the reappearing vortex shedding mechanism. Here the adjacent flame fronts, with no apparent mutual interaction at the lean and ultra-lean level, gradually retract along the flanks of the wake engulfing the hot product vortex zone and it can be conjectured that this separation effectively delays the full reemergence of the Von Karman instability out

of the formation region [11, 19]. These characteristics will be further verified from instantaneous images presented below.

Figures 7(a) and 7(b) display a sequence of LES snapshots of the temperature and the corresponding extinction topology as inferred from two of the criteria discussed previously for a range of fuel settings as we get closer to LBO. The dark areas demarcate reactive regions where both the Karlovitz ($\Lambda$) and the temperature threshold ($\Delta T_c$) criteria are simultaneously valid. Within the present formalism, this is a necessary but not a sufficient condition for sustaining reaction within the indicated areas. From these plots, it appears (Figure 7(a)) that once the side rim vortex loses its stabilizing effectiveness, the flame destabilizes and is unable to recover despite the occasional excursion of reacting fragments into the core of the main recirculation.

The square extinction topology (Figure 7(b)) indicates an efficient and resistant stabilization of the two reacting fronts within the flanking side vortices. However, it appears that when the two reactive layers are *sufficiently* retracted, the inherent intense large-scale instability [12, 17] is reinstated at the flame trailing edges. This has a detrimental effect on the overall stability, and the attribute of the anchoring in the strong flanking vortices is cancelled, leading thereafter to a more rapid approach to LBO in the final stages with respect to the disk behavior. Animation sequences from the simulations, supplied as supplementary material and taken after a sudden fuel reduction to the LBO levels, provide a more detailed illustration of the flow and flame events that occur at the final stages of the blow-out process for the two burners.

The local parameters used to monitor the local extinction behavior are discussed next. The instantaneous scatterplots of the simulated Karlovitz parameter monitored at a position close to the trailing edge (corner) of the two stabilizers, next to the temperature shear layer(s) (at $x/D_b, r/D_b = 0.32, 0.4$

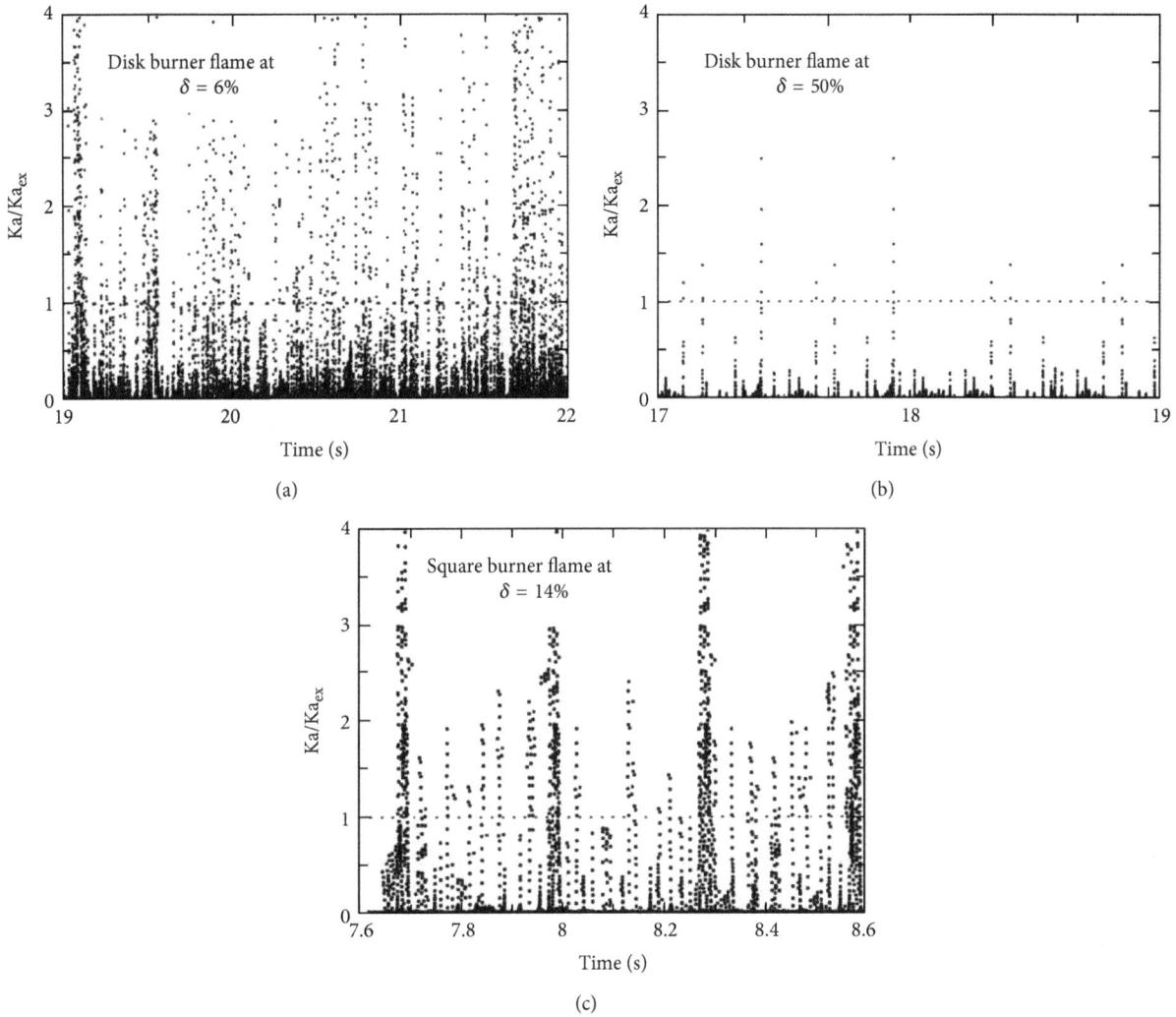

FIGURE 8: Scatterplots of the normalized Karlovitz number for the disk ((a), (b)) and the square (c) at different levels of local extinction ($\Lambda = $ Ka/Ka$_{ex}$ values limited to 4 for clarity of extinction events signified by values greater than 1).

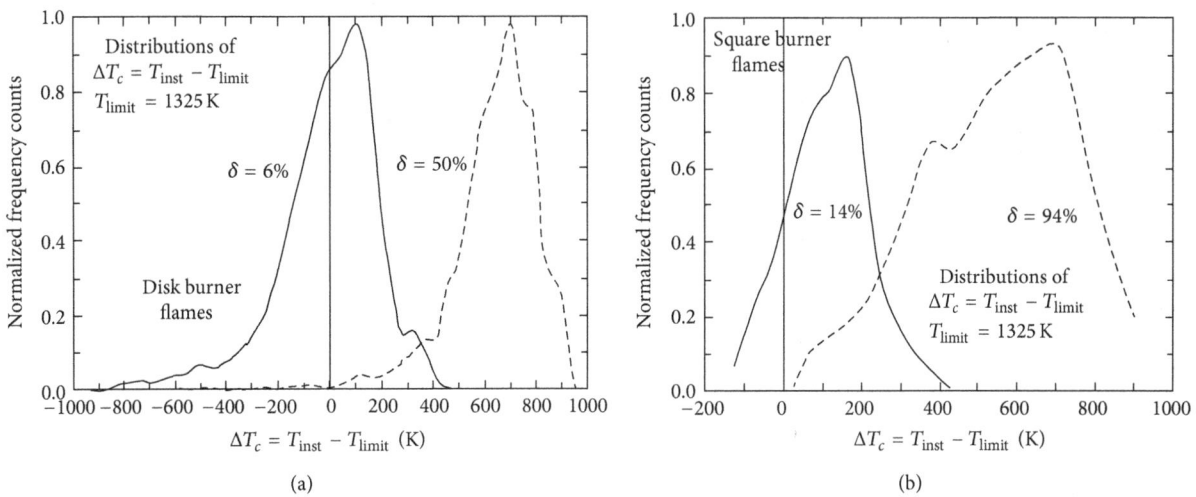

FIGURE 9: Statistical distributions of the instantaneous temperature difference parameter for the disk (a) and the square (b) at different levels of local extinction.

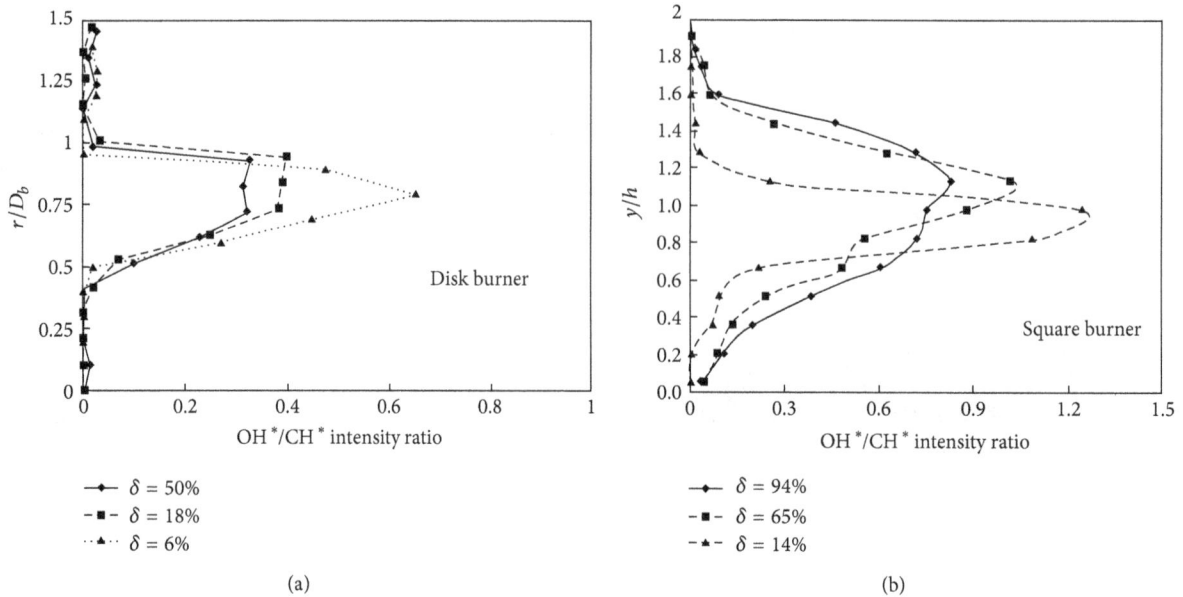

FIGURE 10: Cross-stream profiles of the ratio of the mean $OH^*$ to $CH^*$ chemiluminescence across the flame front within the recirculation zone for three ultra-lean and limiting flame settings, (a) disk and (b) square burner. Chemiluminescence profiles were taken at $x/D_b = 0.44$ for the disk and at $x/h = 1.6$ for the square.

for the disk and at $x/h$, $y/h$ = 0.5, 0.7 for the square), are shown in Figures 8(a), 8(b), and 8(c) for each burner and two levels of fuel reduction. Depending on the level of local extinction, these plots exhibit distinctly different dispersions consistent with the reduction of the local reactivity as determined by the Karlovitz criterion. Interestingly, the scatterplot distribution for the square, depicted in Figure 8(c), displays a recursive appearance of four major extinction events. This corroborates the onset of vortical development at the flame tongues that seem to be reinstated as LBO is approached, as also implied by the low $\delta$ snapshots of Figure 7(b). Such a behavior has been reported previously for fully premixed, for example, [6, 12] and partially premixed configurations [11].

The computed statistical distributions (frequency counts) of the instantaneous temperature threshold parameter, $\Delta T_c$, are displayed for the same monitoring position in Figures 9(a) and 9(b) for the two setups, respectively. A threshold temperature has been previously exploited as a practical and complementary diagnostic parameter for indicating extinction [2, 3, 10]. Based on the present experimental and computational tests and various studies on laminar counterflow [2, 10] and turbulent flames [3], a limit value of 1325 K was here tentatively used for the present conditions. The disposition of the successive distributions of $\Delta T_c$, shown in Figures 9(a) and 9(b), consistently delineates the increased probability for local extinction; since the selected monitoring point lays close to the stabilizing shear layers, these, in combination with further monitoring parameters, could also be considered as indicative of the proximity to *global* extinction.

Figures 10(a) and 10(b) display the cross-stream profiles of the ratio of the time-mean $OH^*/CH^*$ chemiluminescence, within the recirculation regions and across the flame fronts, at $x/D_b$ = 0.44 for the disk and at $x/h$ = 1.6 for the

square, away from the burner face. The $OH^*$ and $CH^*$ disk profiles have been projected onto an axisymmetric plane through an inverse Abel transform for a more consistent comparison of their ratio. The results imply that the flames closer to extinction consistently attain a higher value of this ratio for both burners, and this trend, if further elaborated, could likely be suitable for demarcating limiting conditions. It could be postulated that apart from monitoring the transient behavior of the above or additional parameters, at judiciously chosen locations, the examination of time-mean data could be equally useful in producing useful overall criteria, in a fashion similar to the exploitation of a local Damköhler number, an attribute that merits further verification over a range of flame topologies and conditions.

Figure 11 displays instantaneous $OH^*$ chemiluminescence images obtained at a fuel setting of $\delta$ = 6% for the disk and of $\delta$ = 14% for the square, respectively. The axisymmetric reacting wake structures in the disk images indicate a flame topology with highly variable peripheral liftoff distance with time. These flame front structures form a wide angle relative to the incoming flow due to the effect of the high swirl intensity and seem to occasionally penetrate from the periphery toward the base of the flame within the recirculation region. The overall disposition of the flame front under stratification seems to differ both in terms of length, placement, and anchoring behaviour from both nonpremixed [16, 19] and fully premixed axisymmetric [4, 7, 8] counterpart topologies. The transient $OH^*$ chemiluminescent images for the square verify the two-tongued flame front placement previously implied by the time-mean experimental and simulation results (Figure 3). The two fronts independently reciprocate back and forth in time without any apparent interaction. The images do not provide any indication of

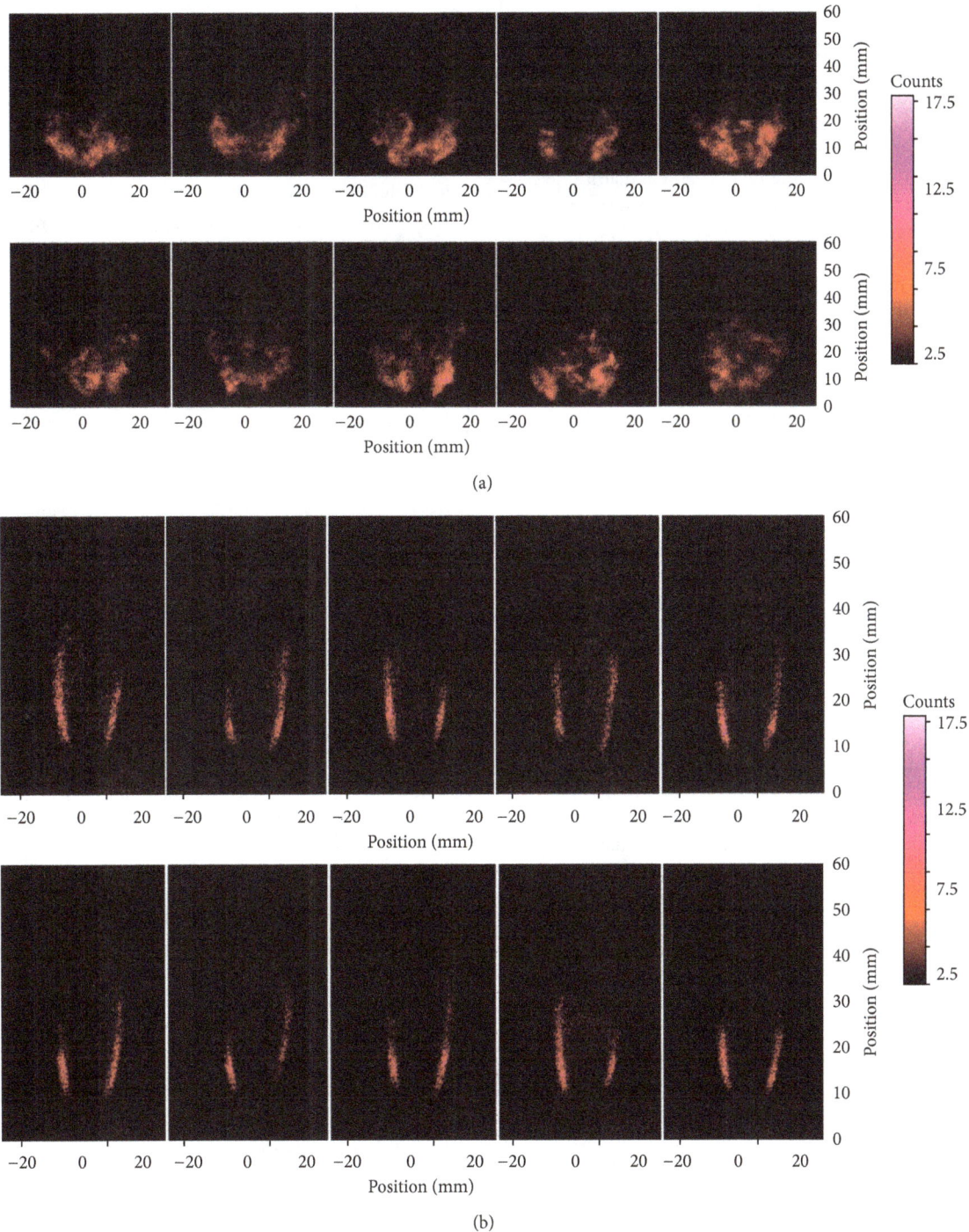

FIGURE 11: Instantaneous OH* images for disk (a) and square (b) burner operating with a mixture level of $\delta = 6\%$ (disk) and $\delta = 14\%$ (square).

reaction within the formation region, which is in stark contrast to the reported behavior [6, 12] of ultralean and limiting slender body stabilization of fully premixed flames.

## 6. Summary and Conclusions

Measurements and simulations of axisymmetric swirling disk and 2D slender square bluff-body reacting wakes, working with an inlet mixture gradient, have been performed under ultra-lean and near blowoff conditions. Results have been presented and compared to highlight some differences and similarities in the flame characteristics of each stabilizer under a variable stratified inlet mixture profile sustained across the leading edge flame front stabilization region. These data can also provide a useful assessment of the ultra-lean operation of these two popular stabilizers under stratified inlet mixture conditions as compared against the more traditional fully premixed operation.

The axisymmetric flames exhibited a more compact, frustum-like shaped near fame region, with significant radial spread dependent on swirl intensity. Recirculating hot products occupied a toroidal region with an extent of about 20 to 40% longer than the isothermal length. This was maintained across the examined mixture strengths and was shorter than that found in counterpart fully premixed axisymmetric cases. Over the ultra-lean range and at the limiting flame conditions reacting, front fragments were identified to penetrate within this recirculation region from the rim vortex stabilization region. The formation of ring vortical structures that has been identified in toroidal disk flames at fully premixed operation was somewhat curtailed under stratified operation, and this aspect requires further investigations since it affects the stability of these flames.

The slender square body flames were stabilized at the rim vortices formed on the square flanks, establishing a two-tongued appearance. The lengths of the two individual reacting fronts were from two to eight square heights depending on mixture strength. Similarly, with reaction, the cold wake formation region was expanded by about four to five times in the lean cases; this was drastically shortened at the weaker inlet mixtures. Peak temperatures for the disk were located both in the toroidal shear layer and extended well within the recirculation zone, while for the square these were clearly positioned along the square flanks. The development of Von-Karman vortical instability was clearly identified to emerge in the square flames as the flame fronts were significantly withdrawn back closer to the burner face; this behaviour is also different from fully premixed configurations that have been reported to maintain periodic symmetric or antisymmetric vortical structures across a wide range of equivalence ratios.

The adopted LES procedure with an appropriate choice of combustion and chemistry submodels reproduced satisfactorily the measured mean velocity and temperature distributions while providing useful qualitative descriptions of the examined species fields which are compared quite favourably with the chemiluminescence imaging measurements. The method also provided some insight into the topology of the leading edge flame front anchoring adjacent to the flanks of axisymmetric or plane geometry stabilizers, in the vicinity of equivalent ratio peaks, within the context of a stratified inlet mixture gradient for ultra-lean and near to blowoff conditions.

## Conflict of Interests

There is no conflict of interests that arises from the methodologies, the instrumentations, or the software used for the production and appraisal of the present research results within the context of the present work performed by the authors.

## Acknowledgment

This work was partly supported by the Research Council of the University of Patras.

## References

[1] B. W. Bilger, S. B. Pope, K. N. C. Bray, and J. F. Driscoll, "Paradigms in turbulent combustion research," *Proceedings of the Combustion Institute*, vol. 30, no. 1, pp. 21–42, 2005.

[2] D. Bradley, "Combustion and the design of future engine fuels," *Proceedings of the Institution of Mechanical Engineers C*, vol. 223, no. 12, pp. 2751–2765, 2009.

[3] G. E. Andrews, N. T. Ahmed, R. Phylaktou, and P. King, "Weak extinction in low NOx gas turbine combustion," in *Proceedings of the ASME Turbo Expo*, pp. 623–638, June 2009.

[4] M. Stohr, I. Boxx, C. Carter et al., "Dynamics of lean blowout of a swirl-stabilized flame in a gas turbine model combustor," *Proceedings of the Combustion Institute*, vol. 33, no. 2, pp. 2953–2960, 2011.

[5] M. S. Sweeney, S. Hochgreb, and R. S. Barlow, "The structure of premixed and stratified low turbulence flames," *Combustion and Flame*, vol. 158, no. 5, pp. 935–948, 2011.

[6] S. G. Tuttle, S. Chaudhuri, K. M. Kopp-Vaughan et al., "Blowoff dynamics of asymmetrically-fueled, bluffbody flames," in *Proceedings of the 49th AIAA Aerospace Sciences Meeting*, AIAA, Orlando, Fla, USA, 2011.

[7] J. R. Dawson, R. L. Gordon, J. Kariuki et al., "Visualization of blow-off events in bluff-body stabilized turbulent premixed flames," *Proceedings of the Combustion Institute*, vol. 33, no. 1, pp. 1559–1566, 2011.

[8] Q. Zhang, S. J. Shanbhogue, T. Lieuwen, and J. O'Connor, "Strain characteristics near the flame attachment point in a swirling flow," *Combustion Science and Technology*, vol. 183, no. 7, pp. 665–685, 2011.

[9] C. Duwig, K.-J. Nogenmyr, C.-K. Chan, and M. J. Dunn, "Large eddy simulations of a piloted lean premix jet flame using finite-rate chemistry," *Combustion Theory and Modelling*, vol. 15, no. 4, pp. 537–568, 2011.

[10] E.-S. Cho, S. H. Chung, and T. K. Oh, "Local karlovitz numbers at extinction for various fuels in counterflow premixed flames," *Combustion Science and Technology*, vol. 178, no. 9, pp. 1559–1584, 2006.

[11] P. Koutmos and K. Souflas, "A study of slender bluff body reacting wakes formed by con-current or counter-current fuel injection," *Combustion Science and Technology*, vol. 184, no. 9, pp. 1343–1365, 2012.

[12] K. M. Kopp-Vaughan, T. R. Jensen, B. M. Cetegen et al., "Analysis of blowoff dynamics from flames with stratified fueling," *Proceedings of the Combustion Institute*, vol. 34, no. 1, pp. 1491–1498, 2013.

[13] C. Z. Xiouris and P. Koutmos, "Fluid dynamics modeling of a stratified disk burner in swirl co-flow," *Applied Thermal Engineering*, vol. 35, no. 1, pp. 60–70, 2012.

[14] O. Colin, F. Ducros, D. Veynante, and T. Poinsot, "A thickened flame model for large eddy simulations of turbulent premixed combustion," *Physics of Fluids*, vol. 12, no. 7, pp. 1843–1863, 2000.

[15] G. Wang, M. Boileau, and D. Veynante, "Implementation of a dynamic thickened flame model for large eddy simulations of turbulent premixed combustion," *Combustion and Flame*, vol. 158, no. 11, pp. 2199–2213, 2011.

[16] P. Koutmos, G. Paterakis, E. Dogkas, and C. H. Karagiannaki, "The impact of variable inlet mixture stratification on flame topology and emissions performance of a premixer/swirl burner configuration," *Journal of Combustion*, vol. 2012, Article ID 374089, 12 pages, 2012.

[17] R. R. Erickson and M. C. Soteriou, "The influence of reactant temperature on the dynamics of bluff body stabilized premixed flames," *Combustion and Flame*, vol. 158, no. 12, pp. 2441–2457, 2011.

[18] J. J. Miau, T. S. Leu, T. W. Liu, and J. H. Chou, "On vortex shedding behind a circular disk," *Experiments in Fluids*, vol. 23, no. 3, pp. 225–233, 1997.

[19] A. G. Bakrozis, D. D. Papailiou, and P. Koutmos, "A study of the turbulent structure of a two-dimensional diffusion flame formed behind a slender bluff-body," *Combustion and Flame*, vol. 119, no. 3, pp. 291–306, 1999.

[20] D. Trimis, *Verbrennungsvorgange in Porosen Inerten Medien, BEV 95.5*, ESYTEC, Elangen, Germany, 1996.

[21] C. Dasch, "One-dimensional tomography: a comparison of Abel, onion-peeling, and filtered back-projection methods," *Applied Optics*, vol. 31, no. 8, pp. 1146–1152, 1992.

[22] ANSYS Fluent, *Release 12.0, Theory Guide*, ANSYS, 2009.

[23] L. Gicquel, G. Staffelbach, and T. Poinsot, "Large Eddy Simulations of gaseous flames in gas turbine combustion chambers," *Progress in Energy and Combustion Science*, vol. 38, no. 6, pp. 782–817, 2012.

[24] A. Briones and B. Sekar, *Effect of Von Kármán Vortex Shedding on Regular and Open-Slit V-Gutter Stabilized Turbulent Premixed Flames*, Spring Technical Meeting of the Central States Section of the Combustion Institute, University of Dayton, Air Force Research Laboratory Dayton, 2012.

# 10

# Assessing the Role of Particles in Radiative Heat Transfer during Oxy-Combustion of Coal and Biomass Blends

**Gautham Krishnamoorthy and Caitlyn Wolf**

*Department of Chemical Engineering, University of North Dakota, 241 Centennial Drive, Stop 7101, Grand Forks, ND 58201, USA*

Correspondence should be addressed to Gautham Krishnamoorthy; gautham.krishnamoorthy@engr.und.edu

Academic Editor: Constantine D. Rakopoulos

This study assesses the required fidelities in modeling particle radiative properties and particle size distributions (PSDs) of combusting particles in Computational Fluid Dynamics (CFD) investigations of radiative heat transfer during oxy-combustion of coal and biomass blends. Simulations of air and oxy-combustion of coal/biomass blends in a 0.5 MW combustion test facility were carried out and compared against recent measurements of incident radiative fluxes. The prediction variations to the combusting particle radiative properties, particle swelling during devolatilization, scattering phase function, biomass devolatilization models, and the resolution (diameter intervals) employed in the fuel PSD were assessed. While the wall incident radiative flux predictions compared reasonably well with the experimental measurements, accounting for the variations in the fuel, char and ash radiative properties were deemed to be important as they strongly influenced the incident radiative fluxes and the temperature predictions in these strongly radiating flames. In addition, particle swelling and the diameter intervals also influenced the incident radiative fluxes primarily by impacting the particle extinction coefficients. This study highlights the necessity for careful selection of particle radiative property, and diameter interval parameters and the need for fuel fragmentation models to adequately predict the fly ash PSD in CFD simulations of coal/biomass combustion.

## 1. Introduction

Radiative transfer from the dispersed phase particulates is important during the combustion of pulverized fuels [1]. The parent fuel, char, fly ash, and soot participate in the radiative transfer process with their radiative properties determined by their size, shape, and composition [2]. The radiative properties of these particulates (absorption and scattering efficiencies, scattering phase functions) are a function of wavelength of light and are often estimated from codes based on the Mie theory [3] or experimental measurements [4, 5]. However, in order to keep the radiation compute time within reasonable limits in combustion simulations, constant absorption and scattering efficiencies or spectrally averaged Planck mean absorption coefficients for the particulates have been proposed. These Planck-averaged properties are dependent on temperature as well as the assumed size distribution of the particles. Furthermore, results from employing these spectrally averaged particle properties have been mixed, with

some investigators confirming their adequacy [6], whereas others have attributed large errors to their usage along with their dependence on the particle size distribution (PSD) [3, 7] which in turn has significant challenges associated with its predictability in pulverized-fuel boilers.

The role of particle radiation has received increased attention in recent years as technologies for carbon capture such as oxy-combustion with flue gas recycle to regulate the temperature are being actively investigated for scale-up and commercialization [8, 9]. The thermal and chemical effects associated with an increased $CO_2$ concentration within the furnace can impact fuel devolatilization and burnout, temperature, radiative heat transfer profiles, and the evolution of the size-segregated fly ash PSD within the boiler. In addition, the volumetric flow rate of the flue gas circulating through the furnace is generally reduced during oxy-combustion and there is the possibility of uncaptured fly ash being recirculated and accumulating within the boiler. While significant efforts have been expended towards developing high fidelity models for the gas radiation properties in these scenarios due to high concentrations of radiatively participating gases [10–13], the role of particle radiation has largely been overlooked and has been modeled by employing either constant absorption and scattering coefficients throughout the domain [14] or constant absorption and scattering efficiencies of the combusting particle [15].

However, it is well known that a combusting particle undergoes significant changes to its absorption and scattering characteristics in a combusting environment as it transitions from a parent fuel to char to an ash particle [1, 2]. Furthermore, the final PSD of the ash particles following char combustion and burnout cannot be explained solely by a shrinking sphere methodology that has been implemented in most CFD codes. For instance, there is a considerable body of literature that discusses fuel fragmentation during devolatilization/char burnout that eventually determines the final PSD of the ash particles [16, 17]. Therefore, the overall goal of this paper is to assess the need for high fidelity particle radiative property and aerosol models that can capture the PSD of fly ash particles under these newer operating conditions. By taking advantage of recent radiative transfer measurements made in a 0.5 MW combustion test facility during combustion of coal-biomass blends under air-fired and oxy-fired conditions [18], a sensitivity study is performed that examines the incident radiative flux variations to the radiative properties of the combusting particle, particle swelling during devolatilization, and the resolution of the fuel PSD at the inlet. In addition, prediction sensitivities to the particle scattering phase function and biomass devolatilization models were also examined.

## 2. Methods

*2.1. Experimental Data.* The simulations reported in this paper are based on a series of experimental measurements of incident radiative heat fluxes during oxy-combustion of coal and biomass blends carried out in a 0.5 MW Combustion Test Facility at RWE npower [18]. The interested reader is referred to the original publication for the experimental details which are only briefly described here. The furnace is approximately 4 m long with an inner cross-section of 0.8 m × 0.8 m. The primary air velocity was maintained at 15 m/s in all of the experiments with the secondary air velocity adjusted to obtain an exit $O_2$ composition of 3%. The fuels PSD along with their proximate and ultimate analysis were also provided in the original reference [18]. Wall incident radiative flux measurements at different locations along the furnace during combustion in air and a once through, dry recycle of flue gas were reported at different recycle ratios (RR) with an experimental uncertainty of ±4%. The RR was defined as

$$RR = \frac{mRFG}{mRFG + mPFG}. \tag{1}$$

mRFG and mPFG correspond to the masses of the recycled flue gas and the product gases, respectively. The simulations in this study correspond to the experimental conditions of oxy-combustion with 75% RR (hereby designated as "oxy-75% RR") and combustion under air-fired conditions. For all

the simulations, the $O_2$ concentration in the primary stream was maintained at 21% (by volume). In the 75% RR scenario, the $O_2$ concentration in the secondary stream was 28% (by volume).

*2.2. CFD Modeling.* The CFD modeling was carried out using the commercial code ANSYS FLUENT (version 15) [19]. The 3D furnace was represented by 2D axisymmetric geometry consisting of 17, 296 cells to enable quick assessments of the different parameters investigated in this sensitivity study. This approach is similar to the 2D axisymmetric representation of this furnace previously carried out by Hu and Yan [14] and is justifiable by the fact that, for the radiative transfer assessments carried out in this study, it is important to adequately capture the temperature distribution within the combustor and maintain an identical volume to surface area ratio (or mean beam length) as the experimental conditions. The highly swirling flames investigated in this study would require millions of computational cells to adequately resolve all the gradients near the burner in a 3D representation thereby making the parametric assessments in this study very compute-intensive [15]. It was deemed critical to numerically capture the internal recirculation zone that was present in the experimental conditions for flame stabilization. This was undertaken in this study by employing the 2D axisymmetric solver with swirl in ANSYS FLUENT [19]. A very fine mesh was employed near the burner along with second order upwind discretization schemes for the momentum and turbulence variable transport equations to ensure that the results reported in this study were grid independent. The inlet PSD of the two coals (Russian and South African (SA)) and two types of biomass particles (Shea meal and saw dust) investigated in this study were fit to a two-parameter Rosin-Rammler distribution function characterized by a most probable particle size and a spread parameter. Figure 1 shows the fuel PSD against Rosin-Rammler curve fits for the different coal and biomass particles employed in this study. The vertical axis in Figure 1 denotes the mass fraction ($Y_d$) of fuel particles that have a diameter that is greater than "$d$." The biomass particles are in general larger than the coal particles at the inlet.

The variables of interest in most radiative transfer analysis are the distributions of radiative heat flux vectors ($\mathbf{q(r)}$) and the radiative source terms ($-\nabla \cdot \mathbf{q(r)}$). Distributions of the radiative heat flux vectors are necessary to perform accurate energy balances on any solid/liquid interfaces within the solution domain and at the boundaries. The radiative source term describes the conservation of radiative energy within a control volume and goes into the energy balance equation associated with the fluid flow. Therefore, these variables couple radiation with the other physical processes that occur in a combustion simulation. In order to determine the distributions of these quantities, a transport equation describing the radiative transfer first needs to be solved.

If "$I$" represents the directional intensity, $k$ the absorption coefficient (which is due to the gases as well as the particulates in the combustion media), $\sigma$ the scattering coefficient (due to the particulates alone), $I_b$ the black body emissive power,

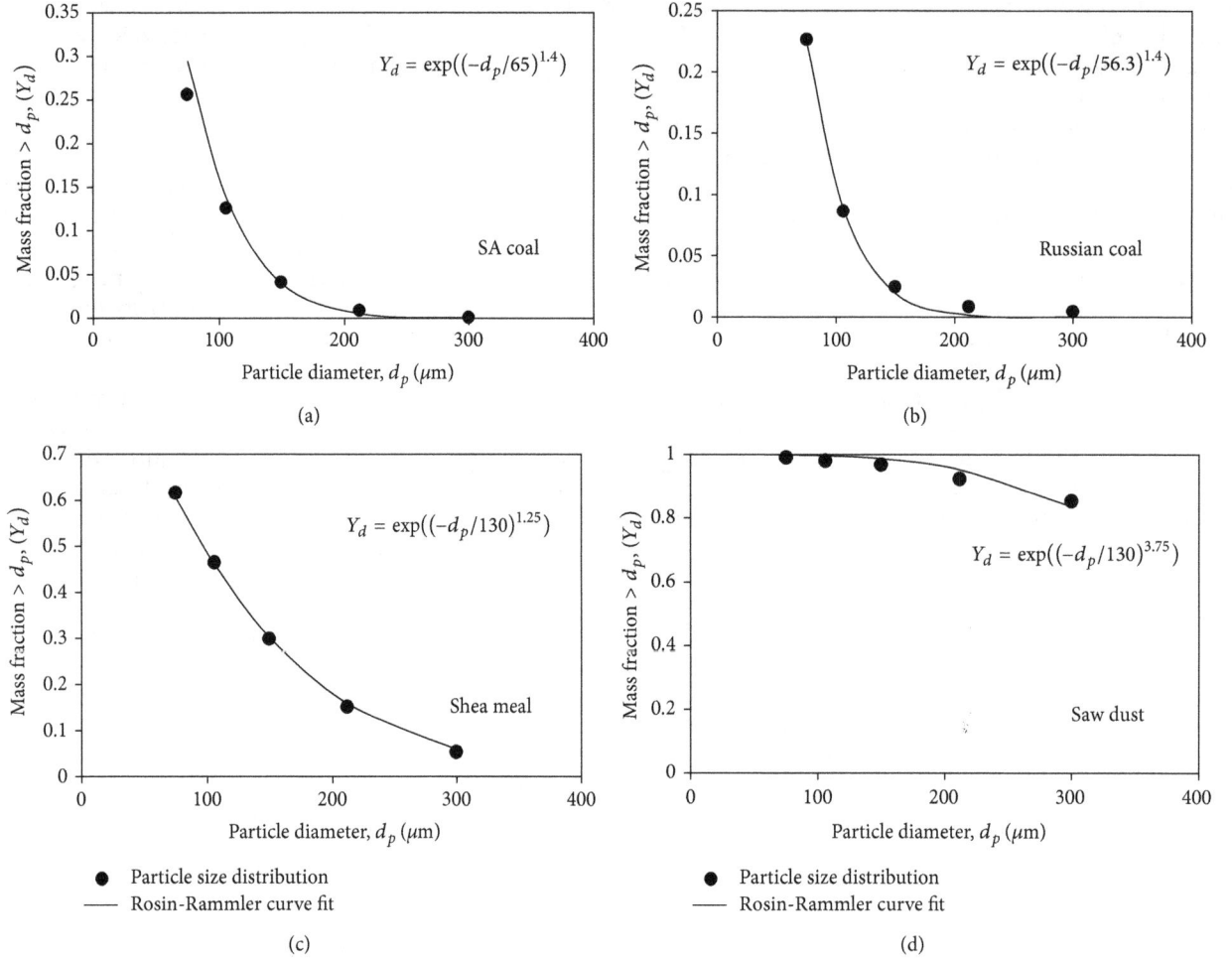

FIGURE 1: Fuel particle size distributions and Rosin-Rammler curve fits (a) SA coal; (b) Russian coal; (c) Shea meal; (d) saw dust.

and $\Phi$ the scattering phase function (assumed to be forward scattering for the bubbles), then the differential equation governing the radiative transfer can be written as

$$
\begin{aligned}
\nabla \cdot \left( I_i \left( \vec{r}, \hat{s} \right) \hat{s} \right) = & -\left( k_{g,i} + k_p + \sigma_p \right) I_i \left( \vec{r}, \hat{s} \right) \\
& + k_{g,i} w_{g,i} I_{bg} \left( \vec{r}, \hat{s} \right) + k_p w_p I_{bp} \left( \vec{r}, \hat{s} \right) \\
& + \frac{\sigma_p}{4\pi} \int_0^{4\pi} I_i \left( \vec{r}, \hat{s} \right) \Phi \left( \hat{s}, \hat{s}' \right) d\Omega'.
\end{aligned}
\tag{2}
$$

In (2), the subscripts "$g$" and "$p$" correspond to the gas-phase and particulate phase, respectively. The subscript "$i$" corresponds to the "gray gas" employed to estimate the gas radiative properties and "$w$" corresponds to the weighting factor associated with the blackbody emissive power. The gas radiative property modeling methodology adopted in this study falls under the category of weighted-sum-of-gray-gases (WSGG) models where the emissivities of gas mixtures are expressed in a functional form consisting of temperature independent absorption coefficients ($k_i$) and temperature dependent weighting fractions $w_i(T)$. The WSGG model

parameters employed in this study have been validated previously through comparisons against benchmark/line-by-line (LBL) data for prototypical problems that are representative of compositions, $H_2O/CO_2$ ratios, and temperatures encountered during combustion in air and oxy-combustion [11, 20]. For further details on the WSGG modeling methodology and the gas radiative property models employed in this study, the reader is referred to [2, 11]. In this study, the particle radiative properties were assumed to be "gray," that is, maintained constant across all gray gases ($i$), and the particle temperature was assumed to be equal to the surrounding gas temperature (i.e., $I_{bg}$ and $I_{bp}$ were equal at all spatial locations) [21].

The radiative heat fluxes within the domain or at a surface and the radiative flux divergence are determined by integrating over the solid angles ($\Omega$) surrounding a point as [2]

$$
\mathbf{q}(\mathbf{r}) = \sum_i \int_{4\pi} I(\mathbf{r}, \hat{s}) \hat{s} d\Omega
\tag{3}
$$

$$
\nabla \cdot \mathbf{q}(\mathbf{r}) = \sum_i \left( k_{g,i}(\mathbf{r}) + k_{p,i}(\mathbf{r}) \right) \left( 4\pi I_b(\mathbf{r}) - G_i(\mathbf{r}) \right).
\tag{4}
$$

$G$, the incident radiation, is calculated as

$$G(\mathbf{r}) = \sum_i \int_{4\pi} I(\mathbf{r}, \hat{\mathbf{s}}) \, d\Omega. \tag{5}$$

The particle absorption and scattering coefficients of the particles in (2) were computed as

$$k_p = \underset{V \to 0}{\text{Limit}} \sum_N \varepsilon_{pn} \frac{A_{pn}}{V} = \underset{V \to 0}{\text{Limit}} \sum_N Q_{abs} \frac{A_{pn}}{V} \tag{6}$$

$$\sigma_p = \underset{V \to 0}{\text{Limit}} \sum_N \left(1 - \varepsilon_{pn}\right)\left(1 - f_{pn}\right) \frac{A_{pn}}{V}$$

$$= \underset{V \to 0}{\text{Limit}} \sum_N Q_{scat} \frac{A_{pn}}{V}. \tag{7}$$

In (6) and (7), the summation is over $N$ particles within the control volume $V$, $\varepsilon_{pn}$ is the particle emissivity, $A_{pn}$ is the projected area of the $n$th particle, and $f_{pn}$ is the scattering factor associated with the $n$th particle. $Q_{abs}$ and $Q_{scat}$ correspond to the absorption and scattering efficiencies, respectively. In this study, $Q_{abs}$ and $Q_{scat}$ were assumed to vary as the combusting particle transitions from the parent fuel to char to an ash particle. This variation adopted in this study is shown in Table 1. The variation in the coal radiative properties is according to Kuehlert [22]. These values indicate that the effective "extinction efficiency factors" are in the range 1.75 to 2.43 which compares well with the effective extinction factors reported by Mengüç et al. [5] for different sized particles. The absorption and scattering efficiencies of the biomass particles were based on the assumption that the absorption efficiency of biomass particles at the temperatures encountered in this study was roughly 50% of those of the coal particles and the scattering efficiencies were about 10% of those of coal particles [3]. In order to avail computational savings, the radiation calculations were performed once every 20 fluid/energy iterations. This approach has no bearing on the final results reported in this study and is justifiable by the fact that the radiation field does not change as rapidly between iterations as the fluid/temperature field. Furthermore, since the results reported in this study correspond to steady state calculations, the convergence in the radiation calculations was assessed at the end by ensuring radiative energy balance throughout the furnace. This was done by confirming that the volume integral of the radiative source term was equal to the surface integral of the net radiative flux through all the domain boundaries.

Other important physical model settings employed in the simulations are given in Table 2 and the interested reader is referred to the ANSYS FLUENT user's guide for further details on these models [19]. The primary modeling options were employed in all of the simulations explored in this study, unless otherwise indicated. The variable properties for the particles and the gas radiative property models were implemented in this study as user-defined functions (UDFs) in ANSYS FLUENT. The heterogeneous coal char combustion was modeled employing the intrinsic model which assumes that the surface reaction rate includes the effects of both

Table 1: Coal and biomass radiative property variations employed in the simulations.

(a)

| Coal | Absorption efficiency | Scattering efficiency | Scattering/absorption efficiency |
|---|---|---|---|
| Parent fuel | 1.13 | 1.3 | 1.15 |
| Char | 0.59 | 1.5 | 2.54 |
| Ash | 0.05 | 1.7 | 34.00 |

(b)

| Biomass | Absorption efficiency | Scattering efficiency | Scattering/absorption efficiency |
|---|---|---|---|
| Parent fuel | 0.565 | 0.13 | 0.23 |
| Char | 0.295 | 0.15 | 0.51 |
| Ash | 0.025 | 0.17 | 6.80 |

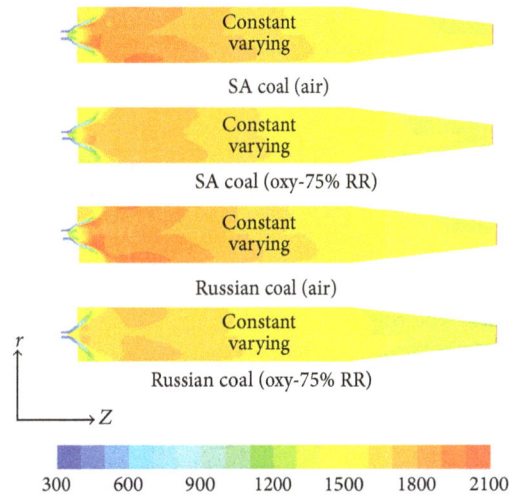

Figure 2: Temperature contours during the combustion of Russian and SA coal in air and in the oxy-75% RR scenario (on the top half of the axisymmetric furnaces are the predictions from constant particle radiative properties; in bottom half are predictions from varying particle radiative properties).

bulk diffusion and surface reaction while the combustion of the biomass was assumed to be diffusion limited [19]. The thermal boundary conditions at the wall were set at emissivity of 0.6 with heat transfer coefficient of 14 W/m$^2$-K and surrounding fluid temperature of 300 K since the furnace was water jacketed [14].

## 3. Results and Discussion

*3.1. Sensitivity to Particle Radiative Property during Coal Combustion.* Figure 2 shows contours of temperature during the combustion of Russian and SA coal in air and in the oxy-75% RR scenario. Two simulations were performed corresponding to each experimental condition: one in which the radiative properties of the combusting particles varied according to Table 1 and the second where the radiative

TABLE 2: Summary of physical models employed in the CFD simulations.

| Physical models | Primary modeling option | Other modeling options explored in the sensitivity analysis |
|---|---|---|
| Coal devolatilization | The two competing rates (Kobayashi) model | |
| Biomass devolatilization | The two competing rates (Kobayashi) model | Constant devolatilization rate (20 sec$^{-1}$) |
| Gas-phase chemistry | Two steps: (1) char oxidation to CO, (2) CO oxidation to $CO_2$ | |
| Gas-phase radiative property | Perry (5 gg) [11]* | |
| Turbulence | Realizable $k$-$\varepsilon$ | |
| Radiative transport equation solver | Discrete ordinates method (angular resolution, theta × phi: 3 × 3) | |
| Particle radiative property | Variable $K_{abs}$ and $K_{scat}$ (Table 1)* | Constant $K_{abs}$ and $K_{scat}$ (corresponding to that of char in cf. Table 2); |
| Particle scattering phase function | Anisotropic (forward scattering) | Isotropic scattering |
| Fuel particle size distribution (PSD) at inlet | Rosin-Rammler distributions (Figure 1) (8 diameter intervals) | Rosin-Rammler distributions (Figure 1) (40 diameter intervals) |
| Particle swelling coefficient (during devolatilization) | 1 (no change in particle diameter during devolatilization) | 1.5 (particle diameter increases by 50% during devolatilization) |

*These models were implemented as User-Defined Functions (UDFs) in ANSYS FLUENT.

properties were held constant (corresponding to that of the coal char in Table 1). These are labeled as "Varying" and "Constant" in the figure legends, respectively. While the temperature contours for the two coals are identical under similar firing conditions, oxy-firing at 75% RR results in reduced temperatures resulting from the higher heat capacities of $CO_2$. Furthermore, there are noticeable differences between the temperature fields predicted by the constant and varying particle properties.

Figure 3 shows contours of $CO_2$ and $H_2O$ mole fractions during the combustion of Russian and SA coal in air and in the oxy-75% RR scenario when employing varying particle radiative properties. The gas compositions are noticeably homogeneous downstream of the flame indicating the completion of devolatilization and combustion. In the simulations explored in this study, the $H_2O$/$CO_2$ ratios at the furnace exit were in the range 0.35–0.65 during combustion in air and in the range 0.08–0.10 during oxy-75% RR.

Figure 4 shows the contours of particle scattering coefficient to particle absorption coefficients ($\sigma_p$/$k_p$). Since the scattering and absorption coefficients are directly proportional to the scattering and absorption efficiencies, respectively ((6) and (7)), this ratio (Table 1) is an indicator of the state of the combusting particles inside the furnace which increases from 1.15 for the parent coal particle to 34 when only ash is present within the furnace. For both coals we observe that complete devolatilization and burn-out occurs within 1.5 m from the burner. Furthermore, a large volume of the furnace within this devolatilization and burnout region consists of a mixture of char and ash particles with significantly different radiative properties (Table 1). Since this region also corresponds to the highest flame temperature and therefore radiative heat fluxes, these variations between the char and ash radiative properties must be accounted for if the particle radiation is deemed to be important in

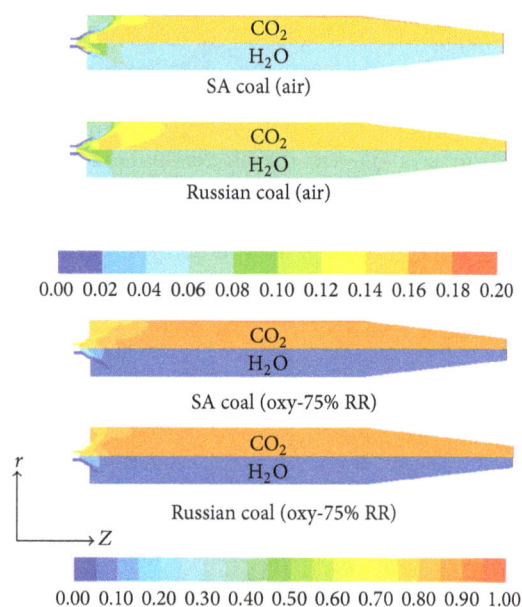

FIGURE 3: Contours of $CO_2$ and $H_2O$ mole fractions during the combustion of Russian and SA coal in air and in the oxy-75% RR scenario (varying particle radiative properties).

this region. Although the contours in Figure 4 are largely identical for combustion in both air and oxy-combustion, differences in the devolatilization and burnout characteristics are observed in the recirculation zone near the burner. Since the temperature in this recirculation zone (circled regions) is generally lower (Figure 2), this can impact the devolatilization and burnout characteristics particularly in the oxy-combustion scenario.

Figure 5 shows the contours of particle concentrations during the combustion of Russian and SA coals. Since

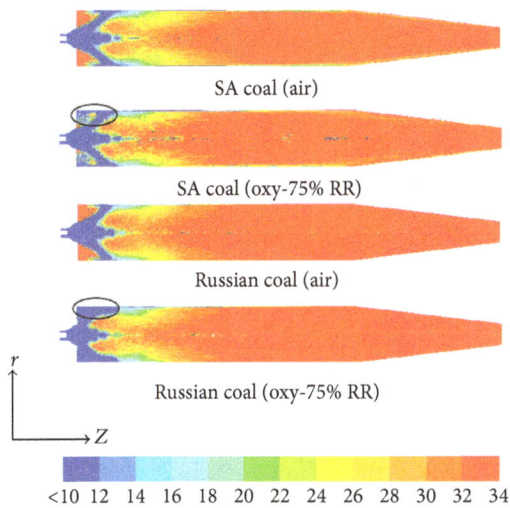

FIGURE 4: Contours of particle scattering coefficient to particle absorption coefficients ($\sigma_p/k_p$) computed by using varying radiative properties for the combusting particle. Differences in the ignition and burnout predictions are seen in the recirculation zone in the oxy-75% RR scenarios as indicated by circles when compared against combustion in air.

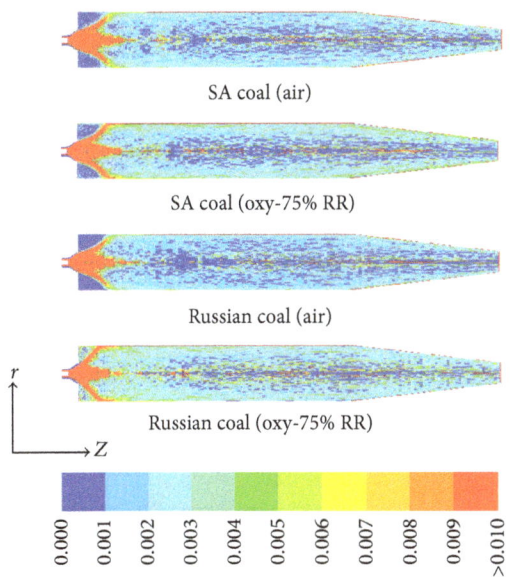

FIGURE 5: Contours of particle concentrations (in kg/m$^3$) during the combustion of Russian and SA coal in air and in the oxy-75% RR scenario.

the particles are tracked in a Lagrangian reference frame and the particle concentrations are proportional to the residence time within a computational cell and therefore to the velocities and volumetric flow rates of the gas within the furnace, the oxy-combustion scenario clearly results in larger particle concentrations due to the lower volumetric flow rates of the flue gas circulating within this furnace. Since the particle radiative properties are directly proportional to the particle concentrations ((6) and (7)), the contours of particle extinction coefficients ($\sigma_p + k_p$) in Figure 6 show higher values associated with the oxy-combustion scenario.

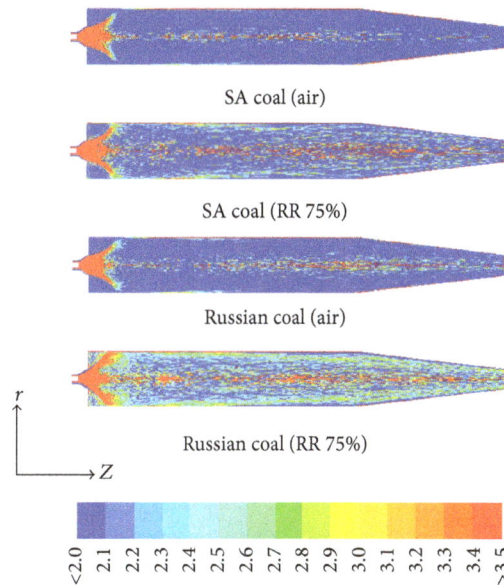

FIGURE 6: Contours of particle extinction coefficients ($\sigma_p + k_p$) (in m$^{-1}$) during the combustion of Russian and SA coal in air and in the oxy-75% RR scenario.

Figure 6 also shows that there are significant variations in the extinction coefficients within the furnace as well as between operating conditions and therefore an assumption of a single constant extinction coefficient cannot be made a priori and is likely to result in incorrect predictions.

Figure 7 shows profiles of incident radiative fluxes during the combustion of the Russian and SA coals. Although reasonable agreement is seen between the experimental measurements and simulation predictions, there are significant variations between the constant and variable particle radiative property predictions. The constant radiative property predictions overestimate the absorption efficiencies and particle emissivities (4) in the high temperature regions containing a mixture of char and ash particles resulting in higher incident radiative fluxes. It is worth noting that it is not uncommon to employ a constant particle emissivity greater than 0.6 throughout the domain in CFD simulations of coal combustion [15, 23]. Due to the lower temperatures encountered in the 75% RR scenario, the corresponding incident radiative fluxes are lower than those encountered during combustion in air. While the numerical predictions are generally greater than the experimental measurements, the predictions in the 75% RR scenarios are lower than the measurements in the near-burner region. This is a result of the inability of the combustion model to predict char ignition characteristics in the recirculation zone (circled regions in Figure 4). This overestimation of the local particle emission when employing constant radiative properties also results in cooling the flame by decreasing the value of the radiative source term (since "$-\nabla \cdot \mathbf{q(r)}$" is a source term in the temperature/enthalpy equation). This explains the slightly cooler flames observed in Figure 2 with the constant particle radiative property simulations when compared against the varying radiative property simulations. These effects are

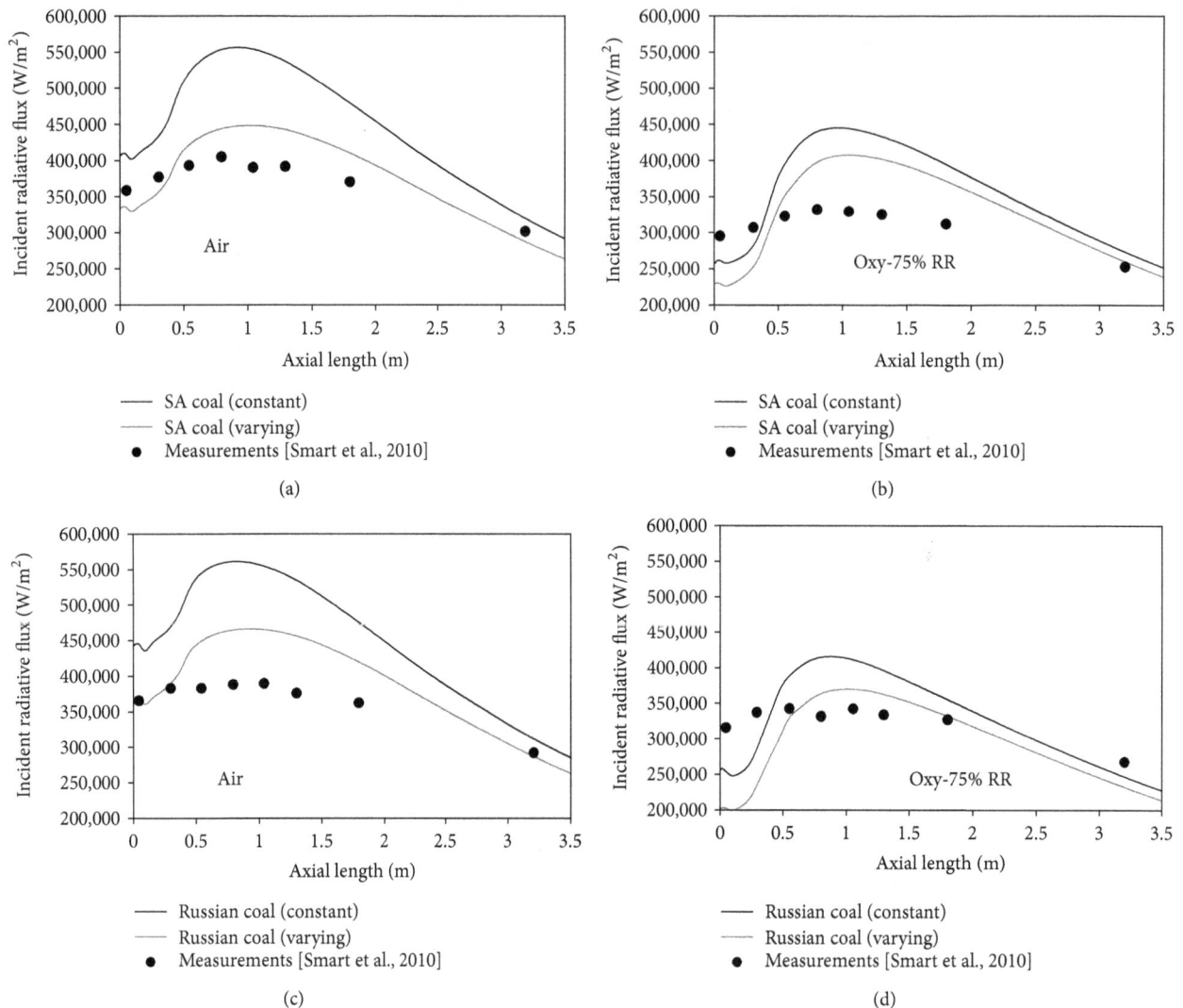

FIGURE 7: Wall incident radiative flux profiles predicted during the combustion of Russian and SA coals in air and in the oxy-75% RR scenario. Prediction sensitivities for employing "constant" and "varying" radiative properties are shown.

further quantified in Table 3 where the observed average incident radiative fluxes along the furnace wall are higher and the furnace temperatures are lower when employing the constant radiative properties. These differences are slightly minimized in the "oxy-75% RR" scenario due to the slightly lower flame temperatures as well as due to the increased contribution of gas radiation.

In Table 3, the net radiative flux which is the difference between the incident radiative flux and the wall emission is seen to be nearly identical in all four scenarios. It is important to recall that, due to the convective boundary conditions imposed at the wall in this study, an increase in incident radiative flux results in an increase in wall temperature which in turn increases the wall emission. Therefore, any increase in incoming radiation at the wall approximately cancels out with a corresponding increase in the outgoing radiation thereby resulting in a constant net radiative flux. However, the total net radiative flux reported in Table 3 is also a

measure of the total heat loss due to radiation from the flames. Considering that the total thermal input of the flames is 500 kW, the radiant fraction (total radiative heat loss/thermal input) of these flames is more than 0.4. Therefore, large errors or differences arising in the radiative transfer calculations (such as the radiative source term) will have strong bearing on the incident radiative fluxes as well as the temperature predictions (as shown in Figures 2 and 7).

### 3.2. Sensitivity to Particle Radiative Property during Coal-Biomass Co-Combustion.
Figure 8, shows the temperature contours during the co-combustion of Russian coal and biomass in air. The weight % of the biomass particles in the fuel while maintaining a constant thermal input of 0.5 MW is also indicated. Increasing the mass fraction of the biomass particles decreases the temperature within the furnace. The moisture content of the SA coal, Russian coal, Shea meal, and

TABLE 3: Radiation and temperature predictions during coal combustion.

| Fuel | Combustion mode | Net radiative flux (kW) | | Average incident radiative flux (W/m$^2$) | | Average furnace temperature (K) | |
| | | Constant property | Variable property | Constant property | Variable property | Constant property | Variable property |
| --- | --- | --- | --- | --- | --- | --- | --- |
| SA coal | Air firing | 222 | 204 | 363 | 313 | 1631 | 1666 |
| SA coal | Oxy-75% RR | 215 | 210 | 296 | 277 | 1570 | 1578 |
| Russian coal | Air firing | 227 | 214 | 363 | 320 | 1643 | 1676 |
| Russian coal | Oxy-75% RR | 210 | 203 | 273 | 249 | 1533 | 1542 |

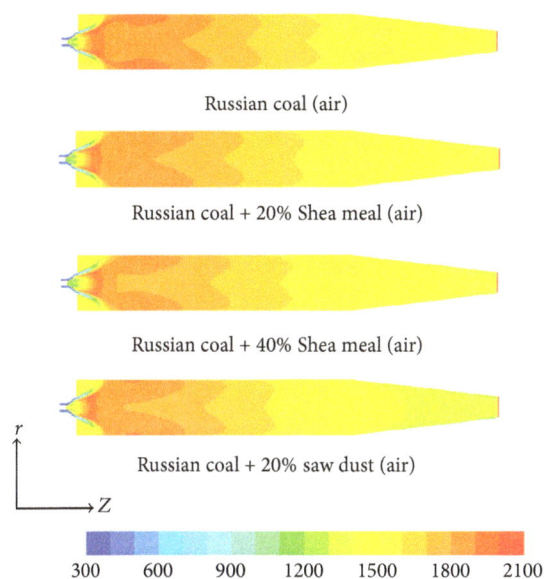

FIGURE 8: Temperature contours during the co-combustion of Russian coal and biomass particles in air.

saw dust was 4.50%, 6.23%, 11.58%, and 30.00%, respectively. The higher moisture content and the lower calorific value of the biomass particles suppress the flame temperatures.

Figure 9 shows the contours of incident radiative heat fluxes in the Russian coal-biomass cofiring scenarios in air with the calculations performed with variable particle radiative properties for both the coal and biomass particles (Table 1). A reasonable agreement is seen between the experimental measurements and the simulation predictions with the trends reflective of the temperature predictions; that is, increasing the mass fraction of the biomass particles in the feed decreases the incident radiative fluxes. Saw dust by virtue of its higher moisture content and lower calorific value when compared to Shea meal results in lowering the flame temperatures and incident radiative fluxes even more for the same mass fraction in the feed. In Figure 9(b), the sensitivities of the predictions to the biomass devolatilization models employed in the simulations were assessed. Since the default devolatilization modeling option (two competing rates (Kobayashi) model) was developed more for coal combustion scenarios, a constant devolatilization rate of $20 \, s^{-1}$ was also employed in the simulations. The incident radiative

flux predictions were found to be largely insensitive to the biomass devolatilization model employed in the simulations as seen in Figure 9(b).

Figure 10 compares the incident radiative flux predictions during the cofiring of SA coal and Shea meal employing constant radiative properties (corresponding to the coal char and biomass char in Table 1) and variable radiative properties for both the coal and biomass particles. Similar to the observations in the coal combustion scenario (Figure 7), there are considerable variations in the incident radiative flux predictions between the two property models. Employing a constant radiative property corresponding to that of the fuel char can therefore lead to a considerable overestimation of the incident radiative fluxes.

Apart from the variability in the absorption and scattering efficiencies of the combusting particle, there can also be a great deal of variability in the scattering phase functions ($\Phi$ in (2)) of the combusting particle. The scattering phase function is strongly dependent on the particle size as well as shape [5]. Both coal and biomass particles have irregular shapes which can change further resulting from fragmentation during the combustion process. While experimental observations have confirmed that coal particles are highly forward scattering [5], there can be considerable variability in the scattering phase function. Furthermore, as noted in Table 1 and Figure 4, the particle scattering coefficients of the ash particles are much higher than the absorption coefficients. Therefore, in order to assess the sensitivity to the scattering phase function, a simulation with isotropic scattering phase function was carried out in addition to the highly forward scattering dirac-delta phase function that was employed in all of the simulations. The predictions are compared in Figure 11 during the combustion of Russian coal during "oxy-75% RR." The results indicate very little sensitivity to the scattering phase function characteristics of the particles. These observations regarding scattering in coal-fired furnaces are similar to those reported in a previous study in a nearly identical sized furnace for combustion in air [24]. In that study, neglecting scattering altogether was suggested as a preferred alternative to isotropic scattering when neglecting the strongly forward scattering behavior of coal particles becomes necessary to reduce the computational effort. However, since "oxy-75% RR" scenario was explored in this study, despite the increase in particle concentrations and scattering coefficients, the increased importance of gas radiation likely suppressed further any

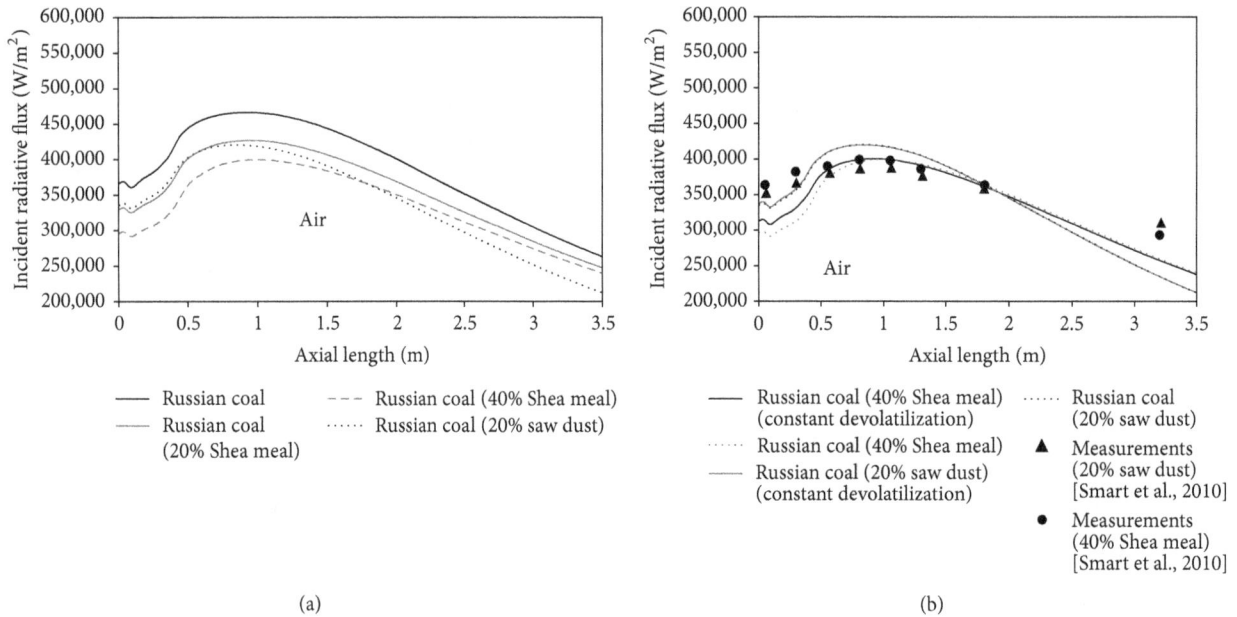

(a)

(b)

FIGURE 9: Wall incident radiative flux profiles predicted during the co-combustion of Russian coal and biomass particles in air.

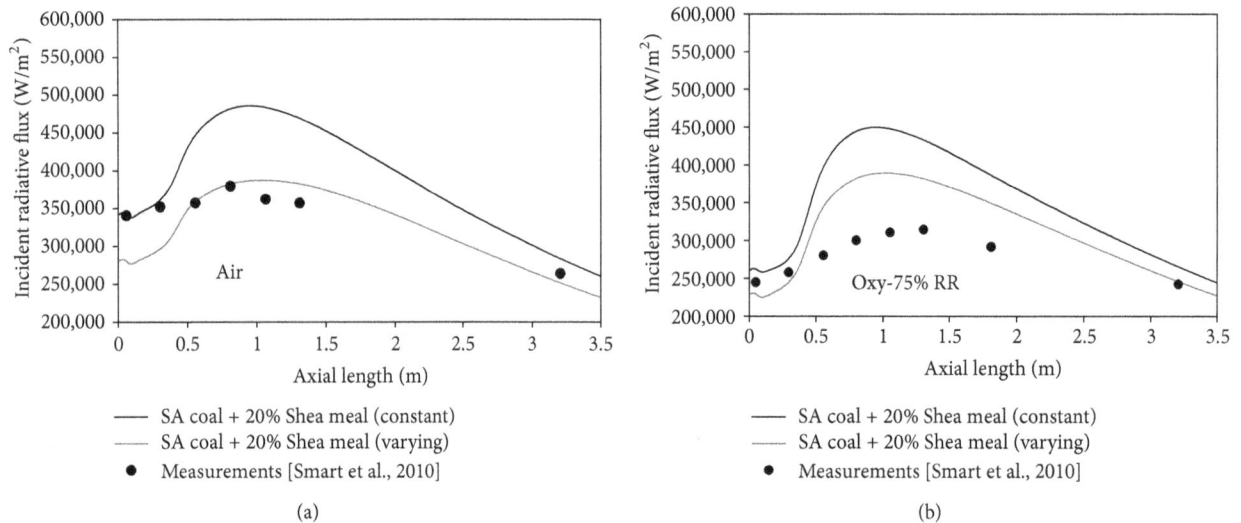

(a)

(b)

FIGURE 10: Wall incident radiative flux profiles predicted during the co-combustion of SA coal and biomass particles in air and oxy-75% RR.

prediction differences arising due to the scattering phase function. This is confirmed in Table 4 where the total radiant powers of the different flames (estimated by employing variable particle radiative properties in the simulations) are reported. This was computed as the surface integral of the net radiative heat fluxes through all the boundaries of the geometry or alternatively as the volume integral of the radiative source term within the furnace. In order to assess the individual contributions of gas and particulates to the total radiant power, the converged temperature and specie concentrations were "frozen," particle radiative properties were set to zero and the radiation field recomputed along with the volume integral of the radiative source term. This new radiant power which accounts for gas radiation alone

is also reported in Table 4. Gas radiation accounts for 60–70% of the total radiant power during combustion in air and more than 80% of the total radiant power during oxy-combustion. However, in the near burner region, particle radiation can dominate and the choice of particle radiative properties can greatly impact the accuracies of the incident radiative flux predictions as shown in Figures 7 and 10. This becomes particularly important when numerical prediction accuracies are assessed against radiometric measurements. It is also important to note that the values of radiant power reported in Table 4 may be slightly different from the values of the net radiative fluxes reported in Table 3 due to the fact that the radiant power estimates were made by accounting for radiative transfer through all boundaries (including inlets

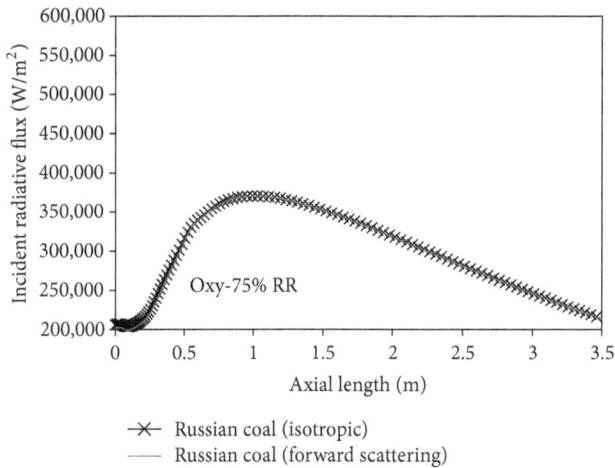

FIGURE 11: Sensitivity to scattering phase function during the combustion of Russian coal during oxy-75% RR.

and outlets) whereas the net radiative fluxes were estimated only along the horizontal walls of the furnace aligned along the axial direction.

*3.3. Sensitivity to Particle Size Predictions.* While the discussion so far has focused on the variability in the particle radiative properties during the combustion process, the absorption and scattering coefficients are also proportional to the particle diameter squared through its dependence on the projected surface area ("$A_{pn}$" in (6) and (7)). Therefore, even modest changes to the particle diameter through swelling and fragmentation during the combustion process may translate to large changes to the absorption and scattering coefficients, the incident radiative flux predictions. In fact, accurate modeling of char fragmentation during burnout has been emphasized as one of the pressing modeling needs not only for radiative transfer but also for carbon-in-ash predictions [25]. One of the theories explaining char fragmentation attributes it to changes to the particle porosity that may result from swelling during the combustion process. Therefore, a swelling factor is also commonly employed in CFD simulations of coal and biomass combustion [26]. Although char fragmentation was not modeled in this study, a set of simulations were carried out employing a swelling factor of 1.5 (which increases the particle absorption coefficient by factor of 2.25). The incident radiative flux predictions are compared in Figure 12. In both combustion in air and 75% RR scenario, particle swelling is observed to have an effect on the incident radiative flux profiles as well as on the location of the peak incident radiative fluxes. Therefore, modeling of swelling and fragmentation during the fuel burnout are deemed to be important for obtaining accurate estimates of radiative transfer.

A second size distribution parameter that was investigated in this study was the number of diameter bins employed in the Rosin-Rammler (RR) size distribution shown in Figure 1. Although the PSD is identical for a given two-parameter RR distribution, it is important to ascertain that adequate resolution (or diameter intervals) is employed in the simulations as there can be some variability in the fraction of mass that is distributed to the different diameter sized particles depending on the number of bins employed. This becomes particularly important when dealing with fuels encompassing a large size range. However, since the computational cost of the Lagrangian particle tracking calculation is directly proportional to the number of bins employed, 6 to 10 bins in RR distribution are generally deemed to be adequate [23, 27].

To investigate the sensitivity to the resolution of the RR distribution, additional simulations were carried out employing 40 bins in the RR distribution. The mass fraction distributions to different particle sizes for particles sampled at the outlet are shown in Figure 13. Since the diameter of the combusting particles in these simulations did not undergo any changes due to swelling/fragmentation, the reported results should have been independent of the number of diameter intervals employed for the fuel PSD at the inlet. While this is largely true when considering the entire size range of the particles (left side images of Figure 13), a closer examination of the larger sized particles reveals that the particle mass fraction distribution in the larger sized particles (which carry the majority of the particle mass) can vary significantly depending on the number of diameter intervals employed at the inlet. This can in turn affect the particle concentration distributions and the particle extinction coefficients and consequently the incident radiative fluxes. The prediction sensitivities of the particle extinction coefficients to the diameter intervals are shown in Figure 14. Since employing fewer diameter intervals overestimate the fraction of mass that is distributed to the larger sized particles (Figure 13), they also overestimate the large particle number densities and concentrations and consequently the particle extinction coefficients as shown in Figure 14. This has an impact on the incident radiative flux predictions as shown in Figure 15. Any further increase in the number of diameter intervals (beyond 40) did not have an effect on the simulation predictions in this study. This again points to the criticality of developing appropriate fragmentation models as even small differences in the mass fractions in the larger sized particles (which carry most of the mass of the fuel particles) can have a big impact on the radiative transfer predictions.

Figure 16 examines the impact on particle swelling and diameter intervals on the temperature predictions within the furnace. While a minimal impact is seen on the temperature field, the differences are not as pronounced as seen in Figure 2 (between the constant and varying radiative property predictions). This indicates that any differences in the radiative source term predictions due to swelling, fragmentation, or the mass distributions in the larger sized particles are likely to be small to have a significant impact on the temperature field but they can influence the incident radiative flux predictions.

# 4. Conclusions

To facilitate the retrofitting of existing coal-fired boilers to oxy-combustion scenarios requires an improved

TABLE 4: Radiant power of the different flames estimated by employing variable particle radiative properties in the simulations.

| Fuel | Combustion mode | Radiant power (kW) | Radiant power (kW) (gas radiation only) |
|------|-----------------|--------------------|-----------------------------------------|
| SA Coal | Air firing | 211 | 141 |
| SA Coal | Recycle 75 | 209 | 183 |
| SA + 20% Shea meal | Air firing | 202 | 138 |
| SA + 20% Shea meal | Recycle 75 | 201 | 168 |
| Russian coal | Air firing | 222 | 145 |
| Russian coal | Recycle 75 | 207 | 175 |
| Russian coal + 20% Shea meal | Air firing | 215 | 145 |
| Russian coal + 40% Shea meal | Air firing | 210 | 146 |
| Russian coal + 20% saw dust | Air firing | 204 | 133 |

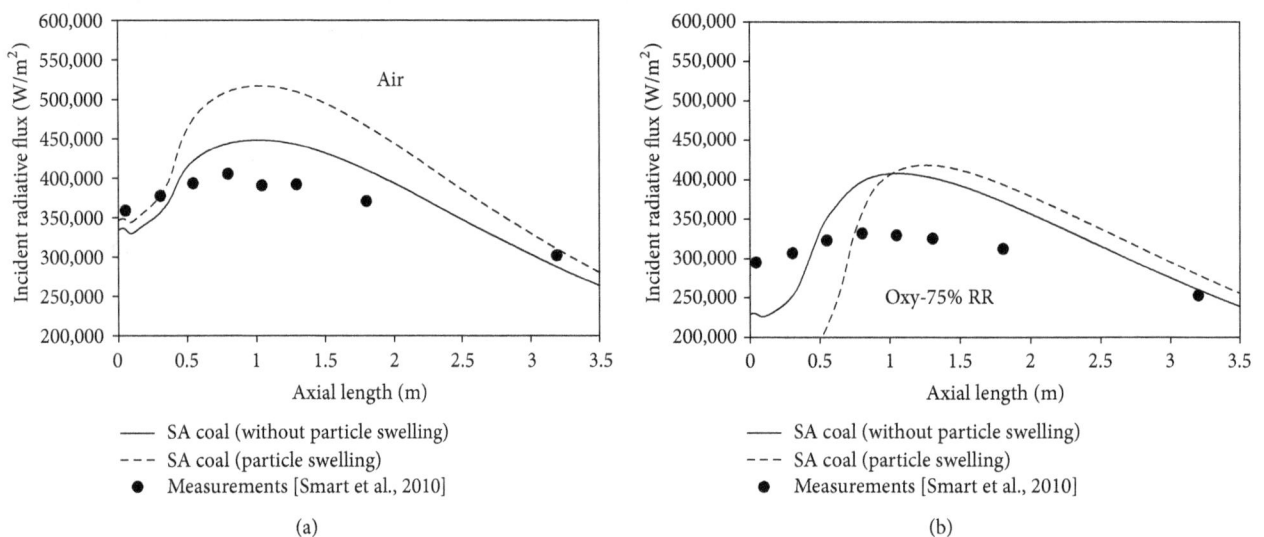

FIGURE 12: Sensitivities of the incident radiative fluxes to particle swelling during combustion of SA coal.

understanding of the role of particle radiation, its contribution relative to that of the radiatively participating gases within the boiler, and how it is influenced by the evolving fly ash particle size distributions (PSDs) under these newer operating conditions. By taking advantage of recent radiative transfer measurements made in a 0.5 MW combustion test facility during combustion of coal-biomass blends under air-fired and oxy-fired conditions, this study assesses the need for and the required fidelity in modeling the evolving PSD and radiative properties of the combusting particles as a function of residence time within the boiler. Two different coals (a South African and a Russian) that were blended with two types of biomass particles (Shea meal and saw dust) were investigated during combustion in air and oxy-combustion at a flue gas recycle ratio of 75%. The fuel and oxidizer compositions were varied to maintain a target exit $O_2$ concentration of 3 mol% (dry basis) in all of the investigated scenarios. The sensitivities of the radiation predictions to the variations in the radiative properties of the combusting particle (absorption and scattering efficiencies), particle swelling during devolatilization, particle scattering

phase function, biomass devolatilization models, and the resolution of the fuel PSD were assessed. The following conclusions may be drawn from the results of this study:

(1) While reasonable agreement with experimental measurements was observed in the investigated scenarios, accounting for the significant variations in the radiative properties (absorption and scattering efficiencies) between the parent fuel, combusting char, and ash particles was deemed important towards improving the prediction accuracies. This was due to the fact that a large volume of the furnace near the radiation measurement locations consisted of combusting char and ash particle mixtures and therefore an assumption of a constant radiative property (corresponding to that of char or the ash) throughout the furnace could result in inaccurate predictions. These variations in the particle radiative properties strongly influenced the incident radiative fluxes as well as the temperature predictions in these strongly radiating flames (which had radiant fractions > 0.4).

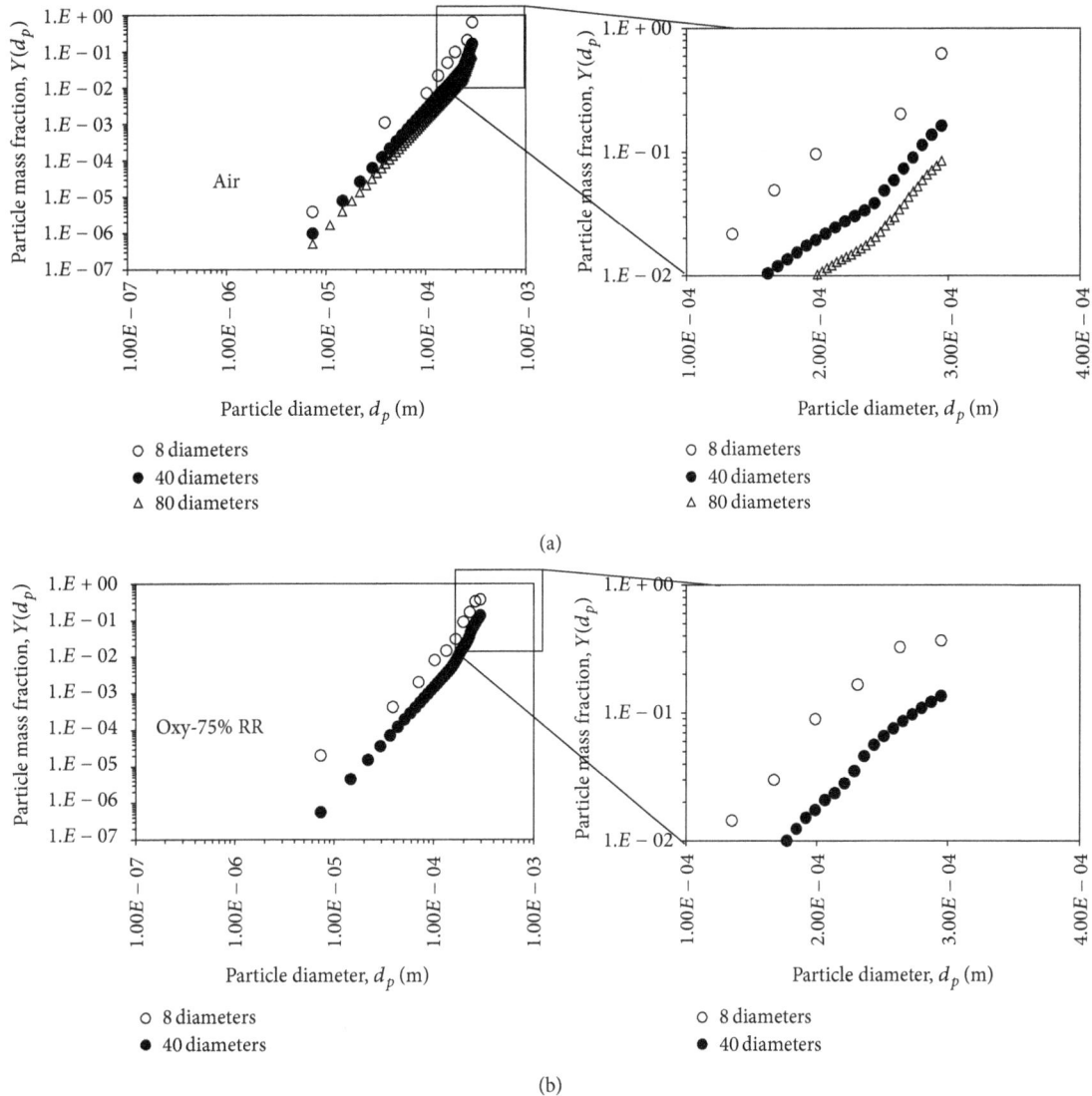

FIGURE 13: Sensitivities of the mass fraction distribution to different particle sizes as a function of the number of diameter intervals (or resolution in the R-R distribution) during combustion of Russian coal.

(2) The lower gas flow rates through the furnace during oxy-combustion resulted in higher particle concentrations and particle extinction coefficients when compared against combustion in air. However, the 75% recycle ratio that was investigated in this study resulted in lower gas and particle temperatures which slightly minimized the differences between the constant particle radiative property predictions and the predictions employing variable particle radiative properties when compared to combustion in air. In addition, the increased contribution of gas radiation in the recycle scenario could have also contributed to this minimization.

(3) Although all the inlet fuel particle size distributions were fit to a Rosin-Rammler (RR) distribution function, the resolution or the diameter intervals employed in the RR function and particle swelling during devolatilization had an impact on the particle extinction coefficients as well as the incident radiative flux predictions. On the other hand, the impact of variations in these parameters on the temperature field was minimal.

(4) Although the modeled scattering efficiencies of particles were significantly larger than their absorption efficiencies, both isotropic and forward scattering phase functions resulted in identical radiative flux profiles. Therefore, any change to the scattering phase function due to the evolving size and shape of the fly ash particle as it moves through the furnace is likely to have a minimal impact on the furnace radiative transfer at these high temperatures. Therefore, it is recommended that the highly forward scattering characteristics attributed to spherical char and ash

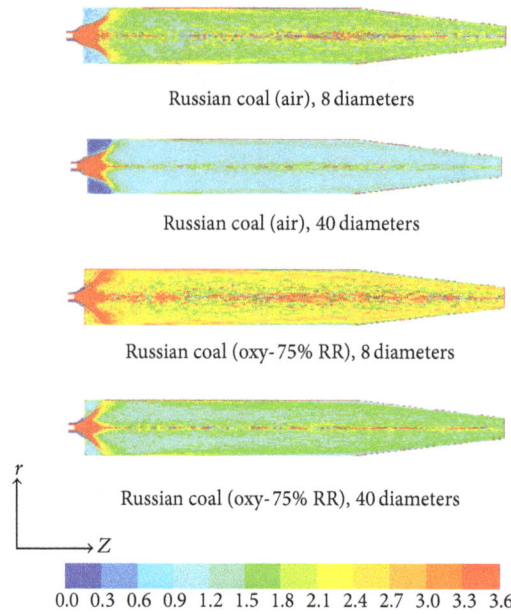

FIGURE 14: Sensitivities of the particle extinction coefficients ($\sigma_p + k_p$) (in m$^{-1}$) to the number of diameter intervals (or resolution in the R-R distribution) during combustion of Russian coal.

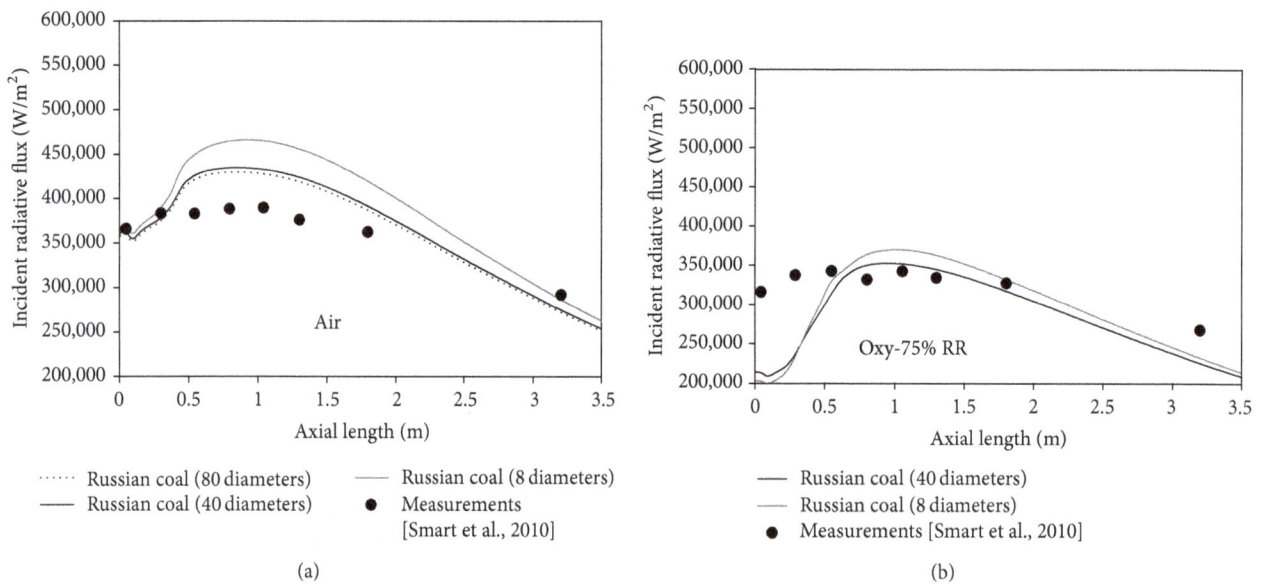

FIGURE 15: Sensitivities of the incident radiative fluxes to the number of diameter intervals (or resolution in the R-R distribution) during combustion of Russian coal.

particles can be applied in these scenarios without too much loss in accuracy.

(5) Although the fuel heating value was maintained at 0.5 MW in the biomass-coal cofiring scenarios as well, the gas temperatures and consequently the wall incident radiative fluxes both decreased with an increase in mass of biomass in the feed due to the higher moisture content and lower calorific values of the biomass particles. The choice of biomass devolatilization models did not significantly impact the predictions in this study since the devolatilization was completed at short residence times from the burner.

This study therefore highlights the necessity for careful selection of particle radiative property and diameter interval parameters in CFD simulations of coal/biomass combustion and the need for fuel fragmentation models to adequately predict the fly ash PSD.

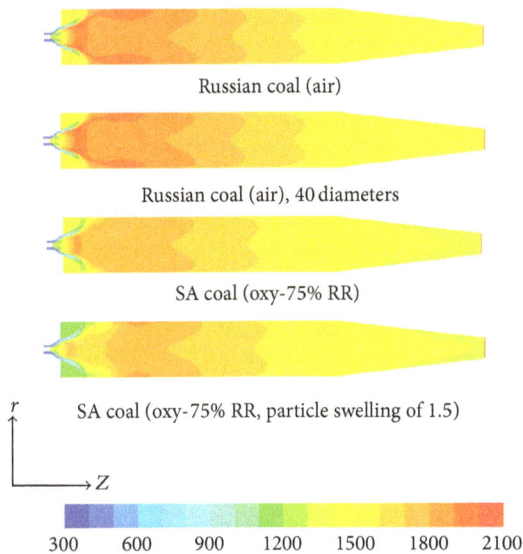

FIGURE 16: Sensitivities of the temperature contours to particle swelling and diameter intervals during coal combustion.

## Nomenclature

$A$:  Projected area of particle ($m^2$)
$f$:  Scattering factor
$G$:  Total incident radiation ($W\,m^{-2}$)
$I$:  Radiative intensity ($W\,m^{-2}\,Sr^{-1}$)
$k$:  Absorption coefficient ($m^{-1}$)
$Q_{abs}$:  Absorption efficiency of particles
$Q_{scat}$:  Scattering efficiency of particles
$q$:  Incident radiative flux ($W\,m^{-2}$)
**r**:  Spatial coordinate
$\widehat{\mathbf{s}}$:  Angular direction
$V$:  Particle volume ($m^3$)
$w$:  Weighting factor associated with black body emissive power
$Y_d$:  Mass fraction of particles having a diameter that is greater than "$d$."

*Greek Letters*

$\varepsilon$:  Particle emissivity
$\Omega$:  Solid angles
$\Phi$:  Scattering phase function
$\sigma$:  Scattering coefficient ($m^{-1}$).

*Subscripts*

$b$:  Blackbody
$g$:  Gas phase
$i$:  Gray gas
$p$:  Particle phase.

## Conflict of Interests

The authors declare that there is no conflict of interests regarding the publication of this paper.

## Acknowledgment

This work was partially supported by a grant to Ms. Caitlyn Wolf administered through the North Dakota EPSCoR's Advanced Undergraduate Research Awards (AURA) program.

## References

[1] R. Viskanta and M. P. Mengüç, "Radiation heat transfer in combustion systems," *Progress in Energy and Combustion Science*, vol. 13, no. 2, pp. 97–160, 1987.

[2] M. F. Modest, *Radiative Heat Transfer*, Academic Press, 2013.

[3] H. Hofgren and B. Sundén, "Evaluation of Planck mean coefficients for particle radiative properties in combustion environments," *Heat and Mass Transfer*, 2014.

[4] D. G. Goodwin and M. Mitchner, "Flyash radiative properties and effects on radiative heat transfer in coal-fired systems," *International Journal of Heat and Mass Transfer*, vol. 32, no. 4, pp. 627–638, 1989.

[5] M. P. Mengüç, S. Manickavasagam, and D. A. D'Sa, "Determination of radiative properties of pulverized coal particles from experiments," *Fuel*, vol. 73, no. 4, pp. 613–625, 1994.

[6] R. P. Gupta, T. F. Wall, and J. S. Truelove, "Radiative scatter by fly ash in pulverized-coal-fired furnaces: application of the Monte Carlo method to anisotropic scatter," *International Journal of Heat and Mass Transfer*, vol. 26, no. 11, pp. 1649–1660, 1983.

[7] F. Liu and J. Swithenbank, "The effects of particle size distribution and refractive index on fly-ash radiative properties using a simplified approach," *International Journal of Heat and Mass Transfer*, vol. 36, no. 7, pp. 1905–1912, 1993.

[8] G. Scheffknecht, L. Al-Makhadmeh, U. Schnell, and J. Maier, "Oxy-fuel coal combustion—a review of the current state-of-the-art," *International Journal of Greenhouse Gas Control*, vol. 5, no. 1, pp. S16–S35, 2011.

[9] P. Edge, M. Gharebaghi, R. Irons et al., "Combustion modelling opportunities and challenges for oxy-coal carbon capture technology," *Chemical Engineering Research and Design*, vol. 89, no. 9, pp. 1470–1493, 2011.

[10] R. Johansson, K. Andersson, B. Leckner, and H. Thunman, "Models for gaseous radiative heat transfer applied to oxy-fuel conditions in boilers," *International Journal of Heat and Mass Transfer*, vol. 53, no. 1–3, pp. 220–230, 2010.

[11] G. Krishnamoorthy, "A new weighted-sum-of-gray-gases model for oxy-combustion scenarios," *International Journal of Energy Research*, vol. 37, no. 14, pp. 1752–1763, 2013.

[12] C. Yin, L. C. R. Johansen, L. A. Rosendahl, and S. K. Kær, "New weighted sum of gray gases model applicable to computational fluid dynamics (CFD) modeling of oxy-fuel combustion: derivation, validation, and implementation," *Energy and Fuels*, vol. 24, no. 12, pp. 6275–6282, 2010.

[13] T. Kangwanpongpan, F. H. R. França, R. Corrêa da Silva, P. S. Schneider, and H. J. Krautz, "New correlations for the weighted-sum-of-gray-gases model in oxy-fuel conditions based on HITEMP 2010 database," *International Journal of Heat and Mass Transfer*, vol. 55, no. 25-26, pp. 7419–7433, 2012.

[14] Y. Hu and J. Yan, "Numerical simulation of radiation intensity of oxy-coal combustion with flue gas recirculation," *International Journal of Greenhouse Gas Control*, vol. 17, pp. 473–480, 2013.

[15] M. Gharebaghi, R. M. A. Irons, L. Ma, M. Pourkashanian, and A. Pranzitelli, "Large eddy simulation of oxy-coal combustion in an industrial combustion test facility," *International Journal of Greenhouse Gas Control*, vol. 5, supplement 1, pp. S100–S110, 2011.

[16] L. L. Baxter, "Char fragmentation and fly ash formation during pulverized-coal combustion," *Combustion and Flame*, vol. 90, no. 2, pp. 174–184, 1992.

[17] N. Syred, K. Kurniawan, T. Griffiths, T. Gralton, and R. Ray, "Development of fragmentation models for solid fuel combustion and gasification as subroutines for inclusion in CFD codes," *Fuel*, vol. 86, no. 14, pp. 2221–2231, 2007.

[18] J. P. Smart, R. Patel, and G. S. Riley, "Oxy-fuel combustion of coal and biomass, the effect on radiative and convective heat transfer and burnout," *Combustion and Flame*, vol. 157, no. 12, pp. 2230–2240, 2010.

[19] Fluent, ANSYS, "15.0 Theory Guide", ANSYS Inc, 2014.

[20] G. Krishnamoorthy, "A new weighted-sum-of-gray-gases model for $CO_2$-$H_2O$ gas mixtures," *International Communications in Heat and Mass Transfer*, vol. 37, no. 9, pp. 1182–1186, 2010.

[21] J. Zhang, T. Ito, S. Ito, D. Riechelmann, and T. Fujimori, "Numerical investigation of oxy-coal combustion in a large-scale furnace: non-gray effect of gas and role of particle radiation," *Fuel*, vol. 139, pp. 87–93, 2015.

[22] K. Kuehlert, *Modellbildung und Berechnung der Wärmestrahlung in Gas- und Kohlenstaubfeuerungen [Ph.D. thesis]*, RWTH Aachen University, Aachen, Germany, 1998.

[23] R. Filkoski, I. Petrovski, and P. Karas, "Optimization of pulverised coal combustion by means of CFD/CTA modeling," *Thermal Science*, vol. 10, no. 3, pp. 161–179, 2006.

[24] J. G. Marakis, C. Papapavlou, and E. Kakaras, "A parametric study of radiative heat transfer in pulverised coal furnaces," *International Journal of Heat and Mass Transfer*, vol. 43, no. 16, pp. 2961–2971, 2000.

[25] R. Weber, M. Mancini, N. Schaffel-Mancini, and T. Kupka, "On predicting the ash behaviour using Computational Fluid Dynamics," *Fuel Processing Technology*, vol. 105, pp. 113–128, 2013.

[26] J. Li, A. Brzdekiewicz, W. Yang, and W. Blasiak, "Co-firing based on biomass torrefaction in a pulverized coal boiler with aim of 100% fuel switching," *Applied Energy*, vol. 99, pp. 344–354, 2012.

[27] R. V. Filkoski, "Pulverised-coal combustion with staged air introduction: CFD analysis with different thermal radiation methods," *Open Thermodynamics Journal*, vol. 4, pp. 2–12, 2010.

# 11

# Predicting Radiative Heat Transfer in Oxy-Methane Flame Simulations: An Examination of Its Sensitivities to Chemistry and Radiative Property Models

**Hassan Abdul-Sater,**[1] **Gautham Krishnamoorthy,**[1] **and Mario Ditaranto**[2]

[1]*Department of Chemical Engineering, University of North Dakota, Harrington Hall Room 323, 241 Centennial Drive, Grand Forks, ND 58202-7101, USA*
[2]*SINTEF Energy Research, 7465 Trondheim, Norway*

Correspondence should be addressed to Gautham Krishnamoorthy; gautham.krishnamoorthy@engr.und.edu

Academic Editor: Dimitrios C. Rakopoulos

Measurements from confined, laminar oxy-methane flames at different $O_2/CO_2$ dilution ratios in the oxidizer are first reported with measurements from methane-air flames included for comparison. Simulations of these flames employing appropriate chemistry and radiative property modeling options were performed to garner insights into the experimental trends and assess prediction sensitivities to the choice of modeling options. The chemistry was modeled employing a mixture-fraction based approach, Eddy dissipation concept (EDC), and refined global finite rate (FR) models. Radiative properties were estimated employing four weighted-sum-of-gray-gases (WSGG) models formulated from different spectroscopic/model databases. The mixture fraction and EDC models correctly predicted the trends in flame length and OH concentration variations, and the $O_2$, $CO_2$, and temperature measurements outside the flames. The refined FR chemistry model predictions of $CO_2$ and $O_2$ deviated from their measured values in the flame with 50% $O_2$ in the oxidizer. Flame radiant power estimates varied by less than 10% between the mixture fraction and EDC models but more than 60% between the different WSGG models. The largest variations were attributed to the postcombustion gases in the temperature range 500 K–800 K in the upper sections of the furnace which also contributed significantly to the overall radiative transfer.

## 1. Introduction

Oxy-fuel combustion, where a fuel is burnt in a mixture of oxygen and recycled flue gas stream (containing primarily $CO_2$ and $H_2O$), is a promising near-zero emission technology that can be adapted by existing and new electric power generation stations to mitigate the human impact on climate change [1]. To facilitate the retrofitting of existing combustors, oxidizer compositions and recycle ratios are currently being optimized with the aim of matching the flame temperature and wall radiative fluxes encountered during air-combustion by carefully taking into account the thermal effects associated with replacing $N_2$ with $CO_2$ [2]. Further, chemical effects associated with the participation of $CO_2$ through the equilibrium: $CO_2 + H \leftrightarrow CO + OH$, have also been identified

[3–7]. Consequently, the need to refine models for gas-phase chemistry and radiative properties that have previously been deemed to be accurate in computational fluid dynamic (CFD) simulations of combustion in air has garnered much attention in recent reviews [8, 9]. This has led to the development and validation of gas-phase radiative property models in the form of weighted-sum-of-gray-gases (WSGG) coefficients based on different spectroscopic/model databases [10–13] and the refinement of kinetic parameters in previously proposed global kinetic mechanisms [14, 15] to improve their prediction accuracy in oxy-combustion scenarios [16–18]. Summarized in Table 1 are the results from CFD simulations of oxy-methane/natural gas combustion employing these refined chemistry and radiative property models. The studies indicate that while refined global kinetics, equilibrium based,

TABLE 1: A summary of previous CFD studies of oxy-methane/natural gas combustion.

| Reference | System | Chemistry models | Radiation modeling | Results and conclusions |
|---|---|---|---|---|
| Prieler et al. [24] | Turbulent high temperature furnace for smelting and annealing applications | Eddy dissipation concept (EDC) with 46 reversible reactions, steady laminar flamelet (SFM) approach with flamelet libraries generated employing 25-step, 46-step mechanisms as well as 325 reactions in GRI-Mech 3.0 | Default WSGGM in ANSYS FLUENT (gray model) employed with P-1 and discrete ordinates (DO) radiation models | Temperature and species concentration predictions from EDC and SFM were very similar. SFM calculations were about 5 times faster than the EDC calculations that employed detailed chemistry. The PI radiation model overestimated emission and predicted lower temperatures than measurements |
| Nemitallah and Habib [25] | Turbulent, diffusion flame in a gas-turbine combustor investigated for a wide range of operational parameters | Modified 2-step chemistry mechanism [16] | DO radiation model; radiative property model was not specified | Overall flame shape and exhaust gas concentrations were well-predicted employing the modified two-step mechanisms |
| Galletti et al. [26] | 3 MW semi-industrial burner | Eddy dissipation model (EDM), EDC, modified global mechanisms | PI/DO, revised WSGGM coefficients [10] | Turbulence chemistry interactions play an important role in determining the temperature and species concentrations; EDC provides satisfactory temperature and species predictions |
| Wheaton et al. [27] | 0.8 MW turbulent burner, high temperature air combustion (HTAC) burner | Nonadiabatic equilibrium PDF | DO radiation model, revised WSGGM [28] employed in nongray simulations | Reasonable agreement of temperature with measurements, high concentrations of radiatively participating gases by themselves are not enough to warrant the use of nongray models. The peak temperatures, temperature gradients, and furnace dimensions also need to be taken into consideration |
| Yin et al. [17] | Semi-industrial furnace | Global 2-step and 4-step reaction mechanisms [14, 15] as well as modified versions of these mechanisms | DO radiation model, revised WSGGM [11] employed in gray simulations | The refined chemistry models were able to better predict the temperature and CO concentrations downstream of the furnace |
| Bhadraiah and Raghavan [29] | Laminar, unconfined | Four global mechanisms, 43-step skeletal mechanism | Optically thin radiation model | Major gases and temperature predictions from the 2-step are closer to the 43-step mechanism. Two-step mechanism predicts the location of the reaction zone accurately. Improvements agree at higher flow rates but only qualitative prediction of CO |
| Kim et al. [30] | 0.78 MW turbulent natural gas furnace | Conservative conditional moment closure for turbulence chemistry interactions with detailed chemistry mechanisms | Radiation model and the determination of radiative properties not specified | A good qualitative agreement is obtained with the temperature overestimated at short radial distances. $CO_2$ was underestimated and CO was overestimated in the high temperature regions |

TABLE 1: Continued.

| Reference | System | Chemistry models | Radiation modeling | Results and conclusions |
|---|---|---|---|---|
| Bennett et al. [31] | Laminar, unconfined diffusion flames | GRI-Mech 3.0 | Optically thin radiation model | Computational and experimental flame lengths and maximum centerline temperatures show excellent agreement. Radial profiles when plotted at fixed values of a dimensionless axial coordinate also show excellent agreement |
| Abdul-Sater and Krishnamoorthy [21] | Laminar, confined diffusion flames | Nonadiabatic equilibrium PDF | DO radiation model, revised WSGGM [28] employed in gray and nongray simulations | Computational and experimental flame lengths and temperature profiles show excellent agreement. Significant variations in the flame radiant fraction predictions between the gray and nongray models |

and reduced chemistry models were deemed to be accurate for predicting the temperature and major gas species concentrations in industrial applications, detailed chemistry models that account for dissociation were found to be necessary to accurately predict pollutant formation such as CO. Similarly, refined WSGG models for the gas-phase radiative properties were determined to be necessary to accurately model radiative transfer. However, a key step towards quantifying CFD prediction uncertainties from employing these refined models requires simultaneous measurements of species concentrations and temperature as well as radiative transfer variables under well-controlled, laminar flow conditions where measurement accuracies are quantified and minimized. To address this gap, experimental measurements of radiative transfer from oxy-methane flames at different oxidizer $O_2$-$CO_2$ dilution ratios were recently published [19]. Measurement accuracies and data correction were ensured by accounting for background wall emission and calculating the gas absorption at several axial locations. A cylindrical shape of the flame was assumed in their analysis and total radiative properties (absorption coefficients) were computed based on measured temperature and specie concentrations. These were employed in line-of-sight (1D) radiative transfer calculations performed using the statistical narrow band (SNB) model RADCAL [20]. A Monte Carlo ray tracing method was employed to determine the view factors of the various radiating elements seen by sensor. The 1D analysis showed that the peak value of the wall radiative fluxes was located at a height corresponding to 60–70% of the flame lengths and that the wall radiative fluxes declined rapidly above the length of the flame. Consequently, the temperature and specie concentrations measurements were also restricted to slightly above the flame lengths of the different flames. Further, Ditaranto and Oppelt [19] observed that while similar heat flux distributions and radiant fraction values were obtained for methane combustion in air as well as 35% $O_2$-65% $CO_2$ oxidizers, these values increased with an increase in oxidizer $O_2$ concentration. They attributed this increase to higher peak flame temperatures and enhanced soot formation rates. The goal of this study is to garner insights into these experimental observations through CFD simulations of the experimental conditions employing chemistry and radiative property models that have previously been deemed to be accurate for simulating these oxy-combustion scenarios (Table 1). Specifically, the goals of this study were to

(1) provide more complete descriptions of the flame shape and the spatial variations in temperature and gas concentrations throughout the furnace to increase the fidelity of the radiative transfer calculation compared to the previous study [19],

(2) quantify the contributions of the radiatively participating gases from the upper sections of the furnace to the overall radiative transfer to improve flame radiant fraction estimates since limited measurements were made in this region,

(3) understand the underlying cause for the observed experimental trends in flame radiant fraction variations with oxidizer compositions,

(4) assess the prediction sensitivities to the choice of modeling options employed in the simulations by considering three chemistry models and four radiative property models. Particular emphasis was placed on evaluating those chemistry models that can predict concentrations of minor species such as H, OH, and O which are critical towards determining the soot concentrations.

It is also important to recognize that the results reported in this study provide estimates of the wall radiative fluxes and total radiant power from oxy-methane flames in an environment where turbulence-radiation interactions (TRI) were minimized [21]. The importance of TRI was recognized in a recent study carried out by Becher et al. [22] where significant TRI caused considerable variability in radiative transfer predictions and precluded the determination of the most accurate radiative property model.

Experimental measurements from confined, laminar (Re 1404) methane flames in oxidizer compositions of 21% $O_2$-79% $N_2$, 35% $O_2$-65% $CO_2$, and 50% $O_2$-50% $CO_2$ are first reported in this study, followed by computational fluid dynamic (CFD) simulations of these flames. The gas-phase radiative properties of $H_2O$ and $CO_2$ were estimated employing different WSGGM for oxy-combustion scenarios that were formulated employing four different spectroscopic databases [10–13]. All of the WSGGM were implemented as user-defined functions (UDFs) and employed in conjunction with the CFD code ANSYS FLUENT [23]. In addition, the modeling of the gas-phase chemistry was undertaken employing the nonadiabatic extensions of the equilibrium probability density function (PDF) based mixture fraction model, a two-step global finite rate chemistry model with modified rate constants, and the Eddy dissipation concept (EDC) employing a 41-step detailed chemistry mechanism, models that have been deemed to be acceptable in oxy-combustion scenarios as listed in Table 1. In the numerical predictions of methane-air flames, the simulations were performed employing models appropriate for combustion in air.

## 2. Methods

*2.1. Experimental Conditions.* Figure 1 provides the geometric details of the furnace investigated in this study. The furnace consists of a cylindrical geometry with a fuel nozzle of diameter 5 mm and an oxidizer nozzle of diameter 100 mm. These are enclosed in a stainless steel walled combustion chamber of diameter 350 mm and length 1000 mm. The large dimension of the furnace relative to that of the flame minimized interaction and perturbation of the flow with the wall. The oxidizer gas was sent through a series of perforated plates to ensure uniform velocity distribution. The inside faces of the stainless steel walls of the reactor were coated with a blackbody paint of emissivity 0.98. The temperature and gas concentration ($CO_2$, $O_2$, and CO) profiles around the flames were measured at different axial heights and in a radial direction from the wall until the vicinity of the outer limit of the flame is defined as when the CO concentration increased sharply. The gas was sampled with a quartz probe

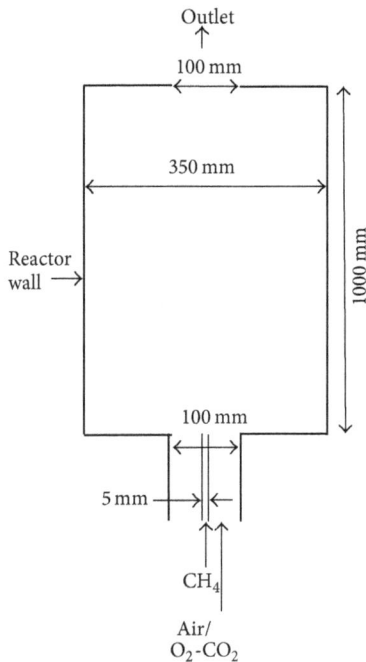

FIGURE 1: Geometric details of the furnace.

and analyzed with conventional gas analyzers and the temperature profiles were obtained by transversing radially four equally spaced $500\,\mu m$ fine bead thermocouples in the gas layer. Obtaining reliable species profiles within these small laboratory flames requires nonintrusive laser diagnostics, which were not applied in this study. The usage of the sampling probe was therefore been limited to regions where its measurements could be trusted.

The coflow velocity and fuel velocity were maintained at 0.25 m/s and 4.6 m/s, respectively. The inlet temperature of the fuel was maintained at 288 K for the fuel densities to match the fuel inlet Reynolds number of 1404. The temperature of the $CO_2$ supply was observed to be strongly dependent on the flow rate and the pressure in the liquid $CO_2$ container due to its sensitivity to Joule-Thomson effects. Consequently, some temperature fluctuations in the coflow plenum were observed and an average coflow temperature of 288 K was therefore employed in the simulations. Although these variations in the coflow temperatures can strongly influence local flame extinction and lift-off characteristics, they were anticipated to play a minimal impact on the results and conclusions pertaining to radiative heat transfer that are reported in this study. For further details regarding the experimental conditions, the interested reader is referred to Ditaranto and Oppelt [19].

### 2.2. CFD Modeling Approach

#### 2.2.1. Mesh and Flow Modeling.
The CFD simulations were carried out using the commercial code ANSYS FLUENT [23]. The furnace was modeled in a 2D axisymmetric domain to take advantage of the symmetry of the problem. The geometry was meshed employing 30,700 hexahedral control volumes. Further refinement of the mesh to 263,000 control

volumes did not change the results reported here. The flames in this study were determined to be in the buoyant regime since their laminar Froude numbers were less than unity [19]. Therefore, the pressure-velocity coupling was accomplished using the SIMPLEC algorithm [32] which we have determined from past experience to perform well in such buoyancy driven enclosure flows. The PRESTO [33] and QUICK [34] schemes were employed for the spatial discretization of the pressure and momentum terms, respectively, since hexahedral cells were employed in the calculations. Strong recirculation patterns were observed numerically and experimentally (via a decrease in gas temperature close to the walls as a result of downward flow). Therefore, obtaining steady-state converged results in the CFD simulations to compare against experimental measurements necessitated the utilization of the standard $k$-$\varepsilon$ turbulence model along with standard wall functions. The small flames examined in this study did not undergo appreciable spreading and were far away from the walls. Previous investigations of TRI in these flames [21] along with the very low values of the mixture fraction variance computed outside the flames in the simulations also reaffirm that the utilization of a turbulence model in the simulations is for numerical stability and to facilitate convergence only and has very little bearing on the results and conclusions reported in this study.

#### 2.2.2. Radiation Modeling and Boundary Conditions.
The radiation was modeled by solving the radiative transport equation (RTE) employing the discrete ordinate (DO) model. The angular discretization was carried out by employing a $3 \times 3$ theta $\times$ phi discretization. The adequacy of this angular resolution was established by determining that the reported variables did not change with any further increase in angular resolution. Unless otherwise mentioned (for instance, in the section where the sensitivity to the spectroscopic database employed in the WSGGM formulations is assessed) the radiative properties of the gas mixtures were all determined employing a recently proposed WSGG model [12] that is based on the SNB RADCAL. The WSGG model (with five gray-gases) accurately calculates the radiative properties of $CO_2$ and $H_2O$ vapor mixtures that are encountered in scenarios encompassing methane, natural gas, or coal combustion under air-fired and oxy-fired conditions.

For boundary conditions for the radiation model, the stainless steel walls of the furnace were assigned an emissivity of 0.98 at the inside walls of the reactor [19]. The external walls (assigned an emissivity of 0.7) were assumed to radiate to the ambient air at 300 K and the furnace wall temperature was established by an energy balance between the net radiative and convective heat fluxes at the inside walls of the furnace and the net emission from the outside wall of the furnace. The nonavailability of the steady-state wall temperature measurements for all the flames necessitated the adopting of a physically realistic thermal boundary condition. The utilization of these boundary conditions has been validated in our previous study [21].

In order to investigate the sensitivity of the predictions to the spectroscopic/model databases employed in the WSGGM formulations, radiative transfer calculations were carried out in a fully coupled manner (with radiative transfer feedback to the energy equation) as well as in a decoupled manner by "freezing" the converged flow field that was obtained with the SNB RADCAL based WSGGM.

The decoupled simulations ensured that all the WSGGM calculations were performed on the same temperature and species concentration fields and any differences in the radiative source term predictions across the models do not translate to any further changes to the thermal field that can further magnify/minimize the differences in the radiative transfer predictions among the models. However, results from the coupled calculations confirmed that the impact of these differences in the radiative source term predictions among the WSGG models on the temperature field was minimal with maximum temperature variations generally less than 25 K across all flames. This is consistent with our observations in our previous study of these flames [21] where we noted significant differences in the volume integrated radiative source term predictions between the gray and five gray gas formulations of the same WSGG model but their impact on the temperature field was minimal. Therefore, results from the decoupled radiative transfer calculations are reported in this study for consistency in quantifying the differences among the WSGG models. Similar comparisons carried out by fully coupling the radiative transfer (by employing different WSGGM) and the flow field in all of our simulations are likely to result in only small changes to the results reported in this study while unaltering our overall conclusions.

The different WSGGM investigated in this study along with their model notations employed in this paper, the number of gray-gases (gg), and the spectroscopic data bases associated with their formulation are listed in Table 2.

*2.2.3. Chemistry Modeling.* Based on the recent studies summarized in Table 1, it is evident that the nonadiabatic extension of the equilibrium PDF based mixture fraction model, the Eddy dissipation concept (EDC), and global finite rate chemistry (FR) have been deemed to be appropriate for simulating the laminar to transitional oxy-methane flames investigated in this study. Therefore, simulations employing all three models were employed in this study to examine the variations in radiative transfer.

The FR chemistry modeling in this study was carried out employing the global kinetic parameters reported for the Westbrook and Dryer (WD) mechanism in Yin et al. [17]. The original unmodified kinetic parameters [14] were employed in the methane-air flame whereas the modified rate constants were employed in the oxy-methane flame simulations.

In a previous study [21], we had deemed the appropriateness of employing the nonadiabatic formulation of the equilibrium PDF based mixture fraction model (denoted as PDF) for these buoyant enclosure flames. The flame length and temperature predictions were found to agree well with the experimental measurements as well as trends at different oxidizer compositions and fuel inlet Reynolds numbers. In the mixture fraction approach, the instantaneous thermochemical state of a fluid is related to its mixture fraction and

TABLE 2: A summary of WSGGM investigated in this study.

| Model notation (number of gray gases) | Spectroscopic database | Reference |
|---|---|---|
| Perry (5 gg) | SNB RADCAL, Perry's Chemical Engineering Handbook | [12] |
| EM2C (5 gg) | EM2C SNB | [10] |
| EWBM (5 gg) | EWBM | [11] |
| HITEMP 2010 (5 gg) | HITEMP 2010 | [13] |

its enthalpy. The benefits of this model are that, by assuming equal species diffusivities, the individual species conservation equations reduce to a single "sourceless" conservation equation for the mixture fraction as a result of the cancellation of the reaction source terms in the species equations due to elemental conservation. This enabled us to estimate the concentrations of minor species such as OH as well as get accurate predictions of the flame temperature in a computationally efficient manner without the need to resort to a detailed chemical mechanism that is applicable to oxy-combustion conditions. Under the assumption of chemical equilibrium, all thermochemical scalars (species fractions, density, and temperature) are uniquely related to the mixture fraction(s) and the value of each mass fraction, density, and temperature were determined from calculated values of mixture fraction, variance in mixture fraction, and the enthalpy. Twenty chemical species were considered in the equilibrium calculations ($CH_4$, $C_2H_2$, $CH_3$, $C_2N_2$, $C_2H_6$, $C_2H_4$, $C_4H_2$, $C_3H_3$, HNC, C(s), CO, $CO_2$, $H_2O$, OH, $N_2$, $O_2$, H, O, $HO_2$, and $H_2$). An assumed shape probability distribution function (PDF) was employed to describe any turbulence-chemistry interactions where the average value of the scalars is related to their instantaneously fluctuating values. In this study, the shape of the PDF was described by the beta function. 80 points in the mixture fraction and variance in mixture fraction space and 121 points in the enthalpy space were employed to carry out the interpolations and integrations within the PDF model.

In the EDC model, chemical reactions are assumed to take place in small turbulent structures referred to as fine scales. This is based on the turbulent energy cascade where large eddies break up into smaller eddies. The fine scales are then treated as constant pressure reactors where the combustion occurs. The concentration of species is then calculated by integrating the chemistry within these fine scales. This is undertaken by modeling the volume fraction of these fine structures in which the reactions take place, and the timescale for mass transfers from the fine structure to the surrounding fluid as functions of the turbulent kinetic energy ($k$) and turbulent dissipation rates ($\varepsilon$). However, the validity of the EDC model is limited to turbulent Reynolds numbers greater than 64 [35, 36]. The turbulent Reynolds number is a function of the fluid density, distance to the near wall, turbulent kinetic energy, and the laminar viscosity of the fluid [23]. The turbulent Reynolds numbers were determined to be greater than 64 in the majority of the furnace and within the flames. However, there was a small localized region outside

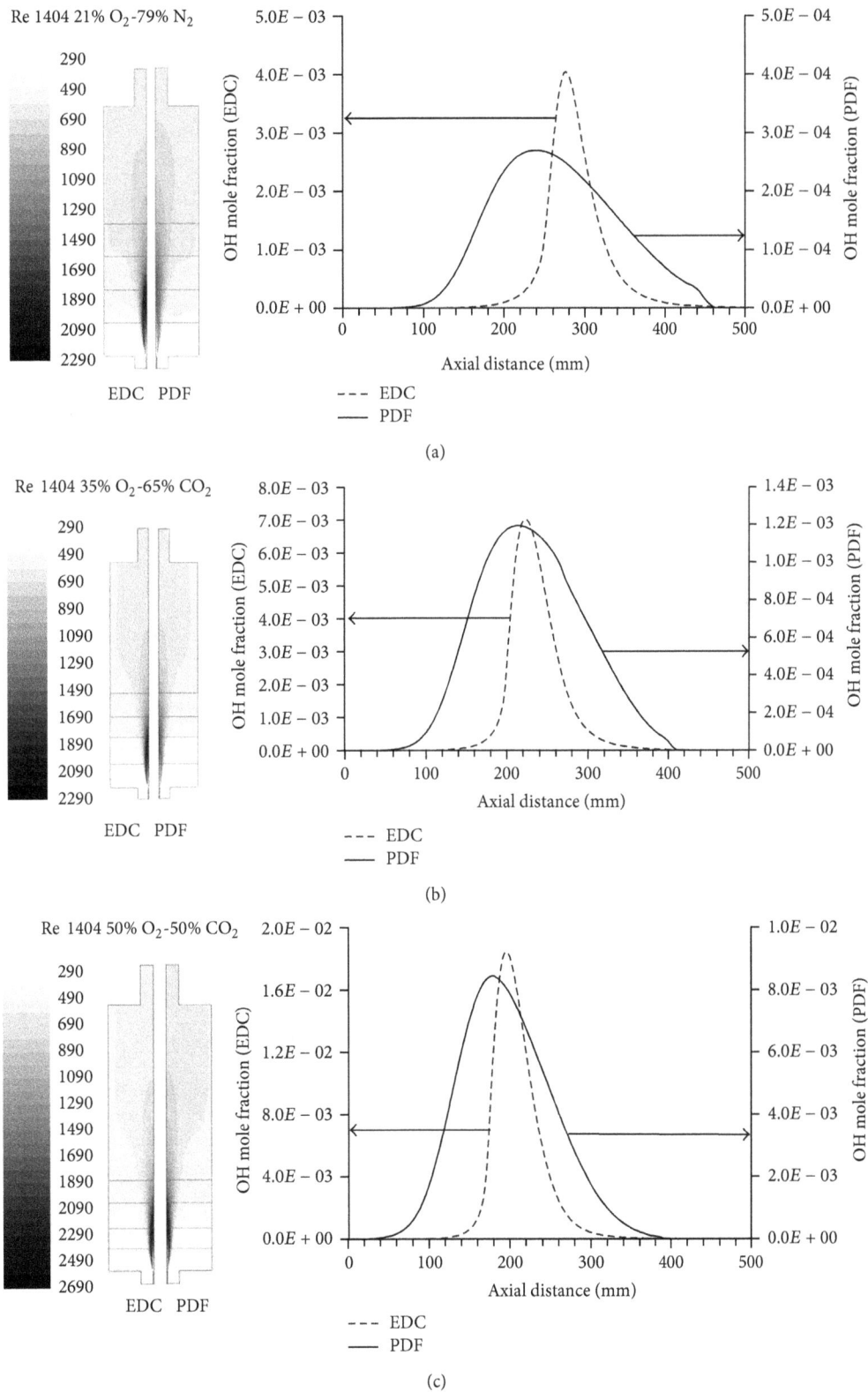

FIGURE 2: Temperature contours from the EDC and PDF model predictions and the corresponding axial variations in the OH concentrations for the three Re 1404 flames examined in this study. Oxidizer is (a) air, (b) 35% $O_2$-65% $CO_2$, and (c) 50% $O_2$-50% $CO_2$.

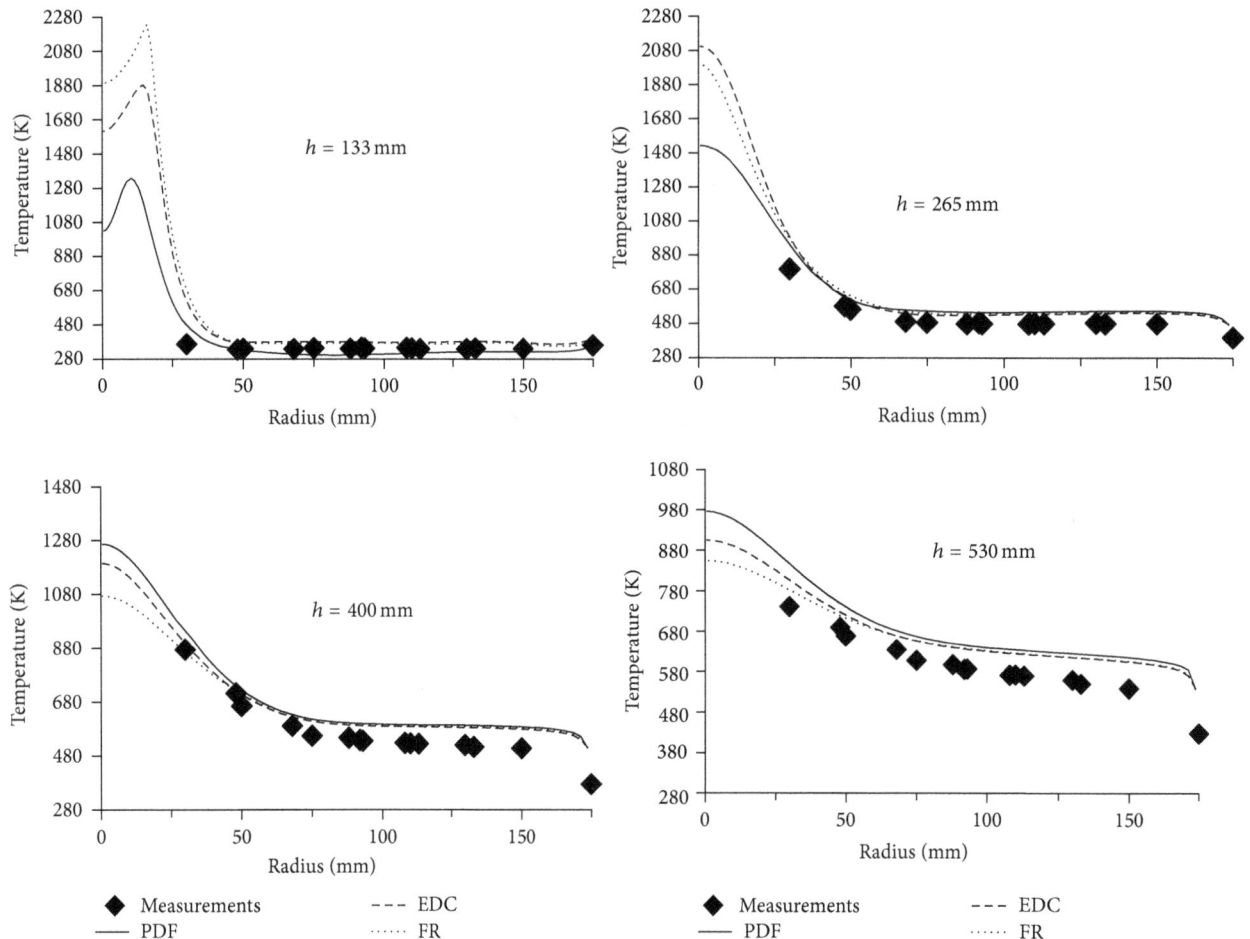

FIGURE 3: Radial temperature profiles at different axial locations for the Re 1404 air-methane flame.

the flame and the measurement locations near the bottom of the furnace corresponding to a recirculation zone near the walls, with temperature ranging from 300 K to 450 K where this requirement was not met. However, these were not expected to impact the results and conclusions reported in this study since the contribution of this region to the overall radiative transfer was minimal. In this study, the skeletal mechanism proposed by Smooke [37] for methane combustion which consists of 33 reactions and 17 species was used to represent the chemistry associated with the EDC model. The thermal and mass diffusion coefficients of the species were determined from kinetic theory and Soret diffusion was also accounted for in the calculations.

## 3. Results and Discussion

Radial measurements of temperature, $CO_2$, and $O_2$ outside the flame region were made at several axial locations in the three flames and are reported in this study. The CFD predictions are then compared against the experimental results.

*3.1. Temperature.* Temperature contours from the EDC and PDF model predictions and the corresponding axial variations in the OH concentrations for the three Re 1404 flames

examined in this study are shown in Figure 2. The axial OH concentrations enable us to estimate the flame lengths [21]. The experimentally measured flame lengths in these flames were reported to be 482 mm, 434 mm, and 373 mm, respectively [19]. The temperature contours and the axial OH concentrations show that the flames resulting from the PDF model are longer (by 25–50 mm) than the flames predicted from employing the EDC chemistry model. Consistent with experimental observations, both models predict a decrease in flame length, an increase in peak flame temperature, and an increase in OH concentration, with an increase in $O_2$ concentration in the oxidizer stream.

Figures 3–5 compare the numerical temperature predictions against experimental measurements along the radial direction at four different axial locations for both air and oxy-combustion cases. The axial centerline ($r = 0$) corresponds to the flame center. In Figure 3, predictions from all chemistry models are seen to agree well against experimental measurements. Variations between the models are more pronounced at lower axial and radial distances closer to the centerline axis. The FR chemistry model by virtue of limited dissociation predicts higher temperature than the equilibrium based mixture fraction and EDC approaches. The EDC based approach incorporates multiple minor species that may have

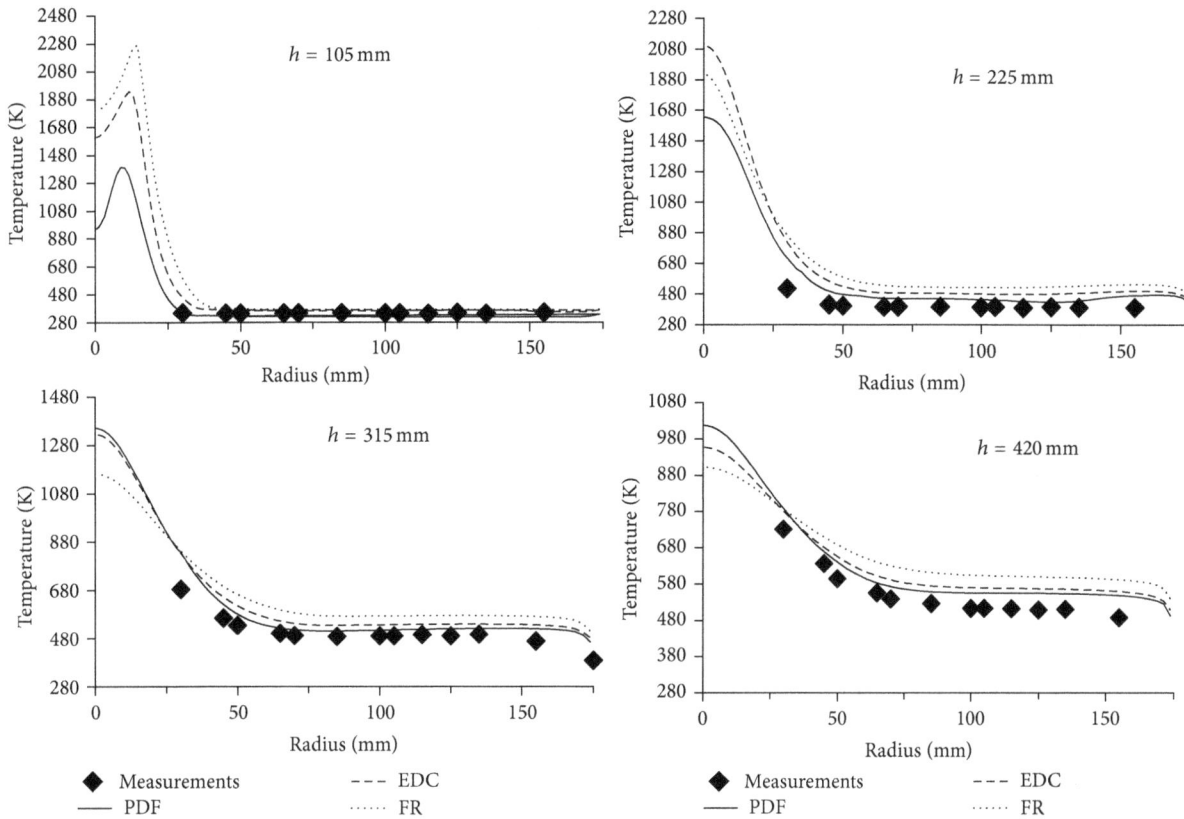

FIGURE 4: Radial temperature profiles at different axial locations for the Re 1404 oxy-methane flame (35% $O_2$-65% $CO_2$).

slower rates of formation and are not yet at their equilibrium concentrations at smaller residence times or lower axial distances. Hence, its temperature predictions are lower than those from the FR model and higher than the equilibrium PDF approach at lower axial distances in all three flames. As noted in Figure 2, the flame length (or the point where the stoichiometric mixture fraction is reached) is displaced further downstream in the PDF model calculations when compared to the EDC model calculations. Therefore, at lower axial locations the mixture fraction is higher in the PDF model predictions when compared against the EDC model predictions. This along with the fact that the PDF model assumes complete dissociation with heat loss (nonadiabatic) results in cooler temperatures at those lower axial locations. However, predictions from all models start converging to the same values at large residence times in the postflame zone when all of the combustion products attain equilibrium, as well as in the fuel lean regions outside the flame. This agreement between the temperature measurements and predictions near the wall further affirms the adequacy of the wall boundary conditions and the wall temperatures predicted by the simulations. The trends in the temperature predictions in the oxy-flames shown in Figures 4 and 5 are consistent with those observed in Figure 3. In the oxy-flames, the three chemistry models converge to the same temperature values at lower axial distances since the flame lengths get shorter with an increase in oxygen concentration in the oxidizer stream [19].

3.2. $CO_2$. Figures 6–8 compare the numerical variations in $CO_2$ concentrations (mol% dry basis) against experimental measurement along the radial direction at four different axial locations for both air and oxy-fuel combustion cases. Corresponding to the temperature plot, the FR chemistry model overpredicts the $CO_2$ at lower axial distances and therefore at small flame residence times whereas the EDC and equilibrium mixture fraction models converge to the correct value at large residence times. At lower axial distances (or small residence times) the PDF model due to the assumption of complete dissociation predicts lower $CO_2$ concentrations when compared against the FR and EDC chemistry models. The finite rate chemistry model with the modified WD rate constants, while performing well in the 35% $O_2$-65% $CO_2$ oxy-flame, overpredicts the $CO_2$ concentrations in the 50% $O_2$-50% $CO_2$ oxy-flame. This is likely due to the fact that the refined kinetic parameters employed in the global finite rate chemistry model have not been validated for flames where oxygen concentration in the oxidizer stream was as high as 50%.

3.3. $O_2$. Figure 9 shows the predicted radial variations in the $O_2$ concentrations (mol% dry basis) against experimental measurements. While reasonable agreement with the experimental predictions is obtained from the equilibrium mixture fraction approach and the EDC model in all three flames, the finite rate chemistry model with the modified WD rate

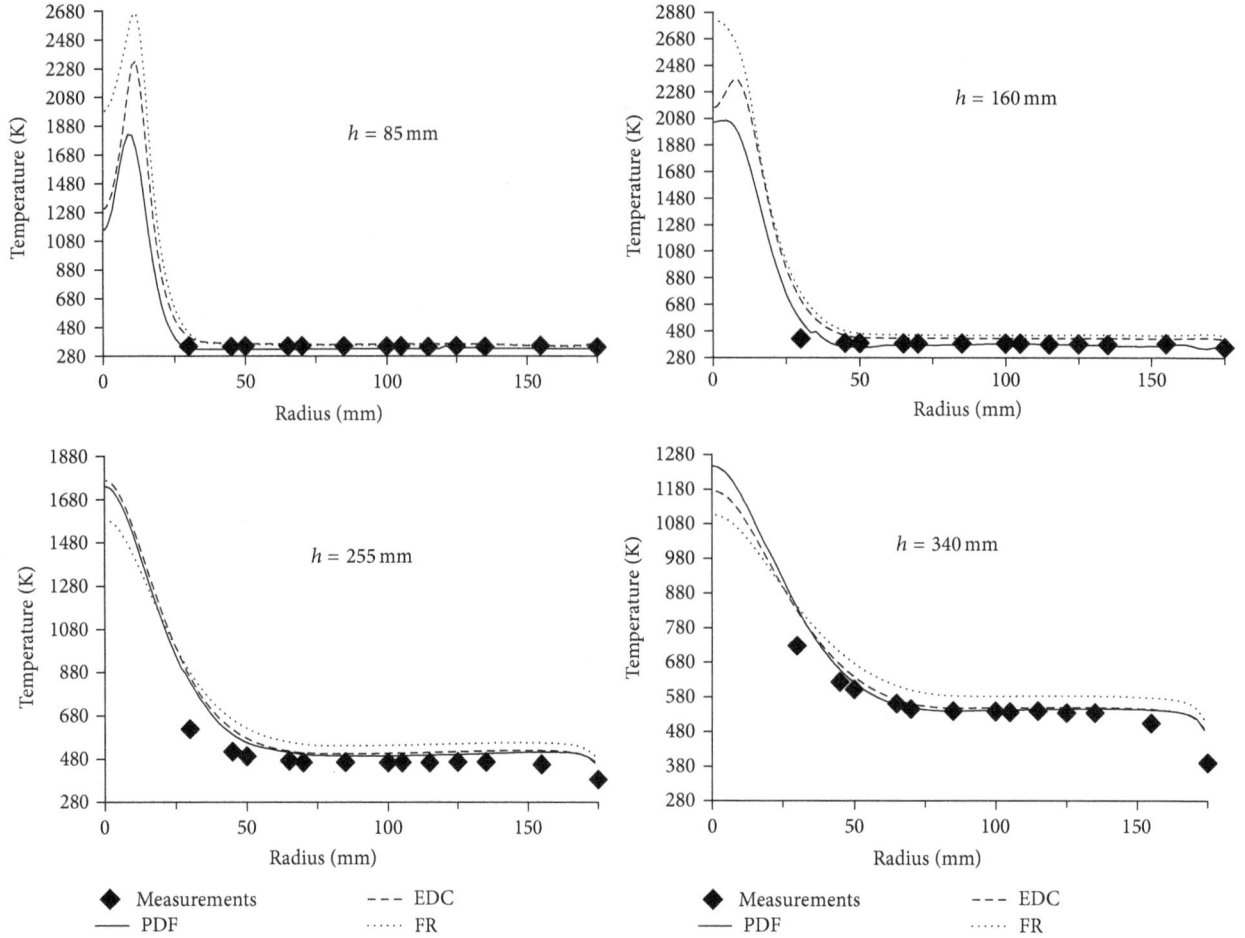

FIGURE 5: Radial temperature profiles at different axial locations for the Re 1404 oxy-methane flame (50% $O_2$-50% $CO_2$).

constants, while performing well in the 35% $O_2$-65% $CO_2$ oxy-flame, underpredicts the $O_2$ concentrations in the 50% $O_2$-50% $CO_2$ oxy-flame.

*3.4. Effect of WSGGM Spectroscopic/Model Database.* The radiative source term $(-\nabla \cdot \mathbf{q}(\mathbf{r}))$ describes the conservation of radiative energy within a control volume and goes into the total energy $(E)$ equation (see (1)), thereby coupling radiation with the other physical processes that occur in a combustion simulation [23]:

$$\frac{\partial}{\partial t}(\rho E) + \nabla \cdot (\vec{v}(\rho E + p))$$

$$= \nabla \cdot \left( \Gamma \nabla T - \sum_j h_j \vec{J}_j + (\overline{\overline{\tau}} \vec{v}) \right) + S_h - \nabla \cdot q. \quad (1)$$

On the left hand side of (1), $E$ is the total energy and $p$ is the pressure. The first three terms on the right hand side represent the "diffusion" contributions to the total energy equation. "$\Gamma$" the thermal conductivity along with the temperature "$T$" represents the conduction contribution. "$h$" and "$J$" represent the enthalpy and diffusion flux of species $j$ and represent energy transfer due to species diffusion, and "$\tau$" and "$v$" are

the fluid shear stress and velocity, respectively, and represent the viscous dissipation contribution. "$S_h$" and "$-\nabla \cdot \mathbf{q}(\mathbf{r})$" are volumetric source terms representing the energy released during chemical reaction and the energy absorbed/emitted through radiative exchange, respectively.

If "$k$" represents the absorption coefficient within a control volume, "$G$" the total incident radiation, and $I_b$ the black body emissive power, then the radiative source term $-\nabla \cdot \mathbf{q}(\mathbf{r})$ can be computed by summing the contributions over all bands "$i$" as

$$-\nabla \cdot \mathbf{q}(\mathbf{r}) = \sum_{i=\text{band}} k_i(\mathbf{r}) \left( G_i(\mathbf{r}) - 4\pi I_{b,i}(\mathbf{r}) \right). \quad (2)$$

$G$, the incident radiation in (2), is calculated by integrating the directional intensities $(I)$ associated with the wavelength band "$i$" over all directions as

$$G_i(\mathbf{r}) = \int_{4\pi} I_i\left(\mathbf{r}, \hat{\mathbf{s}}\right) d\Omega. \quad (3)$$

A negative value of the radiative source term corresponds to a net emission. The volume integral of the radiative source term therefore is a measure of the total radiative energy lost by the flame as a result of emission and absorption and is therefore employed to determine the flame radiant fractions

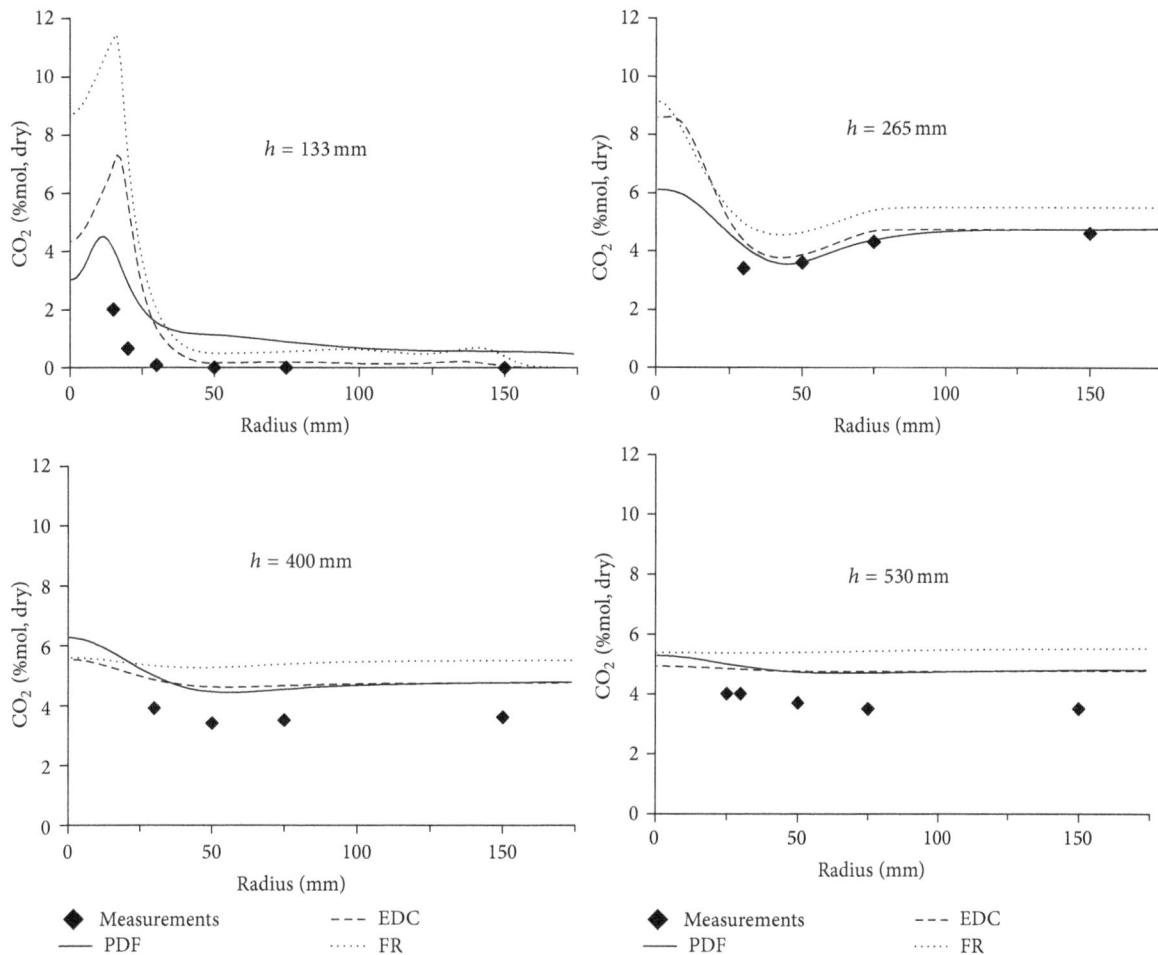

FIGURE 6: Radial $CO_2$ profiles at different axial locations for the Re 1404 air-methane flame.

TABLE 3: Volume integrated radiative source (in W) ($Q_R$) for the flames investigated in this study (sensitivity to WSGG models).

|  | EWBM (5 gg) | Perry (5 gg) | EM2C (5 gg) | HITEMP 2010 (5 gg) | Max % variation |
|---|---|---|---|---|---|
| | | | PDF, Re 1404 | | |
| 21% $O_2$ | −801 | −822 | −889 | −1089 | **−36** |
| 35% $O_2$ | −740 | −755 | −1189 | −788 | **−61** |
| 50% $O_2$ | −813 | −781 | −1277 | −842 | **−64** |
| | | | PDF, Re 2340 | | |
| 21% $O_2$ | −1871 | −1889 | −2061 | −2518 | **−35** |
| 35% $O_2$ | −2049 | −1724 | −2804 | −2210 | **−63** |
| 50% $O_2$ | −2146 | −1794 | −2855 | −2245 | **−59** |

(by dividing the volume integral of (2) by the product of fuel heating value and mass flow rate). Table 3 compares the volume integrated radiant source (or the magnitude of radiant power) predicted by the different WSGGM (by not varying the chemistry model). The fuel inlet Reynolds numbers and the chemistry model employed in the calculations are indicated in bold. The volume integrated radiant source or flame radiant power ($Q_R$) was computed as

$$Q_R = -\iiint_V \nabla \cdot \mathbf{q}(\mathbf{r})\, dV. \qquad (4)$$

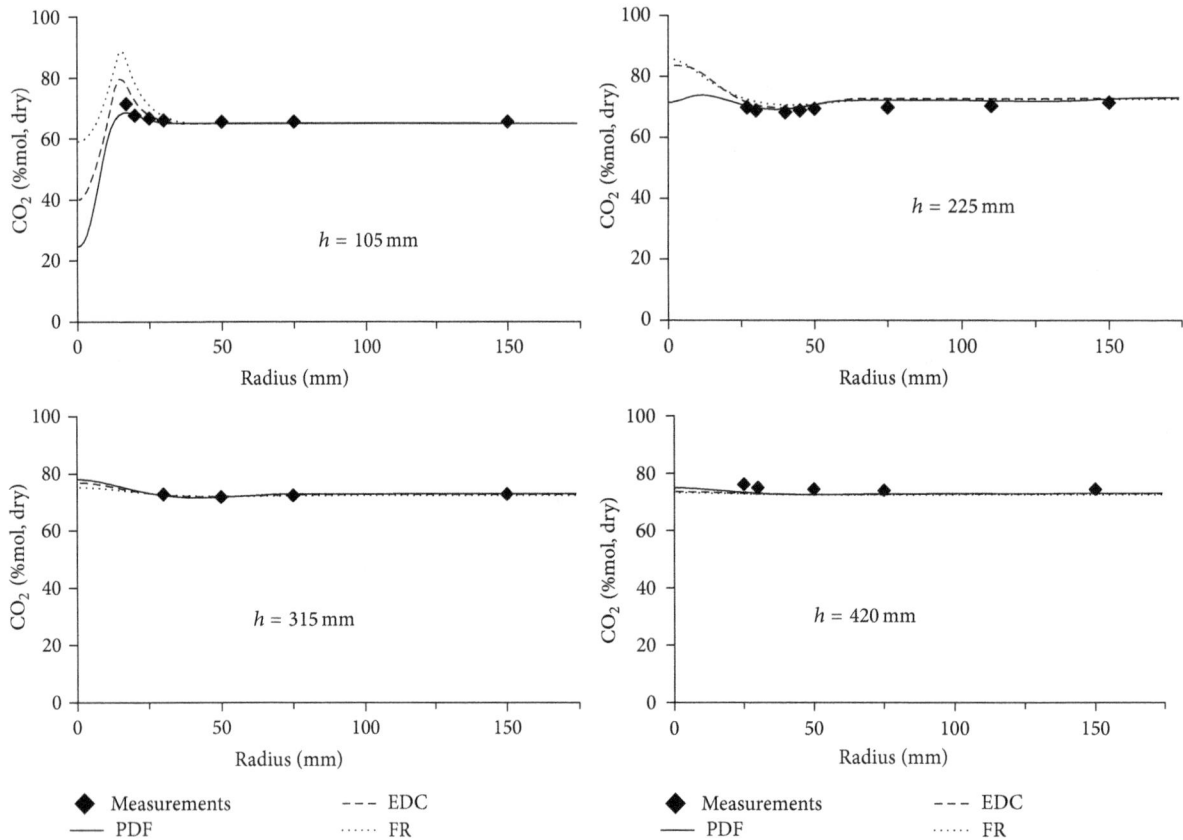

FIGURE 7: Radial $CO_2$ profiles at different axial locations for the Re 1404 oxy-methane flame (35% $O_2$-65% $CO_2$).

The maximum percentage variations in $Q_R$ for all the flames are also shown in Table 3 at two different fuel inlet Reynolds numbers. The simulations reported in Table 3 were all carried out using the PDF chemistry model. The maximum percentage variation for each flame was computed as

$$\frac{Q_{R,high} - Q_{R,low}}{Q_{R,high}} \times 100. \qquad (5)$$

The maximum variations in $Q_R$ are observed in the oxy-flames (greater than 60%) and these are seen to be independent of the fuel inlet Re. While previous studies have attributed high accuracies to both the EM2C (5 gg) [38] and the HITEMP 2010 (5 gg) models [13], Table 3 shows significant differences in the predictions between these two models in both methane-air flames and the oxy-methane flames.

### 3.5. Effect of Chemistry Model.
Table 4 compares the corresponding variations in $Q_R$ for the different chemistry models. The radiative properties for all the flames reported in Table 4 were computed employing the Perry (5 gg) model. The FR chemistry model by virtue of predicting high temperatures as a result of limited dissociation predicts significantly higher $Q_R$ values than the EDC and PDF models. However, despite differences in the flame lengths and the temperature predictions at lower axial locations between the EDC and PDF models, $Q_R$ varies by less than 10% between the chemistry models.

TABLE 4: Volume integrated radiative source (in W) ($Q_R$) for the flames investigated in this study (sensitivity to chemistry models).

| | Re 1404, Perry (5 gg) | | | |
|---|---|---|---|---|
| | FR | EDC | PDF | Max % variation |
| 21% $O_2$ | −1041 | −871 | −822 | −27 |
| 35% $O_2$ | −1069 | −769 | −755 | −42 |
| 50% $O_2$ | −1109 | −867 | −781 | −42 |

### 3.6. Effect of Soot.
The variations in the radiative source term predictions among the WSGG models (Table 3) made it difficult to numerically ascertain the trends in radiant fraction variations with the changes in the oxygen concentration in the oxidizer stream. The EWBM (5 gg) and Perry (5 gg) models in general predicted similar radiant fractions at all three oxidizer compositions (Tables 3 and 4). These results are generally in agreement with the observations of Ditaranto and Oppelt [19] who observed similar heat flux distributions and radiant fraction values for combustion in air as well as 35% $O_2$-65% $CO_2$ oxidizers. The EM2C (5 gg) model predicted a significant increase in radiant fraction with the increase in oxygen composition in the oxidizer stream whereas the HITEMP 2010 (5 gg) model predicted a decrease in radiant fraction in the oxy-flames.

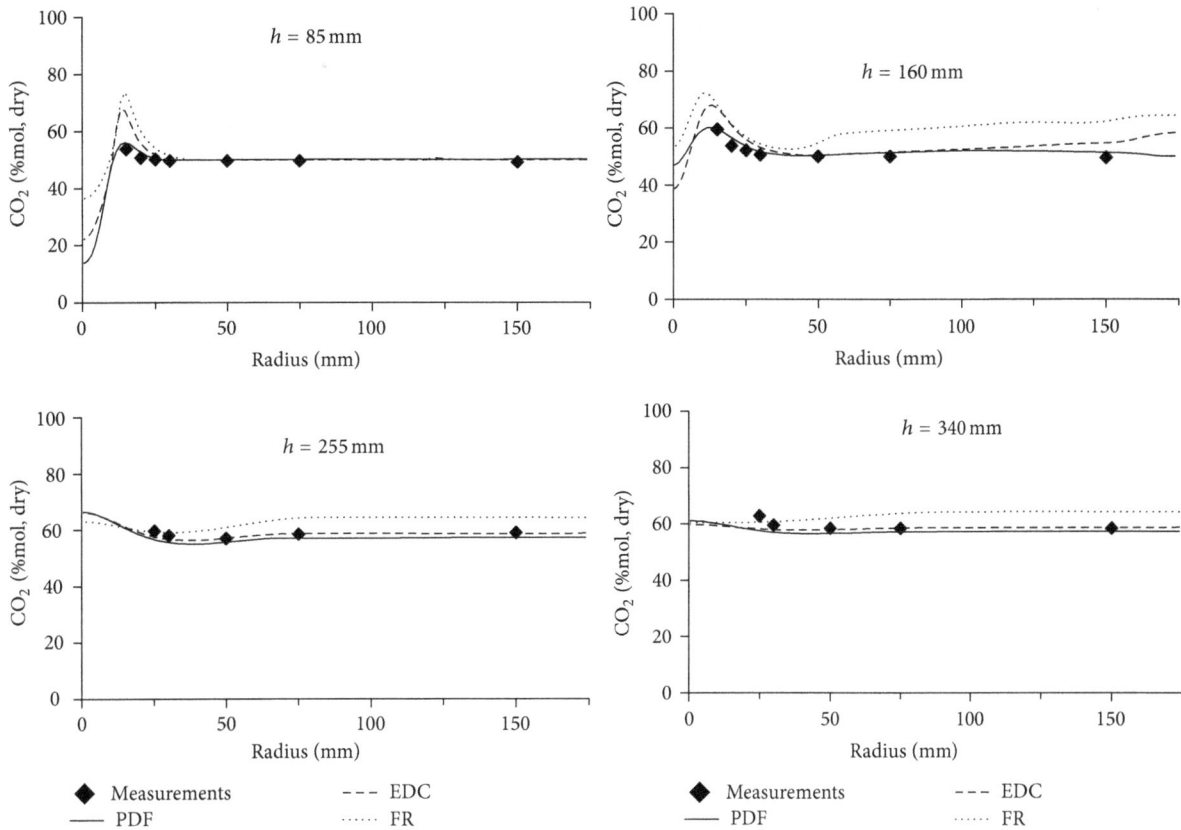

FIGURE 8: Radial $CO_2$ profiles at different axial locations for the Re 1404 oxy-methane flame (50% $O_2$-50% $CO_2$).

However, Ditaranto and Oppelt [19] observed an increase in radiant fraction with the oxygen composition in the oxidizer (beyond 35%) and attributed this to the increase in the soot inception rates at higher oxygen concentrations, resulting in higher soot concentrations and therefore higher soot emissions in the oxygen-rich flames [39]. In order to investigate this effect, the implementation of the Moss-Brookes soot model [40] in ANSYS FLUENT that has mainly been developed and validated for methane flames was employed to predict the soot volume fractions in the different flames. We have previously reported the soot volume fraction predictions employing the Perry (5 gg) WSGG at Reynolds numbers 468 and 2340 for several oxidizer compositions [21]. In this study, similar calculations were performed employing the EDC chemistry model. Contrary to the deductions of Ditaranto and Oppelt [19], we found that the peak soot volume fraction decreases with higher oxygen concentrations in the oxidizer possibly due to the increased oxidation of soot by the OH radicals. The peak soot concentrations values in the Re 1404 flames during combustion in air, 35% $O_2$-65% $CO_2$ and 50% $O_2$-50% $CO_2$, were on the order of $10^{-2}$, $10^{-3}$, and $10^{-4}$ ppm, respectively. Consequently, due its low concentrations, the presence of soot was observed to have a negligible effect on the flame radiant power predictions. Soot formation rates however have been observed to increase with increase in Reynolds numbers due to shorter residence times and incomplete mixing. The effects of soot on radiative transfer during

methane combustion are in general agreement with our previous observations from simulations of methane pool fires where mixing is incomplete. Even at a maximum soot volume fraction of 0.1 ppm, less than 5% change in the flame radiant fraction predictions was observed [41]. Nevertheless, the effect of soot needs further investigation at higher fuel jet velocities.

3.7. Wall Incident Radiative Fluxes. Figure 10 shows the variations in the incident radiative fluxes along the walls of the furnace (in the axial direction) for two oxy-flames (35% $O_2$-65% $CO_2$) from Table 3. The fuel inlet Re and chemistry model employed in the calculations are also indicated in the title. The wall incident radiative fluxes (that would be measured by a sensor for instance) were determined through an integration of the normal component of the directional intensities over a hemisphere as

$$\mathbf{q}(\mathbf{r}) = \int_{4\pi} I\left(\mathbf{r}, \hat{\mathbf{s}}\right) \hat{\mathbf{s}} \, d\Omega. \qquad (6)$$

The variations in $Q_R$ among the WSGGM reported in Table 3 translate to corresponding variations in the wall incident radiative fluxes through the radiative energy balance on this furnace. The radiative energy balance within this furnace implies that the integral of the radiative source term ($Q_R$) within the furnace volume is equal to the surface integral

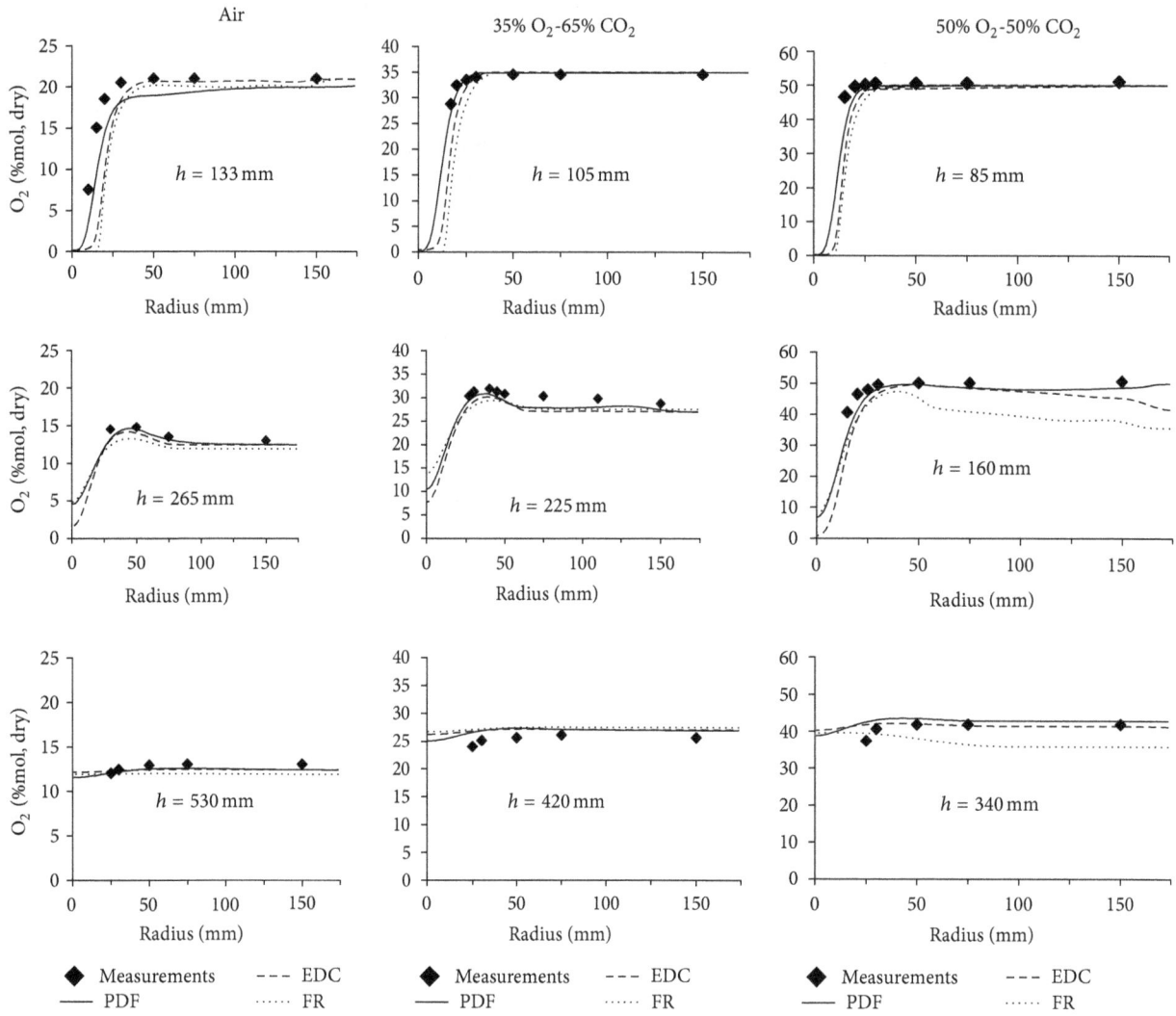

FIGURE 9: Radial $O_2$ profiles at different axial locations for the Re 1404 air and oxy-flames investigated in this study.

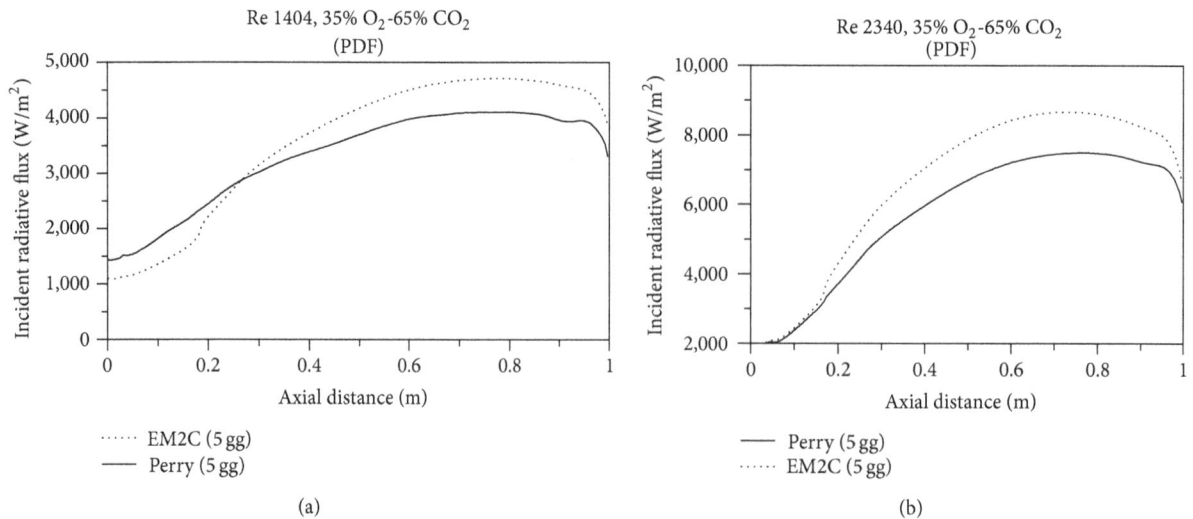

FIGURE 10: Variations in the incident radiative fluxes (in $W/m^2$) at the wall predicted by the different WSGG models: (a) Re 1404, oxy-methane flame (35% $O_2$-65% $CO_2$); (b) Re 2340, oxy-methane flame (35% $O_2$-65% $CO_2$).

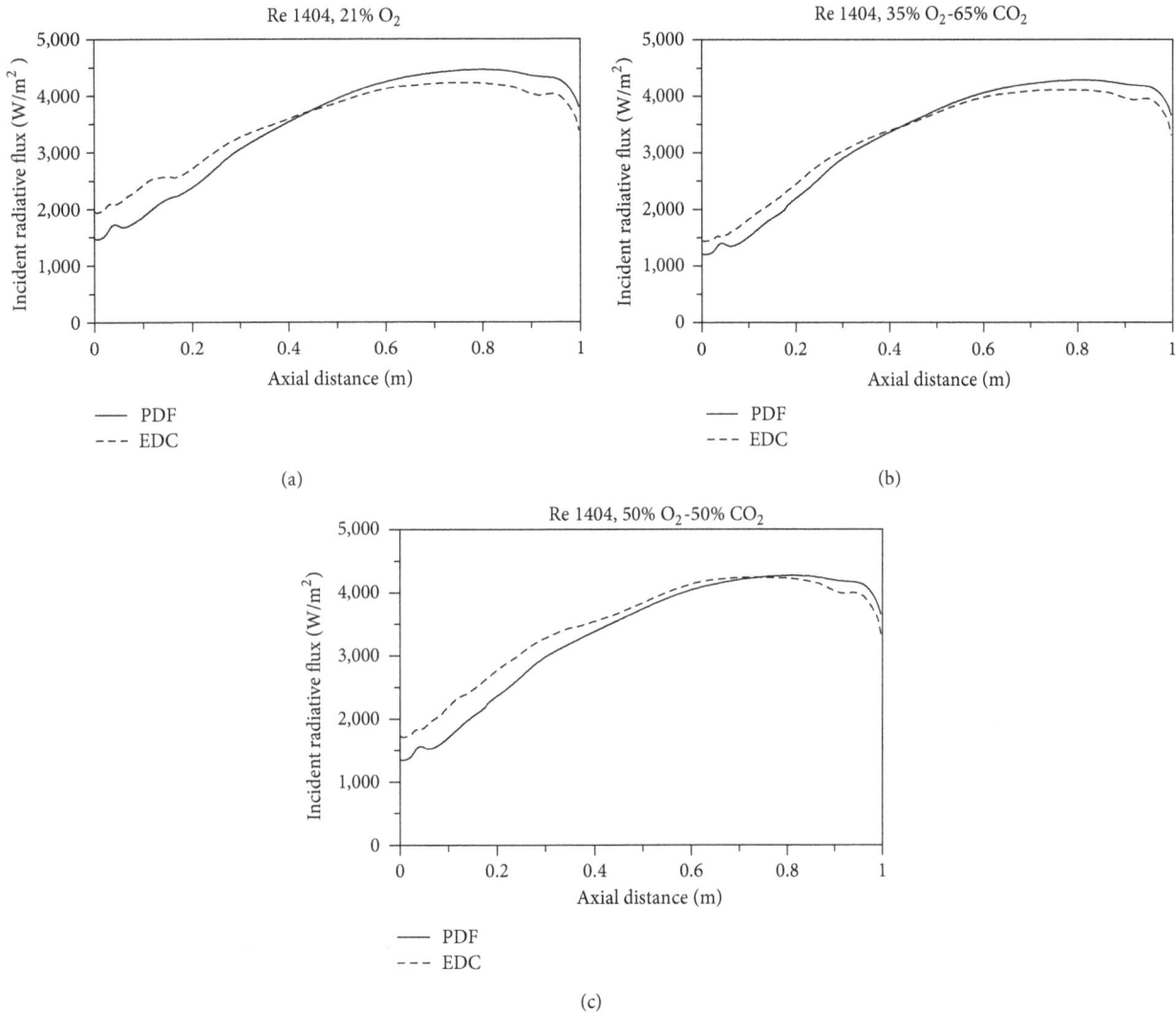

FIGURE 11: Variations in the incident radiative fluxes (in $W/m^2$) at the wall predicted by the different chemistry models: (a) Re 1404, methane-air flame; (b) Re 1404, oxy-methane flame (35% $O_2$-65% $CO_2$); (c) Re 1404, oxy-methane flame (50% $O_2$-50% $CO_2$).

of the net radiative flux lost through the boundaries of the volume. The volume of the furnace is 0.098 m$^3$ whereas the area of the sidewall of the furnace along which the incident radiative fluxes are reported is 1.1 m$^2$. Due to a more than 10-fold difference in the magnitude of these two numbers, a 60% variation in $Q_R$, for instance, will correspond to a 6% difference in the net radiative flux along the sidewall (since this variation gets distributed along the entire surface area). Therefore, the average variations in the surface incident radiative fluxes (which are obtained from the net radiative fluxes and the wall emission) shown in Figure 10 are about 10%.

The variations in the incident radiative fluxes (in $W/m^2$) at the wall predicted from employing the EDC and PDF models chemistry models in the simulations are shown in Figure 11. The Perry (5 gg) model was employed to estimate the radiative properties. Similar to the variations in $Q_R$ observed in Tables 3 and 4, the selection of the WSGG model

has a greater impact on the incident radiative flux predictions than deciding on the chemistry modeling option between the PDF and EDC methods.

*3.8. $Q_R$ Variations by Temperature Region.* In order to isolate the region within the reactor that is the source of the strongest variations among the WSGG models, the following postprocessing methodology was employed on the fully converged flame fields after the radiative source terms had been estimated throughout the domain. The computational domain of the flames was divided into three regions: temperatures less than 500 K (representative of temperatures outside the range of validity of the WSGGM examined in this study), temperatures between 500 K and 800 K (corresponding to temperatures outside the flame consisting mainly of post-combustion gases whose gas composition and the $H_2O/CO_2$ ratios are nearly homogeneous), and temperatures greater than 800 K. The contributions of the different temperature

TABLE 5: Contributions of different temperature ranges to the volume integrated radiative source (in W) ($Q_R$) for the Re 1404 and Re 2340 flames investigated in this study (sensitivity to WSGG models).

| | EWBM (5 gg) | Perry (5 gg) | EM2C (5 gg) | HITEMP 2010 (5 gg) | Max % variation | $H_2O/CO_2$ |
|---|---|---|---|---|---|---|
| | | | PDF, Re 1404, $T < 500$ K | | | |
| 21% $O_2$ | 23 | 19 | 22 | 18 | **22** | 1.94 |
| 35% $O_2$ | 123 | 155 | 136 | 121 | **22** | 0.05 |
| 50% $O_2$ | 175 | 180 | 187 | 182 | **6** | 0.04 |
| | | | PDF, Re 2340, $T < 500$ K | | | |
| 21% $O_2$ | 34 | 31 | 38 | 35 | **18** | 1.93 |
| 35% $O_2$ | 207 | 198 | 239 | 209 | **17** | 0.02 |
| 50% $O_2$ | 252 | 230 | 268 | 246 | **14** | 0.02 |
| | | | PDF, Re 1404, $500$ K $< T < 800$ K | | | |
| 21% $O_2$ | −546 | −504 | −612 | −794 | **−58** | 2.00 |
| 35% $O_2$ | −521 | −481 | −862 | −522 | **−79** | 0.17 |
| 50% $O_2$ | −498 | −443 | −820 | −462 | **−85** | 0.13 |
| | | | PDF, Re 2340, $500$ K $< T < 800$ K | | | |
| 21% $O_2$ | −807 | −821 | −907 | −1177 | **−46** | 2.00 |
| 35% $O_2$ | −1122 | −873 | −1609 | −1279 | **−84** | 0.20 |
| 50% $O_2$ | −1037 | −745 | −1456 | −1121 | **−95** | 0.26 |
| | | | PDF, Re 1404, $T > 800$ K | | | |
| 21% $O_2$ | −278 | −337 | −299 | −313 | **−21** | 2.09 |
| 35% $O_2$ | −342 | −429 | −463 | −387 | **−35** | 0.23 |
| 50% $O_2$ | −490 | −518 | −644 | −562 | **−31** | 0.18 |
| | | | PDF, Re 2340, $T > 800$ K | | | |
| 21% $O_2$ | −1098 | −1099 | −1192 | −1376 | **−25** | 2.02 |
| 35% $O_2$ | −1134 | −1049 | −1434 | −1140 | **−37** | 0.22 |
| 50% $O_2$ | −1361 | −1279 | −1667 | −1370 | **−30** | 0.27 |

ranges to the $Q_R$ in the PDF model simulations are summarized in Table 5 along with the average $H_2O/CO_2$ ratios in the region. In other words, the results reported in Table 3 are broken down by different temperature regions in Table 5.

Figure 12 shows the corresponding variations in $Q_R$ among the WSGG models in the calculations that employed the EDC chemistry model. In Figure 12, the maximum percentage variations in $Q_R$ were determined to be 36%, 63%, and 59% for the flames with the oxidizers: air, 35% $O_2$-65% $CO_2$ and 50% $O_2$-50% $CO_2$, respectively. Therefore, the variations in $Q_R$ were seen to be independent of the chemistry model employed in the simulations as well. However, from Table 5 and Figure 12 we see that the maximum variations among the WSGG models are observed to originate from the 500 K–800 K temperature range corresponding to postcombustion gases in the upper sections of the furnace.

While all of the WSGG models employed in this study have been validated through comparisons against benchmark/line-by-line (LBL) data for prototypical problems [12, 13, 38] in the general temperature range 1000 K–2000 K, our results demonstrate that variations among the underlying spectroscopic databases employed in their formulation can result in corresponding variations in the radiative transfer predictions in coupled combustion calculations. Lallemant et al. [42] observed a significant variation in the emissivity predictions due to the differences in spectroscopic/model databases employed in the emissivity calculations. Recently,

a similar evaluation at conditions representative of oxy-combustion scenarios was carried out by Becher et al. [43], however, at constant $H_2O/CO_2$ ratios and constant temperature conditions (1073 K–2073 K). By comparing the total emissivities predicted by different WSGG model formulations against those obtained from the HITEMP 2010 spectroscopic database, they determined the WSGGM formulation of Johansson et al. [44] to be the most versatile and computationally efficient model, followed by the older model of Johansson et al. [10]. A comparison against the updated WSGGM parameters based on the RADCAL SNB that was published in Krishnamoorthy [12] was not carried out likely due to the close timing of both these publications. However, in a more recent study, Kangwanpongpan et al. [13] published WSGGM correlations based on fitting the coefficients to total emissivities from the HITEMP 2010 LBL database. In their study, the radiative heat fluxes and its divergence were obtained from line-by-line (LBL) calculations for test cases encompassing wide ranges of composition nonhomogeneities, nonisothermal media, and path lengths and were treated as benchmarks. Their proposed WSGGM was then demonstrated to be the most accurate formulation when compared against the benchmarks with large errors associated with the EM2C SNB based model of Johansson et al. [44].

The laminar oxy-methane flames in enclosed environments (such as those examined in this study) as a result of the oxygen enriched combustion environment were seen to

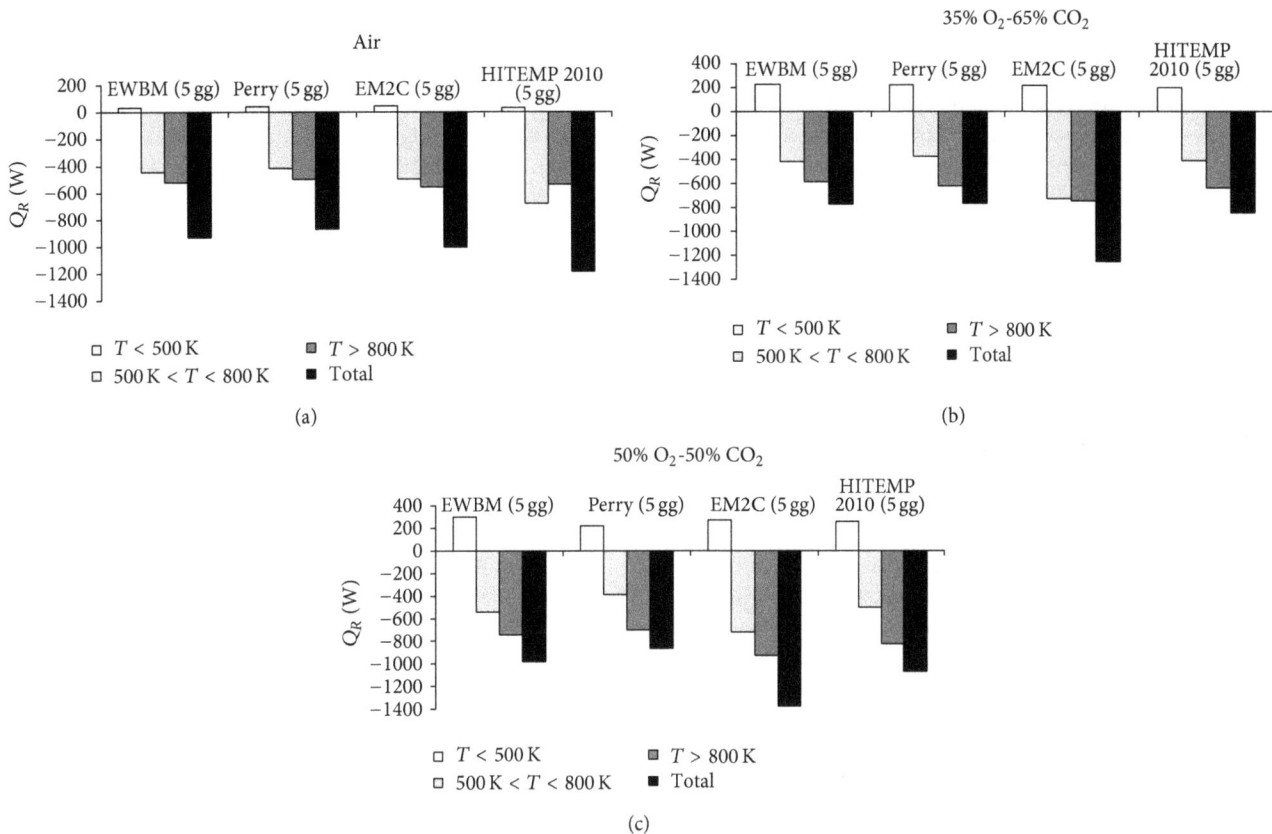

FIGURE 12: Contributions of different temperature ranges to the volume integrated radiative source (in W) ($Q_R$) when the flames were simulated using the EDC chemistry model (sensitivity to WSGG models): (a) Re 1404, methane-air flame; (b) Re 1404, oxy-methane flame (35% $O_2$-65% $CO_2$); (c) Re 1404, oxy-methane flame (50% $O_2$-50% $CO_2$).

attain temperatures greater than 2000 K within the flame and by virtue of their shorter flame lengths, encounter sharper temperature and species concentration gradients, and result in reactor temperatures that are lower than 800 K in a large portion of the reactor volume (cf. Figures 2–5). Since these extremes are well outside the range of conditions examined in the studies of Becher et al. [43] and Kangwanpongpan et al. [13] and total emissivities were compared in one study [22] while directionally integrated quantities (radiative flux and its divergence) was compared in the other [13], this study demonstrates the variations in radiative transfer predictions that might result from employing different WSGGM formulations in multiphysics CFD simulations of flames.

The different WSGGM considered in this study presents model coefficients for different $H_2O/CO_2$ ratios encompassing methane-air and oxy-methane combustion scenarios. However, as seen in Table 5, the nonhomogeneity in the $H_2O/CO_2$ ratio within the flame (corresponding to temperatures > 800 K) or employing the WSGG models outside the lower temperature limit of their validity (temperatures < 500 K) result in only 20–40% variations among the model predictions. However, within the homogeneous region of postcombustion gases encompassing the temperature range 500 K–800 K, maximum variations in $Q_R$ among the models are observed. Furthermore, this region contributes to more

than 60% and 45% of the total $Q_R$ in the Re 1404 and Re 2340 oxy-flames, respectively. Previous studies have not assessed the accuracies of the different WSGG formulations in this temperature region against LBL data since this temperature range is generally not of interest in industrial scenarios. However, in lab-scale enclosed flames where large regions of low temperatures (<800 K) are encountered, modelers must exercise caution when validating model predictions against experimental measurements of radiative transfer since the choice of the validated radiative property models by themselves can cause significant variations in the predictions.

## 4. Conclusions

A knowledge gap currently exists in our understanding of radiative heat transfer from oxy-methane flames particularly at different $O_2$-$CO_2$ dilution ratios employed in the oxidizer stream to regulate the flame temperature and radiative fluxes. The peak flame temperature increases, but the flame length decreases, with an increase in $O_2$ concentration in the oxidizer stream. These two factors not only have opposing impacts on the radiative heat fluxes and flame radiant fractions but also can alter the wall radiative flux profiles and the location of the peak fluxes. Further, recent experimental observations have revealed that similar heat flux distributions and flame radiant fraction values were obtained during combustion of methane in air as well as 35% $O_2$-65% $CO_2$ as the oxidizers. However, these values increased with an increase in oxidizer $O_2$ concentration which was attributed to higher peak flame temperatures and enhanced soot formation rates. A goal of this study was to garner insights into these experimental observations through CFD simulations of the experimental conditions employing chemistry and radiative property models that have previously been deemed to be accurate for simulating these oxy-combustion scenarios. Furthermore, this study also aims to improve estimates of the flame radiant power by providing more complete descriptions of the flame shape and the spatial variations in temperature and gas concentrations and highlight the prediction sensitivities to the choice of chemistry and radiative property modeling options employed in the simulations.

The chemistry was modeled employing the nonadiabatic extension of the equilibrium probability density function based mixture fraction model, the Eddy dissipation concept (EDC) employing a 41-step detailed chemistry mechanism, and two-step global finite rate chemistry (FR) model with modified rate constants proposed to work well under oxy-methane conditions. The gas-phase radiative properties were estimated employing different formulations of the weighted-sum-of-gray-gases models (WSGGM) that were based on four different spectroscopic/model databases. Measurements from and numerical predictions of methane-air flames are also included for comparison, with the simulations performed employing models appropriate for methane combustion in air. Based on the results from this study the following conclusions may be drawn.

(1) Consistent with experimental observations, the mixture fraction and EDC models correctly predicted a decrease in flame length, an increase in peak flame temperature, and an increase in OH concentration, with an increase in $O_2$ concentration in the oxidizer stream.

(2) The mixture fraction and EDC model predictions were also in reasonable agreement against the experimental measurements of $O_2$, $CO_2$, and gas temperatures outside the flames at all oxidizer compositions. However, the global finite rate chemistry model with the modified rate constants, while predicting the $CO_2$ and $O_2$ concentrations well in the 35% $O_2$-65% $CO_2$ oxy-flame, deviated from their measured values in the 50% $O_2$-50% $CO_2$ oxy-flame. This is likely due to the fact that the refined kinetic parameters have not been validated for flames where oxygen concentration in the oxidizer stream was as high as 50%.

(3) The global finite rate chemistry model by virtue of limited dissociation predicted higher temperature than the mixture fraction and EDC approaches. In spite of differences in the temperature predictions between the EDC and mixture fraction models within the flames at lower axial locations, the flame radiant power estimates between these two models varied by less than 10%. This was attributed to facts that

the EDC model predicted higher flame temperatures but shorter flames than the mixture fraction model, temperatures prediction differences between the models were minimized in regions outside the flame and at higher axial locations, and more than 50% of the contribution to the radiant power was from these regions outside the flame where the temperature predictions between the models were nearly identical.

(4) Differences in the spectroscopic/model databases employed in the WSGGM formulations resulted in more than a 60% variation in the volume integrated radiative source term predictions in all of the oxy-flames. The corresponding variations in the methane-air flames were about 35%. These variations were found to be independent of the fuel Reynolds numbers (Re 1404, Re 2340) and the chemistry models (EDC, mixture fraction) employed in the simulations. These variations in the radiant fraction predictions among the WSGG models were identified to be the strongest in the upper sections of the reactor where the postcombustion gas temperature was between 500 K and 800 K and the gas compositions were nearly homogeneous.

(5) While all of the WSGG models employed in this study have been validated through comparisons against benchmark/line-by-line (LBL) data for prototypical problems in the general temperature range 1000 K–2000 K, our results therefore demonstrate that at lower temperatures (500 K–800 K) significant prediction variations among these models can arise due to differences in the underlying spectroscopic databases employed in their formulation and the emissivity curve fitting procedure. The variations in radiative transfer predictions resulting from these variations would depend on the fractional contribution from this temperature range to the overall radiative heat transfer. This fractional contribution can be quite significant (>50%) from small laminar to transitional flames in enclosed domains as observed in this study.

(6) These variations in the radiative source term predictions among the WSGG models made it difficult to numerically ascertain the trends in radiant fraction variations with the changes in the oxygen concentration in the oxidizer stream. The EWBM (5 gg) and Perry (5 gg) models in general predicted similar flame radiant fractions at all three oxidizer compositions. These results are generally in agreement with the observations of Ditaranto and Oppelt [19] who observed similar heat flux distributions and radiant fraction values for combustion in air as well as 35% $O_2$-65% $CO_2$ oxidizers. The EM2C (5 gg) model predicted a significant increase in radiant fraction with the increase in oxygen composition in the oxidizer stream whereas the HITEMP 2010 (5 gg) model predicted a decrease in radiant fraction in the oxy-flames.

(7) While these variations in the radiative source term predictions did not significantly impact the gas temperature predictions (the maximum temperature variation was generally within 25 K among the different WSGGM), they did have a bearing on the incident radiative flux (that would be measured by a sensor) predictions at the walls through the overall radiative energy balance.

(8) In the high temperature regions within the flame ($T >$ 800 K) that exhibited strong temperature gradients and sharp variations in the $H_2O/CO_2$ ratios, the corresponding variations in the volume integrated radiative source term predictions among the WSGGM were considerably less. Therefore, when the radiation is dominated by high temperature gases (>1000 K), the incident radiative fluxes at the walls may not be very sensitive to the choice of WSGGM employed in the simulation as noted in previous simulations of semi-industrial scale furnaces [27].

(9) As a result of the low soot volume fractions predicted in the simulations, the effect of soot on the radiative transfer predictions and conclusions reported in this study was determined to be negligible.

## Conflict of Interests

The authors declare that there is no conflict of interests regarding the publication of this paper.

## Acknowledgments

This research was partly funded by a ND EPSCoR New Faculty Start-Up Award to Dr. Gautham Krishnamoorthy. Mario Ditaranto was funded by the BIGCCS Centre performed under the Norwegian Research Program Centers for Environment-Friendly Energy Research (FME) and acknowledges the following partners for their contributions: ConocoPhillips, Gassco, Shell, Statoil, TOTAL, GDF SUEZ, and the Research Council of Norway (193816/S60).

## References

[1] B. J. P. Buhre, L. K. Elliott, C. D. Sheng, R. P. Gupta, and T. F. Wall, "Oxy-fuel combustion technology for coal-fired power generation," *Progress in Energy and Combustion Science*, vol. 31, no. 4, pp. 283–307, 2005.

[2] T. Wall, R. Stanger, and S. Santos, "Demonstrations of coal-fired oxy-fuel technology for carbon capture and storage and issues with commercial deployment," *International Journal of Greenhouse Gas Control*, vol. 5, supplement 1, pp. S5–S15, 2011.

[3] P. Glarborg and L. L. B. Bentzen, "Chemical effects of a high $CO_2$ concentration in oxy-fuel combustion of methane," *Energy & Fuels*, vol. 22, no. 1, pp. 291–296, 2008.

[4] A. Amato, B. Hudak, P. D'Souza et al., "Measurements and analysis of CO and $O_2$ emissions in $CH_4/CO_2/O_2$ flames," *Proceedings of the Combustion Institute*, vol. 33, no. 2, pp. 3399–3405, 2011.

[5] F. S. Liu, H. S. Guo, G. J. Smallwood, and Ö. L. Gülder, "The chemical effects of carbon dioxide as an additive in an ethylene

[6] K. C. Oh and H. D. Shin, "The effect of oxygen and carbon dioxide concentration on soot formation in non-premixed flames," *Fuel*, vol. 85, no. 5-6, pp. 615–624, 2006.

[7] H. Watanabe, T. Marumo, and K. Okazaki, "Effect of $CO_2$ reactivity on $NO_x$ formation and reduction mechanisms in $O_2/CO_2$ combustion," *Energy and Fuels*, vol. 26, no. 2, pp. 938–951, 2012.

[8] L. Chen, S. Z. Yong, and A. F. Ghoniem, "Oxy-fuel combustion of pulverized coal: characterization, fundamentals, stabilization and CFD modeling," *Progress in Energy and Combustion Science*, vol. 38, no. 2, pp. 156–214, 2012.

[9] P. Edge, M. Gharebaghi, R. Irons et al., "Combustion modelling opportunities and challenges for oxy-coal carbon capture technology," *Chemical Engineering Research and Design*, vol. 89, no. 9, pp. 1470–1493, 2011.

[10] R. Johansson, K. Andersson, B. Leckner, and H. Thunman, "Models for gaseous radiative heat transfer applied to oxy-fuel conditions in boilers," *International Journal of Heat and Mass Transfer*, vol. 53, no. 1–3, pp. 220–230, 2010.

[11] C. Yin, L. C. R. Johansen, L. A. Rosendahl, and S. K. Kær, "New weighted sum of gray gases model applicable to computational fluid dynamics (CFD) modeling of oxy-fuel combustion: derivation, validation, and implementation," *Energy and Fuels*, vol. 24, no. 12, pp. 6275–6282, 2010.

[12] G. Krishnamoorthy, "A new weighted-sum-of-gray-gases model for oxy-combustion scenarios," *International Journal of Energy Research*, vol. 37, no. 14, pp. 1752–1763, 2013.

[13] T. Kangwanpongpan, F. H. R. França, R. C. Da Silva, P. S. Schneider, and H. J. Krautz, "New correlations for the weighted-sum-of-gray-gases model in oxy-fuel conditions based on HITEMP 2010 database," *International Journal of Heat and Mass Transfer*, vol. 55, no. 25-26, pp. 7419–7433, 2012.

[14] C. K. Westbrook and F. L. Dryer, "Simplified reaction mechanisms for the oxidation of hydrocarbon fuels in flames," *Combustion Science and Technology*, vol. 27, no. 1-2, pp. 31–43, 1981.

[15] W. P. Jones and R. P. Lindstedt, "Global reaction schemes for hydrocarbon combustion," *Combustion and Flame*, vol. 73, no. 3, pp. 233–249, 1988.

[16] J. Andersen, C. L. Rasmussen, T. Giselsson, and P. Glarborg, "Global combustion mechanisms for use in CFD modeling under oxy-fuel conditions," *Energy and Fuels*, vol. 23, no. 3, pp. 1379–1389, 2009.

[17] C. Yin, L. A. Rosendahl, and S. K. Kær, "Chemistry and radiation in oxy-fuel combustion: a computational fluid dynamics modeling study," *Fuel*, vol. 90, no. 7, pp. 2519–2529, 2011.

[18] S. Hjärtstam, F. Normann, K. Andersson, and F. Johnsson, "Oxy-fuel combustion modeling: performance of global reaction mechanisms," *Industrial and Engineering Chemistry Research*, vol. 51, no. 31, pp. 10327–10337, 2012.

[19] M. Ditaranto and T. Oppelt, "Radiative heat flux characteristics of methane flames in oxy-fuel atmospheres," *Experimental Thermal and Fluid Science*, vol. 35, no. 7, pp. 1343–1350, 2011.

[20] W. L. Grosshandler, "RADCAL—a narrow-band model for radiation calculations in a combustion environment," NIST Technical Note 1402, 1993.

[21] H. Abdul-Sater and G. Krishnamoorthy, "An assessment of radiation modeling strategies in simulations of laminar to transitional, oxy-methane, diffusion flames," *Applied Thermal Engineering*, vol. 61, no. 2, pp. 507–518, 2013.

[22] V. Becher, J.-P. Bohn, P. Dias, and H. Spliethoff, "Validation of spectral gas radiation models under oxyfuel conditions—part B: natural gas flame experiments," *International Journal of Greenhouse Gas Control*, vol. 5, no. 1, pp. S66–S75, 2011.

[23] *ANSYS FLUENT User's Guide*, Version 12, ANSYS Inc., Canonsburg, Pa, USA, 2009.

[24] R. Prieler, M. Demuth, D. Spoljaric, and C. Hochenauer, "Evaluation of a steady flamelet approach for use in oxy-fuel combustion," *Fuel*, vol. 118, pp. 55–68, 2014.

[25] M. A. Nemitallah and M. A. Habib, "Experimental and numerical investigations of an atmospheric diffusion oxy-combustion flame in a gas turbine model combustor," *Applied Energy*, vol. 111, pp. 401–415, 2013.

[26] C. Galletti, G. Coraggio, and L. Tognotti, "Numerical investigation of oxy-natural-gas combustion in a semi-industrial furnace: validation of CFD sub-models," *Fuel*, vol. 109, pp. 445–460, 2013.

[27] Z. Wheaton, D. Stroh, G. Krishnamoorthy, M. Sami, S. Orsino, and P. Nakod, "A comparative study of gray and non-gray methods of computing gas absorption coefficients and its effect on the numerical predictions of oxy-fuel combustion," *Industrial Combustion*, Article ID 201302, pp. 1–14, 2013.

[28] G. Krishnamoorthy, "A new weighted-sum-of-gray-gases model for $CO_2$–$H_2O$ gas mixtures," *International Communications in Heat and Mass Transfer*, vol. 37, no. 9, pp. 1182–1186, 2010.

[29] K. Bhadraiah and V. Raghavan, "Numerical simulation of laminar co-flow methane-oxygen diffusion flames: effect of chemical kinetic mechanisms," *Combustion Theory and Modelling*, vol. 15, no. 1, pp. 23–46, 2011.

[30] G. Kim, Y. Kim, and Y.-J. Joo, "Conditional moment closure for modeling combustion processes and structure of oxy-natural gas flame," *Energy and Fuels*, vol. 23, no. 9, pp. 4370–4377, 2009.

[31] B. A. V. Bennett, Z. Cheng, R. W. Pitz, and M. D. Smooke, "Computational and experimental study of oxygen-enhanced axisymmetric laminar methane flames," *Combustion Theory and Modelling*, vol. 12, no. 3, pp. 497–527, 2008.

[32] J. P. van Doormaal and G. D. Raithby, "Enhancements of the simple method for predicting incompressible fluid flows," *Numerical Heat Transfer*, vol. 7, no. 2, pp. 147–163, 1984.

[33] S. V. Patankar, *Numerical Heat Transfer and Fluid Flow*, Hemisphere, Washington, DC, USA, 1980.

[34] B. P. Leonard and S. Mokhtari, "ULTRA-SHARP non-oscillatory convection schemes for high-speed steady multidimensional flow," NASATM1-2568 (ICOMP-90-12), NASA Lewis Research Center, 1990.

[35] A. De, E. Oldenhof, P. Sathiah, and D. J. E. M. Roekaerts, "Numerical simulation of delft-jet in-hot-coflow (DJHC) flames using the eddy dissipation concept model for turbulence-chemistry interaction," *Flow, Turbulence and Combustion*, vol. 87, no. 4, pp. 537–567, 2011.

[36] A. Shiehnejadhesar, R. Mehrabian, R. Scharler, G. M. Goldin, and I. Obernberger, "Development of a gas phase combustion model suitable for low and high turbulence conditions," *Fuel*, vol. 126, pp. 177–187, 2014.

[37] M. D. Smooke, *Reduced Kinetic Mechanisms and Asymptotic Approximation for Methane—Air Flames: a Topical Volume*, vol. 384 of *Lecture Notes in Physics*, Springer, Berlin, Germany, 1991.

[38] H. Chu, F. Liu, and H. Zhou, "Calculations of gas thermal radiation transfer in one-dimensional planar enclosure using LBL and SNB models," *International Journal of Heat and Mass Transfer*, vol. 54, no. 21-22, pp. 4736–4745, 2011.

[39] S. Inge and M. Ditaranto, "Soot formation in diffusion flames in oxy-fuel atmospheres," 2015, http://www.sintef.no/.

[40] S. J. Brookes and J. B. Moss, "Predictions of soot and thermal radiation properties in confined turbulent jet diffusion flames," *Combustion and Flame*, vol. 116, no. 4, pp. 486–503, 1999.

[41] G. Krishnamoorthy, S. Borodai, R. Rawat, J. Spinti, and P. J. Smith, "Numerical modeling of radiative heat transfer in pool fire simulations," in *Proceedings of the ASME International Mechanical Engineering Congress and Exposition (IMECE '05)*, pp. 327–337, American Society of Mechanical Engineers, Orlando, Fla, USA, November 2005.

[42] N. Lallemant, A. Sayre, and R. Weber, "Evaluation of emissivity correlations for $H_2O$-$CO_2$-$N_2$/AIR mixtures and coupling with solution methods of the radiative transfer equation," *Progress in Energy and Combustion Science*, vol. 22, no. 6, pp. 543–574, 1996.

[43] V. Becher, A. Goanta, and H. Spliethoff, "Validation of spectral gas radiation models under oxyfuel conditions—part C: validation of simplified models," *International Journal of Greenhouse Gas Control*, vol. 11, pp. 34–51, 2012.

[44] R. Johansson, B. Leckner, K. Andersson, and F. Johnsson, "Account for variations in the $H_2O$ to $CO_2$ molar ratio when modelling gaseous radiative heat transfer with the weighted-sum-of-grey-gases model," *Combustion and Flame*, vol. 158, no. 5, pp. 893–901, 2011.

# 12

# Effects of Electric Field on the SHS Flame Propagation of the Si-C System, Examined by the Use of the Heterogeneous Theory

**Atsushi Makino**

*Institute of Aeronautical Technology, Japan Aerospace Exploration Agency, 7-44-1 Jindaiji-Higashi, Chofu, Tokyo, Japan*

Correspondence should be addressed to Atsushi Makino; amakino@chofu.jaxa.jp

Academic Editor: Peter F. Nelson

Relevant to the self-propagating high-temperature synthesis (SHS) process, an analytical study has been conducted to investigate the effects of electric field on the combustion behavior because the electric field is indispensable for systems with weak exothermic reactions to sustain flame propagation. In the present study, use has been made of the heterogeneous theory which can satisfactorily account for the premixed mode of the bulk flame propagation supported by the nonpremixed mode of particle consumption. It has been confirmed that, even for the SHS flame propagation under electric field, being well recognized to be facilitated, there exists a limit of flammability, due to heat loss, as is the case for the usual SHS flame propagation. Since the heat loss is closely related to the representative sizes of particles and compacted specimen, this identification provides useful insight into manipulating the SHS flame propagation under electric field, by presenting appropriate combinations of those sizes. A fair degree of agreement has been demonstrated through conducting an experimental comparison, as far as the trend and the approximate magnitude are concerned, suggesting that an essential feature has been captured by the present study.

## 1. Introduction

Self-propagating high-temperature synthesis (SHS) process, by virtue of a strong exothermic reaction that passes through a compacted mixture of particles, as a flame, has attracted special interests as a rapid and economical way in synthesizing inorganic and/or intermetallic compounds [1–5]. More than hundreds of kinds of materials, including borides, carbides, and silicides, are reported to be synthesized [1–5] by applying this process, while some of them are quite difficult to synthesize in conventional ways. Materials synthesized are now being considered for use as electronic materials, materials resistant to heat, corrosion, and/or wear, and so forth [1–5]. Production of functionally graded materials (FGMs) [6], composed of different components with continuous profiles, is also intended. Note also that near-net-shape fabrication can be anticipated because this is a kind of powder metallurgy.

The SHS process, applicable to various combinations of solid-solid, solid-liquid, and/or solid-gas systems, as pioneered by the group of Merzhanov and Borovinskaya [7], has also been recognized to present diverse phenomena of the flame propagation, such as pulsating, spinning, and/or repeated combustions, as well as steady propagation, as reviewed [1–5, 8, 9]. Among various systems, synthesis of Ti-C system has extensively been examined, because of its simplicity in chemical reaction, as well as its nature of refractory and hardness. It has been confirmed that not the gaseous component but the molten Ti, spreading over carbon particles, plays an important role in the process. Effects of dominant parameters, such as the preheating, the mixture ratio of reactants, the degree of dilution, and the particle size, on the flame propagation speed and/or the range of flammability have been examined. In particular, preheating for rapid synthesis has been examined from the viewpoint of process innovation, to have higher production rates and/or to facilitate melting during the SHS process. Although rapid synthesis can easily be accomplished, with increasing initial temperature, a supplementary heating can sometimes cause an ignition when the compacted mixture is heated over its self- (or spontaneous) ignition temperature [1, 2, 4, 5, 10], as are the cases for usual combustion.

Even for the spontaneous ignition, its utilization has been considered seriously [11–13], for those systems with weak exothermic reactions, because it is not common for those systems that sufficient preheating can be anticipated for sustaining SHS flame propagation. An external heating, by virtue of the Joule heating with applying electric fields [14–20], has also been taken into account, in order to make the materials synthesis by combustion completed.

In this study, silicon carbide (SiC) has been chosen for the research subject since it has attracted special attention in the field of materials science as the light-weight materials in high-temperature structural applications, relevant to the advanced aviation gas turbines. Focus is put on the SHS flame propagation that proceeds in the compacted mixture of Si-C system, from the viewpoint of combustion engineering, which has been rare, although examinations from the viewpoint of materials science have been conducted extensively [14–20]. Since self-propagation of the SHS flame cannot be anticipated without preheating and/or applying electric field, especially for the Si-C system, here, it is intended to elucidate effects of electric field on the burning velocity and/or the limit of flammability, by the use of the heterogeneous theory [9] for the SHS process, which can satisfactorily account for the premixed mode of the bulk flame propagation supported by the nonpremixed mode of particle consumption. Experimental comparisons are also conducted, which are found to be satisfactory, as far as the trend and approximate magnitude are concerned.

## 2. Formulation

### 2.1. Model Definition.
The problem of interest, as shown in Figure 1, is the one-dimensional, planner, heterogeneous flame propagation under electric field, being applied perpendicular to the direction of the propagation, in accordance with the experiment [19]. It is assumed in the heterogeneous theory [9] that the flame front propagates in the doubly infinite domain of a condensed medium, originally consisting of a mixture of particles of nonmetal (or higher melting-point metal) $N$, lower melting-point metal $M$, and an inert $I$ that can also be the product $P$ of the reaction between $N$ and $M$ according to

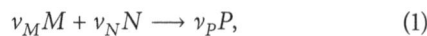

$$\nu_M M + \nu_N N \longrightarrow \nu_P P, \tag{1}$$

where $\nu_i$ is the stoichiometric coefficient of the reaction. For simplicity, it is assumed that the nonmetal particle-size distribution is monodisperse, with an initial radius $R_0$ and number density $n_0$. It is also assumed that there is no reaction between $N$ and $M$ until the mixture has been heated to the melting point $T_m$ of the metal $M$, at which all the metal particles melt instantaneously. The reaction between the solid $N$ and molten $M$ is then assumed to take place at

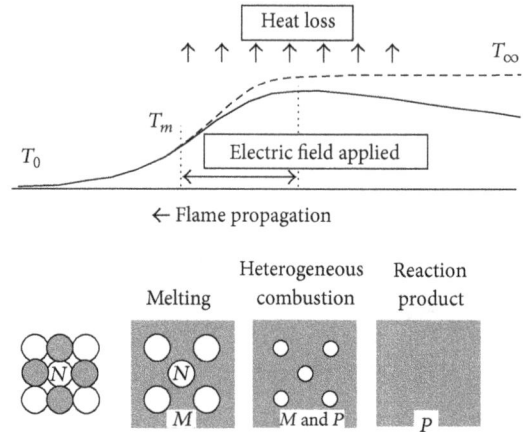

FIGURE 1: Schematic diagram of the flame structure in the propagating heterogeneous flame.

the particle surface at a finite rate and proceed until all of the $N$ particles or $M$ species are consumed. The enthalpy of phase change is neglected because it is much smaller than the heat of combustion. The lateral heat loss to the ambience, occurring throughout the entire flame structure, is also taken into account for the flame propagation considered.

### 2.2. Governing Equations and Boundary Conditions.
By the use of the heterogeneous theory [9] for the SHS flame propagation, which can fairly formulate the situation involving a premixed mode of propagation for the bulk flame, supported by the nonpremixed reaction of the dispersed nonmetal particles in the melt, the governing equations are expressed as follows.

Continuity:

$$\left(\rho_t u\right) = \left(\rho_t u\right)_0 = m, \tag{2}$$

energy conservation:

$$\frac{d}{dx}\left\{\frac{\lambda}{m}\frac{dT}{dx} - c\left(T - T_0\right) + q^0\left(Z - Z_0\right)\right\} = \frac{L - P}{m}, \tag{3}$$

$M$-conservation:

$$\left(\frac{\rho_f D}{m}\right)\frac{dY_M}{dx} = \left(Y_M + f\right)Z - \left(Y_{M,0} + f\right)Z_0, \tag{4}$$

$N$-consumption:

$$\frac{d\left(1 - Z\right)}{dx} = -\frac{4\pi\rho_N n_0 R_0}{m}\frac{u_0}{u}\chi\left(\frac{1 - Z}{1 - Z_0}\right)^{1/3}, \tag{5}$$

where $u$ is the velocity along the $x$ coordinate, $m$ the mass burning rate (per unit area) of the flame, $T$ the temperature, $c$ the specific heat, $q^0$ the heat of reaction per unit mass of $N, Z$ the mass fraction of fluid $(= \rho_f/\rho_t)$, $\rho_f$ the fluid density, $\rho_s$ the mass of solid per unit spatial volume, $\rho_t$ the total density $(= \rho_f + \rho_s)$, $\lambda$ the thermal conductivity, $Y_M$ the mass fraction of $M$ in the fluid, $f$ the stoichiometric mass ratio $[= (\nu_M W_M)/(\nu_N W_N)]$, $W$ the molecular weight, and the subscript 0 designates the unburned state.

The surface regression rate $\chi$, defined by $dR/dt = -\chi/R$, is given [9] as

$$\chi = \left( \frac{\rho_M D}{\rho_N} \right) \tilde{\chi}; \quad \tilde{\chi} = \ln \left[ 1 + \left( \frac{A}{1+A} \right) \frac{Y_M}{f} \right], \quad (6)$$

where $A = Da \exp(-Ta/T)$, $Da = (BR)/D$, $B$, and $Ta$ are, respectively, the reduced Damköhler number, the Damköhler number, the frequency factor, and the activation temperature of the surface reaction. Note here that the nondimensional surface regression rate in (6) is presented by the use of the transfer number in terms of the natural logarithmic term, just like that for the droplet combustion, and that the transfer number here is expressed as that similar to the well-known expression for the combustion rate of solid for the first-order kinetics, suggesting that roles of diffusion and chemical kinetics are both included in the present formulation. In addition, $D$ is the Arrhenius mass diffusivity $[= D_0 \exp(-T_d/T)]$ with an activation temperature $T_d$, being anticipated to increase markedly over a relatively thin, high temperature combustion zone, under the condition that the diffusion-limited situation prevails in the chemical kinetics at high temperatures.

A set of these governing equations is basically the same as that in the previous works [9], except for the term that appears in the energy conservation equation, in order to take account of an effect of electric field. While the heat-loss rate $L$, as expressed in the previous works, is given as

$$L = 4\varepsilon\sigma_{SB} \left( T^2 + T_0^2 \right) \left( T + T_0 \right) \left( T - T_0 \right) \frac{1}{2r}, \quad (7)$$

where $\varepsilon$ is the emissivity and $\sigma_{SB}$ is the Stefan-Boltzmann constant, the energy supply rate $P$ due to the electric field is expressed as

$$P = (1 - Z) \sigma_c E^2, \quad (8)$$

where $E$ is the electric field (V/m) and $\sigma_c$ is the electric conductivity ($\Omega^{-1}m^{-1}$), with the Joule heating being taken into account. In accordance with experimental observations [17, 19, 20] that the Joule heating due to external energy supply has mainly been conducted in the combustion zone, here, it is a priori assumed that it can be made by virtue of the conductible, nonmetal $N$ particles in the melt within the combustion zone, so that the mass ratio $(1 - Z)$ appears in (8). It is also assumed that no energy supplies before arrival of the SHS flame can be expected, because of the extremely high resistance of contact among particles, prior to the appearance of melt. After completion of the materials synthesis, we do not expect any energy supplies, because of the high electric resistance of SiC, which is too high to allow for the electric current.

By introducing nondimensional variables and parameters as

$$\sigma = \frac{(Z_\infty - Z_0) m_a}{(\lambda/c)} x, \qquad \theta = \frac{T - T_0}{T_\infty - T_0},$$

$$\xi = \frac{Z - Z_0}{Z_\infty - Z_0}, \qquad Le_0 = \frac{(\lambda/c)}{(\rho_t D)_0}, \quad (9)$$

$$\Lambda_0 = \left[ \frac{(Z_\infty - Z_0)^2 m_a^2}{4\pi (\rho_M D_0) (\lambda/c) n_0 R_0} \right], \quad (10)$$

$$\Psi = \left( \frac{\lambda}{\rho_t c} \right) \left[ \frac{4\varepsilon\sigma_{SB} \left( T^2 + T_0^2 \right) \left( T + T_0 \right)}{(\rho_t c) \left( u_{0,a} R_0 \right)^2} \right] \left( \frac{R_0^2}{2r} \right), \quad (11)$$

$$H = \frac{(\lambda/c) \sigma_c E^2}{m_a^2 q^0}, \quad (12)$$

the governing equations are as follows:

$$\frac{d}{d\sigma} \left( \frac{d\theta}{d\sigma} - \frac{m/m_a}{Z_\infty - Z_0} \theta + \frac{m/m_a}{\tilde{T}_\infty - \tilde{T}_0} \xi \right)$$

$$= \frac{\Psi \cdot \theta}{(Z_\infty - Z_0)^2} - \frac{H \cdot (1 - \xi)}{(Z_\infty - Z_0) \left( \tilde{T}_\infty - \tilde{T}_0 \right)},$$

$$(13)$$

$$\frac{d\xi}{d\sigma} = \frac{\tilde{\chi}}{(m/m_a) \Lambda_0 (\rho_{t,0}/\rho_t) \exp \left( \tilde{T}_d/\tilde{T} \right)}$$

$$\times \left\{ 1 - \left( \frac{Z_\infty - Z_0}{1 - Z_0} \right) \xi \right\}^{1/3}.$$

In the above, $m_a$ and $\Lambda_0$ are, respectively, the mass burning rate and the mass burning rate eigenvalue in the adiabatic condition. The nondimensional temperature is defined as $\tilde{T} = c_p T/q^0$. Note here that $\Psi$, called the heat-loss parameter and defined as a ratio of the heat-loss and heat-release rates, depends not only on the physicochemical parameters but also on the particle diameter $2R_0$ and representative size $2r$ of a sample specimen [9]. As for the parameter $H$ introduced here, we shall call it a heat-input parameter, hereafter.

Equations (9) through (11), together with the mass burning rate $m$, are to be solved, subject to the boundary conditions

$$\sigma = 0 : \theta = \theta_m, \quad \xi = 0,$$

$$\sigma \longrightarrow \infty : \frac{d\theta}{d\sigma} = -\frac{\Psi \cdot \theta}{(m/m_a) (Z_\infty - Z_0)}, \quad \frac{d\xi}{d\sigma} = 0. \quad (14)$$

Note that the cold boundary difficulty is eliminated by specifying the reaction to be initiated at the melting point of $M$. Thus, the present problem is reduced to be an eigenvalue problem for obtaining $m$ as an eigenvalue, under a nonadiabatic condition in which there exists not only heat loss but also external energy supply.

*2.3. Burning Velocity under Adiabatic Condition.* When there is neither heat loss nor external energy supply, the burning velocity is obtained as [9]

$$u_0 R_0 = \frac{1 - Z_0}{Z_\infty - Z_0} D_0 \sqrt{3 \Lambda_0 Le_0 \frac{\rho_M / \rho_N}{1 - Z_0}}, \qquad (15)$$

once the specific value of the eigenvalue $\Lambda_0$ is determined by conducting a numerical calculation. It is seen that the burning velocity $u_0$ is inversely proportional to the particle diameter $2R_0$. It may be informative to note that the product of $u_0$ and $R_0$ has even been called the SHS rate-constant [9]. The mass ratio of the lower melting-point metal $M$, before and after the combustion, is expressed as

$$1 - Z_0 = \frac{\mu (1 - \kappa)}{\mu + f},$$

$$Z_\infty = \begin{cases} 1 & (\mu \le 1) \\ Z_0 + \dfrac{1 - Z_0}{\mu} & (\mu \ge 1). \end{cases} \qquad (16)$$

Here, $\mu$ is the mixture ratio, defined as the initial molar ratio of $N$ to $M$ divided by the corresponding stoichiometric molar ratio; $\kappa$ is the degree of dilution, defined as the initial mass fraction of diluent.

*2.4. Parameters.* Values of physicochemical parameters employed are those of the Si-C system: $q^0 = 6.23\,\text{MJ/kg}$, $c = 1\,\text{kJ/(kg} \cdot \text{K)}$, $\rho_M = 2.34 \times 10^3\,\text{kg/m}^3$, $\rho_N = 2.25 \times 10^3\,\text{kg/m}^3$, $W_M = 28.1 \times 10^{-3}\,\text{kg/mol}$, $W_N = 12.0 \times 10^{-3}\,\text{kg/mol}$, and $T_m = 1681\,\text{K}$. Although some of the chemical kinetic data are reported in [17], it is not the chemical kinetics but the mass diffusivity, at high temperatures, as explained in Section 2.2. Therefore, use has been made of the mass diffusivity, $D = 3.3 \times 10^{-5} \exp(-3.39 \times 10^4/T)\,\text{m}^2/\text{s}$ [21], so that the representative Lewis number is set to be $Le_0 = 0.36$. The total density before and after the combustion is assumed to be equal. Note that the thermophysical properties used here implicitly account for effects of compact density, gases in void spaces, gas evolution, and so forth. Other thermophysical properties, indispensable for examining the limit of flammability, related to the heat loss, are $\lambda/(\rho c) = 1.2 \times 10^{-6}\,\text{m}^2/\text{s}$ for the thermometric conductivity [22] and $\varepsilon = 0.8$ for the emissivity [23]. In addition, use has been made of the electric conductivity of carbon given by

$$\sigma_c = 1.21 \times 10^4 + 6.25\,(T_r - 450) \quad [\Omega^{-1}\text{m}^{-1}], \qquad (17)$$

being obtained by the use of a curve fitting of data in the literature [24], as a function of the representative temperature $T_r$, defined as an arithmetic mean of the melting point of Si and the maximum flame temperature.

# 3. Results

*3.1. Effects of Heat-Input Parameter in the Adiabatic Condition.* First, effects of heat-input parameter on the mass burning rate

are examined in the adiabatic condition. Here, mass burning rate under electric field, $m_{a,H}$, has been normalized by the use of that without electric field, $m_{a,0}$. Figure 2 shows the normalized mass burning rate $m_{a,H}/m_{a,0}$ for the stoichiometric Si-C system, as a function of the heat-input parameter $H$, with the initial temperature $T_0$ taken as a parameter. With increasing heat-input parameter $H$, the normalized mass burning rate increases first gradually and then rapidly, up to the limiting value determined in (28), to be mentioned later. It is seen that external energy supply is effective in enhancing mass burning rate and that its contribution to $m_{a,H}/m_{a,0}$ becomes large with increasing initial temperature $T_0$.

*3.2. Effects of Heat-Input Parameter under Heat Loss Condition.* Effects of heat loss on the mass burning rate are then examined. In this examination, mass burning rate is normalized by the adiabatic mass burning rate. Figure 3 shows the normalized mass burning rate $m/m_a$, as a function of the heat-loss parameter $\Psi$, with the heat-input parameter $H$ taken as a parameter. When there is no electric field ($H = 0$), the well-known, characteristic extinction turning-point behavior is exhibited, with the upper branch being the stable solution and the turning point designating the state of extinction [9]. In contrast, the trend becomes quite different from this, with increasing heat-input parameter $H$. When $H = 0.1$, although there exists the turning-point behavior, the solution ceases to exist at certain $\Psi$ and $(m/m_a)_H$, because of other restrictions. When $H = 0.2$ or more, there appears the limit of flammability, designated by an open circle, before reaching the turning point. It is seen that, with increasing $H$, the range of $m/m_a$ for the flame propagation contracts, while the range of $\Psi$ expands.

Relevant to the limit of flammability under electric field, an attempt has been made to examine a range for the existence of SHS flame, which yields a relation in (29), to be mentioned later. In Figure 3, those limits are also shown by dotted curves, below which the steady propagation ceases to exist. It should be noted that, in determining the position of the SHS flame, theoretical consideration in the next section is indispensable.

# 4. Theoretical Consideration

*4.1. Flame Position and Temperature.* In the limit of large Zeldovich number, melting, diffusion, and consumption/ convection are confined to a thin layer in the neighborhood of the SHS flame. Outside this layer, the diffusion and consumption/convection terms are exponentially small. If we consider the situation that the flame locates at $\sigma = \sigma_F$ and that the electric field is supplied in the region $0 < \sigma < \sigma_F$, temperature profile outside the flame is expressed as

$$\sigma \le 0: \quad \theta = A_u \exp(\lambda_u \sigma), \quad \xi = 0,$$

$$0 < \sigma < \sigma_F: \quad \theta = A_h \exp(\lambda_u \sigma) + B_h \exp(\lambda_d \sigma)$$

$$+ \frac{H \cdot (Z_\infty - Z_0)}{\Psi \cdot (\tilde{T}_\infty - \tilde{T}_0)}, \quad \xi = 0, \qquad (18)$$

$$\sigma > \sigma_F: \quad \theta = C \exp(\lambda_d \sigma), \quad \xi = 1,$$

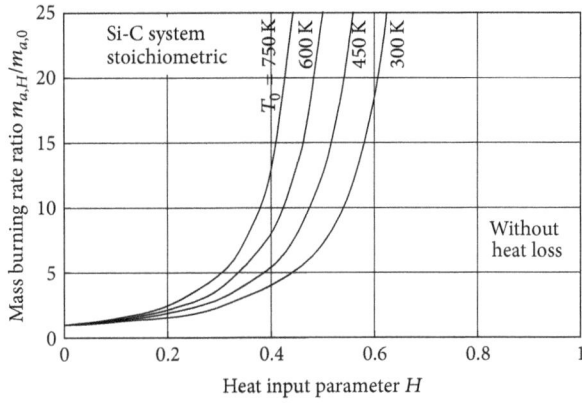

FIGURE 2: Normalized mass burning rate $m_{a,H}/m_{a,0}$ for the stoichiometric Si-C system as a function of the heat-input parameter $H$, with the initial temperature $T_0$ taken as a parameter when there exists no heat loss in the SHS flame propagation.

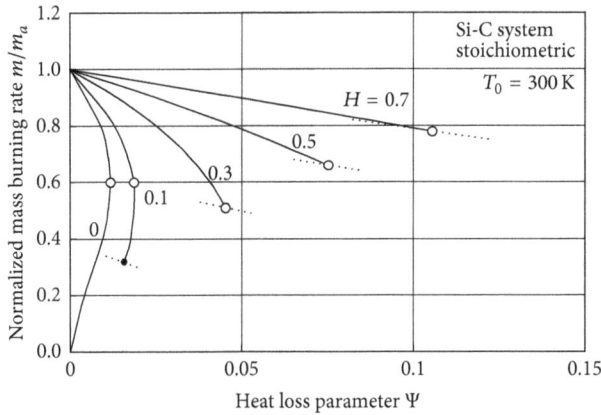

FIGURE 3: Normalized mass burning rate $m/m_a$ for the stoichiometric Si-C system as a function of the heat loss parameter $\Psi$, with the heat-input parameter $H$ taken as a parameter. An open circle designates the limit of flammability.

where

$$\lambda_u = \frac{m/m_a}{2(Z_\infty - Z_0)} \left\{ 1 + \sqrt{1 + \frac{4\Psi}{(m/m_a)^2}} \right\},$$
$$\lambda_d = \frac{m/m_a}{2(Z_\infty - Z_0)} \left\{ 1 - \sqrt{1 + \frac{4\Psi}{(m/m_a)^2}} \right\}. \tag{19}$$

Relations among constants $A_u$, $A_h$, $B_h$, and $C$ are determined by the use of the continuity in temperature and heat flux at the position $\sigma = 0$, the continuity in temperature at the flame position ($\sigma = \sigma_F$), and the jump condition at the flame:

$$\left(\frac{d\theta}{d\sigma}\right)_{\sigma_{F-}} - \left(\frac{d\theta}{d\sigma}\right)_{\sigma_{F+}} = \frac{m/m_a}{\tilde{T}_\infty - \tilde{T}_0}. \tag{20}$$

Then, we have

$$A_u = \frac{1}{\lambda_u - \lambda_d} \frac{m/m_a}{\tilde{T}_\infty - \tilde{T}_0} \exp\left(-\lambda_u \sigma_F\right)$$
$$- \frac{\lambda_d}{\lambda_u - \lambda_d} \frac{H \cdot (Z_\infty - Z_0)}{\Psi \cdot (\tilde{T}_\infty - \tilde{T}_0)} \left[1 - \exp\left(-\lambda_u \sigma_F\right)\right], \tag{21}$$

$$A_h = \left\{ \frac{1}{\lambda_u - \lambda_d} \frac{m/m_a}{\tilde{T}_\infty - \tilde{T}_0} \right.$$
$$\left. + \frac{\lambda_d}{\lambda_u - \lambda_d} \frac{H \cdot (Z_\infty - Z_0)}{\Psi \cdot (\tilde{T}_\infty - \tilde{T}_0)} \right\} \exp\left(-\lambda_u \sigma_F\right), \tag{22}$$

$$B_h = -\frac{\lambda_u}{\lambda_u - \lambda_d} \frac{H \cdot (Z_\infty - Z_0)}{\Psi \cdot (\tilde{T}_\infty - \tilde{T}_0)}, \tag{23}$$

$$C = \frac{1}{\lambda_u - \lambda_d} \frac{m/m_a}{\tilde{T}_\infty - \tilde{T}_0} \exp\left(-\lambda_d \sigma_F\right)$$
$$- \frac{\lambda_u}{\lambda_u - \lambda_d} \frac{H \cdot (Z_\infty - Z_0)}{\Psi \cdot (\tilde{T}_\infty - \tilde{T}_0)} \left[1 - \exp\left(-\lambda_d \sigma_F\right)\right]. \tag{24}$$

If we further note that the temperature at $\sigma = 0$ is the melting point $T_m$, we can determine the flame position $\sigma_F$ as

$$\sigma_F = \frac{1}{\lambda_u} \ln\left[ \frac{(m/m_a) + (\lambda_d/\Psi) H \cdot (Z_\infty - Z_0)}{(\tilde{T}_m - \tilde{T}_0)(\lambda_u - \lambda_d) + (\lambda_d/\Psi) H \cdot (Z_\infty - Z_0)} \right], \tag{25}$$

by the use of (21). The flame temperature $T_\infty$ in the adiabatic condition ($\Psi = 0$), which is indispensable for nondimensionalization, is obtained from (24) as

$$\left(\tilde{T}_\infty - \tilde{T}_0\right) = (Z_\infty - Z_0)\left(1 + \frac{H \cdot \sigma_F}{Z_\infty - Z_0}\right), \tag{26}$$

$$\frac{\sigma_F}{Z_\infty - Z_0} = \ln\left[ \frac{(1 - H)(Z_\infty - Z_0)}{(\tilde{T}_m - \tilde{T}_0) - H \cdot (Z_\infty - Z_0)} \right]. \tag{27}$$

We see that the adiabatic flame temperature is increased due to the application of the electric field.

*4.2. Range of Electric Field.* Even though the electric field is effective in sustaining SHS flame propagation, there exists an indispensable requirement for the flame position to be positive ($\sigma_F \geq 0$), which further requires that the logarithmic term in (27) should be positive. From this requirement, we can determine the range for the valid heat-input parameter $H$ as

$$0 \leq H < \frac{\tilde{T}_m - \tilde{T}_0}{Z_\infty - Z_0} < 1. \tag{28}$$

In the same manner, when there exists heat loss, we have from (25) the following relation between the heat-loss parameter $\Psi$ and the heat-input parameter $H$:

$$0 \leq \Psi < -\frac{1}{2} \left\{ \frac{(m/m_a)^2}{4} - \frac{H(Z_\infty - Z_0)}{\tilde{T}_m - \tilde{T}_0} \right\}$$
$$+ \sqrt{\frac{(m/m_a)^2}{8} \left\{ \frac{(m/m_a)^2}{8} + \frac{H(Z_\infty - Z_0)}{\tilde{T}_m - \tilde{T}_0} \right\}}. \quad (29)$$

## 5. Experimental Comparison

Although an introduction of the heat input parameter $H$ has turned out to be useful in examining the limit of flammability, effects of electric field $E$ on the mass burning rate $m_a$ are to be examined directly because use has been made of $m_a$ in defining $H$, as shown in (12). Figure 4 shows the SHS rate-constant $u_0 R_0$ as a function of the electric field $E$, with the initial temperature $T_0$ taken as a parameter. The SHS rate-constant $u_0 R_0$ expressed in (15) is used, instead of the normalized mass burning rate used in Figures 2 and 3. It is seen that the SHS rate-constant $u_0 R_0$ in the adiabatic condition, shown by solid curves, increases gradually with an increase in the electric field $E$, because of the external energy supply. It is also seen that, at a certain electric field, the SHS rate-constant $u_0 R_0$ becomes high, with increasing initial temperature $T_0$, because of the preheating effect.

A dashed curve in Figure 4 shows the limit of flammability. As the representative size of the cross-sectional area in the compacted specimen, indispensable in determining heat-loss parameter $\Psi$ in (11), use has been made of 9.175 mm for $2r$, determined by use of the relation for a rectangular cross-section as [9]

$$\ell = \frac{2ab}{a+b} \quad (30)$$

with $a = 14.1$ mm and $b = 6.8$ mm, in accordance with the experiment by Feng and Munir [19]. As for the particle size of carbon, use has been made of $2R_0 = 5\,\mu m$, reported. Although the limit of flammability, shown in Figure 4, is only that for $T_0 = 300$ K, it has been confirmed in a preparatory study that it is nearly independent of the initial temperature $T_0$, within the range examined.

Data points in Figure 4 are experimental in the literature [19]; a symbol ($\square$) designates a result for relative density 0.64 and ($\circ$) for relative density 0.52. It is seen that there is no remarkable effect of the relative density and that a fair degree of agreement exists between experimental and theoretical results, as far as the trend and approximate magnitude are concerned. It is also seen that the SHS flame propagation under low electric fields, say, less than about 1000 V/m, proceeds close to the limit of flammability at $T_0 = 300$ K. To the contrary, at high electric fields, the SHS flame propagation proceeds under the condition, free from heat loss, because of the external energy supply.

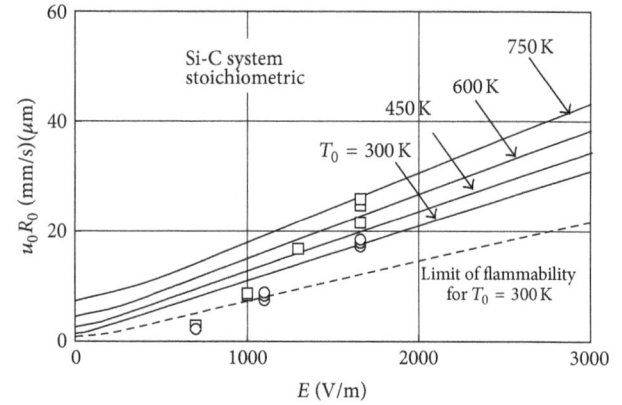

FIGURE 4: SHS rate-constant $u_0 R_0$ as a function of the electric field $E$, with the initial temperature $T_0$ taken as a parameter. Solid curves represent the state when there is no heat loss; a dashed curve represents the limit of flammability. Data points are experimental [19].

## 6. Concluding Remarks

In the present study, flame propagation in the SHS process of the Si-C system has been examined by the use of the heterogeneous theory, with taking account of an effect of electric field. It has been shown that the burning velocity is enhanced by applying an electric field, due to an increase in the external energy supply, as reported in the literature. Furthermore, it has been confirmed that there exists the limit of flammability, even for the SHS flame propagation under electric field, as is the case for usual SHS flame propagation, although its dependence on the heat-loss rate is quite different from that without electric field. A fair degree of agreement between experimental and theoretical results suggests that the present study has captured the essential feature of the SHS flame propagation under electric field, providing useful insight into manipulating the SHS flame propagation, by the use of representative sizes of particles and compacted specimen, which has not been recognized in the previous studies.

## Nomenclature

$A$:   Constant for temperature profile
$a$:   Length of a rectangular sample specimen [m]
$B$:   Constant for temperature profile
$b$:   Width of a rectangular sample specimen [m]
$C$:   Constant for temperature profile
$c$:   Specific heat [J/(kg·K)]
$D$:   Diffusivity [m$^2$/s]
$E$:   Electric field [V/m]
$f$:   Stoichiometric mass ratio[$= (v_M W_M)/(v_N W_N)$]
$H$:   Heat input parameter
$L$:   Heat loss [J/(m$^3$·s)]
$Le$:   Lewis number
$\ell$:   Representative size of the rectangular sample specimen [m]

$m$:  Mass burning rate $[kg/(m^2 \cdot s)]$
$P$:  Energy supply due to electric field $[J/(m^3 \cdot s)]$
$q^0$:  Heat of combustion per unit mass of $N$ $[J/kg]$
$R$:  Particle radius $[m]$
$r$:  Radius of sample specimen $[m]$
$T$:  Temperature $[K]$
$T_d$:  Activation temperature of mass diffusivity $[K]$
$u$:  Burning velocity $[m/s]$
$W$:  Molecular weight
$Y$:  Mass fraction
$Z$:  Mass ratio of species $M$.

*Greek*

$\varepsilon$:  Emissivity
$\theta$:  Nondimensional temperature
$\kappa$:  Degree of dilution
$\Lambda$:  Eigenvalue of the mass burning rate
$\lambda$:  Thermal conductivity $[W/(m \cdot K)]$ or exponent for temperature profile
$\mu$:  Mixture ratio
$v$:  Stoichiometric coefficient
$\xi$:  Normalized mass ratio of $M$
$\sigma$:  Nondimensional length
$\sigma_c$:  Electric conductivity $[\Omega^{-1} m^{-1}]$
$\sigma_{SB}$:  Stefan-Boltzmann constant $[W/(m^2 \cdot K^4)]$
$\rho$:  Density $[kg/m^3]$
$\chi$:  Surface regression rate $[m^2/s]$
$\Psi$:  Heat loss parameter.

*Subscript*

$a$:  Adiabatic condition
$d$:  Downstream
$F$:  Flame
$f$:  Fluid or melt
$H$:  Electric field
$h$:  Hot zone
$M$:  Lower melting-point metal
$m$:  Melting point
$N$:  Nonmetal or higher melting-point metal
$P$:  Combustion product or diluent
$s$:  Solid or surface
$t$:  Total
$u$:  Upstream
$0$:  Initial or unburned state.

*Superscript*

$\sim$:  Nondimensional or stoichiometrically mass weighted.

# References

[1] W. L. Frankhouser, K. W. Brendley, M. C. Kieszek, and S. T. Sullivan, *Gasless Combustion Synthesis of Refractory Compounds*, Noyes, Park Ridge, NJ, USA, 1985.

[2] Z. A. Munir and U. Anselmi-Tamburini, "Self-propagating exothermic reactions: the synthesis of high-temperature materials by combustion," *Materials Science Reports*, vol. 3, no. 7-8, pp. 277–365, 1989.

[3] A. G. Merzhanov, *Self-Propagating High-Temperature Synthesis: Twenty Years of Search and Findings*, Combustion and Plasma Synthesis of High-Temperature Materials, VCH Publishers, New York, NY, USA, 1990, Edited by Z. A. Munir and J. B. Holt.

[4] A. Varma and J.-P. Lebrat, "Combustion synthesis of advanced materials," *Chemical Engineering Science*, vol. 47, no. 9–11, pp. 2179–2194, 1992.

[5] A. Varma, A. S. Rogachev, A. S. Mukasyan, and S. Hwang, "Combustion synthesis of advanced materials: principles and applications," *Advances in Chemical Engineering*, vol. 24, no. C, pp. 79–226, 1998.

[6] W. G. J. Bunk, *Gradient Materials For Structural and Functional Applications*, Advanced Materials '93, III/B, Elsevier, Amsterdam, The Netherlands, 1994.

[7] A. G. Merzhanov and I. P. Borovinskaya, "A new class of combustion processes," *Combustion Science and Technology*, vol. 10, no. 5-6, pp. 195–201, 1975.

[8] S. B. Margolis, "The transition to nonsteady deflagration in gasless combustion," *Progress in Energy and Combustion Science*, vol. 17, no. 2, pp. 135–162, 1991.

[9] A. Makino, "Fundamental aspects of the heterogeneous flame in the self-propagating high-temperature synthesis (SHS) process," *Progress in Energy and Combustion Science*, vol. 27, no. 1, pp. 1–74, 2001.

[10] A. G. Merzhanov, "Nonisothermal methods in chemical kinetics," *Combustion, Explosion, and Shock Waves*, vol. 9, no. 1, pp. 3–28, 1975.

[11] A. S. Rogachev, A. S. Mukasyan, and A. Varma, "Volume combustion modes in heterogeneous reaction systems," *Journal of Materials Synthesis and Processing*, vol. 10, no. 1, pp. 31–36, 2002.

[12] J. P. Lebrat, A. Varma, and A. E. Miller, "Combustion synthesis of Ni$_3$Al and Ni$_3$Al-matrix composites," *Metallurgical Transactions A*, vol. 23, no. 1, pp. 69–76, 1992.

[13] L. Thiers, A. S. Mukasyan, and A. Varma, "Thermal explosion in Ni-Al system: influence of reaction medium microstructure," *Combustion and Flame*, vol. 131, no. 1-2, pp. 198–209, 2002.

[14] V. A. Knyazik, A. G. Merzhanov, V. B. Solomonov, and A. S. Shteinberg, "Macrokinetics of high-temperature titanium interaction with carbon under electrothermal explosion conditions," *Combustion, Explosion, and Shock Waves*, vol. 21, no. 3, pp. 333–337, 1985.

[15] O. Yamada, Y. Miyamota, and M. Koizumi, "Self-propagating high-temperature synthesis of the SiC," *Journal of Materials Research*, vol. 1, no. 2, pp. 275–279, 1986.

[16] V. A. Knyazik, A. G. Merzhanov, and A. S. Shteinberg, "About combustion mechanism of the Ti-C system," *Doklady Akademii Nauk SSSR*, vol. 301, no. 3, pp. 899–902, 1989.

[17] V. I. Gorovenko, V. A. Knyazik, and A. S. Shteinberg, "High-temperature interaction between silicon and carbon," *Ceramics International*, vol. 19, no. 2, pp. 129–132, 1993.

[18] A. Feng and Z. A. Munir, "The effect of an electric field on self-sustaining combustion synthesis: part I. modeling studies," *Metallurgical and Materials Transactions B*, vol. 26, no. 3, pp. 581–586, 1995.

[19] A. Feng and Z. A. Munir, "The effect of an electric field on self-sustaining combustion synthesis: part II. field-assisted synthesis

of $\beta$-SiC," *Metallurgical and Materials Transactions B*, vol. 26, no. 3, pp. 587–593, 1995.

[20] A. Feng and Z. A. Munir, "Relationship between field direction and wave propagation in activated combustion synthesis," *Journal of the American Ceramic Society*, vol. 79, no. 8, pp. 2049–2058, 1996.

[21] Thermophysical Properties Handbook, *Japan Society of Thermophysical Properties*, Yokendo, Tokyo, Japan, 1990.

[22] Thermophysical Properties Handbook, *Japan Society of Thermophysical Properties*, Yokendo, Tokyo, Japan, 1990.

[23] S. Hanzawa, "Refractories of furnaces to reduce environmental impact," *Annual Report of Ceramics Research Laboratory*, vol. 9, pp. 33–42, 2009 (Japanese).

[24] *Thermophysical Properties of High Temperature Solid Materials*, vol. 1, MacMillan, New York, NY, USA, 1965, Edited by: Y. S. Touloukian.

# 13

# The Principal Aspects of Application of Detonation in Propulsion Systems

**A. A. Vasil'ev**

*Lavrentyev Institute of Hydrodynamics SB RAS, Novosibirsk 630090, Russia*

Correspondence should be addressed to A. A. Vasil'ev; gasdet@hydro.nsc.ru

Academic Editor: Eliseo Ranzi

The basic problems of application of detonation process in propulsion systems with impulse and continuous burning of combustible mixture are discussed. The results on propagation of detonation waves in supersonic flow are analyzed relatively to air-breathing engine. The experimental results are presented showing the basic possibility of creation of an engine with exterior detonation burning. The base results on optimization of initiation in impulse detonation engine are explained at the expense of spatial and temporal redistribution of an energy, entered into a mixture. The method and technique for construction of highly effective accelerators for deflagration to detonation transition are discussed also.

## 1. Introduction

The increased interest to use of a detonation process in various technological devices and, in particular, at development of the concept of detonation engine (DE) is stipulated by classical conclusion (e.g., [1]) that from every possible mode of burning of a combustible mixture, the regime of self-sustained detonation (the ideal Chapman-Jouguet wave) is characterized by minimum irreversible losses. The point D (Figure 1) of a tangency of the Michelson-Rayleigh line ODS to a detonation branch 1 of adiabatic curve of energy release $Q$ = const corresponds to minimum growth of an entropy $\Delta S_D$ = min (isentropic curve is tangent to an adiabatic curve from below) in a comparison with any other points. The higher losses are inherent for combustion modes (laminar and turbulent) on a comparison with the C-J detonation mode: the point $F$ of a tangency of the Michelson-Rayleigh line OF to lower deflagration branch corresponds to maximum growth of an entropy (in $F$ isentropic curve is tangent to an adiabatic curve from above: $\Delta S_F$ = max). $\Delta S_D$ = min is the first advantage of detonation mode.

The second advantage is the maximum high pressure of a detonation products $P_D$ on a comparison with traditional combustion, where a final condition is close to point $P_P$ (to be in general agreement with condition $P$ = const) or to point $P_V$ (burning at condition $V$ = const).

Only these two advantages of detonation burning allow to get positive profit at consequent expansion of DW-products.

Additional advantage connects with huge velocities of a mixture burning in a detonation wave (DW) and the highest power (energy in unit time) of detonation energy-release, unattainable for combustion conditions.

## 2. The Basic Scheme of Detonation Engines

To the present time for a realization of idea of burning of a combustible mixture in detonation modes, the set of various devices is offered, including numerous model scheme of a detonation engine (see, e.g., bibliographic lists in the reviews [2, 3]). In the given work, only the basic scheme is considered:

(a) "weapon" schema of a pulsing detonation engine (PDE);

(b) the schema of supersonic pulsing detonation ramjet engine (PDRJE);

(c) the schema of a mixture burning with the help of steady rotating detonation wave.

(A) In the traditional "weapon" schema of an impulse engine, a basic element is the rectilinear tube with one closed and one open pipe ends. The tube is filled periodically by a combustible mixture (as a rule, with separate injection of fuel and oxidizer and their consequent mixture). After filling a tube in mixture near the closed pipe end, the initiation of a detonation wave is carried out, then DW propagates along a tube to an open pipe end. After DW exits from a tube, the phase of an outflow of a detonation products occurs (while the pressure of products in a tube will not decrease to a level permitting to begin a new cycle of filling of a tube by freshen combustible mixture). Thus the tractive force (for a unit shot) is ensured at the expense of action of the raised pressure of a detonation products on the closed pipe end (from the moment of DW initiation near the closed pipe end and during consequent DW distribution along a pipe), and action of flowing out of a detonation products (through an open pipe end in a consequent time interval after DW exits from a pipe). From aforesaid it follows, that if the DW near the closed pipe end is not excited (e.g., insufficient powerful initiator), then it at once will reduce the propulsion performance characteristics. The initiation of a detonation is the one of basic aspects of effectiveness of PDE (alongside with problems of mixture, cooling, etc.).

The frequency of shots is determined by summarized time of tube filling by freshen combustible mixture and time of an outflow of a detonation products. The time of an outflow varies insignificantly at fixed geometry of a tube: in general, it is determined by a sound velocity in reaction products, which in turn insignificantly differs for main conditions of a mixture burning. On Figure 2, the values of a sound velocity in products for typical combustion processes of heptane (as model fuel for kerosene) air mixtures are submitted (detonation: $c_D$, $V$ = const: $c_V$, $P$ = const: $c_P$, deflagration burning with a maximum velocity: $c_F$, initial: $c$). Let us mark, that at shifting along an adiabatic energy-release curve from a detonation point downwards to deflagration point the sound velocity is increased. The time of tube filling is the basic time of a unit cycle of detonation device at traditional injection of freshen mixture from the closed pipe end along tube axes. The filling time is considerably reduced at alternative injection of freshen mixtures into tube: perpendicularly to its axes and simultaneously along whole tube length. However, any information was not possible to find about an engineering realization of the similar schema in the literature of last years. At radial injection of freshen mixture the time of DW-product outflow became as basic time, so the frequency of PDE operation can be increased notably.

Similar schema is used successfully in LIH SB RAS during the latest 30 years in impulse detonation devices for acceleration of micro particles by DW (Figure 3, diameter 20 mm, length 2000 mm) and their consequent drawing on various surfaces (detonation coating) or in impulse device for clearing the technological equipment from dust deposit [2].

Multibarrelled scheme (e.g., system of seven coaxial pipes: 6 peripheral, located in tops of a regular hexagon and one central on an axes) allows to increase total frequency of PDE at the expense of sequential operation of tubes (e.g., in 4 times on 7-barrelled scheme: by first the central tube

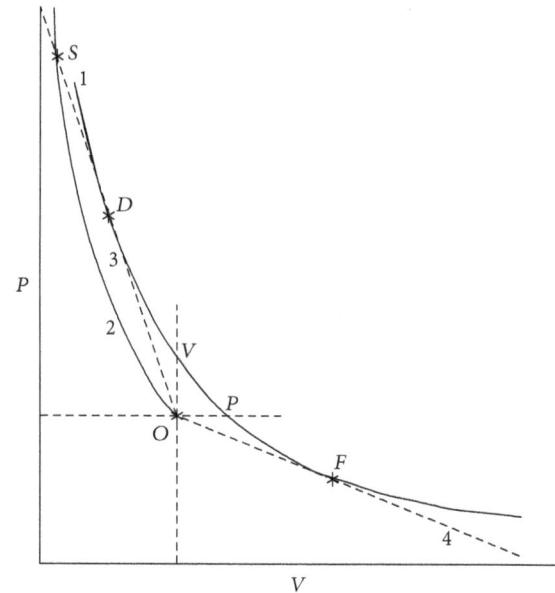

FIGURE 1: Typical ($P$-$V$), diagram of a combustible mixture.

shoots, then peripheral tubes with a defined temporal delay from each other. At this, diametrically located peripheral tubes should work simultaneously). The basic requirement to similar multibarrelled engine is the complete identity of processes in each individual pipe and rigid simultaneity of operation of diametrically pair tubes. Otherwise, it will promote the origin of random rotary moment, operating on an engine. Let us mark, that for identity of impulses, the cross-section of the central tube should twice exceed a cross-section of peripheral tube.

It is necessary to underline that PDE should work cyclically with rather high frequency to compete on effectiveness to a classical engine, in which the stationary combustion is realized. For magnification of tractive force at DW product outflow the open pipe end (usual by whole multibarrelled system) is supplied by additional nozzle.

(B) The schema of supersonic pulsing detonation ramjet engine (PDRJE) represents an alternative type of PDE (Figure 4). The conceptual variant of PDRJE [4] includes an air inlet (index 1 on Figure 4), injection system 2 of fuel in air stream, lengthy combustion chamber 3, exhausted nozzle 6, system of mixture initiation 5 (near to exhaust nozzle), and DW 4. DW in the combustion chamber of such engine in an ideal case (in the judgment of the authors [4]) should cyclically be propagated upwards and downwards along a stream at the expense of cyclical enrichment and depletion of a mixture composition. It would be underlined especially that in the combustion chamber the mixture is moved with a supersonic velocity, and instead of the forward closed pipe end, an air inlet is used.

Traditionally detonation is investigated in a motionless mixture. In a case of subsonic or supersonic flows the boundary layer on walls of an aerodynamic tract is an essential singularity of a stream of a combustible mixture.

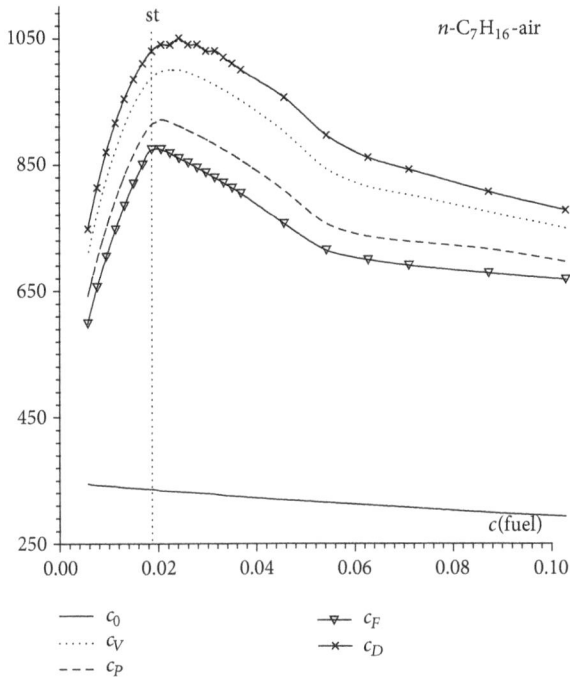

FIGURE 2: Sound velocities on fuel molar concentration for typical regimes of heptane-air mixtures (indexes correspond to characteristic points on Figure 1, vertical dotted line: stoichiometry).

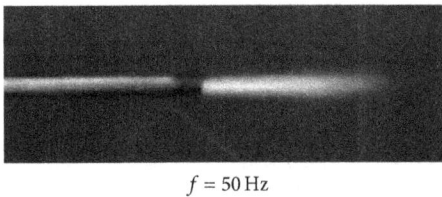

$f = 50\,\mathrm{Hz}$

FIGURE 3: Typical photo of operation of pulsed detonation tube of LIH SB RAS with frequency 50 Hz.

As the outcome, gas-dynamic parameters acquire some characteristic profiles instead of homogeneous on cross-section and length of a pipe (that is typical for motionless mixture). For example, on Figure 5 the profiles for a supersonic mixture stream (driving from left to right) are submitted, on which DW is propagated upstream. The laboratory experimental model of offered PDRJE with a single shot is realized by the authors of idea [4], and in [6], some preliminary results of investigation for stream of the Mach number $M = 3$ were published (DW initiation near output nozzle and DW distribution upstream). The information about experimental realization of cyclical work of such PDRJE is not known.

In [7] the scheme of PDRJE was used as the prototype with essential distinction: the position of DW-initiation point could vary along an aerodynamic tract. The similar modification is important for an evaluation of possibilities of control on propulsion performance characteristics of a real engine (the constant tractive force of an engine at condition of cruising flight is only one point from a range of necessary modes of engine operations). For a hydrogen-air mixture, a possibility of initiation and propagation of self-sustained

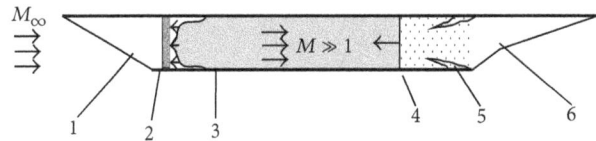

FIGURE 4: Idealized scheme of PDRJE.

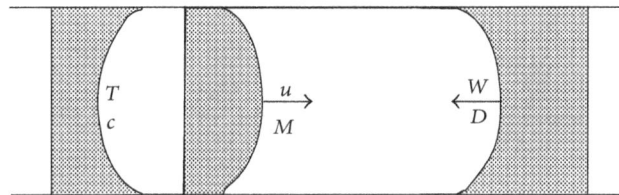

FIGURE 5: Qualitative profiles of gas-dynamic parameters in a supersonic flow of PDRJE.

detonation burning in a supersonic mixture stream of $M = 4$ in range of $\alpha = 0.5 \div 2.5$ ($\alpha$ is the excess air factor) was shown experimentally. Thus, is established that the velocity of DW-propagation upstream of a combustible mixture exceeds the C-J velocity for a motionless mixture and is less than it at DW propagation down (Figure 6). A distance, on which the DDT effect was observed (DDT length), equals some channel diameters and approximately twice less in case of DW-distribution upstream. At removal from stoichiometric ratio of fuel-oxidizer to concentration limits the regimes with an unstable detonation and lengthy zone of forming of detonation front were observed. The hypothesis about probable nature of observable effects was expressed permitting qualitatively to explain not only regularities of given work, but also results of another authors.

(C) The schema of a detonation engine with steady rotating DW (RDE) is original alternative to scheme of pulsing engines.

The essence of idea is the next. In a rectilinear pipe of a round (constant) cross-section at lowering of initial pressure of motionless mixture the regime of stationary DW propagation along pipe axes with the single transversal wave (TW), rotating along an interior concave surface, can be realized (so-called spin detonation). The TW of spinning DW can be stabilized and transformed to rotating wave in some plane (ring schema of mixture burning), if to use a mixture stream with the appropriate velocity (DW stabilization). However it is impossible to realize practically at use of homogeneous mixture, because the stagnation temperature appears much above than the ignition temperature, and mixture will be ignited by uncontrollably near the channel walls. At the same time at separate injection of mixture components (heterogeneous mixture), the given idea can be realized rather simply (on the ring schema), and in this case, DW is automatically stabilized close to injector plane. The given concept was offered and realized in Lavrentyev Institute of Hydrodynamics (LIH SB RAS, Novosibirsk, Russia) almost 55 years back: a mixture burning in steady rotating spinning DW [8, 9], schema of RDE. In Figure 7 the idealized schema of a mixture burning in the ring combustion chamber with

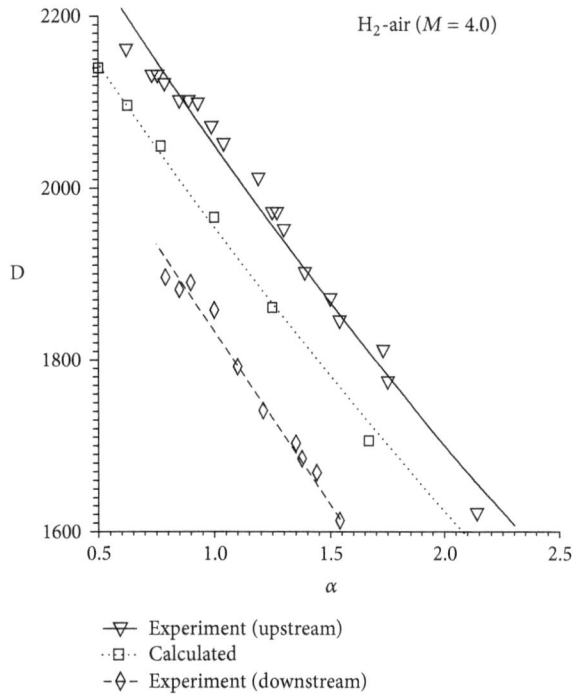

FIGURE 6: DW velocities at propagation upwards (triangles) and downwards (rhombuses) on a supersonic stream of a hydrogen-air mixture ($M = 4$) in a comparison with DW velocities for a motionless mixture (small squares) depending on the excess air factor.

the help of two steady rotated DW and typical streak records of a self-luminescence of rotated waves is demonstrated on Figures 7(b)–7(d) and Figure 8 (from [10, 11]).

The RDE idea of Professor B. Voitsekhovsky was the leading decision during a long time from a practical point of view. The experimental installation in this period was used for realization of new researches (e.g., [10, 11]) or as an operating experimental model at demonstrating a steady rotating detonation to the delegates of various scientific meetings, for example, to participants of the 2-ICOGERS (Novosibirsk, 1969). In Figure 9 the oscillograms of pressure in steady rotating DW (from [11]) are demonstrated, showing stability of process and its good reproduction (for these investigations the special piezoelectric gauges were used with a filter protecting sensitive piezoceramics from long thermal action of high-temperature detonation products).

The new stage of interest to a rotating detonation has appeared in connection with a "euphoria" termination about fast solution of PDE problem. At beginning of 21 centur it has become obvious that the creation of PDE with effectiveness, exceeding existing engines on stationary combustion, is connected directly with successful solution of a problem of DW initiation (see later). For today LIH is the leader of experimental investigations of a steady rotating detonation (see, e.g., review [9]). Recently, it was possible to realize the rotating DW on fuel-air mixtures [12] for liquid and solid (coal) components.

At use of the DW rotation schema two types of this schema represent a practical interest. First-interchamber burning of a mixture, when the concave walls of the combustion chamber are a natural limit boundary of a mixture. Walls ensure rotary propagation of spinning DW in a fixed cross-section of chamber at the expense of constant overdriven and turn of a stream (left schema on Figure 10 at axial injection of a mixture in combustion chamber. Other modifications are radial injection or injection an angle to an axes).

If in the schema of PDE the pressure on closed pipe end is supposed to be uniformly distributed on end surface and identical in any point, then in the RDE schema the pressure is distributed nonuniformly and the zone with increased pressure rotates together with spinning DW. It is necessary take into account at correct estimation of specific impulse. Once again, it should be underlined that at interchamber burning the concave wall of a pipe plays a major role in a turn of TW and constancy of its axial orientation.

The second schema applies to an exterior burning in detonation mode (right schema in Figure 10), when the layer of reactive mixture is created around a solid cylindrical surface and is burned by a rotating detonation wave. Thus the exterior boundary of a reactive layer appears free (without bounding walls). For a free axially symmetric jet (analog of a gas charge in a pipe) the regime of a spin detonation is not observed: transversal wave of spinning configuration damps without support on the external jet boundary and self-sustaining propagation of DW along jet axes becomes impossible [13]. In such free charge jet the limiting regime of stationary DW propagation along charge axes is typical multifront regime with a great number of TWs on detonation front (instead of spin DW with single TW). The diameter of free gaseous charge for a limiting regime of DW propagation is named traditionally as the critical diameter $d_*$: at $d < d_*$; the self-sustaining DW propagation along axes of free charge (without any bounding walls) is not observed. The limiting condition of DW propagation in a pipe with a rigid wall also can be described by "limiting diameter" $d_{lim}$. It is obvious, that $d_{lim} \ll d_*$.

In un rectilinear pipes the concave rigid wall also plays a basic role in a turn of DW at the expense of a regular and irregular DW reflection and over-driven of a stream near a concave wall. A typical case is a DW turn in a curvilinear branch pipe connecting two rectilinear segments of a detonation pipe.

The problem on a possibility of a mixture burning with maximum rates (schema on Figure 10) in a ring charge with the curvilinear free boundaries has scientific and also practical sense. A minimum three questions arise for a solution of a similar problem: (1) about the physical mechanism of a DW turn; (2) about a minimum radius of a curvature of a ring gas charge; (3) about a minimum thickness of a ring mixture layer, necessary for DW propagation. Really, along a ring charge the propagation of classical DW with radial oriented smooth front (perpendicular to boundary) as in Figure 10 is impossible, because for this purpose the velocity of DW propagation on the exterior boundary of a ring charge should exceed a velocity of DW on the interior boundary despite of a constancy of chemical energy release. It is possible to achieve similar effect only on radial stratified mixtures at careful observance of fuel oxidizer ratio along

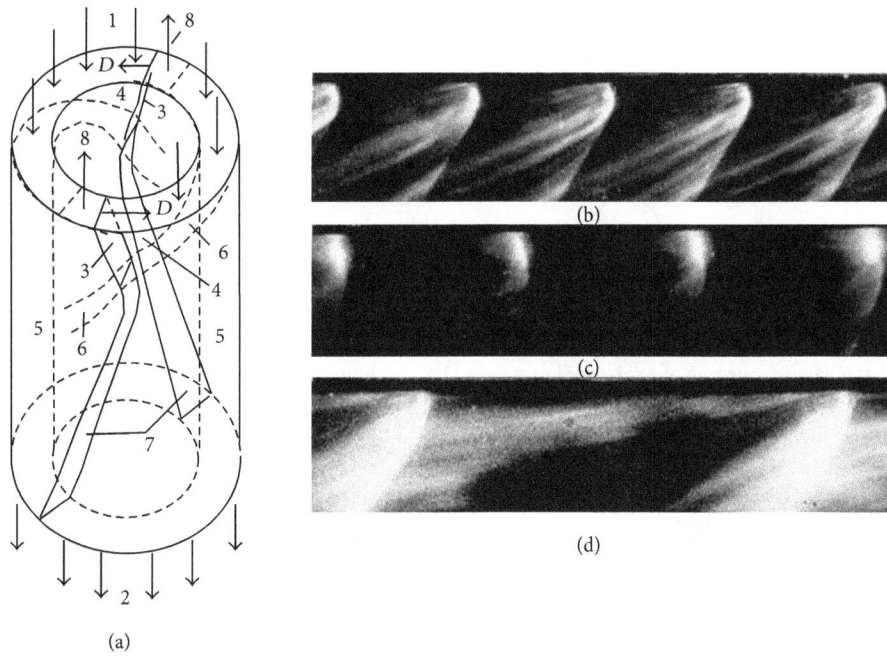

(a)

(b)

(c)

(d)

FIGURE 7: The schema of continuous burning of a mixture in the flowing ring combustion chamber and typical streak records of self-luminosity of rotating DW at registration across slit paralleled to axis.

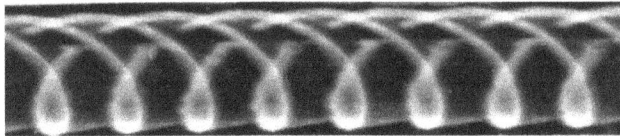

FIGURE 8: Typical streak records of self-luminosity of steady rotating DWs in circle combustion channel at registration along axis (radial mixture flow, three rotating waves on channel circuit).

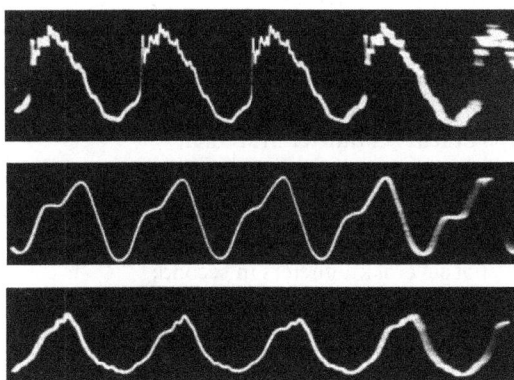

FIGURE 9: Typical pressure oscillograms in steady rotating DW in different points of circle channel (special piezogauges with thermoisolation from hot products [5]).

a radius of a ring charge. Moreover, even for rectilinear free gaseous charge the stationary propagation of multifront DW is possible only in that case, when the diameter of a charge exceeds the "critical" diameter $d > d_*$ (or the thickness of a gas charge as a flat mixture layer exceeds the critical magnitude $h > h_*$). The results of experimental investigations of excitation and quasi-stationary propagation of multifront DW around a cylindrical surface in regime of "exterior" detonation burning were published in [14] at a variation of a radius of a bended surface, thickness of a gas layer, and initial pressure of a mixture (variation of cell size of detonation front).

At initial pressure of a mixture $P_0$ be smaller the critical value $P^*$ (for the chosen radius of a bended surface as interior radius of a gas charge), it is possible to observe in photos the DW failure and transformation to combustion mode, propagated along circle layer. At $P_0 \geq P^*$ the regimes of DW-reinitiation in points A (marked by arrows on Figure 11) on some distance from the bended boundary are observed. The amount of reinitiation centers is increased with growth $P_0$, and they come nearer to a bended surface. Above the value $P^{**}$ (depending on a curvature radius, on mixture composition, on depth of a gap, etc.) DW "rotation" along circle layer happens in a stationary mode.

The photo on Figure 11 illustrates a possibility of a realization of "exterior" burning around a cylindrical surface in a detonation mode. Moreover, photo obviously demonstrates the basic mechanism of a wave turn and quasi-stationary propagation of multifront DW in a ring gas charge: for this purpose it is necessary that new reinitiation centers arise periodically in the weakened wave on some distance from a bended surface. In gas charge with the free boundary the propagation of multifront DW is possible, if the characteristic size of a gas layer around a bended surface is sufficient for constant renewing of the microcenters of reinitiation. In [14], the evaluation of values of a critical radius of a curving and

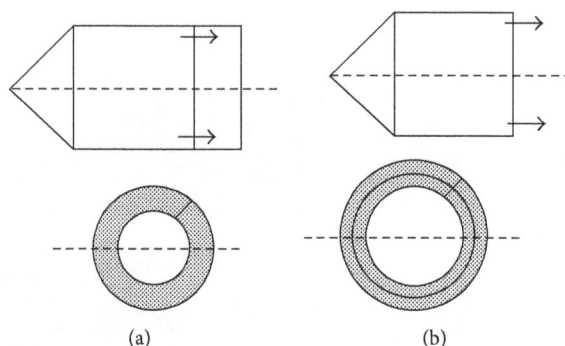

FIGURE 10: The schema of interchamber burning of a mixture in rotating DW ((a) bounded flow) and schema of "exterior" burning in a ring gas layer with the help of rotating DW ((b) unbounded flow, gaseous charge jet is created at axial or radial injection of a mixture).

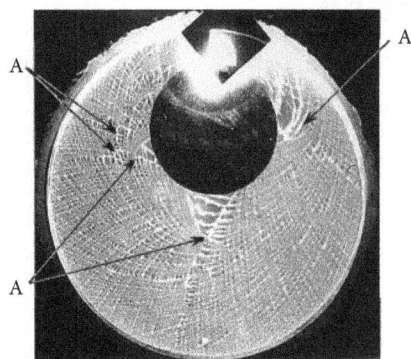

FIGURE 11: A photo of DW "rotation" around of a cylindrical surface. A: Centers of DW reinitiation.

minimum breadth of a ring charge, necessary for a realization of "exterior" burning in detonation mode, was offered: a minimum radius of the interior boundary is $(R_0)_{\min} \approx 0.4l$, thus a necessary breadth of a gas layer is $\Delta R > 0.9l$. These relations for $(R_0)_{\min}$ and $\Delta R$ can be recommended for practical estimations.

Many important problems for PDE (multicyclical DW initiation in an individual tube of multibarrelled system, the high-frequency regimes, full identity of processes in each cycle, etc.) became less important in the concept of RDE. For example, probable frequency of the PDE schema is limited by length of an individual tube and can be enlarged at the expense of multibarrelled system. At the same time, the frequency in RDE schema is determined by a circle length of a tube only (or diameter). Moreover, in RDE schema the mixture is initiated only in the process beginning, and then it is burned continuously by steady rotated spin DWs. And the self-organizing (autotuning) of a system of spin waves ("multiheaded spin") at a wide variation of flow parameters is an essential virtue of the RDE schema, because it allows to inspect and to operate effectively the DW in RDE. At the same time, the self-organizing of a detonation regime is not inherent in the PDE schema. Consequently, the constant rigid monitoring on engine parameters should be carried out at variation of mixture parameters, that it has not left the operating regime.

## 3. Initiation: Ignition, Deflagration to Detonation Transition, Detonation

The theoretically justified advantages of a detonation burning could be realized in PDE on active fuel-oxygen mixtures (FOM), as FOM are characterized by rather small magnitudes of initiation energies (under condition of ideal intermixing of a mixture at separate injection of components). However, the major condition of practical application of any engine (including those operated on detonation) is the use of an atmospheric air and work on fuel-air mixtures (FAM). For example, using hydrogen as most perspective fuel (from power and ecological points of view) for ignition of stoichiometric FAM at normal conditions 0.017 mJ is required approximately, and for direct initiation of a detonation regime about 4000 J is required accordingly (about one gram of TNT) [15]. It is obvious that it is unreal from a practical point of view to realize similar initiation of a detonation for high-frequency mode of operations of multibarrelled schema of PDE.

According to a modern classification the excitation of a combustible mixture is achieved by three basic modes:

(i) the weak initiation (ignition), when only laminar burning is raised with velocities of propagation at a level of tens centimeter in second;

(ii) the strong (direct) initiation, when self-sustained DW is formed in immediate proximity from the initiator and then is spreaded on a mixture with a velocity at a level of several kilometers in second;

(iii) an intermediate case, when the mixture only is ignited on the initial stage and then the front of a flame is accelerated by virtue of the natural or artificial reasons up to visible velocities at a level of hundred meters in second. Under certain conditions further, the deflagration to detonation transition (DDT) can be realized even.

The effect of excitation of combustion or detonation usually carries a "threshold" character ("yes" or "not") for any initiator (Figure 12). The minimum energy ensuring 100-percentage excitation at the given condition traditionally refers to the critical energy. The type of symmetry is very

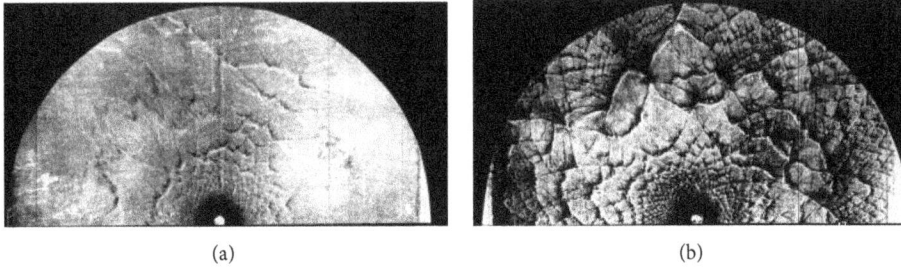

FIGURE 12: Smoked-foil imprints illustrated the threshold character of initiation of cylindrical DW: (a) DW destroyed at $E < E^*$ (subcritical regime), (b) successful DW initiation at $E \geq E^*$.

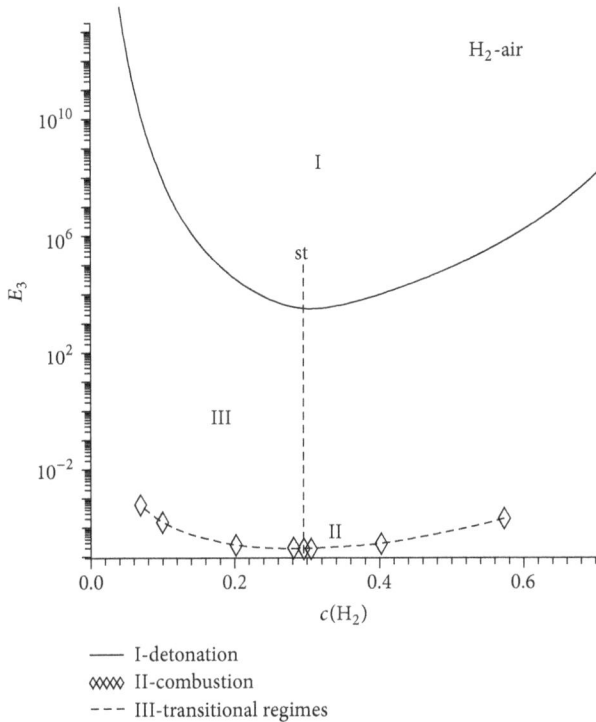

- —— I-detonation
- ⬦⬦⬦ II-combustion
- - - - III-transitional regimes

FIGURE 13: The graph of critical energies of ignition and initiation of detonation on molar fuel concentration for hydrogen-air mixtures.

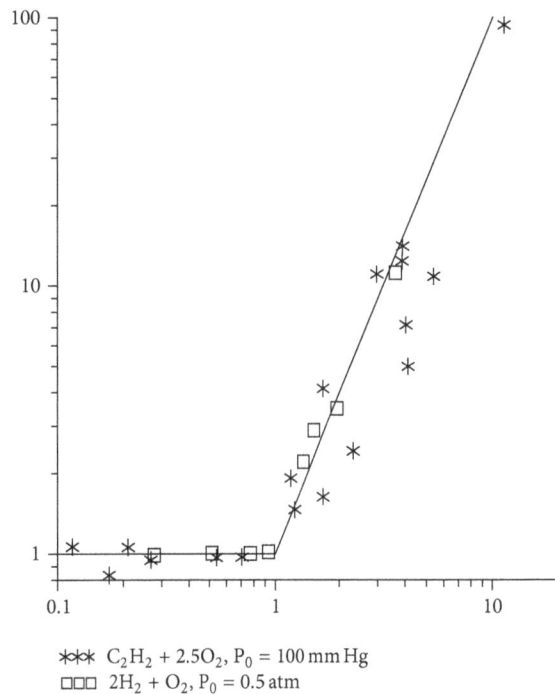

✳✳✳ $C_2H_2 + 2.5O_2$, $P_0 = 100$ mm Hg
□□□ $2H_2 + O_2$, $P_0 = 0.5$ atm

FIGURE 14: Critical initiation energy ($y$-axis) on duration of discharge ($x$-axis), both normalized on values for ideal initiator.

important for initiation: expanding waves (cylindrical or spherical) or quasi-plane (propagation in a rectilinear tube). In expanding waves of laminar burning without artificial action on a wave the basic mechanism of acceleration is autoturbulization of a flame front. The question on a possibility of self-dependent DDT in expanding waves until now is debatable and has not precise experimental confirmation. In pipes alongside with autoturbulization of a flame front the processes of interaction with side walls play rather important role and the principal possibility of DDT is well known (especially for active FOM). Despite an establishment of the DDT fact the data about DDT in pipes are rather inconsistent, especially for FAM.

Usually the critical energy is characterized by some curve which has been carried out on boundary values of energies in area "yes" or "not" ("go" or "no go") on the graph of an initiation energy in dependences on any parameters of

a mixture (initial pressure or temperature, duration of a discharge, inter-electrode distance, etc.). For example, on Figure 13 the critical excitation energy of a spherical flames (dashed curve II and above area) and direct initiation of a spherical detonation (solid line I and area above it) for mixtures of hydrogen-air dependent on molar concentration of hydrogen ($[E_3]$—J) are presented. The vertical dashed straight line corresponds to stoichiometric ratio of hydrogen with air. The area III between lines I and II, corresponds to excitation of regimes, intermediate between low-velocity laminar flame and high-speed self-sustaining detonation. Between lines I and II the boundary line, corresponded to the DDT processes, should be located.

The critical ignition energy $E_{flame}$ (anyway, at spark ignition) traditionally appears as a basic parameter of fire hazards of a mixture (first case of the previously-mentioned classification). The critical energy of initiation of a detonation

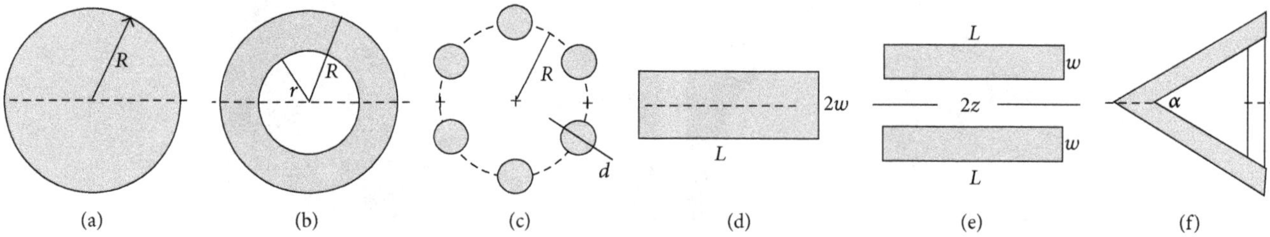

FIGURE 15: Simplest schema of space distribution of input energy.

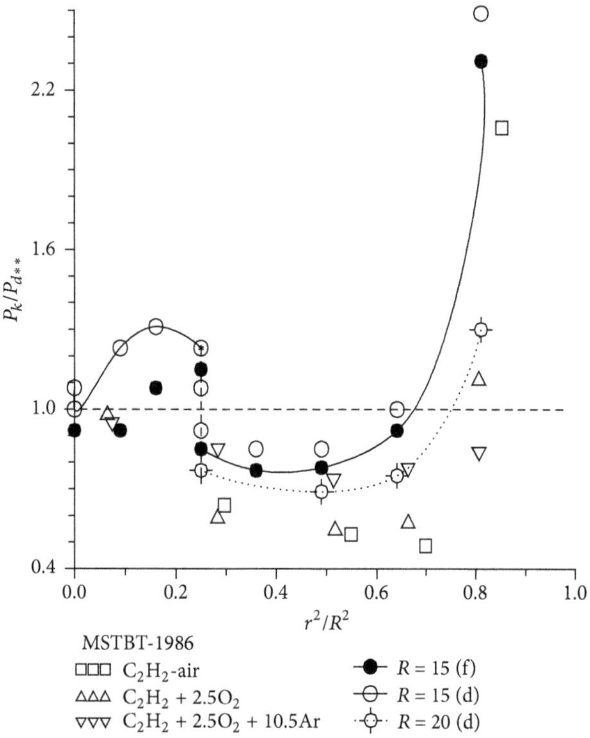

MSTBT-1986

| | |
|---|---|
| ⊡⊡⊡ $C_2H_2$-air | ● $R = 15$ (f) |
| △△△ $C_2H_2 + 2.5O_2$ | ○ $R = 15$ (d) |
| ▽▽▽ $C_2H_2 + 2.5O_2 + 10.5$Ar | ⦶ $R = 20$ (d) |

FIGURE 16: Initial pressure of mixture for critical initiation of quasispherical DW by annular charge Figure 15(b) (normalized on pressure for ring charge of the same diameter Figure 15(a)) on blockage ratio (relation among inner and external radii of annular charge).

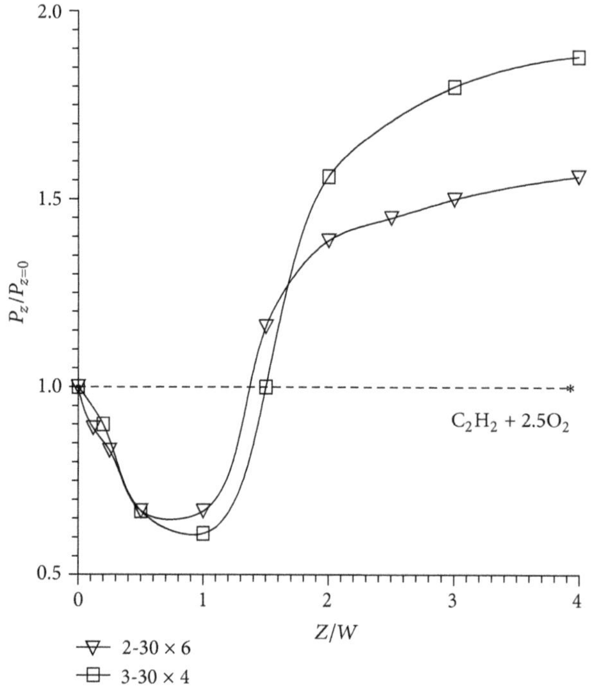

▽ 2-30 × 6
⊟ 3-30 × 4

FIGURE 17: Initial pressure of mixture for critical initiation of quasi-cylindrical DW by two or three parallel charges Figure 15(f) (normalized on pressure for the summary charge Figure 15(d)) on distance among charges (normalized on character width of individual charge).

$E_*$ of ideal initiator (from point of view of spatial-temporary characteristics of the initiator) serves as a measure of detonation danger of mixtures: the $E_*$ smaller, the mixture more dangerous.

The ideality of the initiator is understood in the following sense. Each mixture at given conditions (pressure, temperature, composition,...) are characterized by some characteristic spatial and temporal scales $r_*$ and $t_*$ (e.g., induction size and duration). At the same time, at given conditions, the explosive mixture absorbs some energy $E_\nu$ from the initiator (during a finite time interval $t_0$ in finite area of space $V_0 = f(r)$), part $\eta$ of an energy $E_0$, initially accumulated in the initiator:

$$E_\nu = \int_0^{t_0} \int_0^{V_0} \varepsilon(t, V) \, dt \, dV = \eta E_0, \quad (1)$$

where $\varepsilon(r, t)$ is a function, circumscribing the spatial-temporary law of an entered energy, $\nu$ is index of dimensionality ($\nu = 1, 2, 3$ for plane, cylindrical and spherical symmetries accordingly). Generally $E_\nu$ is a composite function from characteristic scales of a mixture and initiator (more exactly from their relations). And only under conditions $t_0 \le t_*$ and $r_0 \le r_*$, the excitation of multifront DW will be determined by $E_*$ only (see later). These conditions represent the criterions of an "instantaneity" and "punctual" of the used initiator and given mixture.

It is necessary to notice that with formal mathematical point of view the previously-stated integrated equality represents a typical functional in a variational task about minimization of an energy $E_\nu$. At mutual influence of the spatial and temporal factor the requirement of minimization of an entered energy up to magnitude $E_{\min}$ is reduced

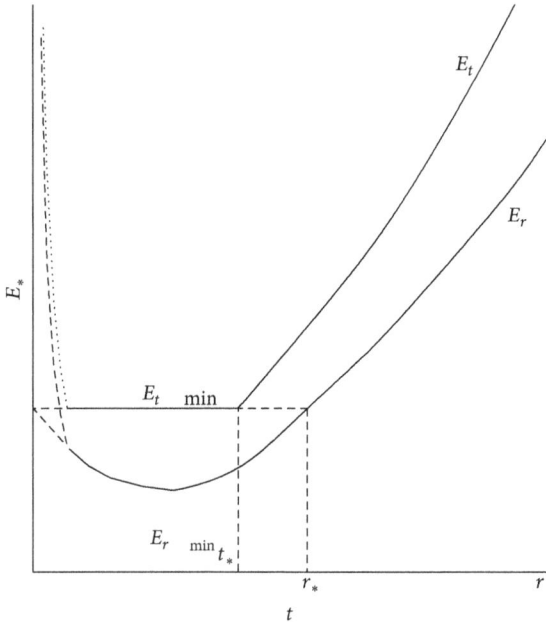

FIGURE 18: Spatial and time components of critical initiation energy on spatial-time parameters of initiator.

simultaneously to minimization of a power density up to $\varepsilon_{min} = \varepsilon(r^*, t^*)$. At the account only of temporal factor $(E_t)$ simultaneously with $E_{min}$ it is required to optimize a power of input energy $\varepsilon_{min} = \varepsilon(t^*)$. The influence of a spatial component of input energy $(E_r)$ should be reduced to optimization of a density of an entered energy $\varepsilon_{min} = \varepsilon(r^*)$ simultaneously with $E_{min}$.

The knowledge of spatial-temporary characteristics of energy release of any individual initiator is necessary for the correct experimental definition of a critical energy and its optimization.

## 4. Optimization of an Initiation: Role of the Spatial and Temporal Factors

The problem of initiation is extremely important not only for PDE, but also as the interdisciplinary task from a point of view of scientific, practical and ecological aspects of safety. The main aim is the determination of the critical initiation energy and optimization of initiation from a point of view of spatial distribution of an energy, entered into a mixture, and temporal performances.

The following gas-dynamic and kinetic parameters of an initiation task are determinanted:

medium: $r$, $t$, $P_0$, $\rho_0$, $\gamma_0$ (coordinate, time, pressure, density, ratio of specific heats);

initiator: $t_0$, $r_0$, $E_y$ (temporal and spatial parameters of an energy, entered into a mixture, energy, type of a symmetry);

combustible mixture: $t_{ind}$, $t_{react}$, $E_{act}$, $Q$, ... (induction and reaction times, activation energy, specific energy release (chemical)).

Traditionally used simplifications and obtained conclusions are the next:

(1) inert medium: $t_{ind} = 0$, $t_{react} = 0$, $E_{act} = 0$, $Q = 0$, ...). Then at additional conditions of $P_0 = 0$, $t_0 = 0$, $r_0 = 0$ the self-similar analytical solution is obtained $r = f(t, E_y, \rho_0, \gamma_0)$. At $t_0 > 0$, $r_0 > 0$, $P_0 > 0$: the non-self-similar solution is obtained (and the numerical procedures);

(2) combustible mixture: at $t_{ind} = 0$, $t_{react} = 0$, $E_{act} = 0$, $Q > 0$ we have the model of a wave with an instant chemical reaction at the front and the self-similar analytical solution as steady propagated flame or detonation, even at a zero excitation energy; if $t_{ind} = 0$, $t_{react} > 0$, $E_{act} > 0$, $Q > 0$, ..., non-self-similar solution is obtained.

The variational task about an energy $E_y$ of the above-mentioned integrated relation can experimentally be investigated by several paths.

(1) If there is a variation of duration of energy input at conservation of spatial sizes of area of energy input: $V_0 = $ const, $t_0 = $ var, then a power density is transformed to a power: $\varepsilon(t, V) \Rightarrow \varepsilon(t)$. Nondimensional critical energies of initiation of a cylindrical detonation independent from dimensionless duration of an electrodischarge (the data [16]) are submitted on Figure 14 for stoichiometric mixtures of hydrogen and acetylene with oxygen. The critical initiation energy is a unique criteria parameter of a mixture only for a case, when the duration of an initiating discharge does not exceed some critical magnitude; otherwise, the more large energy is required for DW initiation, then the discharge is more long. In other words, at $t_0 < t^0$ ($t^0$: the characteristic temporal scale of a mixture) it can be spoken about "quasi-ideal" initiator: only in this case; the initiation is defined by a unique parameter, the critical energy $E_y^*$!. At $t_0 > t^0$ the initiator should be referred to as imperfect. At anyone $t_0$ the critical initiation carries a threshold character: damping at $E < E^*$ and DW excitation at $E > E^*$.

(2) Conservation of duration of energy input and both variations of a size and form of area of energy input: $t_0 = $ const, $V = $ var, $\varepsilon(t, V) \Rightarrow \varepsilon(V)$, power density is transformed to an energy density.

The simple forms of the initiators at investigations of influence of spatial distribution of an entered energy for the critical initiation energy of spherical DW are submitted on Figure 15. Initiator charges can be modeled in the following geometric forms: (a) a circular disk with radius $R$; (b) an annulus with radii $r$ and $R$; (c) a multipoint scheme represented by $n$ charges of diameter $d$ spatially distributed, for example, on a circle of radius $R$; (d) a single rectangular plane charge with $L \cdot 2w$ sizes; (e) a system of parallel line charges of $L \cdot w$ sizes, separated by a distance $2z$; or (f) a system of nonparallel line charges, for example, shaped with an angle $\alpha$ or closed triangle.

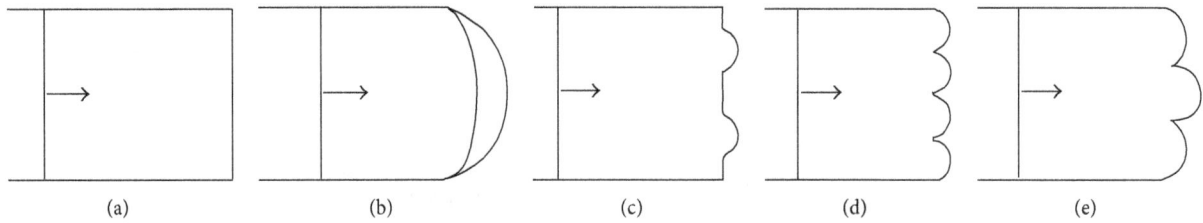

FIGURE 19: The typical forms of concave surfaces for optimization of initiation at SW reflection: (a) plane reflector; (b) single ring or elliptical reflector on whole cross-section or single ring or elliptical reflector on limited part of cross-section; (c) two ring or elliptical reflectors shifted one another; (d) some neighboring elliptical reflectors located along plane; (e) some ring or elliptical reflectors located along concave elliptical line (neighboring or with shift).

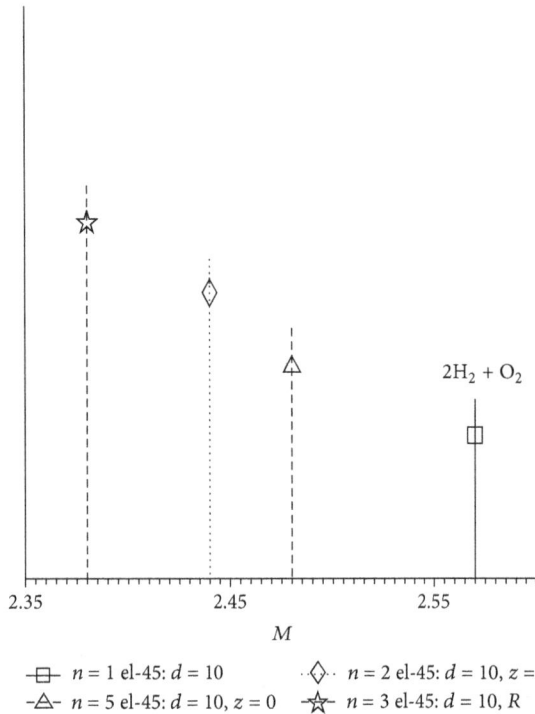

FIGURE 20: The critical (lowest) Mach number for excitation of a detonation at interaction of an incident SW with various reflectors: identical mixture conditions and variation of SW Mach number for individual reflector. The area from the right of vertical line corresponds to successful initiation ("go", "yes"), from the left to unsuccessful initiation ("no go").

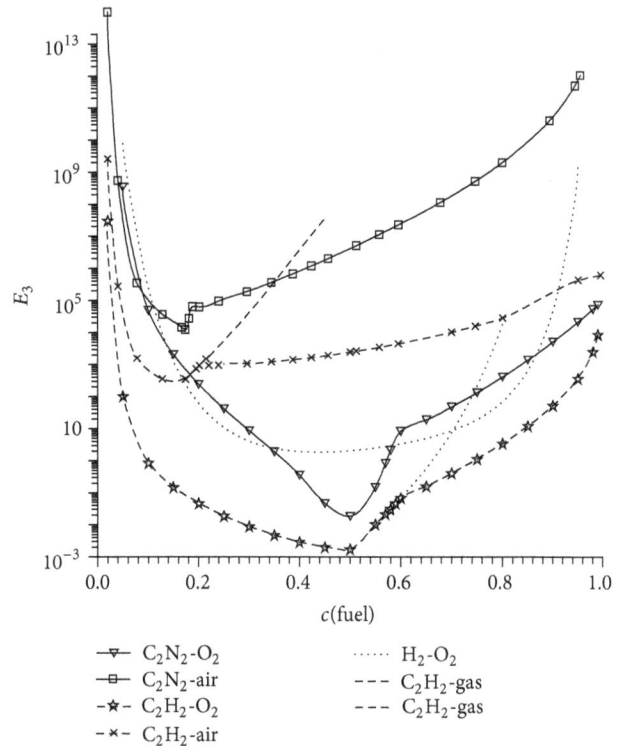

FIGURE 21: The graphical BANK of data about critical initiation energy of spherical DW (J) on fuel molar concentration for different mixtures, which allow to compare mixtures on its detonation hazards. Doubled lines for $C_2H_2$ correspond to cases of gaseous and solid C atoms in detonation products.

The critical conditions of initiation of a spherical detonation by ring charges are submitted on Figure 16, $r$ and $R$: interior and exterior radii of a charge. The area of a below dashed horizontal line corresponds to the greater effectiveness of ring charges in comparison with a compact charge in form of disk of radius $R$. The similar effect of the greater effectiveness of the distributed charges in a comparison with a compact charge is observed and at initiation of cylindrical DW by two or three parallel linear charges—Figure 17 ($z$-distance between charges).

On Figure 18 the influence of a temporal component to an initiation energy is submitted: at variation of duration of initiating impulse, the value $E_t$ = const at $t_0 < t^0$,

at $t_0 > t^0$ $E_t$ will increase. Spatial component $E_r$ is characterized by the U-figurative form with optimum magnitude; that is, the critical initiation energy can be reduced at the expense of spatial distribution. For last case, the problem on a choice of the characteristic scale of a mixture $r^0$ is very important.

The optimization of initiation can be reached with the help of many methods, for example;

(a) spatial initiation, including the multicharge schema with a variation of an amount of charges and their spatial disposition from each other;

(b) initiation by a series of impulses with a variation of amplitude and duration of individual impulse and also their off-duty factor;

(c) initiation at reflection of shock wave (SW) from a focusing surface;

(d) initiation by jets of hot and active substances, including ionized components;

(e) application of promoters;

(f) using mixtures with gradients of parameters (density, temperature, and composition).

For example, in Figure 19, the initiation with the help of SW reflections from concave surface is illustrated. Figures 19(a) and 19(b) are classical schema of SW reflection from a plane surface and from cylindrical (or elliptic-dashed profile) surface (e.g., [17, 18]); the symbols (in order) correspond to one elliptic reflector with an effective diameter of 10 mm; to five adjoining elliptic reflectors with a similar size (as on Figure 19(d)); to two elliptic reflectors of an identical effective diameter that moved from each other on 10 mm (the schema Figure 19(c)); and to three elliptic reflectors located along a concave elliptic curve, "double" focusing of SW (as on Figure 19(e)). The additional positive effect in decreasing the critical Mach number of the incident SW can be reached with the help of multifocused system (MFS), when a few concave reflectors are located on the round or elliptic surface (Figures 19(c)–19(e)). In this case the focusing of the reflected waves creates new "hot spots" as the micro-initiators, which initiate mixture immediately after SW focusing, or later on the following stage of interaction by secondary SWs, generated by "hot spots." Centers of individual reflectors can place as along a direct line (Figure 19(d)), as and along concave curve (Figure 19(e)), and also with some shift $z$ from each other (Figure 19(c)). Three last schemas are more effective at DW initiation comparison with cases Figures 19(a)-19(b). In Figure 20, the experimental data about limiting Mach number of an incident shock wave are presented, at which the initiation of a mixture is observed: decrease of the critical Mach number of the incident SW for DW initiation is precisely fixed.

## 5. BANK of Detonation Danger

The critical energy (as a unique parameter describing excitation of a mixture by an ideal initiator) attracts an attention of many investigators until now. Three types of models are known for estimations of the critical initiation energy:

(1) model of one-dimensional initiation for numerical accounts as a system of the one-dimensional equations of a gas dynamics and kinetics;

(2) semi-empirical models of initiation of one-dimensional DW (about 20 variants);

(3) model of initiation of multifront DW.

The one-dimensional numerical calculations were carried out for the limited amount of mixtures (about ten), while the practice interest requires the analysis of hundreds of

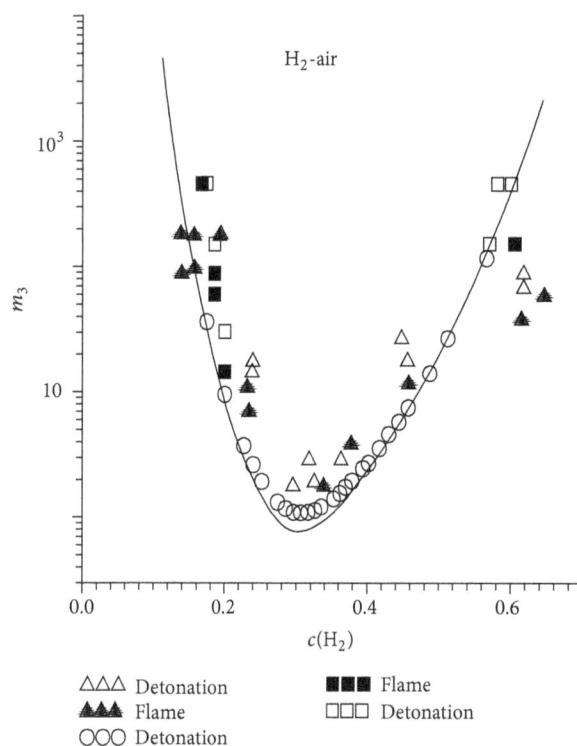

FIGURE 22: Correlation of experimental and calculated data about critical mass of TNT charge for initiation of spherical DW ($[m_3]$ = g) on fuel molar concentration of $H_2$-air mixtures $c(H_2)$.

various mixtures. In LIH (Novosibirsk, Russia), the code "SAFETY" is developed for the similar purpose, in which all known models of a type (2)-(3) (e.g., [15, 19]) are included. The calculated data allow to create the BANK of detonation danger of various combustible systems Figure 21. As an example, on previously-mentioned Figure 13 the graphic interpretation of calculated data about critical energies for initiation of a detonation process and mixture ignition for hydrogen-air mixtures is presented. The critical mass of TNT charge (in gram) for initiation of a spherical detonation in hydrogen-air mixtures independent from molar concentration c of fuel in a mixture is submitted in Figure 22. The good correlation of calculated and experimental values is visible. The detailed analysis of experimental data on different detonation parameters was published in review [20].

## 6. DDT Optimization

The prospects of application of a detonation in various technological processes (as fastest regime of a mixture burning) are stroke on difficulty of its practical realization, because the direct initiation of DW in FAM, as a rule, is achieved with the help of high explosive (HE) charges. The ignition of a mixture by the initiator of low power and consequent artificial acceleration of a flame with the help of highly effective accelerators down to DDT is natural alternative to HE charges.

It is necessary especially to underline that the DDT accelerator should guarantee full identity of DDT process in

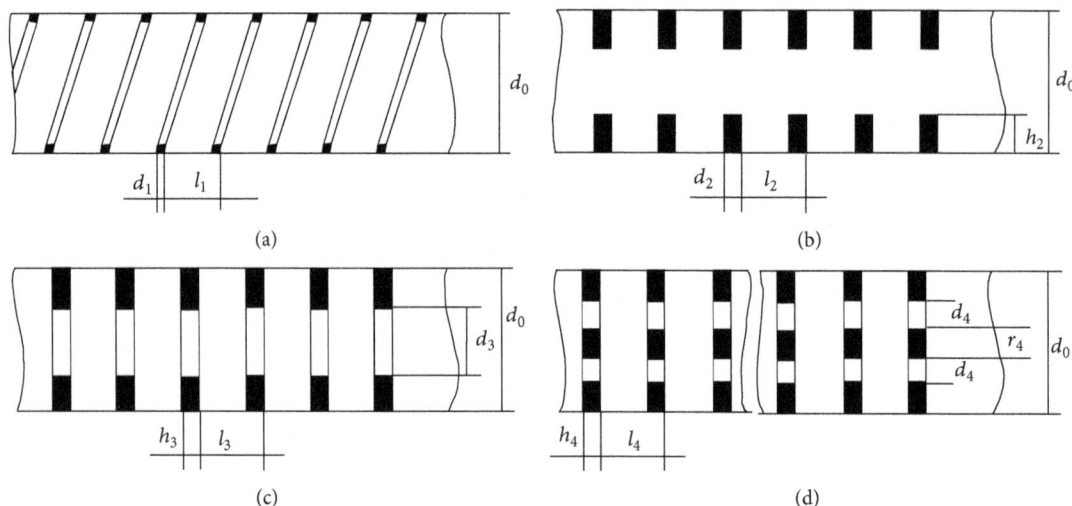

FIGURE 23: Typical schema of use of passive elements for acceleration of "plane" flames: helixes (Tshelkin spiral), each helix is characterized by three parameters with the dimensions of length: tube diameter $d_0$, thickness $d_1$, and pitch $l_1$ of spiral (a); singly standing rods ("forest" schema) (b) (in addition to the dimensions of an individual rod such as diameter $d_2$ and length $h_2$, the rod-type accelerator is characterized by the number $n$ of the used rods and a law of their distribution in the tube, for example, by the step along the axis, azimuthal displacement in neighboring planes, etc.); disks with holes coaxial (c) (the characteristic dimensions are thickness $h_3$ and displacement $l_3$ of disk, and hole diameter $d_3$; the number of holes is of importance also), or with displacement (the dimensions are thickness $h_4$ and displacement $l_4$ of disk, hole diameter $d_4$ and distance $r_4$ among holes) (d).

each individual tube for each cycle of PDE work. Moreover, the initiation should be effectively realized in a high-speed stream of a mixture (subsonic or supersonic) instead of classical initiation of a motionless mixture. It is not evident that the conditions of DW initiation for a motionless mixture and for chemically active stream are identical. Therefore, the additional experimental researches of initiation of high-speed streams of FAM, for example, with the help of wind tunnel techniques in a wide range of the flow Mach numbers are necessary. The similar investigations on optimization of DDT accelerators for FAM are necessary also.

On a today the limited number of correct experimental measurements of the critical ignition energy is known from the literature (it is rather strange!).

We would like to pay attention once again to a possibility of a decrease on some order of the critical initiation energy at simultaneous optimization of the spatial and temporal factors of an entered energy. But it is necessary to underline especially that such decrease is reached at quite defined relations between spatial and temporal parameters of an initiator and spatial: temporal parameters of a combustible mixture. The high effectiveness is reached at an optimum relation between these values, otherwise special effectiveness is not observed. In the literature, more often, only the Tshelkin spiral is used as a basic element for acceleration of DDT process. But even such spiral is characterized by three parameters with dimensionality of length: by an exterior diameter of a spiral (usually equals diameter of a pipe), thickness of a wire, and pitch of a spiral. And only at optimum relations of these parameters with a spatial parameter of a combustible mixture the action on DDT will be highly effective, and at an arbitrary relation the influence of a spiral on DDT is insignificant.

In LIH during the previous years the effective accelerators of DDT for FOM and also the accelerators for excitation of spinning DW in FAM were developed: after ignition of a mixture by an automobile spark the action of the accelerator leads to formation of the spinning DW on a distance of 16 calibers in equimolar hydrogen-air mixture and in stoichiometric methane-air mixture on distance of 25 calibers.

The turbulence plays an important role in intensify of burning (at expense of an increase of a flame surface). Two basic modes of a flame turbulization are known: natural (autoturbulization) and artificial turbulization with the help of various obstacles. Usually an autoturbulization produces a slow acceleration of flame front, and interactions with obstacles and pipe walls are more effective, especially with concave boundary. Some typical schema of use of passive elements for acceleration of "plane" and divergent flames were presented on Figures 23-24. For forming of multifront DW it is necessary that the elements of the accelerator were distributed uniformly on a cross-section of a detonation pipe. Moreover, the elements must repeat the action on combustion front at its acceleration along a detonation tube over and over again. The given idea was realized in [21] with attraction of the theory of turbulent jets.

At a flow of a streamline body by gas behind a body the turbulent layer is formed with the extending boundaries: exterior from a stream axes and interior on direction to axes (Figure 25). The interior boundaries are closed on axes on some distance from streamline body (points 1 on Figure 25), and the point of intersection is defined by a body size and by expansion angle of the interior boundary. If in the given cross-section the screen ("lattice") from a few body is located,

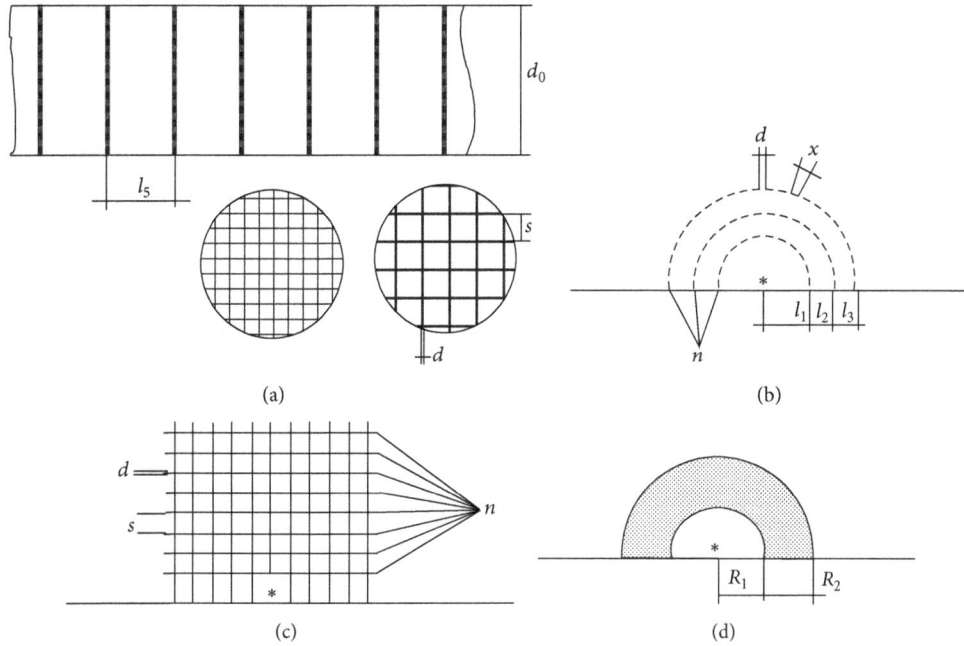

(a)

(b)

(c)

(d)

FIGURE 24: Typical schema of use of passive elements for acceleration of "plane" and divergent flames: wire or rod screens (a) (the characteristic parameters are the diameter of the wire $d$, the permeability $s$ of the screen, and the distance between the screens $l_5$); hemispherical shells with holes (b) (parameters are hole diameter $d$ and holes displacement $x$, number of hemispheres $n$, and their spatial arrangement $l_1, l_2, l_3, \ldots$); 3D rod-structures (c) (parameters are rod diameter $d$ and permeability$s$, number of rods $n_i$ (total and in the planes); obstacles from porous materials (d) (radii $R_1$ and $R_2$, porosity, number of porous layers,$\ldots$), and so forth.

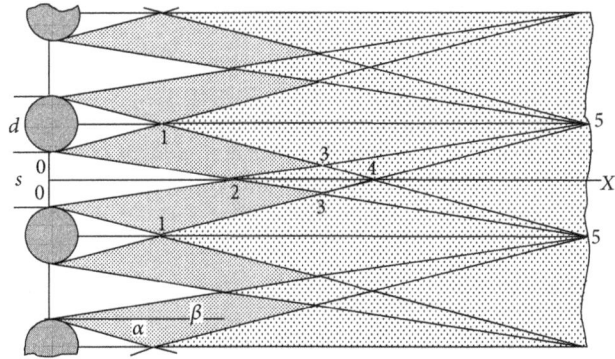

FIGURE 25: Scheme of flow turbulization behind the obstacles.

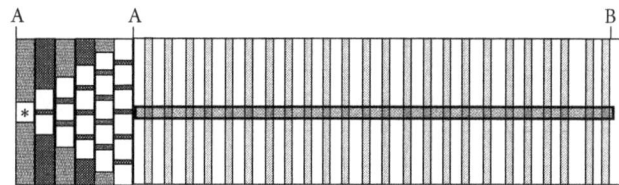

FIGURE 26: Simplest scheme of generator of plane wave and DDT accelerator.

then along with intersections of the interior boundaries on axes of each body, the exterior boundaries of turbulent layers from the neighboring bodies will be closed also. The intersection point of the exterior boundaries is defined as a distance between bodies and by expansion angle of the exterior boundary (points 2 on Figure 25). The optimization condition was formulated as an equidistantness of intersection points of the interior (1) and exterior (2) boundaries of turbulent layers from a plane of "lattices." This condition allows to establish an optimum relation between geometric parameters of individual bodies (as elements of the accelerator) in the given cross-section and also to determine the coordinates of following "lattice" with turbulization elements. Such relations represent an engineering technique for purposeful projection of turbulent device for DDT accelerators.

In [21] with the help of similar DDT accelerator (Figure 26) the record of effectiveness was obtained: at pressure 1 atm for stoichiometric of mixtures of acetylene-air and hydrogen-air the multifront DW in tube of diameter of 250 mm was formed on a distance of about 2 calibers (Figure 27). The advantage of the highly effective DDT accelerator is also that it allows to transfer the investigations of fuel-air mixtures from target ground conditions (with using of high explosive) to laboratory.

It would mark especially that the spatial distribution of ignition centers allows under certain conditions considerably (on orders) to reduce the DDT distance.

## 7. Difficulties in a Realization of a Detonation Engine

(1) "Weapon" scheme of PDE: problem of direct DW initiation ДВ in fuel-air mixtures: it is necessary especially to underline that the characteristic times of

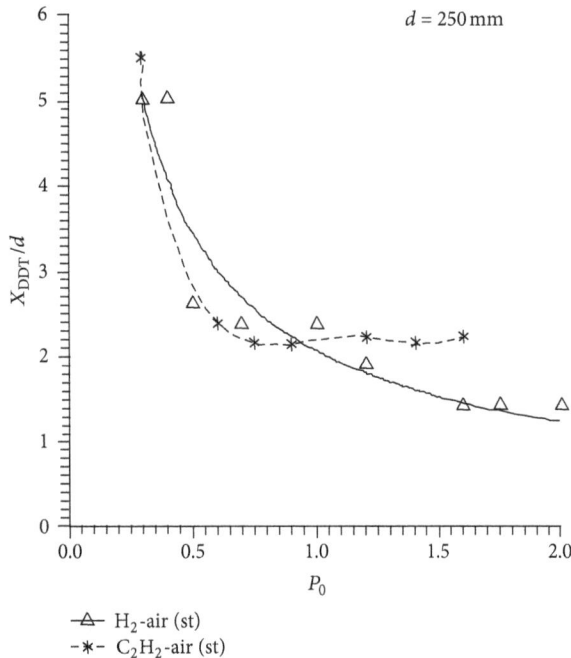

FIGURE 27: DDT distance (normalized on tube diameter) on initial mixture pressure (atm) for some fuel-air mixtures.

DW forming at direct initiation of DW lay in microsecond range, and times of ignition in milliseconds. The times, for which the flame will be transferred in detonation, will make one hundred microseconds at the best. And if, for immovable mixture, it does not play the special role, it becomes defining for flow conditions.

(2) PDRJE: problem of cyclical operation: though the work of such engine on FAM is confirmed experimentally in single cycle, there are doubts in its stability at change of direction of DW propagation and in DW conservation at such change in case of cyclical operation. If DW is broken on some cycle (e.g., at run up to nozzle entering) and is turned to combustion wave, it will be affected on work of an engine and will require immediate operations on restoration of detonation process, for example, with the help of additional system of DW initiation (such system must be in condition of "constant readiness"). It can complicate the control process on an engine.

(3) The scheme of inchamber mixture burning by steady rotating DW removes a set of problems intrinsic to PDE, but the problem on realization of a classical detonation regime remains. During latest years there are some attempts to name any regimes as detonation, if the velocities of which exceed a sound velocity of an initial mixture (in literature, devoted to detonation engine). Such attempts reduce to terminological dissonance. Point is that if the regime has a smaller velocity on a comparison with an ideal C-J velocity (regime of quasi-detonation), the point of contact

of the Mikhelson line with an adiabatic curve of energy release at decreasing of DW velocity will be displaced in region of smaller $P$-$V$, and so pressure in such wave will be lower than pressure in ideal DW. Naturally, it will reduce the integrated work at expansion of products of such "detonation" and the question on the effectiveness degree of such regime requires additional researches and arguments.

(4) Exterior burning: the realization on FOM requires confirmation of serviceability on FAM. Common note to the analysis of effectiveness of various engines and its characteristic cycles, the clearness in definition of process of product expansion (with the frozen or equilibrium state), is necessary: it can essentially influence the obtained values.

The problem of Pulse Detonation Engine and Rotated Detonation Engine is widely investigated in the latest years in many countries. In [22–25], for example, one can read the opinion of the most known specialists of PDE and RDE problems about some specific aspects and perspectives of future investigations of PDE and RDE.

## 8. Conclusion

The basic problems of use of a detonation in propulsion systems with impulse and continuous burning of a combustible mixture are discussed. The basic results on propagation of detonation waves in supersonic streams are presented with references to PFRJE. The results of researches showing a principal possibility of creation of an engine with exterior detonation burning are presented. The base results (concerning PDE) on optimization of initiation are explained at the expense of spatial and temporal redistribution of an energy, entered into a mixture. The technique for projection of highly effective accelerators of DDT is discussed.

## References

[1] K. I. Tshelkin and Ya. K. Troshin, *Gas-Dynamic of Combustion*, in Russian, USSR Academy of Sciences, Moscow, Russia, 1963.

[2] Yu. A. Nikolaev, A. A. Vasil'ev, and V. Yu. Ulianitsky, "Gaseous detonation and its application in technique and technologies," *Combustion, Explosion, and Shock Waves*, vol. 39, no. 4, pp. 22–54, 2003.

[3] G. D. Roy, S. M. Frolov, A. A. Borisov, and D. W. Netzer, "Pulse detonation propulsion: challenges, current status, and future perspective," *Progress in Energy and Combustion Science*, vol. 30, no. 6, pp. 545–672, 2004.

[4] V. G. Aleksandrov, G. K. Vedeshkin, A. N. Kraiko et al., "Supersonic pulsing detonation ram-jet engine (SPDRJE) and method of its functioning" (Russian), The patent of Russian Federation no 2157909, October 2000, priority from 26.05.1999.

[5] A. A. Vasil'ev, *Investigation of a stationary detonation in the ring channel [Ph.D. thesis]*, Novosibirsk State University, 1968, in Russian.

[6] V. G. Alexandrov, A. A. Baskakov, G. K. Vedeshkin et al., "Supersonic pulse detonation ramjet engine: new experimental and theoretical results," in *Application of Detonation to Propulsion*,

G. Roy, S. Frolov, and J. Shepherd, Eds., pp. 277–282, Torus Press, Moscow, Russia, 2004.

[7] A. A. Vasil'ev, V. I. Zvegintsev, and D. G. Nalivaichenko, "Detonation waves in supersonic flow of reacting mixture," *Combustion, Explosion and Shock Waves*, vol. 42, no. 5, pp. 85–100, 2006.

[8] B. V. Voitsekhovsky, "A stationary detonation," *Reports of USSR Academy of Sciences*, vol. 129, no. 6, pp. 1254–1256, 1959 (Russian).

[9] B. V. Voitsekhovsky, "A spin stationary detonation," *Applied Mechanics and Technical Physics*, vol. 3, pp. 157–164, 1960 (Russian).

[10] F. A. Bykovskii, A. A. Vasil'ev, E. F. Vedernikov, and V. V. Mitrofanov, "Detonation burning of a gas mixture in radial ring chambers," *Combustion, Explosion, and Shock Waves*, vol. 30, no. 4, pp. 111–119, 1994.

[11] S. A. Zhdan and F. A. Bykovsky, "Investigations of continuous spin detonations at Lavrentyev Institute of Hydrodynamics," in *Pulse and Continuous Detonation Propulsion*, G. Roy and S. Frolov, Eds., pp. 181–202, Torus Press, Moscow, Russia, 2006.

[12] F. A. Bykovskii, S. A. Zhdan, and E. F. Vedernikov, "A continuous spin detonation in fuel-air mixtures," *Combustion, Explosion and Shock Waves*, vol. 42, no. 4, pp. 107–115, 2006.

[13] A. A. Vasil'ev and D. V. Zak, "Detonation of gas jets," *Combustion, Explosion, and Shock Waves*, vol. 22, no. 4, pp. 463–468, 1986.

[14] A. A. Vasil'ev, "Characteristic conditions of propagation of a multi-front detonation along a convex surface," *Combustion, Explosion and Shock Waves*, vol. 35, no. 5, pp. 86–92, 1999.

[15] A. A. Vasil'ev, *Near-Limiting Regimes of Detonation*, in Russian, Lavrentyev Institute of Hydrodynamics, Novosibirsk, Russia, 1995.

[16] J. H. S. Lee, "Initiation of gaseous detonation," *Annual Review of Physical Chemistry*, vol. 28, pp. 75–104, 1977.

[17] H. Gronig and B. Gelfand, Eds., *Shock Wave Focusing Phenomena in Combustible Mixtures: Ignition and Transition to Detonation of Reactive Media Under Geometrical Constraints*, Shaker, Aachen, Germany, 2000.

[18] O. V. Achasov, S. A. Labuda, O. G. Penyazkov, P. M. Pushkin, and A. I. Tarasov, " Initiation of a detonation at a reflection of a shock wave from a concave surface," *Journal of Engineering Physics and Thermophysics*, vol. 67, no. 1-2, pp. 66–72, 1994.

[19] A. A. Vasil'ev, "Estimation of the critical initiation energy for ignition and detonation in gaseous mixtures," in *Application of Detonation to Propulsion*, G. Roy, S. Frolov, and J. Shepherd, Eds., pp. 49–53, Torus Press, Moscow, Russia, 2004.

[20] A. A. Vasil'ev, "Dynamic parameters of detonation," in *Shock Wave Science and Technology Reference Library*, F. Zhang, Ed., vol. 6, pp. 213–279, Springer, 2012.

[21] A. A. Vasil'ev, "Optimization of accelerators of deflagration-to-detonation transition," in *Confined Detonations and Pulse Detonation Engines*, G. Roy, S. Frolov, R. Santoro, and S. Tsyganov, Eds., pp. 41–48, Torus Press, Moscow, Russia, 2003.

[22] W. H. Heiser and D. T. Pratt, "Thermodynamic cycle analysis of pulse detonation engines," *Journal of Propulsion and Power*, vol. 18, no. 1, pp. 68–76, 2002.

[23] F. Falempin, "Continuous detonation wave engine," in *Advances on Propulsion Technology for High-Speed Aircraft*, RTO-EN-AVT-150, Paper 8, NATO, 2008.

[24] D. A. Schwer and K. Kailasanath, "Numerical investigation of the physics of rotating detonation engines," *Proceedings of the Combustion Institute*, vol. 33, no. 2, pp. 2195–2202, 2011.

[25] M. Hishida, T. Fujiwara, and P. Wolanski, "Fundamentals of rotating detonations," *Shock Waves*, vol. 19, no. 1, pp. 1-10, 2009.

# 14

# A Reduced Order Model for the Design of Oxy-Coal Combustion Systems

**Steven L. Rowan, Ismail B. Celik, Albio D. Gutierrez, and Jose Escobar Vargas**

*West Virginia University, Morgantown, WV 26505, USA*

Correspondence should be addressed to Steven L. Rowan; rowan.steve@gmail.com

Academic Editor: Satyanarayanan R. Chakravarthy

Oxy-coal combustion is one of the more promising technologies currently under development for addressing the issues associated with greenhouse gas emissions from coal-fired power plants. Oxy-coal combustion involves combusting the coal fuel in mixtures of pure oxygen and recycled flue gas (RFG) consisting of mainly carbon dioxide ($CO_2$). As a consequence, many researchers and power plant designers have turned to CFD simulations for the study and design of new oxy-coal combustion power plants, as well as refitting existing air-coal combustion facilities to oxy-coal combustion operations. While CFD is a powerful tool that can provide a vast amount of information, the simulations themselves can be quite expensive in terms of computational resources and time investment. As a remedy, a reduced order model (ROM) for oxy-coal combustion has been developed to supplement the CFD simulations. With this model, it is possible to quickly estimate the average outlet temperature of combustion flue gases given a known set of mass flow rates of fuel and oxidant entering the power plant boiler as well as determine the required reactor inlet mass flow rates for a desired outlet temperature. Several cases have been examined with this model. The results compare quite favorably to full CFD simulation results.

## 1. Introduction

The anthropogenic emission of greenhouse gases (GHG) from the combustion of fossil fuels for the generation of electrical power has been considered as one of the driving factors for global climate change [1]. In order to meet future targets for the reduction of GHG emissions, a number of developing technologies for carbon capture and storage are currently under development; including pre- and postcombustion capture and oxy-fuel combustion [2].

Oxy-fuel combustion utilizes a combination of pure oxygen and recycled flue gas (RFG) as an oxidant for the combustion of the fuel source, producing a gas that consists mostly of $CO_2$ and water vapor [2]. The high concentrations of $CO_2$ in the flue gas enable easier and more economical $CO_2$ separation and compression and have been shown to reduce gaseous emissions of $NO_x$ and $SO_x$ from pulverized coal (PC) power plants [3]. Because of the higher specific heat of $CO_2$ as opposed to nitrogen in conventional air-fired combustion, oxy-coal combustion has been found to lead to significant changes in flame temperatures, chemical species concentrations, and radiation heat transfer within the combustor [3].

As a consequence of this, many researchers have turned to CFD modeling of oxy-combustion as a design tool for the design of new power plants and retrofitting of existing facilities utilizing oxy-coal combustion [3–7]. These simulation tools are capable of providing detailed information about the flow field, turbulent mixing, temperature, and species concentration profiles within a reactor. However, these simulations depend upon a number of model parameters and assumptions. More importantly, each of these simulations requires a considerable amount of time investment and computational costs [7]. If the purpose of the simulations is to obtain a set of operating conditions such as fuel and oxidant flow rates that will lead to reactor temperatures within a certain narrowly defined range, then a large number of simulations may be required, multiplying the costs by a significant margin.

In an effort to reduce the time and effort that must be spent to determine acceptable reactor operating conditions, a simplified reduced order model (ROM) has been developed. The ROM utilizes a combination of a commercially available coal devolatilization model based upon lattice percolation theory, a chemical equilibrium code developed by NASA, and an iterative energy balance calculation spreadsheet using Microsoft Excel. Each of these elements will be described in greater detail in the following sections, as well as a comparison between the ROM predictions and CFD simulations using the Ansys-Fluent software package.

## 2. Description of the Reduced Order Model

For a given coal feedstock, the ultimate and proximate analysis data and higher heating value are used to calculate the required flow rates of fuel (coal) and oxidant (as a mixture of $O_2$ and $CO_2$) for a desired level of input energy, expressed in kW, and equivalence ratio, where the equivalence ratio is expressed as the ratio of available oxygen to that required for stoichiometric combustion. In order to account for the devolatilization of the coal as it is heated, ultimate and proximate analysis data for the selected coal feedstock, as well as a heating rate, is provided as inputs to the Chemical Percolation Devolatilization (CPD) model software developed at Brigham Young University [8–10]. The CPD software enables prediction of the mass fractions of the coal tar and volatile gasses that are released when the coal is heated. The model also predicts the mass fraction of the remaining solid material, which is assumed for the purposes of the ROM to consist of pure carbon (graphite). These mass fractions are then converted into flow rates (moles/s) and, along with the molar flow rates of fuel and oxidant, are provided as inputs for the NASA Chemical Equilibrium Applications (CEA) software. The CEA software is then used to determine the adiabatic flame temperature, total enthalpy, and equilibrium concentrations of the combustion products. This information is then used to calculate an energy balance which accounts for the energy contained within the outlet gas and the energy transferred to the ash materials, as well as an assumed amount of heat lost to the surroundings. The mass and energy balance equations lead to an iterative process in which the CEA software is run repeatedly at different specified temperatures until the energy and mass conservations are satisfied, resulting in an average outlet temperature and mass fractions of prescribed species. In what follows, each of the steps discussed above are described in greater detail.

*2.1. Fuel/Oxidant Stoichiometry Calculations.* The first step in determining the required stoichiometric amounts of oxidant required for complete combustion is to calculate the mass flow rate of fuel (or coal) that is needed to produce the desired amount of power output during the combustion process. This is done by solving (1), where $\dot{m}_f$ is the mass flow rate of fuel in kg/s, $H_{HV}$ is the higher heating value of the feedstock in kJ/kg, and PO is the power output in kW:

$$\dot{m}_f = \frac{PO}{H_{HV}}. \tag{1}$$

Once the mass flow rate of the fuel is known, it is necessary to calculate the required mass flow rate of oxidant for complete

combustion. This is done by solving for the stoichiometric coefficients of the coal combustion reaction shown in (2) for a single kg of coal:

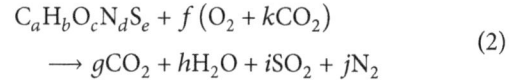

$$C_aH_bO_cN_dS_e + f\left(O_2 + kCO_2\right)$$
$$\longrightarrow gCO_2 + hH_2O + iSO_2 + jN_2 \tag{2}$$

where $f, g, h, I,$ and $j$ are stoichiometric coefficients, and the subscripts $a, b, c, d,$ and $e$ are the number of moles of carbon, hydrogen, oxygen, nitrogen, and sulfur present in the coal (as found by dividing the dry ash-free percent mass of each element, determined via ultimate analysis, by the molecular weight of each element). The number of moles of oxygen per mole of fuel required to complete combustion is equal to the value of $f$. Finally, $k$ is the number of moles of carbon dioxide required to obtain a desired percentage of oxygen in the total oxidant and is calculated from

$$k = \frac{100 - \%O_2}{\%O_2}. \tag{3}$$

In order to determine the required mass flow rate of the oxidant, the number of moles of oxidant per kilogram of dry, ash-free (daf) fuel must then be multiplied by the molecular weight of the oxidant and the total mass flow rate of the fuel:

$$\dot{m}_{ox} = fM_{ox}\dot{m}_f\left(1 - \%_{moisture} - \%_{ash}\right), \tag{4}$$

where the expression in parentheses accounts for the difference between the "as-received" and "daf" coal. It should be noted here that the mass flow rate of oxidant obtained via (4) is that required by stoichiometry for complete combustion. If instead a fuel-lean (or fuel-rich) mixture is desired, then the right-hand side of (4) can be multiplied by the equivalence ratio, $\lambda$, to obtain the desired conditions, where

$$\lambda = \frac{\text{actual moles of oxygen}}{\text{stoichiometric moles of oxygen}}. \tag{5}$$

An equivalence ratio less than one signifies that the combustion will occur in a fuel-rich environment and greater than one will be fuel-lean, and a ratio of one signifies stoichiometric combustion.

*2.2. Coal Devolatilization Model.* The devolatilization characteristics of the coal are modeled using the Chemical Percolation Devolatilization (CPD) model developed by Fletcher et al. [8–11]. Utilizing the ultimate and proximate analysis data of a given type of coal and a specified heating rate, the CPD model is able to predict the evolution of volatile gases, tars, and residual solids during coal pyrolysis (which is considered to be one of the early stages of coal combustion). The CPD model outputs the mass fractions of CO, $CO_2$, $CH_4$, water vapor, (unspecified) tars, solids (i.e., carbon and ash generating materials), and other (unspecified) volatile gases as a function of time and temperature.

For the purposes of the ROM, it is then assumed, for brevity, that the tar consists mainly of phenol ($C_6H_5OH$) and that the balance of the released gasses not in the form of CO, $CO_2$, $CH_4$, or water vapor (i.e., the unspecified gases referred to above) will consist of the following gas species:

TABLE 1: Coal properties (Decker Coal).

| Proximate analysis (wt% as received) | | | VM (% daf) | Ultimate analysis (wt% daf) | | | | | HHV |
|---|---|---|---|---|---|---|---|---|---|
| Moisture | Ash | Combustibles | | C | H | N | S | O | $H_i$ (MJ/kg) |
| 10.2 | 5.0 | 84.8 | 59.4 | 69.9 | 5.4 | 0.6 | 1.0 | 23.1 | 20.9 |

$C_2H_6$ (ethane), $H_2S$, $SO_2$, $O_2$, $N_2$, and $H_2$. The amounts of these individual gas species are then estimated by calculating an elemental mass balance based upon the mass fractions of carbon, hydrogen, nitrogen, sulfur, and oxygen provided by the ultimate analysis of the coal. The selection of the assumed gas and tar species listed above is primarily driven by the available selection of fuel and oxidant compounds within the NASA CEA software, as described below.

*2.3. Chemical Equilibrium Calculations.* The ROM utilizes the NASA Chemical Equilibrium with Applications (CEA) computer code [12] for the calculation of combustion-related parameters, such as adiabatic flame temperature, total enthalpy, and equilibrium gas compositions. The CEA code was developed by the NASA Glenn Research Center and contains modules for constant-pressure or constant-volume combustion, rocket performance based on finite- or infinite-chamber-area models, shock wave calculations, and Chapman-Jouguet detonations. In addition, the code contains databases of thermodynamic and transport properties for more than 2000 chemical species. The CEA code is available both as a web-based application and as downloadable source code written in the Fortran programming language. The CEA code is first used to determine the adiabatic flame temperature, as well as the total enthalpy and chemical composition of the reactor flue gases. This is done using the "combustion" module, in which the reactor pressure and molar flow rates and inlet temperatures of the previously determined fuel and oxidant species are specified as inputs. Unless otherwise specified, the temperature of the calculated oxidant is assumed to be at standard conditions, while the temperature of the devolatilization products is specified at the final temperature specified during devolatilization of the coal via the CPD model. The resulting flame temperature, enthalpy, and gas composition are then recorded and used as the inlet conditions for the energy balance calculations that will be discussed in the next section.

Additionally, during the iterative steps of the energy balance calculations, the CEA code is run again with the specified flow rates, but at specified temperatures and pressures. This step is completed in an iterative manner (with changing temperature) until the energy balance calculations, described below, are satisfied.

*2.4. Energy Balance Calculations.* For the energy balance calculations component of the ROM, the oxy-combustion reactor is treated as a control volume over which a steady-state energy balance is performed. In this treatment, the specified inlet conditions used with the NASA CEA code provides the total enthalpy (energy) entering the control volume. It is assumed that there is no change in energy within the control volume so that the energy leaving the control volume must equal that which is entering the control volume. The total energy leaving the control volume is a combination of the temperature-dependent enthalpy contained within the flue gas, conduction heat transfer losses through the walls of the theoretical reactor, and the energy required to raise the temperature of the remaining solids (ash) to the outlet temperature. The first iteration uses the adiabatic flame temperature data, and subsequent iterations gradually reduce temperatures until the difference between the inlet and outlet energy values is negligibly small, thus satisfying the principle of conservation of energy.

# 3. Results and Discussion

A number of simulations were carried out using the ROM and the results of these simulations were then compared to previously validated CFD simulations for similar combustion conditions. In these simulations, coal and various mixtures of $O_2$ and $CO_2$ are fed into a 100 kW downfired oxy-coal reactor (currently being designed by the authors), assuming an overall heat loss of 25 kW, and the predicted outlet temperatures provided by the ROM and CFD simulations are compared. In total, three different operating conditions are considered; these conditions correspond to the oxidant consisting of 25%, 50%, and 75% $O_2$, with the remaining balance consisting of $CO_2$. The coal properties for each case are shown in Table 1.

*3.1. CFD Simulations.* As previously stated, the reactor used for the comparison of the CFD simulations and the ROM predictions is currently being developed by the authors. This Very High Temperature Entrained Reactor (VHTER) is a vertical downfired design with a combustion chamber that is 150 mm in diameter and 2 m in length. The burner is composed of two concentric ports. The first port is for injection of pulverized coal using a small portion of the oxidant gas as carrier. The remaining balance of oxidant is injected through a second port, which is swirled with a fin angle of 45°. In addition, a pair of tangential injection ports can be used for injection of chemical reagents for desired chemical processes within the reactor. Schematics of this reactor are shown in Figure 1.

For each of the simulations used in this comparison, an assumed power input of 100 kW, corresponding to a mass flow rate of $4.78e - 03$ kg/s of pulverized coal, is used. In addition, three different $O_2/CO_2$ mixtures, with $O_2$ concentrations of 25, 50, and 75%, were considered. These cases are named OF25, OF50, and OF75, respectively. The heat losses through the wall were assumed to be 25% of the power input, or 25 kW. The mass flow rates of oxidant were calculated using an equivalence ratio of 0.85 and are shown in Table 2. The coal and oxidant are assumed to enter the reactor at standard

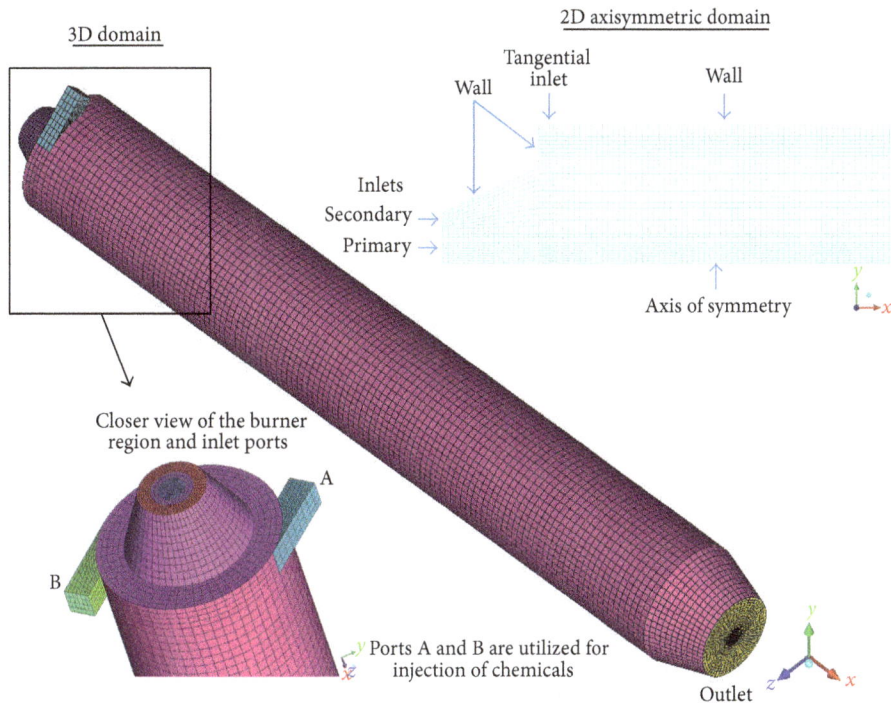

FIGURE 1: CFD computational domain for the VHTER reactor design.

TABLE 2: CFD simulation flow rates (kg/s).

| CASE | $O_2$ | $CO_2$ | Coal |
|------|-------|--------|------|
| OF25 | $8.70e-03$ | $3.59e-02$ | $4.54e-03$ |
| OF50 | $8.70e-03$ | $1.20e-02$ | $4.54e-03$ |
| OF75 | $8.70e-03$ | $3.99e-03$ | $4.54e-03$ |

conditions. Finally, the tangential injection ports were not used in these simulations.

*3.1.1. Solution Method (CFD).* The simulations were performed using the ANSYS-Fluent software package. The SIMPLE scheme [13] was used for pressure-velocity coupling in conjunction with the PRESTO scheme [14] for pressure coupling as it is recommended for swirling flows [15]. Turbulent flow, transport of species, and radiative heat transfer equations were solved using the first order UPWIND scheme [13]. The governing equations for turbulent flow, combustion, heat, and mass transfer for the continuous and discrete phases are solved using the finite volume method in structured grid systems. Both 2D and 3D axisymmetric domains were used; however, only the results for the 2D axisymmetric domain are presented here.

The combustion process is modeled using the Eulerian-Lagrangian approach for the continuous and discrete phases, respectively. The flow dynamics of the continuous media is determined by solving the Navier-Stokes equations by using the Reynolds Average approach (RANS). The Reynolds stresses are modeled by utilizing the standard $k$-$\varepsilon$ model of Launder and Spalding [16]. The modeling of turbulence close to the wall region is performed by using standard wall functions for near-wall treatment from Launder and Spalding [17].

Combustion and species transport for the gas phase is modeled using the well-known time averaged energy equation and species transport equations.

The energy equation is formulated in terms of the specific energy accounting for conduction, convection, and viscous dissipations with effective thermal and diffusive properties modified by turbulent effects. The effect of combustion, radiation heat transfer, and additional sources are also included. The Radiative Transfer Equation (RTE) is solved by using the Discrete Ordinates (OD) Radiation Model (Siegel et al. [18]), which can account for the effect of radiation exchange between the gas and particle phases as well as for scattering. For turbulent-chemistry interaction the turbulent-chemistry model of Hjertaer and Magnussen [19] is assumed, that is, the Eddy-dissipation model.

Coal devolatilization is modeled using the single kinetic rate devolatilization model of Badzioch and Hawksley [20], where the volatiles are defined by a two-step reaction mechanism where the species "volatiles" is a generic hydrocarbon obtained from the proximate and ultimate analysis of coal. Finally, the oxidation of char is modeled by utilizing the kinetics/diffusion-limited surface reaction model of Baum and Street [21] and Field [22]. The models utilized in these simulations were tested for oxy-coal combustion in the exploratory study performed by the coauthors [23].

The boundary conditions for the primary and secondary ports were set as inlet flow conditions. For the secondary port, a swirling flow component was given in order to model the $45°$ fins swirler. The walls were modeled assuming a constant heat transfer rate and a constant emissivity of 0.5.

TABLE 3: Outlet gas composition (mole fraction).

| Case | Volatiles | $O_2$ | $CO_2$ | $H_2O$ | CO | $SO_2$ | $N_2$ |
|------|-----------|-------|--------|--------|-----|--------|-------|
| OF25 | $4.29e-06$ | $3.59e-02$ | $8.58e-01$ | $1.04e-01$ | $2.80e-05$ | $3.98e-04$ | $1.17e-03$ |
| OF50 | $2.51e-04$ | $6.63e-02$ | $7.38e-01$ | $1.92e-01$ | $5.95e-04$ | $7.28e-04$ | $2.09e-03$ |
| OF75 | $6.93e-04$ | $1.05e-01$ | $6.30e-01$ | $2.60e-01$ | $1.09e-03$ | $1.00e-03$ | $2.21e-03$ |

TABLE 4: Reduced order model (ROM) case study NASA CEA inputs.

| | | Feed rate (mole/s) | | |
|---|---|---|---|---|
| | Species | 25% $O_2$ | 50% $O_2$ | 75% $O_2$ |
| Fuel | Carbon (graphite) | 0.1681717 | 0.1681717 | 0.1681717 |
| | $CH_4$ | 0.0110667 | 0.0110667 | 0.0110667 |
| | $C_6H_5OH$ (phenol) | 0.0037952 | 0.0037952 | 0.0037952 |
| | $C_2H_6$ | 0.003525 | 0.003525 | 0.003525 |
| | $H_2$ | 0.0 | 0.0 | 0.0 |
| | $H_2S$ | 0.0 | 0.0 | 0.0 |
| | $N_2$ | 0.0014144 | 0.0014144 | 0.0014144 |
| | $SO_2$ | 0.0004807 | 0.0004807 | 0.0004807 |
| Oxidant | $O_2$ | 0.2724017 | 0.2724017 | 0.2724017 |
| | $CO_2$ | 0.817942 | 0.2732708 | 0.0918949 |
| | CO | 0.003273 | 0.003273 | 0.003273 |
| | $H_2O$ | 0.0824674 | 0.0824674 | 0.0824674 |

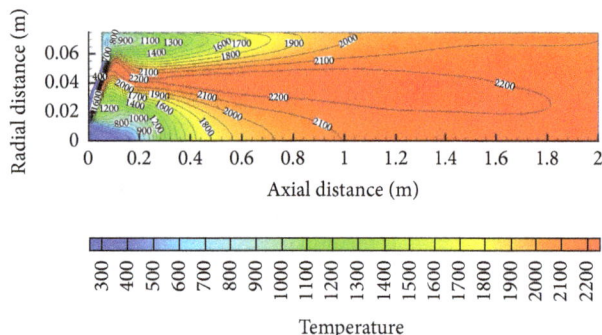

FIGURE 2: Predicted temperature contours for OF50 case.

TABLE 5: Comparison of CFD and ROM results for average outlet temperature.

| Case | Outlet temp K (CFD) | Outlet temp K (ROM) | Difference | % error |
|------|---------------------|---------------------|------------|---------|
| OF25 | 1459 | 1412 | 47 | 3.22 |
| OF50 | 2140 | 2074 | 66 | 3.08 |
| OF75 | 2497 | 2410 | 87 | 3.48 |

The outlet condition is set as pressure outlet with standard pressure and back flow temperature close to the average outlet temperature.

*3.1.2. Simulation Results.* Results for combustion of coal indicate that a complete coal devolatilization and char burnout percentages close to 99% are attained in all cases. Results for a typical temperature contour are shown in Figure 2. The predicted outlet gas composition in mole fractions is shown in Table 3. The predicted outlet temperatures from the CFD simulations are given in Table 5.

*3.2. ROM Predictions.* The three simulation cases described in the previous section were also considered with the reduced order model (ROM) approach. Following the process described previously, the products of coal devolatilization and the corresponding required amount of oxidant were broken down into molar flow rates for each of the various individual chemical compounds and used as input for the NASA CEA code. The resulting values for each of the three

simulation cases are provided in Table 4. As can be seen in the table, all of the values are constant across each case except for the molar flow rate of $CO_2$, which is used as a diluent for regulation of reactor temperature. For these ROM cases, the oxidant is assumed to be entering the reactor at room temperature, while the products of the coal devolatilization are considered to be entering at a temperature of 2000 K. The predicted outlet temperatures are given in Table 5. As expected, as the concentration of oxygen in the oxidant gas stream increases the average outlet temperature decreases due to the reduced amount of diluent $CO_2$.

The resulting outlet gas composition for each of the three cases, calculated through Gibb's free energy minimization by the NASA CEA code, is shown in Table 6. Once again, as expected, the mole fraction of $CO_2$ in the product gas decreases with increasing oxygen percentage within the initial oxidant gas injected into the reactor. These results will be discussed in greater detail in the following section.

*3.3. Comparison to CFD Simulations.* The temperatures obtained from the ROM were compared to the average outlet temperatures obtained from CFD simulations. The resulting comparisons are shown in Table 5. As can be seen from Table 5, the CFD and reduced order model ROM exhibit close agreement with respect to the estimated outlet gas

Table 6: ROM-predicted outlet gas composition (mole fractions).

| Gas species | OF25 | OF50 | OF75 |
|---|---|---|---|
| CO | 0.00001 | 0.00679 | 0.04083 |
| $CO_2$ | 0.85825 | 0.73049 | 0.58303 |
| H | 0.0 | 0.00005 | 0.00090 |
| $HO_2$ | 0.0 | 0.0 | 0.00001 |
| $H_2$ | 0.0 | 0.00036 | 0.00293 |
| $H_2O$ | 0.10550 | 0.19008 | 0.24832 |
| NO | 0.00001 | 0.00026 | 0.00077 |
| $N_2$ | 0.00117 | 0.00202 | 0.00252 |
| O | 0.0 | 0.00029 | 0.00289 |
| OH | 0.00003 | 0.00336 | 0.01566 |
| $O_2$ | 0.03463 | 0.06556 | 0.10116 |
| SO | 0.0 | 0.0 | 0.00001 |
| $SO_2$ | 0.00040 | 0.00073 | 0.00097 |

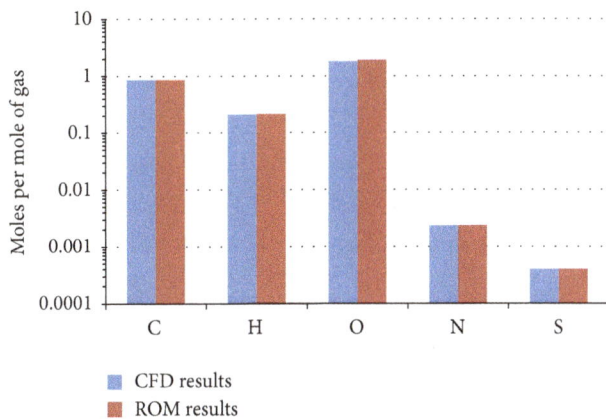

Figure 3: Elemental Balance for OF25.

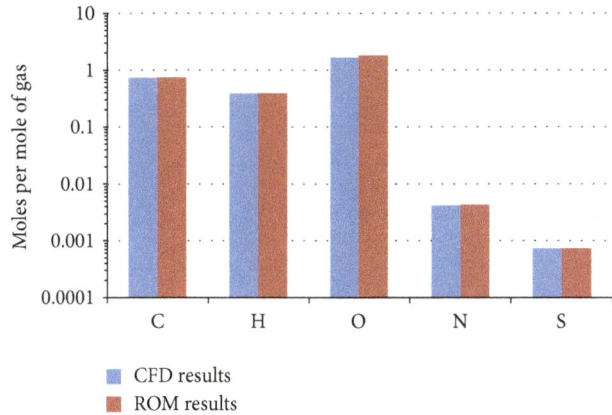

Figure 4: Elemental Balance for OF50.

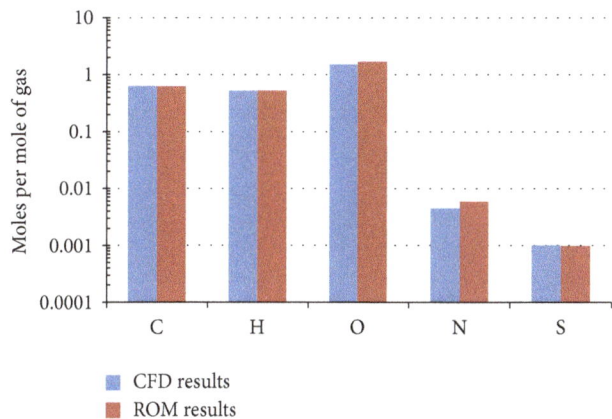

Figure 5: Elemental Balance for OF75.

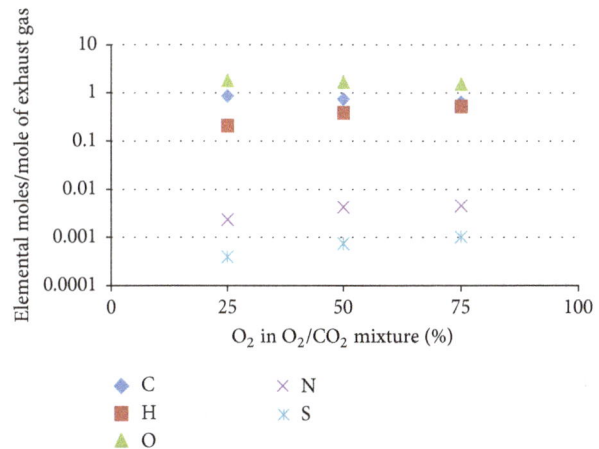

Figure 6: Effect of $O_2$ percentage in oxidant mixture on outlet gas composition (CFD simulations).

temperature. For each of the cases presented, the percent error between the ROM and CFD results falls within the range of 3.0 and 3.5 percent.

A comparison of the outlet gas composition predicted by the ROM (Table 6) and the Fluent CFD simulations (Table 3) shows fairly close agreement with respect to some of the chemical species of interest, while showing significant difference in others. For example, the resulting mole fractions of $CO_2$, $H_2O$, $N_2$, and $O_2$ for both models show good agreement, whereas the ROM predicts CO concentrations that are approximately an order of magnitude larger than that seen in the CFD results. The authors attribute this, at least in part, to the fact that the NASA chemical equilibrium code considers a far larger set of chemical species than what is considered in the Fluent CFD simulations. For instance, the NASA code also includes H, $HO_2$, NO, O, OH, and SO in its determination of chemical equilibrium, whereas Fluent lumps all of the "extraneous" chemical species into a single generic volatile compound of the form $C_{1.17}H_{2.90}O_{0.75}N_{0.0635}S_{0.0107}$.

Because of this difference between the two, it is important to check to make sure that the results from both simulations incorporate a similar amount of each element present (i.e., C, H, O, N, and S) in a single mole of exhaust gas. The results of this analysis are shown in Figures 3–5, and it can be seen

from these figures that there is good agreement between the two methods.

Finally, Figure 6 depicts the behavior of the composition of the predicted exhaust gas composition as a function of the percentage of oxygen in the oxidant mixture for the CFD cases. For the simulations discussed in this work, the flow rate

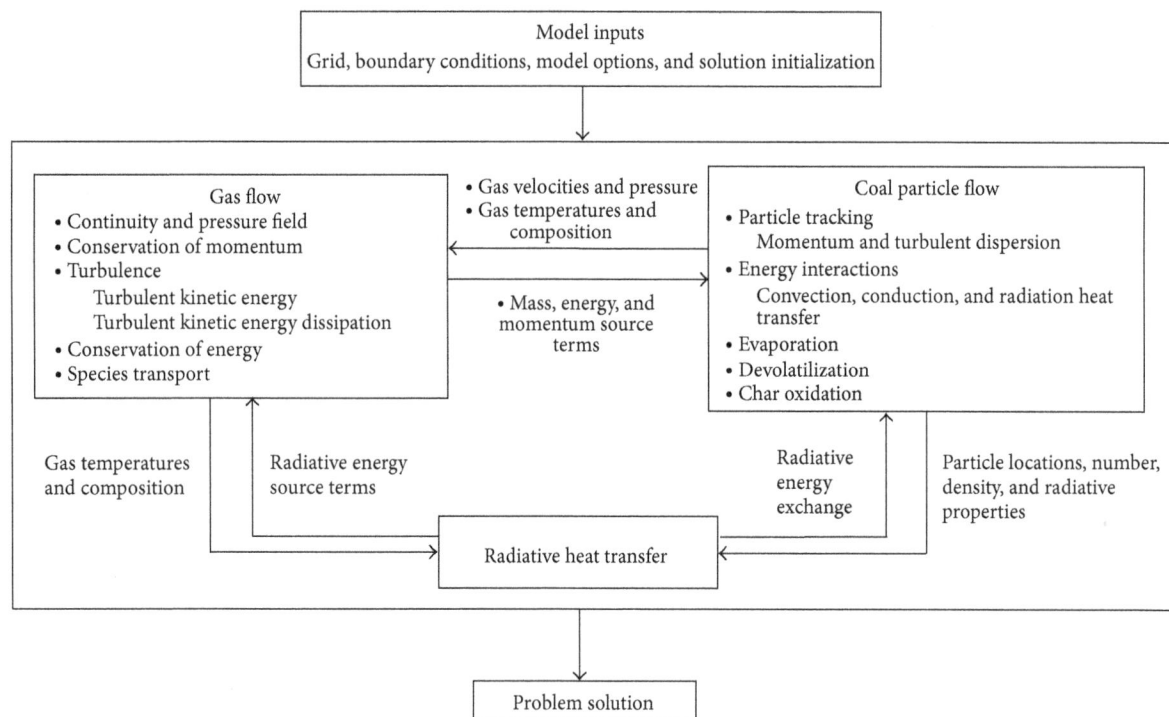

FIGURE 7: CFD model flowchart.

of oxygen was held constant, and the oxygen percentage was obtained by varying the amount of carbon dioxide added to the oxidant. Therefore, increasing percentages of oxygen in Figure 6 are the result of reducing the amount of carbon dioxide. As a consequence, Figure 6 shows a decrease in oxygen and carbon with increasing $O_2$ percentage, as expected. By comparison, there are slight increases in hydrogen, nitrogen, and sulfur with increasing $O_2$ percentage. This is also to be expected, due to the fact that higher $O_2$ percentages mean that there is less oxidant overall to act as a diluent. While not provided here, similar trends can be seen if one looks at the outlet gas composition of the ROM cases as well.

## 4. Conclusions

A reduced order model (ROM) for oxy-coal combustion has been presented. This model, which utilizes the Chemical Percolation Devolatilization (CPD) model developed by Fletcher et al. and the freely-available NASA Chemical Equilibrium with Applications (NASA CEA) code can be quickly used to help identify the outlet gas temperature and chemical composition for an oxy-coal combustion reactor without the requirement of exhaustive CFD simulations. The main advantage of this model is that it provides reactor designers and operators with the ability to quickly obtain preliminary estimates of the required flow rates of fuel and oxidant for a desired operating condition, or a quick estimate of the outlet gas temperature and composition for a given set of initial flow rates. The fact that it can do so without the need for time-intensive CFD simulations can lead to reductions

in computational cost and time requirements. However, it should be noted that CFD simulations will still be required for prediction of gas composition and temperature variations within the reactor vessel itself.

## Appendix

In response to reviewer comments, a brief discussion on CFD and ROM model complexities is included. Figure 7 shows the components integral to the CFD simulations. These simulations consist of gas and solid phase models, as well as a radiative heat transfer model. The flowchart in Figure 7 lists all of the submodels that make up each of the primary models, as well as the interactions between them. The models presented in this study were carried out utilizing the NIFTY computer cluster at West Virginia University, which consists of 96 GB RAM, 2TB Network Storage 6 Dual Quad-Core Xeon 2.66 GHz Processor Nodes, each node containing a total of 8 processors. A single simulation case running on a single one of these nodes took approximately 10 hours of computational time to complete. In comparison, the reduced order model (ROM) presented in this work can be carried out on a single personal computer, in this case a Dell Inspiron Laptop with an Intel i5 2.5 GHz processor with 8 GB RAM, in approximately 30 minutes. In addition, if the flow rates of fuel (i.e., coal) are held constant, and only the flow rates of the oxygen/carbon dioxide oxidant mixture are varied, different cases can be computed in approximately 10–15 minutes. In either case, this represents significant resources savings over full-blown CFD simulations. A flowchart depicting the complexity of the ROM procedure is shown in Figure 8.

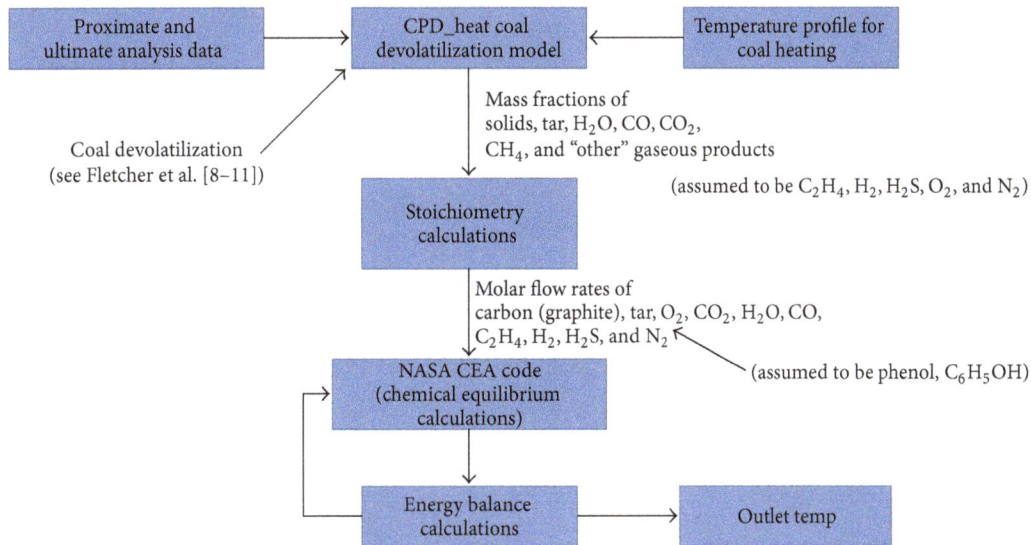

FIGURE 8: Reduced order model (ROM) computational flowchart.

## Conflict of Interests

The authors declare that there is no conflict of interests regarding the publication of this paper.

## Acknowledgment

This technical effort was performed in support of the US-China Clean Energy Research Center–Advanced Coal Technology Consortium's (CERC-ACTC) ongoing research in clean coal conversion processes under the DOE Contract DE-PI0000017.

## References

[1] T. Wall, Y. Liu, C. Spero et al., "An overview on oxyfuel coal combustion—state of the art research and technology development," *Chemical Engineering Research and Design*, vol. 87, no. 8, pp. 1003–1016, 2009.

[2] G. Scheffknecht, L. Al-Makhadmeh, U. Schnell, and J. Maier, "Oxy-fuel coal combustion—a review of the current state-of-the-art," *International Journal of Greenhouse Gas Control*, vol. 5, supplement 1, pp. S16–S35, 2011.

[3] A. H. Al-Abbas, J. Naser, and D. Dodds, "CFD modelling of air-fired and oxy-fuel combustion of lignite in a 100 KW furnace," *Fuel*, vol. 90, no. 5, pp. 1778–1795, 2011.

[4] M. Costa, P. Costen, and F. C. Lockwood, "Pulverized-coal and heavy-fuel-oil flames: large-scale experimental studies at imperial college, London," *Journal of the Institute of Energy*, vol. 64, no. 459, pp. 64–76, 1991.

[5] H. Nikzat, H. Pak, T. Fuse et al., "Characteristics of pulverized coal burner using a high-oxygen partial pressure," *Chemical Engineering Research and Design*, vol. 82, no. 1, pp. 99–104, 2004.

[6] E. H. Chui, A. J. Majeski, M. A. Douglas, Y. Tan, and K. V. Thambimuthu, "Numerical investigation of oxy-coal combustion to evaluate burner and combustor design concepts," *Energy*, vol. 29, no. 9-10, pp. 1285–1296, 2004.

[7] A. D. Gutierrez, S. L. Rowan, and I. B. Celik, "An integrated approach for the design of a pilot scale oxy-coal combustion reactor using CFD and chemical equilibrium software," in *Proceedings of the Fall Technical Meeting of the Eastern States Section of the Combustion Institute*, Clemson, SC, USA, October 2013.

[8] D. M. Grant, R. J. Pugmire, T. H. Fletcher, and A. R. Kerstein, "Chemical model of coal devolatilization using percolation lattice statistics," *Energy & Fuels*, vol. 3, no. 2, pp. 175–186, 1989.

[9] T. H. Fletcher, A. R. Kerstein, R. J. Pugmire, and D. M. Grant, "Chemical percolation model for devolatilization. 2. Temperature and heating rate effects on product yields," *Energy & Fuels*, vol. 4, no. 1, pp. 54–60, 1990.

[10] T. H. Fletcher, A. R. Kerstein, R. J. Pugmire, M. S. Solum, and D. M. Grant, "Chemical percolation model for devolatilization. 3. Direct use of 13C NMR data to predict effects of coal type," *Energy & Fuels*, vol. 6, no. 4, pp. 414–431, 1992.

[11] T. H. Fletcher and R. J. Pugmire, "Chemical Percolation Model for Coal Devolatilization," 2012, http://www.et.byu.edu/~tom/cpd/cpdcodes.html.

[12] M. Zehe, *Chemical Equilibrium with Applications*, NASA Glenn Research Center, 2010, http://www.grc.nasa.gov/WWW/CEAWeb/.

[13] H. K. Versteeg and W. Malalasekera, *An Introduction to Computational Fluid Dynamics: The Finite Volume Method*, Pearson Education Limited, London, UK, 2nd edition, 2007.

[14] S. V. Patankar, *Numerical Heat Transfer and Fluid Flow*, Hemisphere, Washington, DC, USA, 1980.

[15] Ansys Inc, *Ansys Fluent 12.0 Theory Guide*, Ansys Inc, 2009.

[16] B. E. Launder and D. B. Spalding, *Lectures in Mathematical Models of Turbulence*, Academic Press, London, UK, 1972.

[17] B. E. Launder and D. B. Spalding, "The numerical computation of turbulent flows," *Computer Methods in Applied Mechanics and Engineering*, vol. 3, no. 2, pp. 269–289, 1974.

[18] J. Siegel, R. Menguck, and M. Howell, *Thermal Radiation Heat Transfer*, CRC Press, Boca Raton, Fla, USA, 5th edition, 2011.

[19]  F. Hjertaer and B. H. Magnussen, "On the mathematical models of turbulent combustion with special emphasis on soot formation and combustion," in *Proceedings of the 16th International Symposium on Combustion*, Combustion Institute, Cambridge, Mass, USA, August 1976.

[20]  S. Badzioch and P. G. W. Hawksley, "Kinetics of thermal decomposition of pulverized coal particles," *Industrial & Engineering Chemistry: Process Design and Development*, vol. 9, no. 4, pp. 521–530, 1970.

[21]  M. M. Baum and P. J. Street, "Predicting the combustion behavior of coal particles," *Combustion Science & Technology*, vol. 3, no. 5, pp. 231–243, 1971.

[22]  M. A. Field, "Rate of combustion of size-graded fractions of char from a low-rank coal between 1200°K and 2000°K," *Combustion and Flame*, vol. 13, no. 3, pp. 237–252, 1969.

[23]  A. Gutierrez, A. Posada, and I. Celik, "CFD study of oxy coal combustion in a 100 kW down-fired furnace," in *Proceedings of the International Pittsburgh Coal Conference (PCC '12)*, Pittsburgh, Pa, USA, October 2012.

# 15

# Measurements of Gasification Characteristics of Coal and Char in $CO_2$-Rich Gas Flow by TG-DTA

**Zhigang Li,**[1] **Xiaoming Zhang,**[2] **Yuichi Sugai,**[1] **Jiren Wang,**[2] **and Kyuro Sasaki**[1]

[1] *Department of Earth Resources Engineering, Faculty of Engineering, Kyushu University, Fukuoka 819-0395, Japan*
[2] *College of Mining Engineering, Liaoning Technical University, Fuxin 123000, China*

Correspondence should be addressed to Zhigang Li; zhiganglee2009@hotmail.com

Academic Editor: Constantine D. Rakopoulos

Pyrolysis, combustion, and gasification properties of pulverized coal and char in $CO_2$-rich gas flow were investigated by using gravimetric-differential thermal analysis (TG-DTA) with changing $O_2$%, heating temperature gradient, and flow rate of $CO_2$-rich gases provided. Together with TG-DTA, flue gas generated from the heated coal, such as CO, $CO_2$, and hydrocarbons (HCs), was analyzed simultaneously on the heating process. The optimum $O_2$% in $CO_2$-rich gas for combustion and gasification of coal or char was discussed by analyzing flue gas with changing $O_2$ from 0 to 5%. The experimental results indicate that $O_2$% has an especially large effect on carbon oxidation at temperature less than 1100°C, and lower $O_2$ concentration promotes gasification reaction by producing CO gas over 1100°C in temperature. The TG-DTA results with gas analyses have presented basic reference data that show the effects of $O_2$ concentration and heating rate on coal physical and chemical behaviors for the expected technologies on coal gasification in $CO_2$-rich gas and oxygen combustion and underground coal gasification.

## 1. Introduction

As the increased fossil fuels consumption such as coal, oil, and gas leads to rapid deterioration of global environment, nowadays low-carbon economy is getting more and more attention. Low-carbon economy mostly linked greenhouse gases emissions and energy usage together [1, 2]. The economic growth of energy consumption countries impels intensive use of energy and other natural resources; thus, more residues and wastes discharged in the nature lead to environmental aggravation. China has been the second largest energy consumption country in the world, where the total energy consumption increased from 302 million tons of standard coal equivalent in 1960 to 2850 million tons in 2008 [3]. Coal as an energy source plays an important and indispensable role on future energy mix due to its proven stability in supply and its low cost. Coal has improved its long-term position as the world's most widely available fossil energy source with a very large resource base and economically recoverable reserves that are much greater than those of oil and gas. Coal is the most abundant fossil fuel in China.

Present recoverable reserves occupied about 11.67% of global coal reserves based on Key World Energy Statistics 2010 [4], ranked third in the world, with potential total reserves far in excess of this amount. Chinese coal consumption by the year 2020 will be nearly 4.8 billion tons per year with the bulk being consumed through the combustion processes. Therefore, present recoverable reserves are adequate to meet the national coal needs for many decades and potentially much longer. Moreover, most of coal consumptions are for electric power generations, with industrial consumptions of coal for steam and heat and for chemical and metallurgical processes being other major uses [5].

Carbon dioxide ($CO_2$) is regarded to be the main source of greenhouse gases emission that is a major threat of global warming and climate change [6]. According to the Intergovernmental Panel on Climate Change (IPCC), approximately 75% of the increase in atmospheric $CO_2$ is attributable to the consumption of fossil fuels [7, 8]. According to statistics of the IEA (2011) [9], $CO_2$ emission from fossil energy consumption in China was accounted for about 19% of global $CO_2$ emission, of which coal-fired power plants occupied

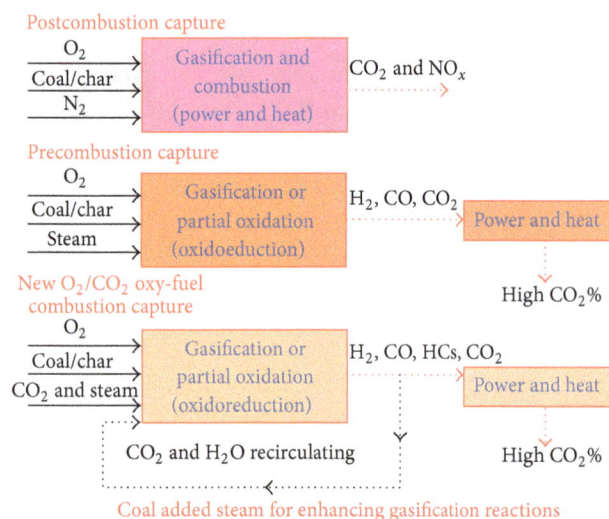

FIGURE 1: Schematic diagram of three kinds of power generation with $CO_2$ capture.

about 52.6% of total $CO_2$ emission in China. International Energy Agency (IEA) predicted that in 2030 China would emit twice as much carbon dioxide as that in 2007, provided that $CO_2$ emissions increase by 2.9% each year [10].

As stationary sources emitting large amounts of $CO_2$, pulverized coal fired power plants could be the best candidates to install $CO_2$ capture system which can be classified into three categories in general: precombustion capture, postcombustion capture, and oxy-fuel strategy as shown in Figure 1 [11, 12]. The traditional coal fired boilers use air for combustion in which $N_2$ gas is 79% in volume ratio. Its flue gas includes only about 15% $CO_2$; therefore, the $CO_2$ capture efficiency by post-combustion system is not high [13, 14]. Furthermore, $CO_2$ capture cost from the flue gas using amine scrubbing is expected to be relatively high [15]. In the case of pre-combustion capture, although calorific value of oxy-fired coal boiler is higher than that of air-fired coal boiler, there is a major disadvantage for oxygen-blown gasifier that is to build an oxygen plant. In general, an oxygen plant consumes about 5% of the gross power generated, which is the main reason why the total of plant investment for an oxygen-blown plant is somewhat higher than that of an air-blown plant [5].

As an alternative, a zero-emission power plant of pulverized coal-fired power generation in a nitrogen-free atmosphere, most known as oxy-fuel or $O_2/CO_2$ combustion technology for pre-combustion capture, is one of new promising methods to approach the problem of $CO_2$ separation and capture. In this technology, $CO_2$ gas substitutes the role of $O_2$ gas to improve and stimulate coal conversion and reduce $O_2$ consumption. Recently, coal gasification with $CO_2$ and oxygen combustion technology has been investigated for next coal fired power [16, 17]. In addition, this type of pulverized coal fired power plant is mainly composed of gasifier and combustor as shown in Figure 1. Moreover, gasification process of pulverized coal in the gasifier is the core part of the technology because it determines synthesis gas product and thermal efficiency. This process is also the focus of the research in this paper; moreover, it has been verified that the processes of coal in $CO_2$-rich gas atmosphere mainly are divided into two temperature ranges for coal devolatilization, char formation, and gasification.

The implementation of these improved combustion technologies for replacing $N_2$ with $CO_2$ in feeding gas requires further understanding of physical and chemical characteristics in the process of oxidation combustion and gasification of coal with gradually increasing temperature in $CO_2$-rich atmosphere. In particular, reaction characteristics of coal gasification in a $CO_2$-rich atmosphere are required for coal seam underground coal gasification (UCG) projects. Li et al. presented the comparisons in TGA experiments with bituminous coal at high temperature of 1000°C with heating rate of 10 to 30°C·$min^{-1}$ in the mixture of $O_2/N_2$ or $O_2/CO_2$ with various oxygen concentrations (21, 30, 40, and 80%) [18], and Liu presented the properties of coal chars prepared from UK high-volatile bituminous and anthracite coals by using TGA with heating rate of 2.5 to 12.5°C·$min^{-1}$ in mixtures of $O_2/CO_2$ and $O_2/N_2$ with $O_2$ concentrations of 3, 6, 10, 21, and 30% [19]. However, researchers did not measure flue gas and heat generation by coal combustion and gasification. The unburned carbon content in $CO_2$ rich atmosphere is expected to be higher than those in air environment due to $O_2$ concentration.

Authors (Li et al., 2012) [20] have presented the combustion and gasification properties of Datong coal and char in $CO_2$-rich gas flow (5 or 10% $O_2$) by rapid heating with temperature gradient of 50 to 200°C·$s^{-1}$ using a $CO_2$ laser beam. In the experiments, the coal conversion ratio to gases was measured for different coal temperature time gradient with monitoring of CO and HC gases generated from heated coal particles. Based on experimental results by the rapid heating of dry, moist coal, and mixing coal-water samples, it has been clarified that coal moisture (internal water) and external water of coal particles have the same function to increase HC-gas production and decrease CO-gas amount by promoting chemical reactions between carbon or CO and $H_2O$. Consequently, a possibility has been shown to accomplish coal gasification in $CO_2$-rich atmosphere including enough water vapor to carry out low-cost $CO_2$ capture.

Most researchers presented the results in TGA experiments with coal and char at high temperature; however, the experimental results were restricted to HCs and CO gasified gases analysis and heat generation after coal gasification. Investigation of gasification and combustion reactivity of coal in $CO_2$-rich atmosphere at high temperature, HCs, CO, and so forth gasified gases generations is essential for the development of gasification technology in the future. In contrast with gasification furnace in commercial process, pyrolysis, combustion, and gasification properties of pulverized bituminous coal were investigated at high temperature of 1400°C by TG-DTA measurements. On the other hand, temperature gradient was set up from 20 to 40°C·$min^{-1}$ in order to discuss gasification and combustion ratio of coal conversion. In addition, Lu et al. and Xie et al. presented that the critical $O_2$ concentration of oxidation combustion at low temperature is around 5–10% [21, 22]; furthermore,

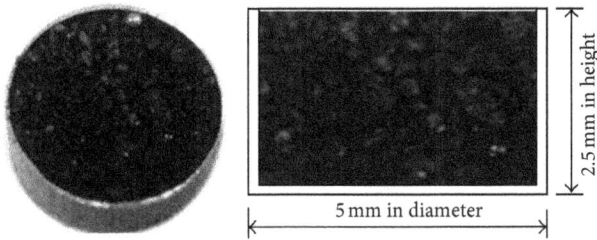

FIGURE 2: Coal sample in platinum container ($\approx$30 mg coal).

TABLE 1: Analysis values of coal (air dried basis).

| Proximate analysis | Weight (%) |
| --- | --- |
| Ash | 12.70 |
| Moisture | 2.42 |
| Fixed carbon | 54.49 |
| Volatile matter | 30.39 |

high temperature combustion with low oxygen concentration ($\leq$5%) is regard as a new generation of high temperature air combustion technology [23]. In other words, even if $O_2$ is controlled as very low concentration ($\leq$5%) in the flow provided to coal sample, accumulated $O_2$ gas amount is mostly enough to complete oxidation combustion during the heating process. Consequently, $CO_2$-rich atmosphere in the experiment was controlled by changing $O_2$ concentration from 0 to 5%. After that, the flue gas generated from the heated coal, such as CO, $CO_2$, $O_2$, and HCs, were analyzed in the combustion and gasification process.

The weight reduction ratio after the measurements, $x$ (%), of coal samples was measured against $O_2$% with increasing temperature in the atmospheric pressure. In addition, same experiments and flue gas analyses were conducted for pulverized char samples in the $CO_2$-rich atmosphere to compare with the measurement results of coal samples.

## 2. Analytical Approach and Experimental Conditions

### 2.1. Coal and Char Samples.
Coal samples, used for the exper-iments, were taken from the 8103 face of Tashan colliery in Shanxi province, China. The properties of which were summarized in Table 1. The samples were crushed into particles 0.25 to 0.5 mm in diameter and dried in a vacuum desiccator. The volume of crushed coal particles was less than that of the platinum container placed in TG-DTA which is almost equal to 49.1 mm$^3$ as shown in Figure 2.

Sample weight placed was about 30 mg, and its porosity was evaluated as 37%. In order to compare coal and char, char samples were made from the same coal samples by heating for 7 minutes in a sealing volatile matter crucible at the temperature of (900 $\pm$ 10)$°$C based on ISO 562:1998 for hard coal and coke determination of the volatile matter. In the same manner, the char particles placed in the container were adjusted into the same diameter range of coal.

### 2.2. Reaction Mechanism.
As reported by Luo and Zhou [24] and Huang et al. [25], the processes of coal combustion and gasification are expressed by the following reactions.

(a) Combustion Reaction:

$$C + O_2 \longrightarrow CO_2 \qquad (1)$$

$$C + \frac{1}{2} O_2 \longrightarrow CO \qquad (2)$$

(b) Gasification Reaction:

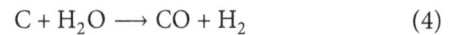

$$C + CO_2 \longrightarrow 2CO \qquad (3)$$

$$C + H_2O \longrightarrow CO + H_2 \qquad (4)$$

$$C + 2H_2 \longrightarrow CH_4 \qquad (5)$$

Reaction equations (1) and (2) are exothermic processes, and reaction equations (3) to (5) are endothermic processes. Carbon in char matrix reacts with oxygen to form CO and $CO_2$. However, it still has not been unified which of them is the favoured product. In general, the proportion of CO/$CO_2$ in products increases gradually with increasing temperature. Thus, CO is the main product when the reaction temperature is over around 1030$°$C and other parameters are constant [26].

### 2.3. Experimental Apparatus and Procedure.
The thermal analysis system from 20 to 1400$°$C in temperature (TG-DTA) used for the present experiments is shown in Figure 3. The thermogravimetric (TG) analyzes sample mass under changing temperature or elapsed time at a temperature program. Differential thermal analysis (DTA) measures the temperature difference between the analyzed sample and a reference material (a substance with no thermal effect in the measured temperature range, such as $Al_2O_3$, as shown in Figure 3(b)) at a sample temperature. TG curves of samples reflect the relationship between changes in the sample mass, temperature, and ambient gas. Injected gas species, gas flow rate, and heating rate are shown in Table 2. The flue gas generated from the heated coal, such as CO, $CO_2$, $O_2$, and HCs, were analyzed by an emission gas analytical system and a gas chromatography system in the combustion and gasification process.

### 2.4. Definition of Weight Loss Ratio on TG Curves.
Thermogravimetric analysis is an established method to study coal oxidation reactions. Based on the fundamental principle of chemical dynamics, we set the reaction model as $f(x)$, in which $x$ is the reaction conversion rate of coal weight loss:

$$x = \frac{m - m_0}{m_0}, \qquad (6)$$

where $m_0$ is the initial sample mass in mg and $m$ is the sample mass in mg at elapsed time $t$ in min.

The TG output showing the ratio of coal sample to reference material ($Al_2O_3$) needs to be adjusted before commencement of TG-DTA experiments. The empty weight

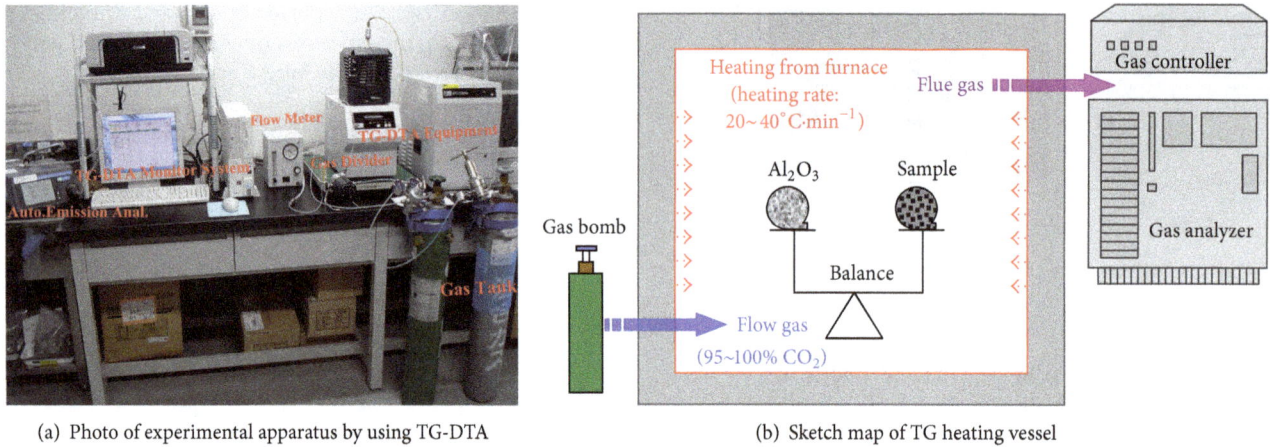

(a) Photo of experimental apparatus by using TG-DTA

(b) Sketch map of TG heating vessel

FIGURE 3: Schematic diagram of experimental apparatus.

TABLE 2: Experimental conditions of temperatures and gas flow rates for the TG-DTA analysis.

| Sample | Gas injected to system | Flow rates (mL min$^{-1}$) | Heating rates $\lambda$ (°C min$^{-1}$) |
|---|---|---|---|
| Coal | Air | 200 | 20 |
| | 95% $CO_2$ + 5% $O_2$ | 100~200 | 20~40 |
| | 96% $CO_2$ + 4% $O_2$ | 100 | 20 |
| | 97% $CO_2$ + 3% $O_2$ | 100 | 20 |
| | 98% $CO_2$ + 2% $O_2$ | 100 | 20 |
| | 99% $CO_2$ + 1% $O_2$ | 100 | 20 |
| | 100% $CO_2$ | 100~200 | 20~40 |
| Char | Air | 100 | 20 |
| | 95% $CO_2$ + 5% $O_2$ | 100 | 20 |
| | 96% $CO_2$ + 4% $O_2$ | 100 | 20 |
| | 97% $CO_2$ + 3% $O_2$ | 100 | 20 |
| | 98% $CO_2$ + 2% $O_2$ | 100 | 20 |
| | 99% $CO_2$ + 1% $O_2$ | 100~200 | 20~40 |
| | 100% $CO_2$ | 100 | 20 |

of a platinum cup for the coal sample was calibrated manually to 0 when the output was stable.

Temperature ($\Theta$) was set with a linear gradient against time $t$ in the measurements:

$$\Theta = \Theta_0 + \lambda t, \qquad (7)$$

where $\Theta$ is the cell temperature in °C, $\Theta_0$ is the initial temperature in °C, and $\lambda$ is the temperature gradient in °C·min$^{-1}$.

*2.5. Conversion Factor for Heat Generation from DTA Output.* At low temperatures less than 200°C, water evaporated from coal sample, and sample mass $m$ reduced from the initial mass $m_0$ ($m \leq m_0$, i.e., $x \leq 0$) as shown in Figure 4. According to the DTA principle, the DTA voltage output, $Q^*$ ($\mu V$), is proportional to heat generation rate per unit mass, $q$ (J·min$^{-1}$·mg$^{-1}$), as the following:

$$q = \beta \frac{Q^*}{m_0}, \qquad (8)$$

where $\beta$ is a conversion factor from $\mu V$ to J·min$^{-1}$. Heat of the Datong coal combustion was previously measured as $H = 30300$ kJ·kg$^{-1}$ = 30.3 J·mg$^{-1}$. Since the DTA curve reached a constant value after 40 min heating, heat of coal combustion was expressed by integrated DTA output from 0 to 40 min using the following:

$$H = \int_0^\infty q(t)\, dt = \frac{\beta}{m_0} \int_0^{40} Q^*(t)\, dt = \frac{\beta}{m_0} \sum_{i=0}^{40} Q_i^* \cdot \Delta t, \quad (9)$$

where $\Delta t$ (=1 min in present experiments) is interval time of the DTA output. The relationship between cumulative heat from time 0 to $t$, and $t$ is shown in Figure 4 with TG curve ($x$-$t$). The conversion factor was calculated as $\beta = 0.17$ J·min$^{-1}$·$\mu V^{-1}$ from the value of $H$ at 40 min.

## 3. TG-DTA Analyses of Coal Combustion and Gasification

*3.1. Effects of $O_2$ Concentration and Gas Flow Rate on Coal Reaction.* In the experiments, the termination temperature was set to 1400°C with a temperature gradient of $\lambda = 20$°C·min$^{-1}$ in order to reduce the unburned carbon content. The injected gas species, gas flow rate, and heating rate are shown in Table 2.

According to the coal TG curves by injecting air (see Figure 5), the processes of pyrolysis, combustion, and gasification of coal in flow air can be divided into three temperature stages (temperature value is coal body temperature):

(1) 25 to 108°C: calefactive-evaporated-alleviative process;

(2) 108 to 276°C: calefactive-adsorption-weight incremental process ($O_2$ absorption);

Measurements of Gasification Characteristics of Coal and Char in CO2-Rich Gas Flow by TG-DTA

189

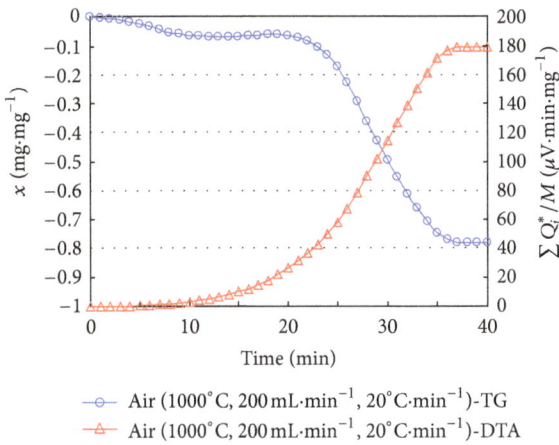

FIGURE 4: A typical TG-DTA curve of coal combustion in air (Air flow: 200 mL·min$^{-1}$; max. temperature: 1000°C; heating rate: 20°C·min$^{-1}$).

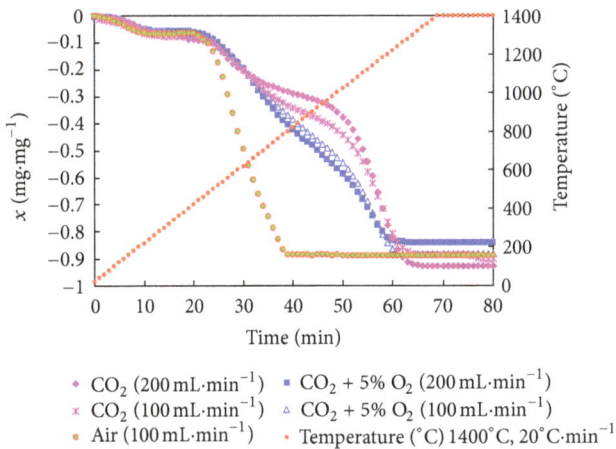

FIGURE 5: TG curves of different gas flow rates and O$_2$ concentrations in flow gas.

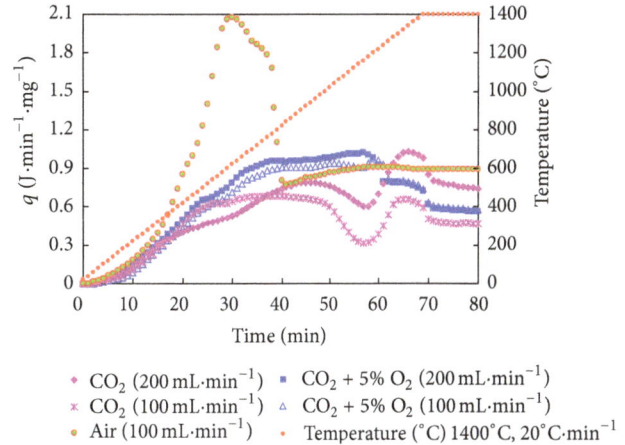

FIGURE 6: DTA curves of different gas flow rates and O$_2$ concentrations in flow gas.

(3) Over 276°C: calefactive-oxidation and combustion-alleviative process [27].

However, the processes of coal in the CO$_2$-rich atmosphere can be divided into four temperature stages:

(1) 25 to 108°C: calefactive-evaporated-alleviative process;

(2) 108 to 276°C: calefactive-adsorption-weight incremental process (CO$_2$ absorption);

(3) 276 to 650°C: calefactive-devolatilization-alleviative process (refer to volatile matter);

(4) Over 650°C: calefactive-char formation and gasification-alleviative process.

TG results in the CO$_2$-rich atmosphere with different gas flow rates and O$_2$ concentrations are shown in Figure 5. When atmospheric temperature is higher than 360°C (at >360°C, the impact of volatile loss on mass is negligible), coal mass reduces with increasing coal body temperature with a

linear line, especially in air, coal conversion was completed when atmospheric temperature reached 800°C. In the CO$_2$-rich atmosphere, the coal burning rate for 95% CO$_2$ + 5% O$_2$ gas mixture is faster, and its conversion time is shorter than that of injected 100% CO$_2$ gas. On the other hand, for the case of 95% CO$_2$ + 5% O$_2$, the coal burning rate increases by increasing gas flow rate from 100 mL·min$^{-1}$ to 200 mL·min$^{-1}$, because provided O$_2$ amount increases in unit of time and its reaction time decreases. These phenomenons suggest that O$_2$ amount is the main working factor for coal conversion rate under the same condition. However, for the case of 100% CO$_2$, the coal burning rate decreases by increasing gas flow rate from 100 mL·min$^{-1}$ to 200 mL·min$^{-1}$ when atmospheric temperature is lower than 1100°C. It can be assumed that increasing CO$_2$ gas flow rate (amount in unit of time) makes the flame propagation speed and the flame stability decline. However, when atmospheric temperature reached 1100°C, the effects of gas flow rate on coal conversion disappeared, because CO$_2$ gas participated in coal gasification reactions. The phenomenon suggests that CO$_2$ gas substitutes the role of O$_2$ gas to improve and stimulate coal conversion in the higher temperature range from 1100 to 1400°C.

As shown in Figure 6, the heat generation rate of coal is the highest by providing air flow. This is due to coal oxidation and combustion being an exothermic process; on the contrary, the reaction between coal and CO$_2$ is an endothermic process (refer to Section 2.2). Moreover, flame stability and coal temperature in CO$_2$ gas-rich flow are lower than those in air flow environment. Additionally, when vessel temperature is lower than 1300°C, the heat generation rate of coal in flow gas of 95% CO$_2$ + 5% O$_2$ is larger than that of 100% CO$_2$. However, the one of a larger flow rate (200 mL·min$^{-1}$) of 100% CO$_2$ gas got the highest heat generation after the temperature reached 1300°C (see Figure 6). In other words, even if 100% CO$_2$ gas was provided, the heat was generated by complex gasification reactions between coal and CO$_2$ gas including a small amount of H$_2$O in the high-temperature range. In addition, a dip and a peak come out on the DTA curve of the 100% CO$_2$ gas. The minimum point of the dip appears

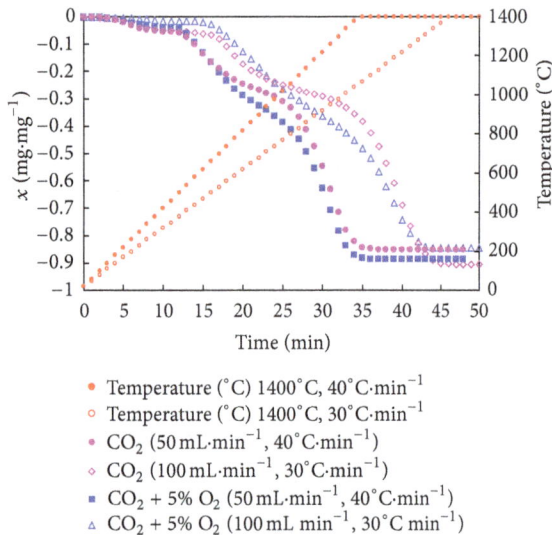

FIGURE 7: TG curves of coal under different temperature gradients.

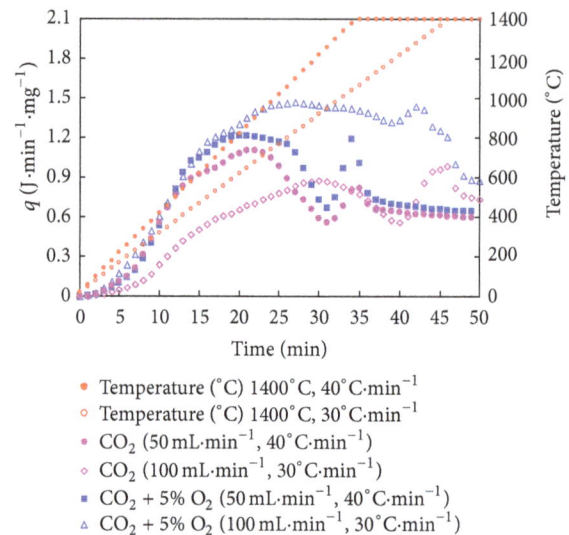

FIGURE 8: DTA curves of coal under different temperature gradients.

at around 1160°C, and the maximum point of the peak comes forth at around 1300°C. The trough may reflect the known phenomenon that gasification absorbs heat and generates a differential thermal drop. In other words, char residues generated from coal are further converted by the gasification reaction over 1100°C. It can be assumed that coal gasification with $CO_2$ gas mainly occurs in higher temperature range from 1100 to 1400°C. In addition, present results show that the chemical reaction of coal in the 100% $CO_2$ differs from air or gas flow containing $O_2$ over 5%. Those results suggest that the flame propagation speed, the flame stability, and the reaction between unburned carbon and gas have been improved in the high-temperature range.

*3.2. Effect of Temperature Gradient on Coal Reaction.* In the measurements, the terminal (or maximum) temperature was set to 1400°C with the temperature gradient of $\lambda = $ 30°C·min$^{-1}$ or 40°C·min$^{-1}$ and gas flow rate of 100 mL·min$^{-1}$ or 50 mL·min$^{-1}$, and TG-DTA measurement results in the $CO_2$-rich atmosphere with different gas flow rates are shown in Figures 7 and 8. Comparing the results of coal weight reduction with the previous results shown in Figure 5, the coal weight reduction ratio is not sensitive to the temperature gradient, because it is mainly affected by $O_2$ concentration and the terminal temperature. However, coal conversion time shortens, and differential thermal peak takes place in advance, and heat generation values increases with increasing temperature gradient from 20°C·min$^{-1}$ to 40°C·min$^{-1}$. Especially for the case of 95% $CO_2$ + 5% $O_2$, as shown in Figures 6 and 8, the troughs of the DTA curves with temperature gradients of 20°C·min$^{-1}$ and 30°C·min$^{-1}$ are unobvious in higher temperature range; however, the trough and the peak of 40°C·min$^{-1}$ are very evident. The phenomenon suggests that even if 95% $CO_2$ + 5% $O_2$ gas was provided, the intensity of gasification reaction instead

of oxidation combustion was enhanced by increasing temperature gradient from 40°C·min$^{-1}$.

In addition, heat generating rates of coal by injecting 95% $CO_2$ + 5% $O_2$ ($q = 0.97$ J·min$^{-1}$·mg$^{-1}$) and 100% $CO_2$ ($q = 0.66$ J·min$^{-1}$·mg$^{-1}$) are increased around 50% with increasing temperature gradient from 20°C·min$^{-1}$ to 30°C·min$^{-1}$. Furthermore, the heat generation for 95% $CO_2$ + 5% $O_2$ with temperature gradient of 30°C·min$^{-1}$ is higher than those of others. The value of coal heat generation decreases with increasing temperature gradient from 30°C·min$^{-1}$ to 40°C·min$^{-1}$ due to endothermic gasification process. In the high-temperature range, the coal conversion is mainly implemented by coal gasification instead of coal combustion, and the temperature gradient is an essential parameter for improving and stimulating coal gasification reactions.

*3.3. Effects of $O_2$ Concentration on Residual and Differential Thermal.* Based on TG-DTA results described, the differences of the effects of 100% $CO_2$ and 95% $CO_2$ + 5% $O_2$ gas flow on weight reduction ratio, $x$, and differential thermal of coal and reaction products are relatively prominent. Therefore, to further investigate temperature range of coal gasification and the effect of $O_2$ concentration on coal gasification, the TG-DTA measurements of the coal were carried out by setting different termination temperatures of 1000°C, 1200°C, and 1400°C with injected gases of 0 (100% $CO_2$) to 5% in $O_2$ concentration, $\lambda = 20$°C·min$^{-1}$, and gas flow rate of 100 mL·min$^{-1}$. The detailed contents are shown in Table 2.

The TG-DTA results indicate that $O_2$%, 0 to 5%, contained in $CO_2$-rich gas flow has relatively strong effect on coal conversion, heat generation, and reaction products for the different termination temperatures as shown in Figures 9, 10, and 11. Coal weight reduction ratio increases with increasing $O_2$ concentration in the $CO_2$-rich atmosphere under the same conditions; moreover, heat generation reduces with increasing $CO_2$ concentration due to intensifying

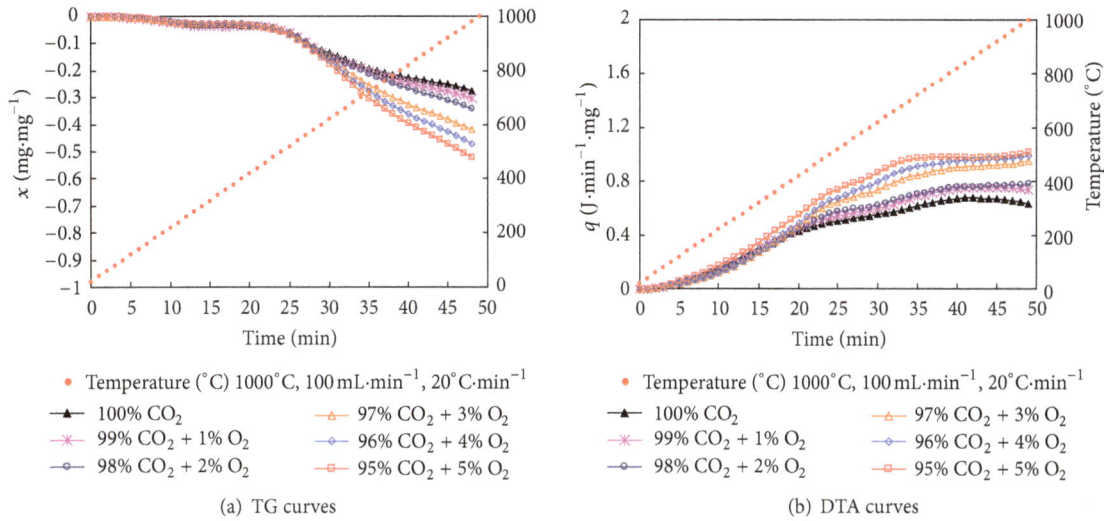

FIGURE 9: TG-DTA curves of different $O_2$ concentrations with a temperature of 1000°C.

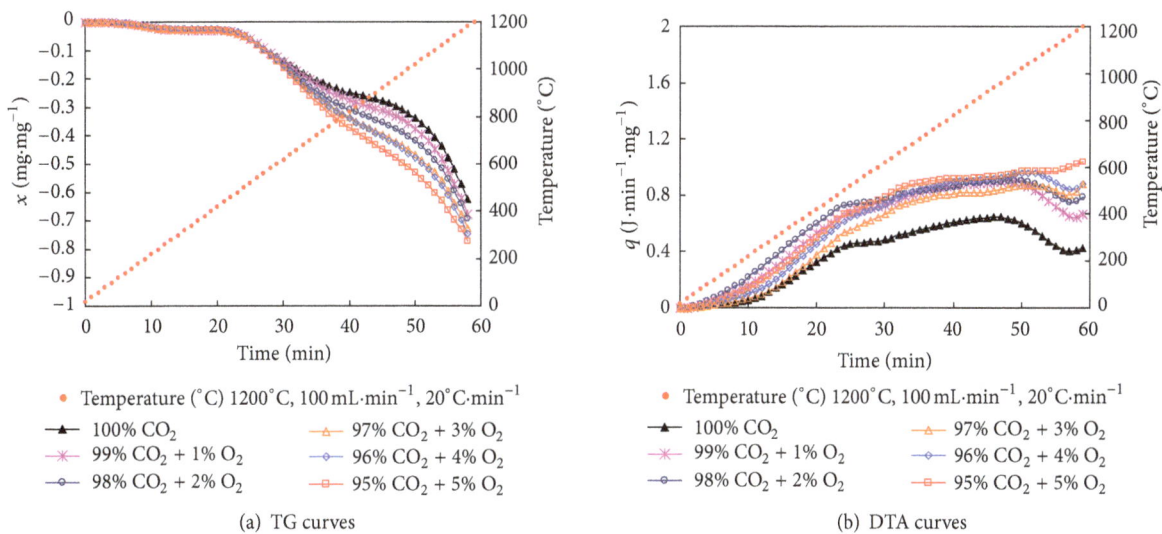

FIGURE 10: TG-DTA curves of different $O_2$ concentrations with a temperature of 1200°C.

endothermic gasification reaction between coal and $CO_2$ gas. Especially, it is clear from the DTA curves that the gasification reaction of coal with $CO_2$ gas mainly occurs in temperature range from 1100 to 1300°C. In addition, the DTA curves in Figures 10(b) and 11(b) indicate that the greater the atmospheric $CO_2$%, the larger trough radian of curves in higher temperature range, that is, intension of gasification reaction. The phenomenon suggests that coal conversion to gases will no longer depend on $O_2$% in the high-temperature range due to gasification reactions.

Residual or ash that remained in the container was analyzed by an Energy Dispersive Spectrum (EDS) analyzer after TG-DTA experiments. The photos and EDS images of residuals or ashes for different $CO_2$ concentrations are shown in Figures 12, 13, and 14. The results of carbon molecular ratios, C% were investigated by EDS analyzer as shown in Figure 15.

The analytic results indicated that carbon molecular ratios at 1000°C before gasification and 1400°C after gasification do not show strong dependency on $O_2$%. On the other hand, the trend of C% against temperature shows that the greater was the atmospheric $O_2$%, the less residual value of C% was measured in the gasification stage with the termination temperature of 1200°C. In particular, carbon molecular ratios of 4% and 5% $O_2$ contained in $CO_2$-rich gas flow are essentially coincident with those of the termination temperature of 1400°C; moreover, carbon molecular ratio is nearly constant with gas flow containing $O_2$ over 4%; in other words, 4% $O_2$ contained in $O_2/CO_2$ gas flow reaches to the saturation ratio of coal combustion and gasification reaction in high-temperature range. It can be verified from the EDS images of residuals or ashes at terminal temperature of 1200°C that coal particle surfaces generated many pores in the 100% $CO_2$ gas flow due to coal gasification with $CO_2$

(a)  TG curves

(b)  DTA curves

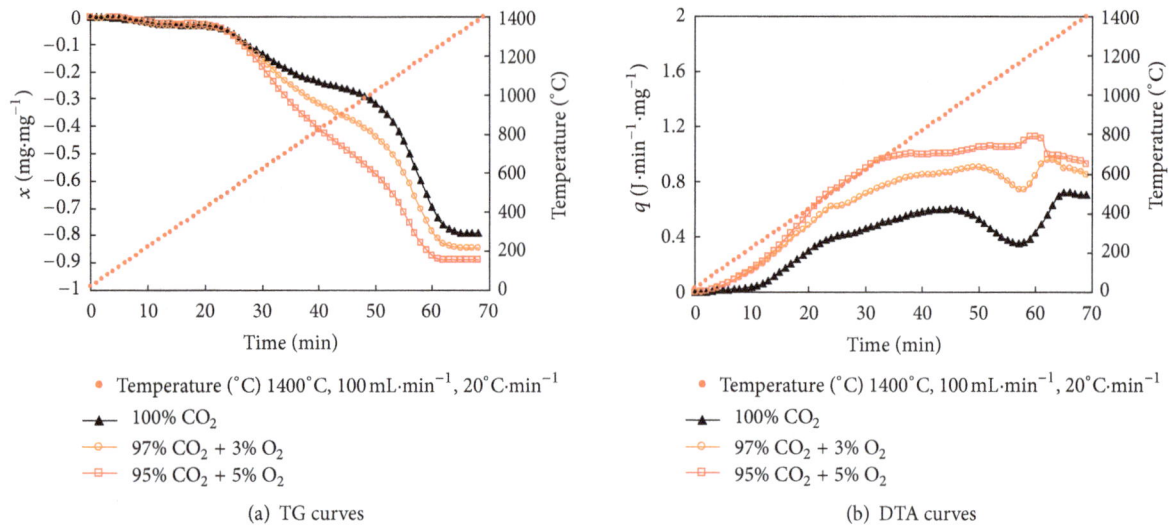

FIGURE 11: TG-DTA curves of different $O_2$ concentrations with a temperature of 1400°C.

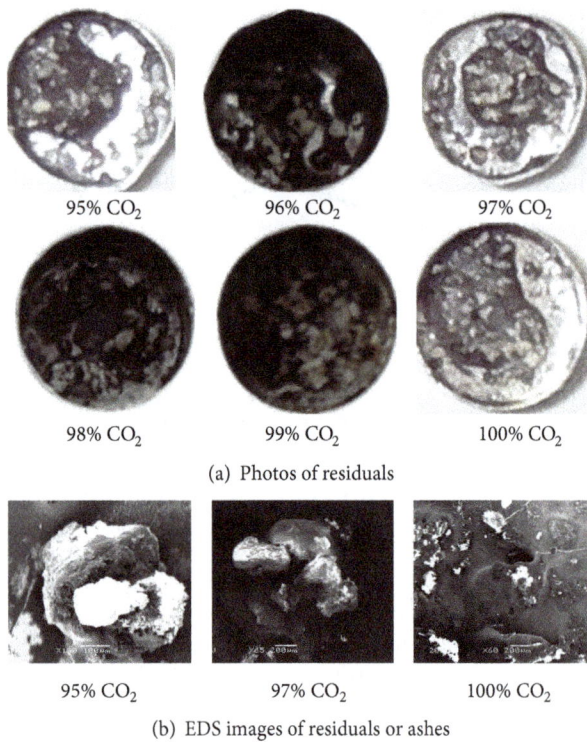

(a)  Photos of residuals

(b)  EDS images of residuals or ashes

FIGURE 12: Photos of residuals at various $CO_2$ concentrations with a temperature of 1400°C.

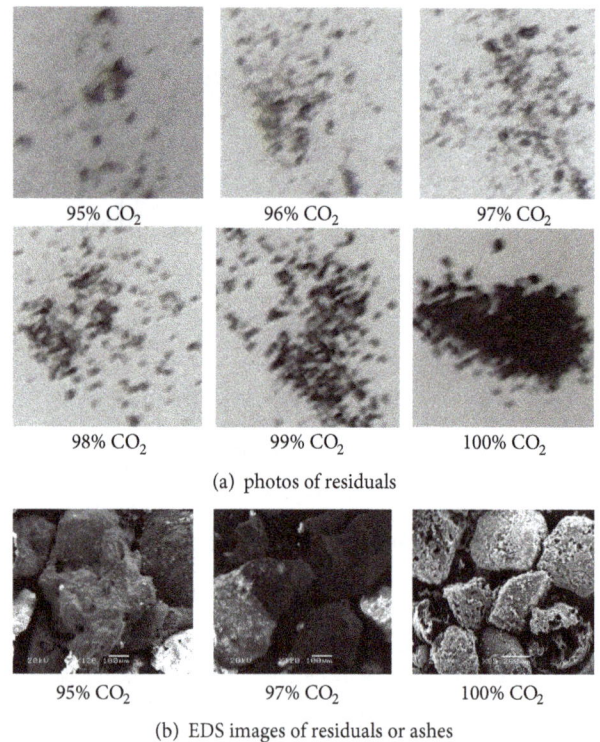

(a)  photos of residuals

(b)  EDS images of residuals or ashes

FIGURE 13: Photos of residuals at various $CO_2$ concentrations with a temperature of 1200°C.

gas. On the other hand, the difference of carbon molecular ratio with the terminal temperature of 1200°C is large from 0 to 3% $O_2$. It can be assumed that coal combustion to gases is not sufficient with reaction (2) instead of (1) after the temperature reached 1100°C; furthermore, $O_2$ amount in unit of time is insufficient for coal conversion within limited heating time. On the other hand, the heat provided by the terminal temperature of 1200°C is not enough to complete

coal gasification with $CO_2$ gas. Therefore, $O_2$ concentration is the key factor for coal conversion to gases with the terminal temperature of 1200°C. However, for the case of terminal temperature of 1400°C, coal conversion to gases is completed; it is clear from the photos that residuals form molten state as shown in Figure 12(a), because the melting point of coal ash is around 1250°C. In addition, as shown in Figure 11, the value of TG-DTA curve with 95% $CO_2$ + 5% $O_2$ gas flow was constant

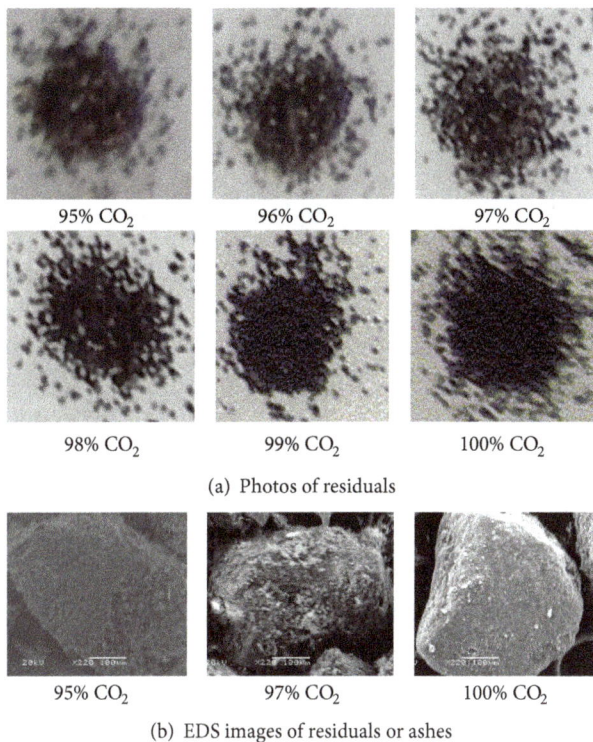

95% $CO_2$            96% $CO_2$            97% $CO_2$

98% $CO_2$            99% $CO_2$            100% $CO_2$

(a) Photos of residuals

95% $CO_2$            97% $CO_2$            100% $CO_2$

(b) EDS images of residuals or ashes

FIGURE 14: Photos of residuals at various $CO_2$ concentrations with a temperature of 1000°C.

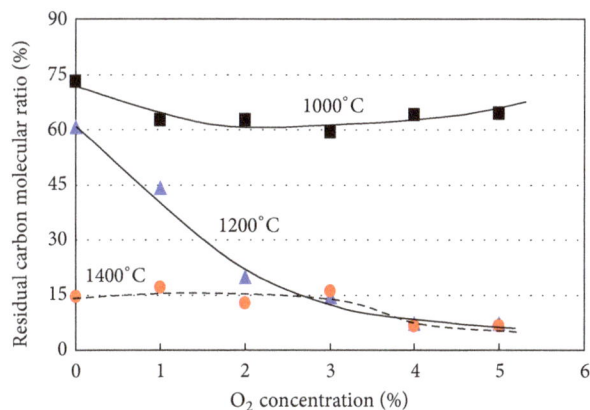

FIGURE 15: EDS analysis of residual carbon molecular ratios against $O_2$% in $CO_2$-rich gas flow.

after the temperature reached 1200°C; however, the ones of 0 and 3% $O_2$ contained in $CO_2$-rich gas flow came to constant when the temperature reached 1300°C. Consequently, coal conversion to gases may break away from $O_2$ gas and promote $CO_2$ reduction reactions when atmospheric temperature is over 1300°C.

### 3.4. Coal Weight Loss Rate versus Temperature.
As shown in Figure 16(a), the relationship between coal weight loss rate

(equal to conversion rate of coal to gasses) and temperature can be expressed by the following Arrhenius formula:

$$\frac{\Delta V_m}{\Delta t} = A_0 \cdot \exp\left(-\frac{E}{RT}\right),$$

$$\Delta V_m = \frac{m_0 - m}{m_0},$$

(10)

where, $\Delta V_m/\Delta t$ is the average rate of coal weight loss at unit time in $s^{-1}$, $m_0$ is the initial coal mass in mg, $m$ is the coal mass at elapsed time $\Delta t$, in mg, $R(=8.314\,\text{J·K}^{-1})$ is the gas constant, $T(K)$ is the absolute temperature, $A_0\,(s^{-1})$ is the pre-exponential factor, and $E$ (kJ·mol$^{-1}$) is the activation energy.

The measurement results indicate that pre-exponential factor is almost constant; however, activation energy is mainly dependency of $O_2$ concentration as shown in Figure 16(b); moreover, it decreased gradually with increasing $O_2$ concentration. Consequently, the Arrhenius equation can be expressed as the following:

$$\frac{\Delta V_m}{\Delta t} = 0.00045 \cdot \exp\left(-\frac{-0.96Y_{O_2,S} + 135.93}{RT}\right),$$

(11)

where $Y_{O_2,S}$ is the $O_2$ concentration at surface of coal particle, in mole fraction.

### 3.5. Coal Conversion and Heat Generation Rates.
Pyrolysis, combustion, and gasification of coal can be clarified from three peaks generated by analyzing time differential values of coal mass denoted by

$$\frac{dx}{dt} = \frac{d\left[(m - m_0)/m_0\right]}{dt}.$$

(12)

As shown in Figure 17, the range of 20 to 230°C is dominated by evaporation processes. The value of $dx/dt$ decreases gradually at temperature range from 230 to 360°C and becomes near to zero. The process corresponds to gas adsorption onto coal internal surface pores after releasing moisture. Volatile matter and HCs gases separate out from coal matrix in 360 to 650°C and the porous chars form in 650 to 900°C. In the temperature range of 900 to 1400°C, the $dx/dt$ shows large values due to gasification and combustion reactions of chars in the $CO_2$ rich gas flow containing a small percentage of $O_2$.

Heat generation rates in unit of coal mass against time $t$ or vessel temperature, denoted by $s = q/dx/dt$ (J·mg$^{-1}$), are shown in Figure 18. Heat generation rate of coal can be classified into four stages based on changes with temperature.

(1) 20 to 230°C: coal drying by water evaporation.

(2) 230 to 360°C: $O_2$ and $CO_2$ gases adsorption before oxidation and combustion. The value of $s$ jumps dramatically since its mass change is small against heat generation.

(3) 360 to 1100°C: coal oxidation and combustion. Maximum peaks of heat generation rate in unit of coal mass are observed, but the value of $s$ reduces gradually with the formation of porous chars.

(a) Arrhenius plots of coal weight loss rate versus $T^{-1}$

(b) The effects of $O_2$ concentration on $A_0$ and $E$

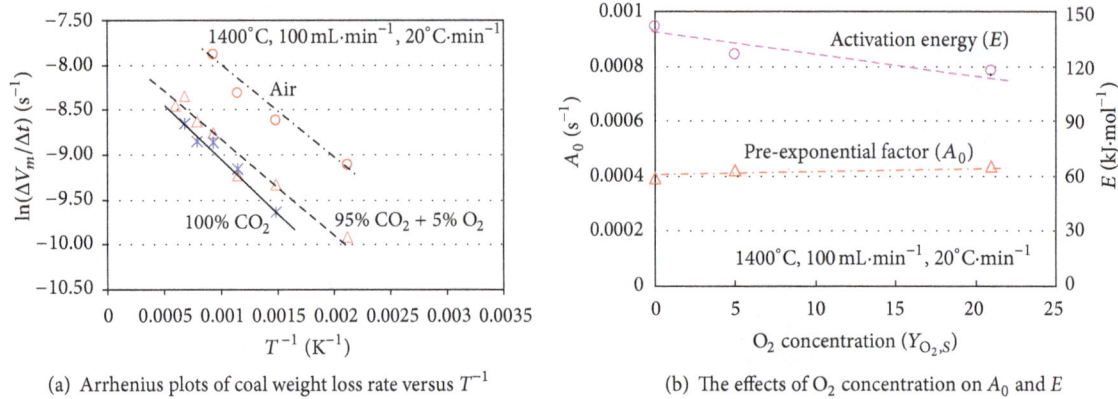

FIGURE 16: Coal weight loss rates with different temperatures and $O_2$ concentrations.

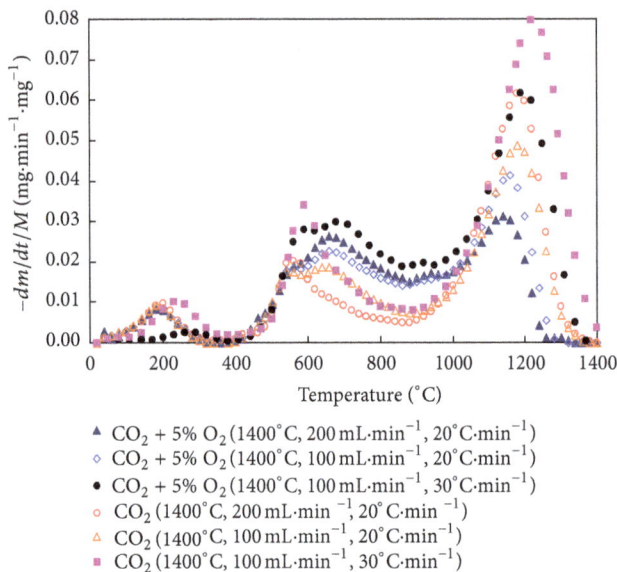

▲ $CO_2 + 5\% O_2$ (1400°C, 200 mL·min$^{-1}$, 20°C·min$^{-1}$)
◇ $CO_2 + 5\% O_2$ (1400°C, 100 mL·min$^{-1}$, 20°C·min$^{-1}$)
● $CO_2 + 5\% O_2$ (1400°C, 100 mL·min$^{-1}$, 30°C·min$^{-1}$)
○ $CO_2$ (1400°C, 200 mL·min$^{-1}$, 20°C·min$^{-1}$)
△ $CO_2$ (1400°C, 100 mL·min$^{-1}$, 20°C·min$^{-1}$)
■ $CO_2$ (1400°C, 100 mL·min$^{-1}$, 30°C·min$^{-1}$)

FIGURE 17: Coal conversion rate versus atmospheric temperature.

▲ $CO_2 + 5\% O_2$ (1400°C, 200 mL·min$^{-1}$, 20°C·min$^{-1}$)
◇ $CO_2 + 5\% O_2$ (1400°C, 100 mL·min$^{-1}$, 20°C·min$^{-1}$)
● $CO_2 + 5\% O_2$ (1400°C, 100 mL·min$^{-1}$, 30°C·min$^{-1}$)
○ $CO_2$ (1400°C, 200 mL·min$^{-1}$, 20°C·min$^{-1}$)
△ $CO_2$ (1400°C, 100 mL·min$^{-1}$, 20°C·min$^{-1}$)
■ $CO_2$ (1400°C, 100 mL·min$^{-1}$, 30°C·min$^{-1}$)

FIGURE 18: Heat generation rate versus atmospheric temperature.

(4) 1100 to 1300°C: coal gasification with the reaction between residual carbon and $CO_2$ gas.

# 4. Comparisons of Coal and Char

In view of coal properties of oxidation, combustion, and gasification in the $CO_2$-rich atmosphere, comparisons of TG-DTA results of coal and char were also conducted by providing the $CO_2$-rich gas flow. The contents of experimental temperature and ambient gas are shown in Table 2. In addition, during the experiments, flue gases generated from the heated coal or char particles were simultaneously analyzed by the gas analytical system transferred from TG-DTA using an air pump (100 mL·min$^{-1}$) in order to discuss the optimum $O_2$% in the $CO_2$-rich gas for coal or char combustion and gasification. Gases of $O_2$, HCs, and CO in the $CO_2$-rich gas were measured with a time interval of 1 second by the gas analytical system.

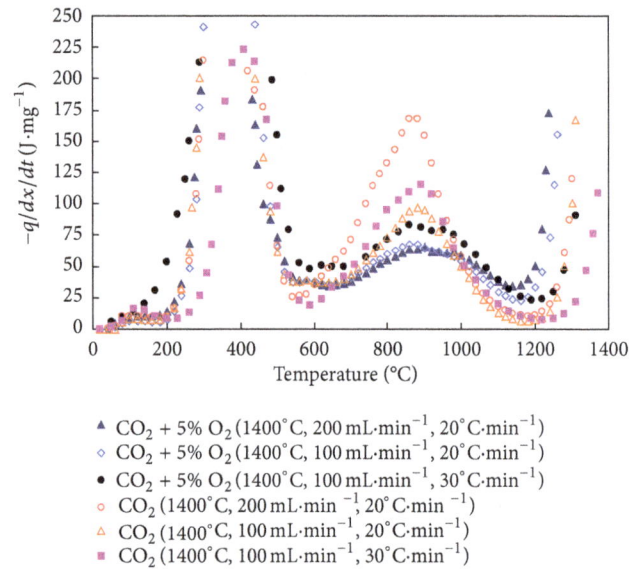

## 4.1. Comparisons of TG-DTA Curves.
In the TG-DTA measurements, the termination temperature was set to 1400°C with a temperature gradient of $\lambda = 20°C \cdot min^{-1}$, and ambient gas was supplied by 100% $CO_2$ or 95% $CO_2 + 5\%$ $O_2$ gas with gas flow rate of 100 mL·min$^{-1}$. Sample mass used in the experiments was around 30 mg (see Figure 2).

The TG-DTA measurement results of coal and char in the $CO_2$-rich gas flow are shown in Figures 19 and 20. Difference of conversion rate between coal and char is obvious with water and volatile matter evaporations in the initial stage of 20 to 230°C as shown in Figure 19. The water and volatile matter separate out from coal particles, but there is no change for char during the period. After that, the stage transfers to the next common stage of forming porous matrix. Coal or char gasification stage is verified at 1100°C by char mass loss in 100% $CO_2$ atmosphere (see Figure 19). In addition, heat generation rate of char in 95% $CO_2 + 5\%$ $O_2$ gas flow is higher than those in other gas flows as shown in Figure 20. It can be assumed that 5% $O_2$ in the $CO_2$-rich gas flow was

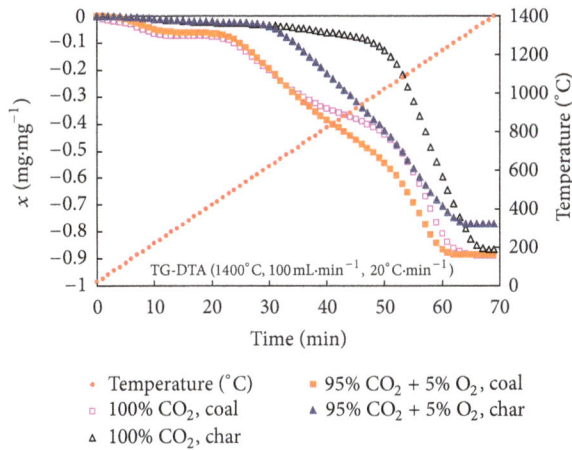

FIGURE 19: TG comparisons of coal and char in the $CO_2$-rich atmosphere.

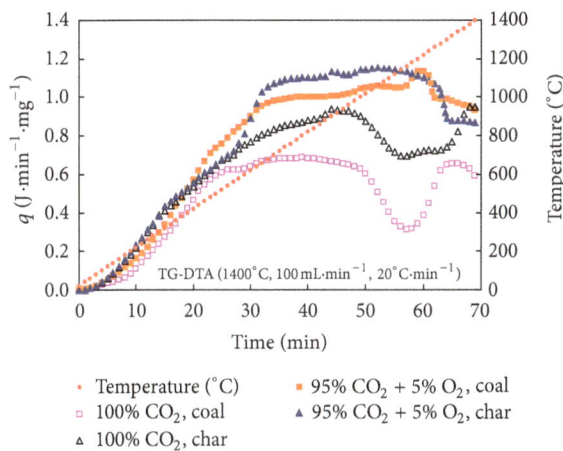

FIGURE 20: DTA comparisons of coal and char in the $CO_2$-rich atmosphere.

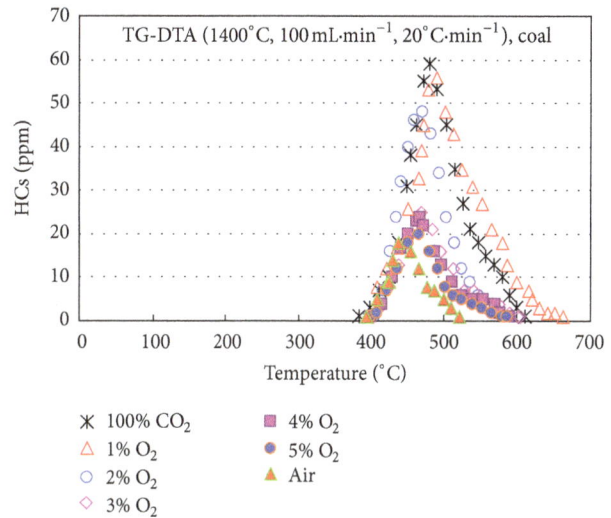

FIGURE 21: HCs gas generation from coal in air and $CO_2$-rich gas flows.

sufficient to complete char combustion. On the other hand, heat generation of char gasification in 100% $CO_2$ gas flow is larger than that of coal; the measurement result suggests that the heat of adsorption with char gasification in the gas flow is smaller than that of coal, because of heat consumption from evaporation of volatile matter in coal and the formation of porous chars.

*4.2. Gas Generation from Coal or Char by Various $O_2$ Concentrations.* In the TG-DTA heating process, the flue gases generated from the heated coal, such as CO and HCs gases, have been analyzed by the emission gas analytical system. HCs gas amount generated from coal was measured by providing gas flow with various $O_2$%. As shown in Figure 21, the gases were generated from coal samples in the temperature range from 400 to 650°C. In the case of char, there is no HCs gas generation in the same condition. It can be determined that HCs gas is formed with the moisture or volatile matter of coal. In addition, HCs generation rate in air is lower than those of other gases; furthermore, HCs generation from coal

in 100% $CO_2$ is evaluated as roughly 1.2 mL·g$^{-1}$-coal that is higher than those of other gases containing $O_2$. In other words, under the same condition, low $O_2$% can promote HCs generation in the $CO_2$-rich gas flow.

Figures 22 and 23 show CO gas generation from the heated coal or char in air or $CO_2$-rich gas flow. In the case of coal in air flow, CO gas concentration is less than 500 ppm at temperature lower than 700°C. On the other hand, CO gas generation from coal in 100% $CO_2$ gas flow at temperature over 900°C is roughly 235 mL·g$^{-1}$-coal that is the largest among the $CO_2$-rich gas flows, although the maximum peak concentration is recorded in 99% $CO_2$ + 1% $O_2$ gas flow. Moreover, the peak time of CO concentration matches the trough bottom of DTA heat generation curves (refer to Figures 18 and 20) which correctly verifies endothermic processes of coal gasification. It is intuitively confirmed by CO generation area with increasing temperature from 900 to 1400°C as shown in Figure 22.

Additionally, CO generation amount of coal gradually decreases with increasing $O_2$ concentration from 0 to 5% in the $CO_2$-rich gas. Furthermore, CO generation concentrations from coal for 4% and 5% $O_2$ contained in $CO_2$-rich gas flow arealmost lower than 300 ppm in TG-DTA heating process which are smaller than that in air flow. The measurement results suggest that the optimum $CO_2$-rich gas flow for coal gasification and combustion is evaluated with 96% $CO_2$ + 4% $O_2$ gas from the present TG-DTA experiments. The analytical result exactly matches the EDS analysis of Section 3.3. Moreover, it can be assumed that CO generation from coal in air partly formed by reaction (2); however, rich $CO_2$ gas inhibited coal conversion to gases in the temperature range from 20 to 600°C.

On the other hand, CO generation concentrations from char for 2 to 5% $O_2$ contained in $CO_2$-rich gas flow areapproximately lower than 300 ppm in TG-DTA heating process. Moreover, nothing but CO gas generation in 99% $CO_2$ + 1% $O_2$ gas flow is obvious in the $CO_2$-rich gas flow. Therefore,

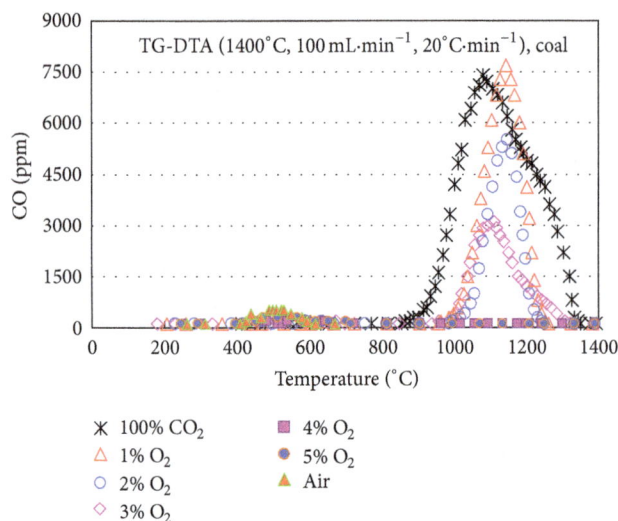

FIGURE 22: CO gas generation of coal in air and $CO_2$-rich gas flows.

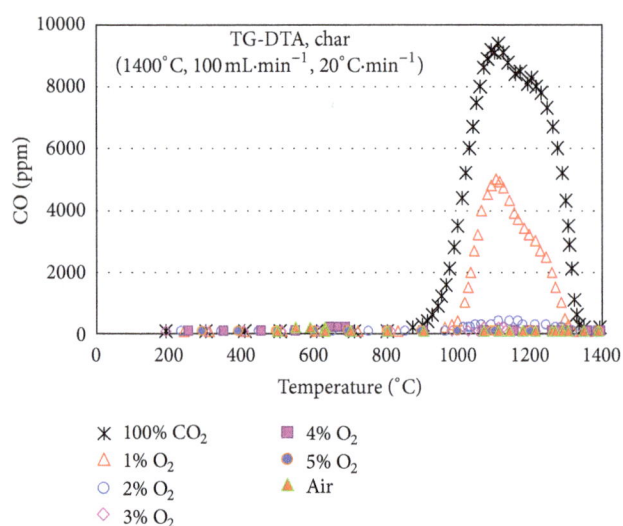

FIGURE 24: HCs gas generation of coal with different temperature gradients and gas flow rates.

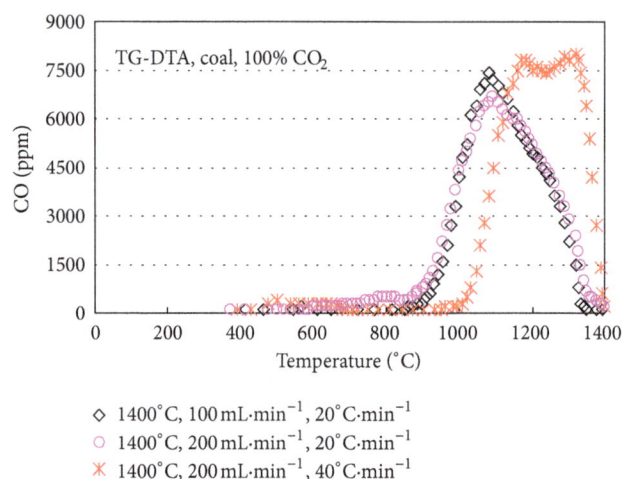

FIGURE 23: CO gas generation of char in air and $CO_2$-rich gas flows.

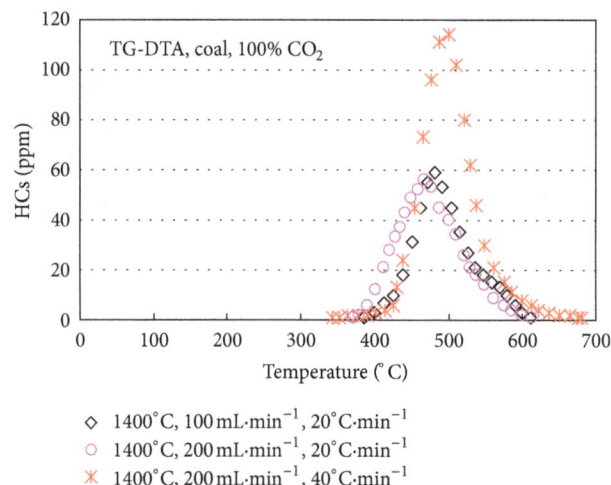

FIGURE 25: CO gas generation of coal with different temperature gradients and gas flow rates.

the optimum $CO_2$-rich gas flow for char gasification and combustion is evaluated with 98% $CO_2$ + 2% $O_2$ gas flow as shown in Figure 23. Similarly, CO generation amount from char samples in 100% $CO_2$ gas is roughly 460 mL·g$^{-1}$-char that is also higher than those of other gases. Comparing the results of coal and char samples in 100% $CO_2$ gas flow, CO gas generation amount of char samples is higher than that of coal samples, because the volatile matter of coal participates in carbon gasification with $CO_2$ gas and decreases CO gas generation.

*4.3. Effects of Temperature Gradient and Flow Rate on Flue Gas Generation.* The measurement results of flue gases generated from coal samples in 100% $CO_2$ gas with different gas flow rates and temperature gradients set to TG-DTA are shown in Figures 24 and 25. In the measurements of coal samples,

the termination temperature was set to 1400°C. Both of CO and HCs gases generations are not sensitive to the gas flow rate because it is mainly controlled by $O_2$% and temperature range. However, HCs gas generation becomes approximately double by increasing temperature gradient from 20°C·min$^{-1}$ to 40°C·min$^{-1}$. Furthermore, CO gas generation amount also increases with increasing temperature gradient, and the peak extent of CO gas concentration is extended against temperature. These measurement results suggest that high temperature gradient accelerates coal gasification and stimulates gasified gases generation. On the contrary, low temperature gradient promotes slow oxidation of coal and gas generation of $CO_2$.

In the measurements of char samples, the termination temperature was set to 1400°C with a temperature gradient of 20°C·min$^{-1}$ and 40°C·min$^{-1}$; ambient gas was supplied by injected 99% $CO_2$ + 1% $O_2$ gas with gas flow rate of 50 mL·min$^{-1}$ and 100 mL·min$^{-1}$. As shown in Figure 26, CO

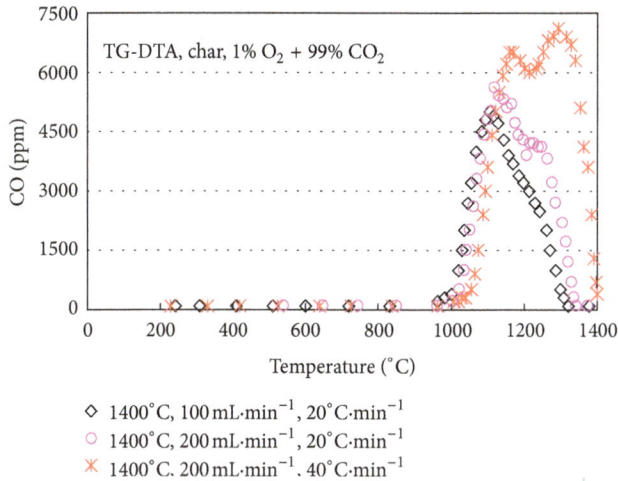

FIGURE 26: CO gas generation from char in 99% $CO_2$ + 1% $O_2$ gas flow with different temperature gradients and gas flow rates.

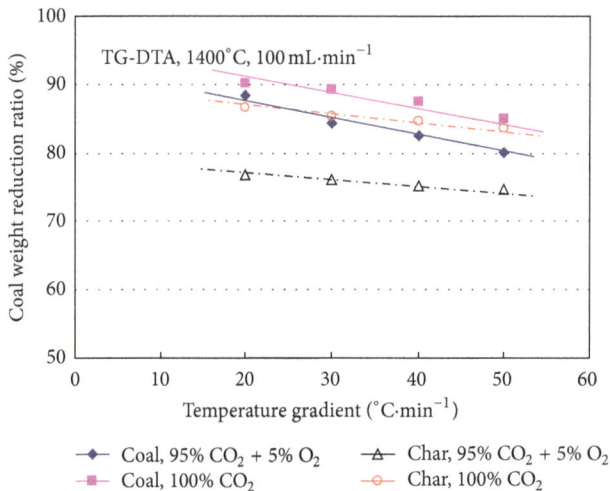

FIGURE 27: Comparisons of weight reduction ratio versus temperature gradient between coal and char.

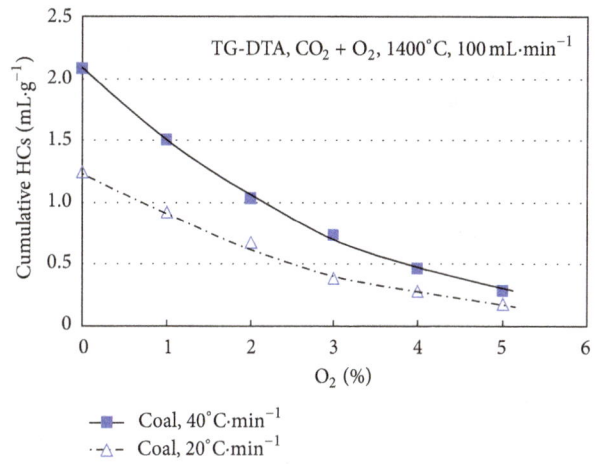

FIGURE 28: Cumulative HCs gas versus $O_2$% for different temperature gradients during 20 to 700°C (0 to 35 min).

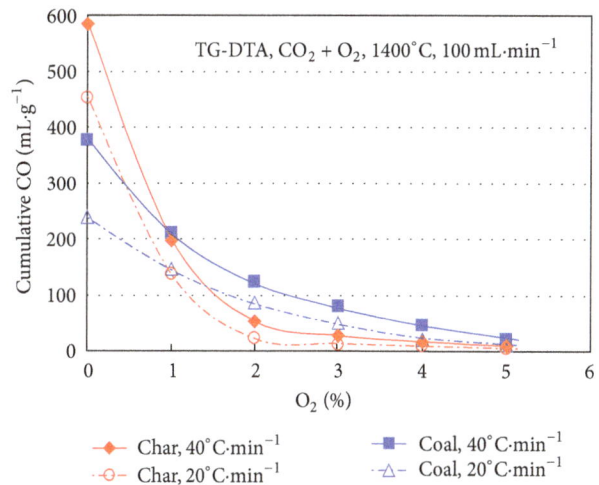

FIGURE 29: Cumulative CO gas versus $O_2$% for different temperature gradients during 20 to 1400°C (0 to 70 min).

gas generation is not sensitive to gas flow rate; however, the amount of generating CO gas becomes nearly double by increasing temperature gradient from 20°C·min$^{-1}$ to 40°C·min$^{-1}$. In addition, one phenomenon occurs in which temperature area of CO gas generated from coal with the temperature gradient of 40°C·min$^{-1}$ is almost uniform to the area of CO generation by the heated char in the $CO_2$-rich gas flow under low $O_2$%. It shows that the condition of high temperature gradient in a $CO_2$-rich gas flow with low $O_2$% is beneficial to coal gasification with $O_2$ incomplete combustion, and $CO_2$ circulation technology further improves the utilization of $CO_2$ gas on coal-fired power generation technologies.

*4.4. Effects of Temperature Gradients on Coal Weight Reduction Ratio and Gases Generation.* As shown in Figure 27, both of weight reduction ratios of coal and char decrease with

increasing temperature gradient at various $O_2$% in $CO_2$-rich gas flow. However, coal is more sensitive to the temperature gradient, and its weight decreases more than that of char, since volatile matter of coal participates in converting coal to gases.

In the TG-DTA measurements, the effects of temperature gradient on cumulative HCs and CO gases are shown in Figures 28 and 29. The experimental results indicate that cumulative HCs and CO gases from coal and char are increased by increasing temperature gradient from 20 to 40°C·min$^{-1}$. Furthermore, the cumulative CO gas volume generated from the char is the largest at temperature gradient of 40°C·min$^{-1}$. It is clear that the effect of temperature gradient on HCs and CO gases generations decreases with increasing $O_2$% from 0 to 5% $O_2$, especially for CO gas. In addition, cumulative CO gas generated from coal in 99% $CO_2$ + 1% $O_2$ gas flow with different temperature gradients is essentially coincident with that of char. It can be assumed that carbon in volatile matter of coal almost converted to $CO_2$ gas. The measurement results suggest that $O_2$% is the primary parameter of coal gasification reactions with low temperature gradient. Consequently, the condition of higher temperature gradient and low $O_2$% less than 1% makes stimulate and enhances generations of gasified gases such HCs and CO.

## 5. Conclusions

In this study, characteristics of pyrolysis, combustion, and gasification of Datong coal and char were investigated at temperature range from 20 to 1400°C with heating temperature gradient of 20 to 40°C·min$^{-1}$ in a $CO_2$-rich gas flow by TG-DTA analyses. The TG-DTA results were discussed to make clear the effects of $O_2$% in $CO_2$-rich gas and heating temperature gradient on coal physical and chemical behaviors related to coal gasification with $CO_2$ and oxygen combustion and underground coal gasification.

The main findings are summarized as follows.

(1) The processes of coal in $CO_2$-rich gas atmosphere mainly are divided into two temperature ranges for coal devolatilization, char formation, and gasification.

(2) Coal mass reduces with increasing coal matrix temperature with a linear line at various $O_2$ concentrations that is the main impact factor for coal conversion rate at temperature lower than 1100°C. Moreover, coal weight loss against temperature followed the Arrhenius equation.

(3) There are dip and peak on the DTA curves for 100% $CO_2$ gas flow. The minimum point of the dip appears at around 1160°C, and the maximum point is at around 1300°C. Coal gasification in 100% $CO_2$ gas flow is generated mainly in the temperature range from 1100 to 1300°C. Therefore, coal conversion to gases may break away from $O_2$ gas and promote $CO_2$ reduction reactions at temperature over 1300°C.

(4) For the case of terminal temperature of 1200°C, the higher $O_2$% was in $CO_2$-rich gas flow, the less C% in was measured in residuals or ashes.

(5) HCs gas generated from coal was measured at temperature range from 400 to 650°C in the $CO_2$-rich gas flow, and it can be formed from moisture and volatile matter in coal.

(6) CO gas generation amount gradually decreases with increasing $O_2$% during 0 to 5% in $CO_2$-rich gas flow. Additionally, the peak time on CO gas concentration or generation matched with the time showing the trough bottom of heat generation curves measured by DTA.

(7) The optimum $CO_2$-rich gas flow for gasification and combustion reactions of coal and char is evaluated with 96% $CO_2$ + 4% $O_2$ and 98% $CO_2$ + 2% $O_2$ gas from the present TG-DTA experiments, respectively.

(8) Temperature gradient per unit time for heating coal and char samples is a secondary essential parameter to improving and stimulating coal and char gasification.

Coal conversion factor is mainly implemented by coal gasification instead of coal combustion when the temperature gradient is over 40°C·min$^{-1}$.

(9) The higher temperature gradient accelerates coal and char gasification reaction with $CO_2$ gas; on the contrary, low temperature gradient promotes coal and char slow oxidation with low gases generation rate.

## Acknowledgments

This study was partly supported by the NEDO (P08020) project on Innovative Zero-emission Coal Gasification Power Generation, JSPS KAKENHI Grant-in-Aid for Scientific Research (B) no. 25303030, and the cooperative research project between Kyushu University and Liaoning Technical University on "$CO_2$ geological storage and utilization for coal."

## References

[1] W. Zhang and Z. Wu, "A study on establishing low-carbon auditing system in china," *Low Carbon Economy*, vol. 3, no. 2, pp. 35–38, 2012.

[2] S. Zeng and S. Zhang, "Literature review of carbon finance and low carbon economy for constructing low carbon society in China," *Low Carbon Economy*, vol. 2, no. 1, pp. 15–19, 2011.

[3] R. Lei, Y. Zhang, and S. Wei, "International technology spillover, energy consumption and $CO_2$ emissions in China," *Low Carbon Economy*, vol. 3, no. 3, pp. 49–53, 2012.

[4] Key World Energy Statistics, International Energy Agency (IEA), "Clearly-presented data on the supply, transformation and consumption of all major energy sources," Stedi Media, 2010.

[5] A. Williams, M. Pourkashanian, J. M. Jones, and N. Skorupska, *Combustion and Gasification of Coal*, Applied Energy Technology Series, Taylor & Francis, New York, NY, USA, 1999.

[6] S. Hossain, "An econometric analysis for $CO_2$ emissions, energy consumption, economic growth, foreign trade and urbanization of Japan," *Low Carbon Economy*, vol. 3, no. 3, pp. 92–105, 2012.

[7] C. Ramírez and J. González, "Contribution of finance to the low carbon economy," *Low Carbon Economy*, vol. 2, no. 2, pp. 62–70, 2011.

[8] IPCC Working Group II, "Climate change 2007: impacts, adaptation and vulnerability," Assessment Report of the Intergovernmental Panel on Climate Change, Cambridge University Press, 2007.

[9] Key World Energy Statistics, International Energy Agency (IEA), "Evolution from 1971 to 2010 of World $CO_2$ emissions by region," Stedi Media, 2011.

[10] IEA, "$CO_2$ emissions from fuel combustion," 2011, http://www.iea.org/CO2highlights.

[11] H. Herzog and D. Golomb, "Carbon capture and storage from fossil fuel use," *Encyclopaedia of Energy*, vol. 1, pp. 277–287, 2004.

[12] K. Jordal, M. Anheden, J. Y. Yan, and L. Strömberg, "Oxyfuel combustion for coal-fired power generation with $CO_2$ capture-opportunities and challenges," *Greenhouse Gas Control Technologies 7*, vol. 1, pp. 201–209, 2005.

[13] Office of Fossil Energy, "Carbon capture & separation," U.S. Department of Energy, 2004, http://fossil.energy.gov/programs/sequestration .

[14] J. C. Chen, Z. S. Liu, and J. S. Huang, "Emission characteristics of coal combustion in different $O_2/N_2$, $O_2/CO_2$ and $O_2/RFG$ atmosphere," *Journal of Hazardous Materials*, vol. 142, no. 1-2, pp. 266–271, 2007.

[15] D. Singh, E. Croiset, P. L. Douglas, and M. A. Douglas, "Techno-economic study of $CO_2$ capture from an existing coal-fired power plant: MEA scrubbing versus $O_2/CO_2$ recycle combustion," *Energy Conversion and Management*, vol. 44, no. 19, pp. 3073–3091, 2003.

[16] E. S. Hecht, C. R. Shaddix, M. Geier, A. Molina, and B. S. Haynes, "Effect of $CO_2$ and steam gasification reactions on the oxy-combustion of pulverized coal char," *Combustion and Flame*, vol. 159, pp. 3437–3447, 2012.

[17] B. J. P. Buhre, L. K. Elliott, C. D. Sheng, R. P. Gupta, and T. F. Wall, "Oxy-fuel combustion technology for coal-fired power generation," *Progress in Energy and Combustion Science*, vol. 31, no. 4, pp. 283–307, 2005.

[18] Q. Li, C. Zhao, X. Chen, W. Wu, and Y. Li, "Comparison of pulverized coal combustion in air and in $O_2/CO_2$ mixtures by thermo-gravimetric analysis," *Journal of Analytical and Applied Pyrolysis*, vol. 85, no. 1-2, pp. 521–528, 2009.

[19] H. Liu, "Combustion of coal chars in $O_2/CO_2$ and $O_2/N_2$ mixtures: a comparative study with non-isothermal thermo-gravimetric analyzer (TGA) tests," *Energy and Fuels*, vol. 23, no. 9, pp. 4278–4285, 2009.

[20] Z. H. Li, X. M. Zhang, Y. Sugai, J. R. Wang, and K. Sasaki, "Properties and developments of combustion and gasification of coal and char in a $CO_2$-rich and recycled flue gases atmosphere by rapid heating," *Journal of Combustion*, vol. 2012, Article ID 241587, 11 pages, 2012.

[21] P. Lu, G. X. Liao, J. H. Sun, and P. D. Li, "Experimental research on index gas of the coal spontaneous at low-temperature stage," *Journal of Loss Prevention in the Process Industries*, vol. 17, no. 3, pp. 243–247, 2004.

[22] J. Xie, W. M. Cheng, and F. Q. Liu, "Technology and effect of open nitrogen injection at fully mechanized face," *Journal of Safety in Coal Mines*, vol. 3, pp. 33–35, 2007.

[23] H. Y. Qi, Y. H. Li, C. F. You, J. Yuan, and X. C. Xu, "Emission on $NO_x$ in high temperature combustion with low oxygen concentration," *Journal of Combustion Science and Technology*, vol. 8, no. 1, pp. 17–22, 2002.

[24] C. X. Luo and W. H. Zhou, "Coal gasification technology & its application," *Sino-Global Energy*, vol. 1, pp. 28–35, 2009.

[25] J. J. Huang, Y. T. Fang, and Y. Wang, "Development and progress of modern coal gasification technology," *Journal of Fuel Chemistry and Technology*, vol. 30, no. 5, pp. 385–391, 2002.

[26] Y. F. Liu and X. K. Xue, "Thermal calculation methods for oxy-fuel combustion boilers," *East China Electric Power*, vol. 36, pp. 355–357, 2008.

[27] J. R. Wang, C. B. Deng, Y. F. Shan, L. Hong, and W. D. Lu, "New classifying method of the spontaneous combustion tendency," *Journal of the China Coal Society*, vol. 33, no. 1, pp. 47–50, 2008.

# 16

# Eucalyptus-Palm Kernel Oil Blends: A Complete Elimination of Diesel in a 4-Stroke VCR Diesel Engine

**Srinivas Kommana,[1] Balu Naik Banoth,[2] and Kalyani Radha Kadavakollu[3]**

[1]*Department of Mechanical Engineering, VRSEC, Vijayawada 520 007, India*
[2]*Department of Mechanical Engineering, JNTU, Hyderabad 500 085, India*
[3]*Department of Mechanical Engineering, JNTU, Anantapur 515 002, India*

Correspondence should be addressed to Srinivas Kommana; kommanasrinivas@yahoo.com

Academic Editor: Sergey M. Frolov

Fuels derived from biomass are mostly preferred as alternative fuels for IC engines as they are abundantly available and renewable in nature. The objective of the study is to identify the parameters that influence gross indicated fuel conversion efficiency and how they are affected by the use of biodiesel relative to petroleum diesel. Important physicochemical properties of palm kernel oil and eucalyptus blend were experimentally evaluated and found within acceptable limits of relevant standards. As most of vegetable oils are edible, growing concern for trying nonedible and waste fats as alternative to petrodiesel has emerged. In present study diesel fuel is completely replaced by biofuels, namely, methyl ester of palm kernel oil and eucalyptus oil in various blends. Different blends of palm kernel oil and eucalyptus oil are prepared on volume basis and used as operating fuel in single cylinder 4-stroke variable compression ratio diesel engine. Performance and emission characteristics of these blends are studied by varying the compression ratio. In the present experiment methyl ester extracted from palm kernel oil is considered as ignition improver and eucalyptus oil is considered as the fuel. The blends taken are PKE05 (palm kernel oil 95 + eucalyptus 05), PKE10 (palm kernel oil 90 + eucalyptus 10), and PKE15 (palm kernel 85 + eucalyptus 15). The results obtained by operating with these fuels are compared with results of pure diesel; finally the most preferable combination and the preferred compression ratio are identified.

## 1. Introduction

Continuous rise in fuel price, increase in number of vehicles on road, depletion of petroleum resources, and rapid increase in greenhouse gases are the main reasons for the search of alternative fuels. Up till now many alternative fuels are identified from different resources like vegetables, plants, animal fat, and so forth; these are successfully tested over engine with slight modifications in engine or without any modifications. Alternative fuels are typically produced through the reaction of a vegetable oil or animal fat with alcohol in the presence of the catalyst to produce their esters [1]. These are eco-friendly since these are extracted from plants and animals have the advantage of being biodegradable, less polluting, and renewable [2]. With environmental laws becoming more and more stringent these days, fortunately this renewable source of energy will help to protect the world from the effects of pollution, such as global warming and acid rain [3]. In general, oils extracted from plants are classified into two categories. They are triglyceride oils (TG oils) and terpene oils (light oils). In present study triglyceride oils are used. Triglyceride oils are extracted from plant seeds but eucalyptus oil is taken from the leaves and young twigs of plant. Present study involves two triglyceride oils, namely, eucalyptus oil and methyl ester of palm kernel oil. Eucalyptus oil is prepared from leaves and young twigs of plant, whereas palm kernel oil is prepared from palm kernel. Eucalyptus oil alone cannot replace diesel in diesel engine since the cetane number of eucalyptus oil is insufficient to ignite; but the blended form of eucalyptus oil and methyl ester of palm kernel oil can replace diesel to a maximum extent since the properties of blends are nearer to the properties of diesel fuel. These blends can be used over diesel engine without any further modifications. In the present work performance, emission, and combustion characteristics of these biofuel blends are examined over a 4-stroke direct injection diesel engine by varying the compression ratio.

## 2. Energy Scenario

Energy and agriculture are always on the same path, but unfortunately the relation between these has been changing over the time period. Agriculture is the main source of energy and energy is the main input for agriculture production. Until the nineteenth century, animals provided almost all the horse power used for transport and farm equipment and in many parts of the world they still do. Agriculture produces the fuel to feed these animals; two centuries ago, around 20 percent of the agricultural area in the United States of America was used to feed draught animals. At the same time relation over input market increased as agriculture became more dependent on chemical fertilizers extracted from fossils and machinery operated over diesel. Processing and distribution involve higher energy costs, therefore implying a direct and strong impact on agricultural production costs and food costs. The recent advances and researches over biofuels depending on agricultural crops which are used as transport fuels involve the relation between agriculture and energy output markets. The world's total primary energy demand amounts to about 11400 million tons of oil equivalent (Mtoe) per year (IEA, 2012); biomass including forest and agricultural products and organic wastes and residues accounts for 10 percent of this total.

## 3. Specification for Biodiesel

Specifications and standards are the major important parameters for any biodiesel producers, suppliers, and users of biodiesel. Standards are required for the evaluation of safety, risks, environmental problems approval, and warranty of vehicles. Creation of standards shall help expand the market for renewable sources of energy in the world. Generally, standards and codes are developed by examining the existing standards and codes in different countries and then writing standards for the own country. In December 2001, American society of testing and materials (ASTM) issued a specification (D6751) for biodiesel (B100) which is presented in Table 1 and corresponding properties of experimental blends are shown in Tables 2 and 3 and also test rig specifications are depicted in Table 4. The physical and chemical properties of the test fuels were determined earlier in accordance with the ASTM standards [4].

## 4. Experimental Setup

A single cylinder, four-stroke, constant speed, variable compression ratio, water cooled, direct injection diesel engine is used for the present study. The loading is by means of Eddy current dynamometer. Water cooling system and various sensors and instruments integrated with computer data acquisition. A five-gas analyzer is used to obtain the exhaust gas concentrations in exhaust. Setup enables the

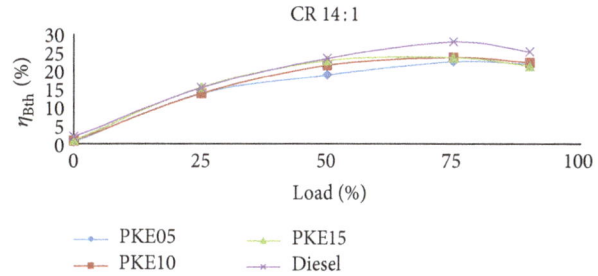

Figure 1: Load versus brake thermal efficiency at 14 : 1 CR.

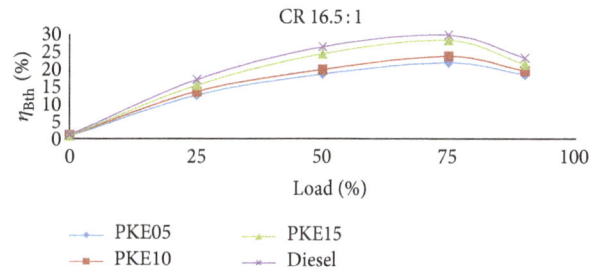

Figure 2: Load versus brake thermal efficiency at 16.5 : 1 CR.

evaluation of performance emissions constituents. Performance parameters include brake thermal efficiency, indicated thermal efficiency, mechanical efficiency, brake specific fuel consumption, and exhaust gas temperature.

## 5. Results and Discussion

*5.1. Engine Performance Analysis.* The performance of an engine is mainly studied with the help of the operating conditions. The characteristics obtained by operating the single cylinder diesel engine with the blends of palm kernel oil and eucalyptus oil are discussed below. The obtained results were compared with the results obtained when operating with diesel, palm kernel + diesel mixture, and eucalyptus oil + diesel mixture at various compression ratios. Figures 1–21 are used to study various characteristics of engine like brake thermal efficiency, mechanical efficiency, engine outlet temperature, and so forth. Comparison of performance was done for different values of compression ratio [5].

*5.1.1. Brake Thermal Efficiency ($\eta_{Bth}$), %.* As shown in the graphs (Figures 1, 2, and 3), at all compression ratios, the brake thermal efficiency of diesel is higher than the remaining blends, but at the full load at the same compression ratio the efficiency is nearer to the efficiency obtained by remaining blends.

Next to diesel PKE15 (palm kernel 85 + eucalyptus 15) is higher. This is mainly due to the presence of high volatile eucalyptus oil in the blend used. Eucalyptus oil mainly consists of cineole. Cineole decomposes easily at low temperature; it releases intermediate compounds in a heavy manner as soon as the fuel is injected. The reduced viscosity leads to improved atomization, fuel vaporization, and combustion. Eucalyptus oil's presence increases the ignition delay period

TABLE 1: ASTM specification (D6751) for biodiesel (B100).

| Property | ASTM method | Limits | Units |
|---|---|---|---|
| Flash point | D93 | 130 min | Degree C |
| Water & sediment | D2709 | 0.050 max | % volume |
| Kinematic viscosity | D445 | 1.9–6.0 | Mm$^2$/sec |
| Copper strip corrosion | D130 | Number 3 max | — |
| Cetane | D613 | 47 min | — |
| Cloud point | D2500 | Report | Degree C |
| Carbon residue(100% sample) | D4530 | 0.050 max | % mass |
| Acid number | D664 | 0.80 max | Mg KOH/gm |
| Free glycerine | D6584 | 0.020 max | % mass |

TABLE 2: Properties of experimental blends being done at ETA laboratories, Chennai, India.

| Oils/blends | Viscosity (cSt) | Calorific value (Kcal/Kg) | Density (kg/mm$^3$) |
|---|---|---|---|
| Diesel | 3.25 | 42,700 | 0.84 |
| Palm kernel oil | 4.839 | 37,250 | 0.883 |
| Eucalyptus oil | 2.024 | 43,270 | 0.991 |
| Palm kernel 95 + eucalyptus 05 | 4.246 | 38,158 | 0.892 |
| Palm kernel 90 + eucalyptus 10 | 3.986 | 40,350 | 0.889 |
| Palm kernel 85 + eucalyptus 15 | 3.826 | 41,793 | 0.88 |

TABLE 3: Properties of base fuels.

| Properties | Diesel | Eucalyptus oil | Palm kernel oil |
|---|---|---|---|
| Specific gravity, kg/m$^3$ | 0.83 | 0.918 | 0.883 |
| Viscosity, cSt | 3.25 | 2.024 | 4.839 |
| Flash point, °C | 74 | 53 | 167 |
| Lower heating value, KJ/Kg | 42700 | 43270 | 37,250 |
| Autoignition temperature, °C | 250 | 300–330 | 400–450 |

TABLE 4: Test rig specifications.

| Engine | Type: single cylinder, four-stroke diesel, water cooled |
|---|---|
| Rated power | 3.7 kW at 1500 rpm |
| Stroke | 110 mm |
| Bore | 87.5 mm |
| Compression ratio range | 12 to 24 |
| Dynamometer | Type eddy current, water cooled, with loading unit |
| Load indicator | Digital, range 0–50 Kg, supply 230 VAC |
| Load sensor | Load cell, type strain gauge, range 0–50 Kg |
| Piezosensor | Range 5000 psi, with low noise cable |
| Air flow transmitter | Pressure transmitter, range 0–250 mm |
| Software | "EngineSoftLV" Engine performance analysis software |
| Exhaust gas analyzer | Make: Indus Scientific, five-gas analyzer |

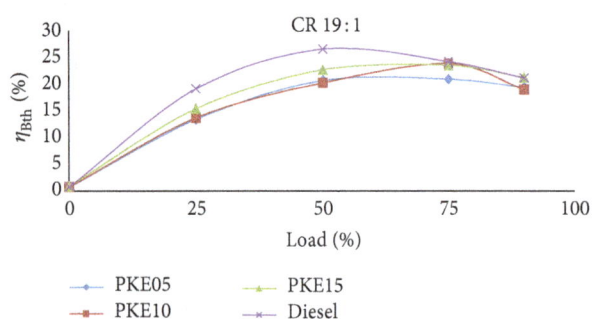

FIGURE 3: Load versus brake thermal efficiency at 19 : 1 CR.

and it causes the heavy accumulation of fuel which results in large heat release and higher brake thermal efficiency and high cylinder pressure.

5.1.2. Mechanical Efficiency ($\eta_{me}$). As shown in the graphs (Figures 4, 5, and 6), the mechanical efficiency of diesel is very high at compression ratio 14 when compared to the remaining experimental blends at all loads.

At compression ratios 16.5 : 1 and 19 : 1 the mechanical efficiency is equal for all the experimental fuels and diesel at all the loads. Comparatively, at compression ratio 19 : 1 CR, full load efficiency for PKE15 is higher than remaining experimental blends. Mechanical efficiency is the result of both brake thermal efficiency and indicated thermal efficiency. The brake thermal efficiency and indicated thermal efficiency of diesel are higher than remaining blends. Next to the diesel PKE15 blend is having higher efficiency. Mechanical

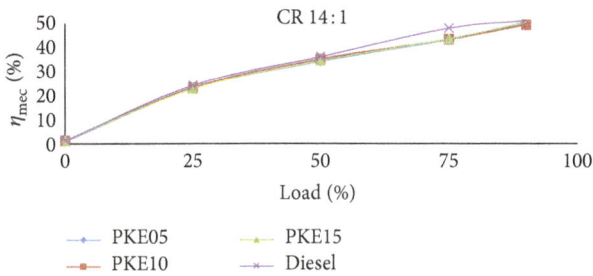

FIGURE 4: Load versus mechanical efficiency at 14 : 1 CR.

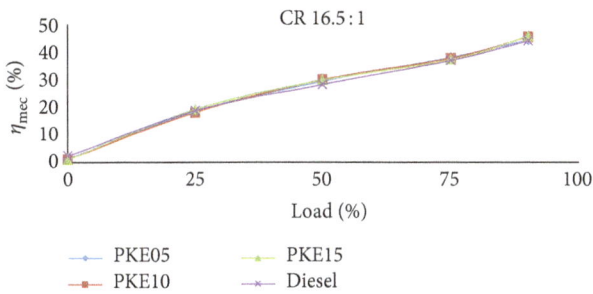

FIGURE 5: Load versus mechanical efficiency at 16.5 : 1 CR.

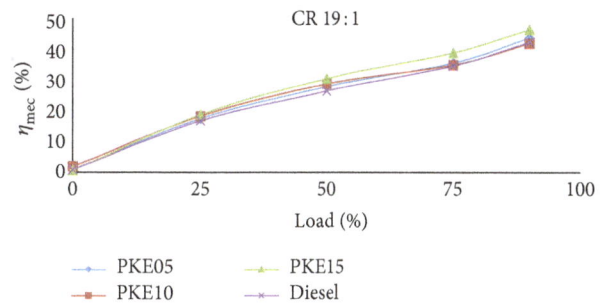

FIGURE 6: Load versus mechanical efficiency at 19 : 1 CR.

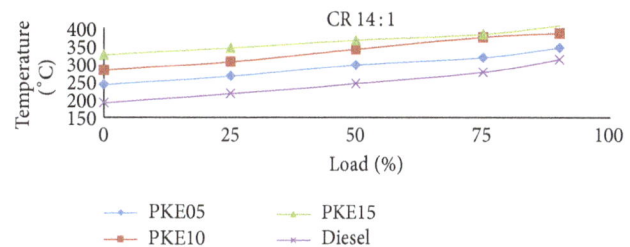

FIGURE 7: Load versus exhaust gas temperature at 14 : 1 CR.

efficiency of diesel and the experimental blends are almost equal with a slight variation.

*5.1.3. Exhaust Gas Temperature.* As shown in the graphs (Figures 7, 8, and 9) the exhaust temperature of diesel is lower than the remaining experimental blends. This is mainly due to the presence of eucalyptus oil in the blends which has lower cetane number and high percentage of oxygen which leads to longer ignition delay and rapid combustion which results in higher temperature and high cylinder pressure. Next to diesel PKE05 is having lower exhaust temperature at all the compression ratios. It has lower temperature compared to remaining blends since the quantity of eucalyptus oil present in this blend is smaller than remaining blends.

*5.2. Emission Analysis.* From these reports, the effects of biofuel on regulated emissions were studied [6].

*5.2.1. Hydrocarbon (HC) Emissions.* As shown in the graphs (Figures 10, 11, and 12) the hydrocarbon emissions of diesel are higher than all the experimental blends. At all compression ratios and different loads diesel is having the high emission of hydrocarbons. In our present experimental blends PKE05 is having higher HC emissions. HC emissions for biofuels are low because of the complete combustion of the fuel because of more availability of oxygen in biodiesel. As the ignition period increases due to the lower cetane number, at some areas of the combustion chamber the mixture becomes too lean which leads to the lower HC emissions.

*5.2.2. Carbon Dioxide (CO$_2$) Emissions.* As shown in the graphs (Figures 13, 14, and 15) the carbon dioxide emissions of all the experimental fuels are higher when compared to

diesel. This is mainly because of the oxygen enrichment in the eucalyptus oil and addition of palm kernel oil leads to the oxidation of CO produced during the exhaust process. Rather, diesel PKE05 is having lesser emissions in the present experimental fuels.

*5.2.3. Oxides of Nitrogen Emissions.* As shown in the graphs (Figures 16, 17, and 18) nitrous oxide emissions of diesel are lesser when compared to experimental fuels. This is mainly because of the heavy existence of oxygen in palm kernel oil and further more existence of oxygen in eucalyptus oil and elevated temperatures derived because of complete combustion. NO$_x$ is recorded higher than diesel fuel due to high oxygen content in biodiesel and lower cetane number (CN) of eucalyptus oil, also lengthening the ignition delay period [7].

*5.2.4. Oxides of Sulphur Emissions.* As shown in the graphs (Figures 19, 20, and 21) sulphur oxide emission of diesel is high at all compression ratios at all the loads. In our experimental fuels PKE05 is having higher emissions and PKE15 is having lower emissions.

## 6. Conclusion

The effective compression ratio can be fixed based on the experimental results obtained in the engine since the findings of the present research work infer that the biofuel obtained from palm kernel oil and eucalyptus oil blend is a promising alternative fuel for four-stroke VCR engine [8].

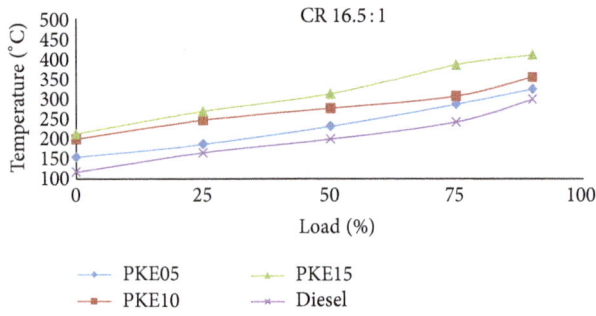

FIGURE 8: Load versus exhaust gas temperature at 16.5 : 1 CR.

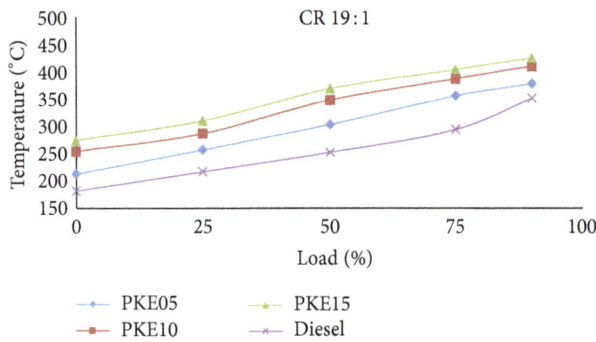

FIGURE 9: Load versus exhaust gas temperature at 19 : 1 CR.

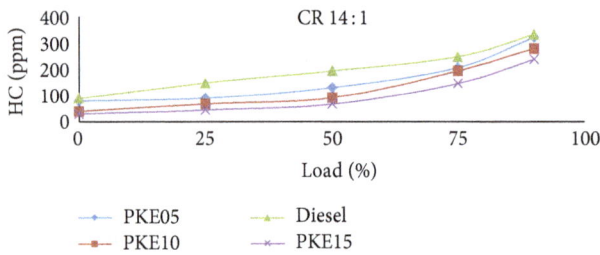

FIGURE 10: Load versus HC emissions at 14 : 1 CR.

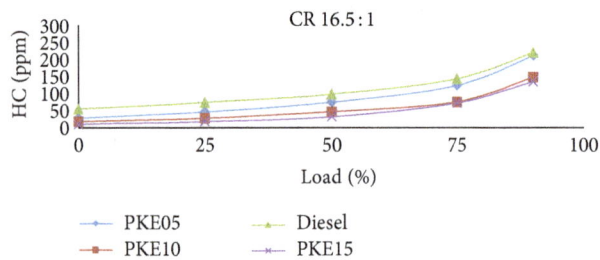

FIGURE 11: Load versus HC emissions at 16.5 : 1 CR.

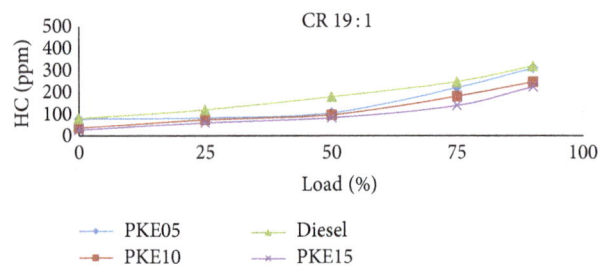

FIGURE 12: Load versus HC emissions at 19 : 1 CR.

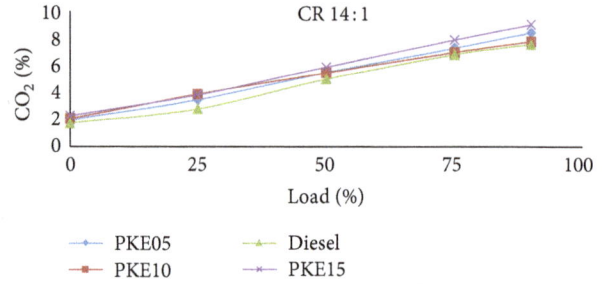

FIGURE 13: Load versus $CO_2$ emissions at 14 : 1 CR.

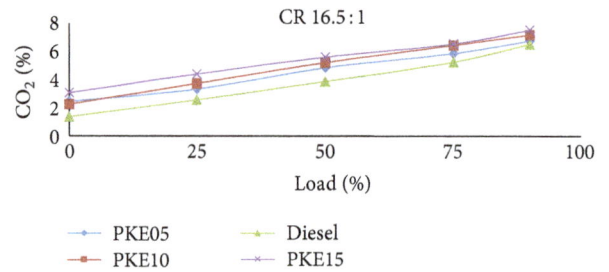

FIGURE 14: Load versus $CO_2$ emissions at 16.5 : 1 CR.

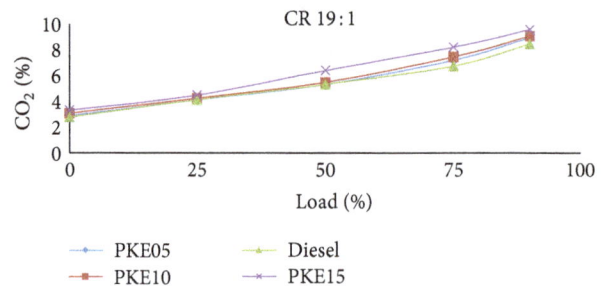

FIGURE 15: Load versus $CO_2$ emissions at 19 : 1 CR.

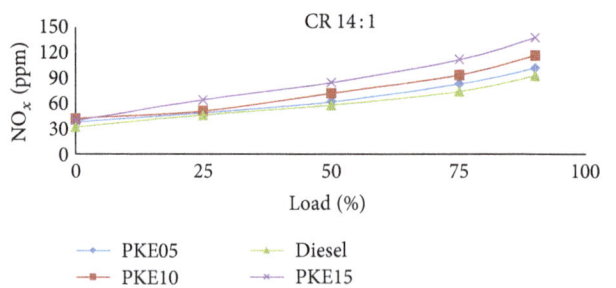

FIGURE 16: Load versus $NO_x$ emissions at 14 : 1 CR.

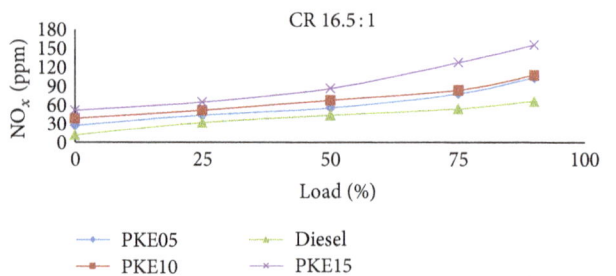

FIGURE 17: Load versus $NO_x$ emissions at 16.5 : 1 CR.

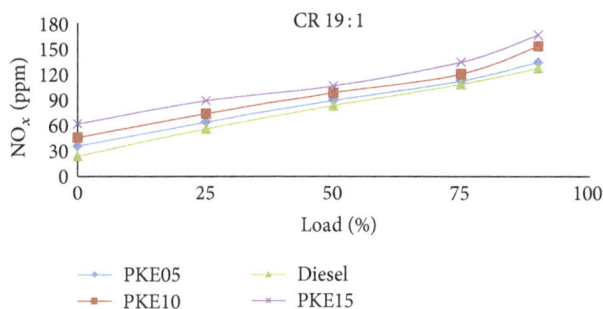

FIGURE 18: Load versus $NO_x$ emissions at 19:1 CR.

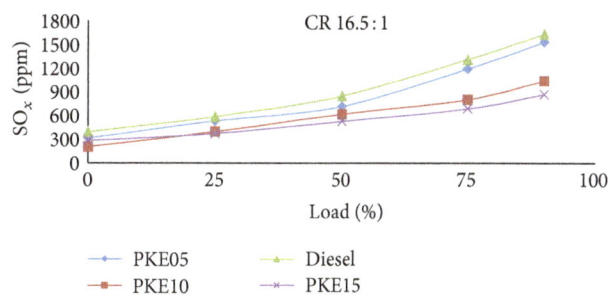

FIGURE 20: Load versus $SO_x$ emissions at 16.5:1 CR.

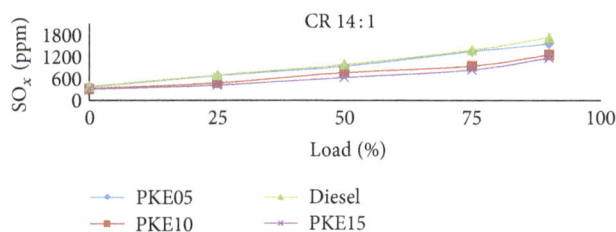

FIGURE 19: Load versus $SO_x$ emissions at 14:1 CR.

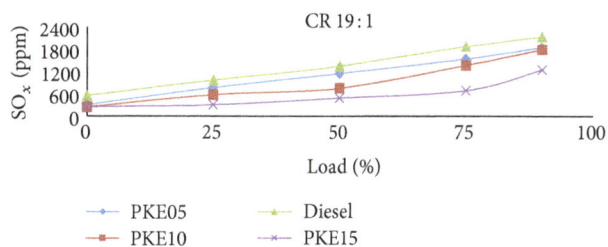

FIGURE 21: Load versus $SO_x$ emissions at 19:1 CR.

### 6.1. Performance Analysis

(1) Brake thermal efficiency of all the considered blends at all the compression ratios for minimum loads is equal; as the load increases efficiency increases and diesel has higher efficiency compared to remaining blends. Compared to remaining compression ratios, at CR 16.5:1, maximum efficiency is obtained but as the load approaches full load the efficiency gradually decreases.

(2) Mechanical efficiency is higher at compression ratio 14:1 when compared with remaining compression ratios. At CR 14:1 the efficiency of diesel is higher but at remaining compression ratios with minute variations PKE15 (palm kernel 85 + eucalyptus 15) it has higher mechanical efficiency when compared to diesel.

(3) Exhaust gas temperature of diesel is lower than all the considered experimental fuels. Comparatively next to diesel PKE05 has lower exhaust gas temperature.

### 6.2. Emission Analysis

(1) HC emissions are high for PKE05 compared to the remaining experimental blends. By looking at overall results, at compression ratio 14:1, HC emissions are high for all the blends. Compared to all the considered blends, PKE15 emits lower HC emissions than diesel at all the compression ratios.

(2) Carbon dioxide emissions are higher at compression ratio 19:1 for all the considered blends. As the compression ratio increases $CO_2$ emissions increase.

At Cr 14:1 lower $CO_2$ emissions are observed. As the load increases the emission increases. PKE15 has higher emissions compared to all the considered fuels.

(3) At minimum loads the oxygen emissions are high; as the load increases oxygen emissions decrease. Oxygen emitted by PKE10 (palm kernel 90 + eucalyptus 10) is almost equal to the oxygen content emitted by diesel at all the loads for all the compression ratios. PKE15 emits the least oxygen emissions.

(4) As the compression ratio increases the emissions increase. Diesel emits higher $SO_x$ emissions when compared to the considered experimental blends.

(5) As the compression ratio increases $NO_x$ emissions increase. At CR 19:1 $NO_x$ emissions are higher; as the load increases $NO_x$ emissions increase. For PKE15 (palm kernel 85 + eucalyptus 15) $NO_x$ is higher than remaining experimental fuels.

The results proved that the blending of palm kernel oil with eucalyptus oil can be used as an alternative fuel in diesel engine by complete elimination of diesel. The emissions and performance are of considerable range for all blends especially for PKE15 blend. It can be concluded that palm kernel 85 + eucalyptus 15 can be used in diesel engine without any major modifications to the engine.

## Conflict of Interests

The authors declare that there is no conflict of interests regarding the publication of this paper.

# References

[1] R. Prakash, S. P. Pandey, S. Chatterji, and S. N. Singh, "Emission analysis of CI engine using rice bran oil and their esters," *Journal of Engineering Research and Studies*, vol. 2, pp. 173–178, 2011.

[2] S. O. Eze, M. O. Ngadi, J. S. Alakali, and C. J. Odinaka, "Quality assessment of biodiesel produced from after fry waste Palm Kernel Oil (PKO)," *European Journl of Natural and Applied Science*, vol. 1, no. 1, pp. 38–46, 2013.

[3] A. S. Rocha, M. Veerachamy, V. K. Agrawal, and S. K. Gupta, "Jatropha liquid gold—the alternative to diesel," in *Proceedings of the 1st WIETE Annual Conference on Engineering and Technology Education*, pp. 91–96, Pattaya, Thailand, February 2010.

[4] B. P. Anand, C. G. Saravanan, and C. A. Srinivasan, "Performance and exhaust emission of turpentine oil powered direct injection diesel engine," *Renewable Energy*, vol. 35, no. 6, pp. 1179–1184, 2010.

[5] M. Venkatraman and G. Devaradjane, "Experimental investigation of effect of compression ratio, injection timing and injection pressure on the performance of a CI engine operated with diesel-pungam methyl ester blend," in *Proceedings of the Frontiers in Automobile and Mechanical Engineering (FAME '10)*, pp. 117–121, Chennai, India, November 2010.

[6] J. Xue, T. E. Grift, and A. C. Hansen, "Effect of biodiesel on engine performances and emissions," *Renewable and Sustainable Energy Reviews*, vol. 15, no. 2, pp. 1098–1116, 2011.

[7] M. S. Shehata, "Emissions, performance and cylinder pressure of diesel engine fuelled by biodiesel fuel," *Fuel*, vol. 112, pp. 513–522, 2013.

[8] M. T. Raj and M. K. K. Kandasamy, "Tamanu oil-an alternative fuel for variable compression ratio engine," *International Journal of Energy and Environmental Engineering*, vol. 3, no. 1, article 18, 2012.

[9] B. T. Tompkins, H. Song, J. A. Bittle, and T. J. Jacobs, "Efficiency considerations for the use of blended biofuel in diesel engines," *Applied Energy*, vol. 98, pp. 209–218, 2012.

[10] B. S. Chauhan, N. Kumar, H. M. Cho, and H. C. Lim, "A study on the performance and emission of a diesel engine fueled with Karanja biodiesel and its blends," *Energy*, vol. 56, pp. 1–7, 2013.

# Permissions

# List of Contributors

**K. F. Mustafa, S. Abdullah and K. Sopian**
Department of Mechanical and Materials Engineering, Faculty of Engineering and Built Environment, Universiti Kebangsaan Malaysia (UKM), 43600 Bangi, Selangor, Malaysia

**M. Z. Abdullah**
School of Mechanical Engineering, Universiti Sains Malaysia Engineering Campus, Seri Ampangan, 14300 Nibong Tebal, Penang, Malaysia

**Selvakumaran Palaniswamy, M. Rajavel, A. Leela Vinodhan, B. Ravi Kumar and A. Lawrence**
Bharat Heavy Electricals Limited, Tiruchirappalli, Tamil Nadu 620 014, India

**A. K. Bakthavatsalam**
National Institute of Technology, Tiruchirappalli, Tamil Nadu 620015, India

**José C. F. Pereira, José M. C. Pereira, André L. A. Leite, and Duarte M. S. Albuquerque**
IDMEC, Instituto Superior Tecnico, Universidade de Lisboa, Avenida Rovisco Pais, 1049-001 Lisbon, Portugal

**Ahmad El Sayed**
Department of Mechanical and Mechatronics Engineering, University of Waterloo, 200 University Avenue West, Waterloo, ON, Canada N2L 3G1
Institut de Combustion Aèrothermique Rèactivitè et Environnement (CNRS), 1C avenue de la Recherche Scientifique, 45071 Orlèans Cedex 2, France

**Roydon A. Fraser**
Department of Mechanical and Mechatronics Engineering, University of Waterloo, 200 University Avenue West, Waterloo, ON, Canada N2L 3G1

**Joseph Kalman**
University of Illinois at Urbana-Champaign, 1206 W. Green Street, Urbana, IL 61801, USA
Naval Air Warfare Center Weapons Division, 1 Administration Circle, China Lake, CA 93555, USA

**Nick G. Glumac and Herman Krier**
University of Illinois at Urbana-Champaign, 1206 W. Green Street, Urbana, IL 61801, USA

**Zhiwei Li and Hongzhou He**
Cleaning Combustion and Energy Utilization Research Center of Fujian Province, Jimei University, 9 Shigu Road, Xiamen 361021, China

**Seyed Ehsan Hosseini, Ghobad Bagheri, Mostafa Khaleghi and Mazlan Abdul Wahid**
High Speed Reacting Flow Laboratory, Faculty of Mechanical Engineering, Universiti Teknologi Malaysia, 81310 Skudai, Johor, Malaysia

**R. M. Davies**
Department of Agricultural and Environmental Engineering, Niger Delta University, PMB 071, Yenagoa, Bayelsa State, Nigeria

**O. A. Davies**
Department of Fisheries and Aquatic Environment, Rivers State University of Science and Technology, PMB 5080, Port Harcourt, Rivers State, Nigeria

**G. Paterakis, K. Souflas, E. Dogkas and P. Koutmos**
Laboratory of Applied Thermodynamics, Department of Mechanical Engineering and Aeronautics, University of Patras, 26504 Patras, Greece

**Gautham Krishnamoorthy and Caitlyn Wolf**
Department of Chemical Engineering, University of North Dakota, 241 Centennial Drive, Stop 7101, Grand Forks, ND 58201, USA

**Hassan Abdul-Sater and Gautham Krishnamoorthy**
Department of Chemical Engineering, University of North Dakota, Harrington Hall Room 323, 241 Centennial Drive, Grand Forks, ND 58202-7101, USA

**Mario Ditaranto**
SINTEF Energy Research, 7465 Trondheim, Norway

**Atsushi Makino**
Institute of Aeronautical Technology, Japan Aerospace Exploration Agency, 7-44-1 Jindaiji-Higashi, Chofu, Tokyo, Japan

**A. A. Vasil'ev**
Lavrentyev Institute of Hydrodynamics SB RAS, Novosibirsk 630090, Russia

**Steven L. Rowan, Ismail B. Celik, Albio D. Gutierrez and Jose Escobar Vargas**
West Virginia University, Morgantown, WV 26505, USA

**Zhigang Li, Yuichi Sugai and Kyuro Sasaki**
Department of Earth Resources Engineering, Faculty of Engineering, Kyushu University, Fukuoka 819-0395, Japan

**Xiaoming Zhang and JirenWang**
College of Mining Engineering, Liaoning Technical University, Fuxin 123000, China

**Srinivas Kommana**
Department of Mechanical Engineering, VRSEC, Vijayawada 520 007, India

**Balu Naik Banoth**
Department of Mechanical Engineering, JNTU, Hyderabad 500 085, India

**Kalyani Radha Kadavakollu**
Department of Mechanical Engineering, JNTU, Anantapur 515 002, India

www.ingramcontent.com/pod-product-compliance
Lightning Source LLC
Chambersburg PA
CBHW080652200326
41458CB00013B/4829

9 781632 385109